Designing Effective Public Policies to Address Global Warming

Ömer Uğur
Gumushane University, Turkey

Kadir Caner Doğan
Gumushane University, Turkey

IGI Global
Scientific Publishing
Publishing Tomorrow's Research Today

Vice President of Editorial Melissa Wagner
Director of Acquisitions Mikaela Felty
Director of Book Development Jocelynn Hessler
Production Manager Mike Brehm
Cover Design Jose Rosado

Published in the United States of America by
IGI Global Scientific Publishing
701 East Chocolate Avenue
Hershey, PA, 17033, USA
Tel: 717-533-8845
Fax: 717-533-7115
Website: https://www.igi-global.com E-mail: cust@igi-global.com

Library of Congress Cataloging-in-Publication Data

LCCN: 2025013673 (CIP Data Pending)
ISBN13: 9798337313108
Isbn13Softcover: 9798337313115
EISBN13: 9798337313122

British Cataloguing in Publication Data
A Cataloguing in Publication record for this book is available from the British Library.

Table of Contents

Detailed Table of Contents

As one of the most urgent, multidimensional and deepening problems of our time, global warming represents one of the greatest challenges in human history. So why does this crisis require such an urgent response? How are rising temperatures, melting glaciers and increasing extreme weather events affecting our planet's ecosystems and human communities? Is climate change merely an environmental disaster, or is it an issue that deepens social inequalities, shakes economic systems and tests global cooperation? Moreover, what steps should governments, societies and individuals take to build an effective response to this threat? These questions reveal that global warming is not only a scientific reality, but also a complex challenge that requires the shared responsibility and solidarity of humanity.

Climate change necessitates innovative approaches to sustainable urban development. Conventional housing models often exacerbate environmental and social challenges through resource inefficiency, high carbon emissions, and social fragmentation. This chapter investigates collaborative housing (co-housing) as a sustainable alternative, emphasizing shared spaces, participatory governance, and ecological design. Focusing on case studies in Vienna and in Lower Austria, it analyzes how energy-efficient construction, renewable energy systems, and sustainable mobility reduce ecological footprints while enhancing social cohesion. Additionally, the chapter examines Vienna's progressive housing policies that facilitate co-housing through targeted subsidies and incentives. While challenges such as finance and scalability remain, co-housing offers significant potential for climate change mitigation and urban resilience, with transferable insights for global sustainability efforts.

Climate justice emphasizes the moral imperative to address the disproportionate impacts of climate change on marginalized communities while striving for global sustainability. As local communities often bear the brunt of environmental degradation, their needs must be prioritized to achieve equitable solutions. This concept acknowledges that climate issues intersect with social inequities, demanding actions that not only reduce carbon emissions but also address poverty, health, and access to resources. Effective climate justice strategies integrate diverse perspectives, fostering collaboration across borders and sectors. By aligning grassroots initiatives with global climate goals, we can create comprehensive solutions that enhance resilience and promote social justice. Ultimately, climate justice advocates for a holistic approach to sustainability, recognizing that local actions are fundamental to achieving global climate objectives.

Global warming represents an urgent crisis requiring immediate action. According to the IPCC's latest report, human-induced climate change is causing widespread disruption to nature and affecting billions worldwide, with vulnerable populations bearing the heaviest burden. While multiple feasible solutions exist to reduce emissions and adapt to climate change, immediate action is crucial for securing a liveable future. The challenge of addressing climate change is intrinsically linked to social justice and sustainable development. Marginalized communities, often having the least resources to adapt, face disproportionate impacts from rising sea levels, extreme weather events, and food insecurity. Effective climate policies must therefore embrace climate justice principles while ensuring a just transition for workers in carbon-intensive industries. Success requires integrating climate action with broader sustainable development goals, including poverty alleviation, food security, and gender equality. By pursuing these synergies, policymakers can maximize benefits.

China faces its own complications in adapting to the new realities in the energy sector. The distinct strategic vision of the country was boosted by an on-target assortment of practical tools for energy readjustment. The motivation to transform the national energy mix triggered the efforts to join the global trend for advancement in green energy. In the light of these issues, the paper explores the principal dimensions of the integration of the green pillar in the energy strategy and policy of PRC, in order to evaluate their repercussions in the domestic and the international realm. The research showcases the essence of the promoted concept for "ecological civilization", defined by many experts as a form of sustainable development with Chinese characteristics, and its role in the rise of the country as a renewable energy superpower. The study is oriented towards outlining the main dimensions of the metamorphoses in the Asian country's energy policy and presenting its potential for the use of renewable energy to generate green growth.

This study focuses on the fiscal aspect of the Carbon Border Adjustment Mechanism (CBAM) introduced by the European Union (EU) and aims to examine the use of revenues from the mechanism for developed and developing countries. The CBAM aims to prevent carbon leakage and promote environmental sustainability through carbon pricing, while at the same time generating significant public revenue for the EU. The envisaged use of these revenues in strategic areas such as green transformation, sustainable infrastructure projects and similar is an important step in contributing to the EU's environmental goals. However, given the risk that underdeveloped and developing countries are disproportionately affected by CBAM, the need for effective use of these revenues in these countries becomes even more evident. In this context, the effective use of CBAM revenues in less developed and developing countries is considered as an important tool to support not only environmental sustainability but also the economic and social transformation processes of these countries.

Green finance plays a pivotal role in aligning financial systems with climate policy objectives to mitigate global warming. This chapter examines strategies that integrate green finance with regulatory frameworks, ensuring that capital flows support climate resilience and sustainability. It explores mechanisms such as green bonds, sustainability-linked loans, carbon pricing, and public-private partnerships to mobilize investments toward low-carbon technologies. Additionally, the chapter highlights the role of financial institutions in promoting climate disclosure and risk assessment while addressing challenges like greenwashing and policy misalignment. Case studies illustrate successful implementations of green finance policies in different jurisdictions, offering insights into best practices and regulatory advancements. By fostering collaboration between governments, financial markets, and international organizations, green finance can accelerate the transition toward a net-zero economy.

Governance explores the pivotal role of local governance in addressing the urgent challenges posed by climate change. As the effects of climate disruption become increasingly localized, cities, towns, and communities are on the front lines, facing the immediate impacts while also offering innovative solutions. This chapter examines the strategies that enable local governments to integrate climate resilience into their policies, governance structures, and day-to-day decision-making processes. It discusses the importance of building local capacity, fostering collaboration among stakeholders, and utilizing data-driven approaches for adaptive management. Through case studies and real-world examples, the chapter highlights how local leadership can empower citizens, engage in sustainable urban planning, and implement climate-mitigation practices. By enhancing the capacity for climate-responsive governance, this work underscores the need for a bottom-up approach in achieving global climate goals and creating resilient communities.

This study examines the causal relationship between environmental taxes and carbon footprint in the context of climate change mitigation, using data from ten major EU economies (Austria, Belgium, France, Germany, Ireland, Italy, Netherlands, Poland, Spain, Sweden) over 1994–2022. Employing panel causality analysis, the study first applies tests for cross-sectional dependence, slope heterogeneity, and panel unit roots. The Dumitrescu and Hurlin test results indicate a significant causality from environmental taxes to per capita carbon emissions in at least one country. The findings suggest that well-designed environmental taxes can play a critical role in reducing emissions. The study highlights the importance of revenue allocation and policy coherence and recommends future research on the effect of such taxes on ecological footprint components for more comprehensive sustainability assessment.

The relationship between environmental protection expenditures and income inequality is undoubtedly a critical issue in terms of social justice and sustainable development. In societies with high income inequality, environmental protection expenditures are generally lower. This is because low-income individuals tend to have less motivation to develop awareness of environmental issues or to support environmental protection policies. On the other hand, in societies where income is more equitably distributed, environmental protection expenditures tend to increase. A key factor behind this is that a larger segment of society views environmental sustainability as a public good and a necessary component of inclusive economic development, thereby generating collective demand in this direction. In this study, the relationship between income inequality and environmental protection expenditures in eight OECD countries was investigated using the Bootstrap Panel Rolling Window causality test.

The increasing frequency of global environmental disasters is forcing the world to join the fight against climate change. The solution, whether on a global or local, lies in governance system that protects the world. However, with over half of the world's population living in cities, climate change poses increasingly challenging problems for urban. Due to globalization, local governments are becoming important. This study focuses on local practices and examining climate actions that integrate local governments with international climate policies governing the world. The study examines the foundations linking local government actions to global climate change in terms of political, technological, and cooperative dimensions. The political dimension defines climate change governance, the cooperative dimension is defined by international alliances, and the technological dimension is defined by international coalitions. Without collaboration among local governments, states, international agreements, alliances, and coalitions, sustainable solutions to climate change cannot be found.

Climate change poses one of the most urgent challenges of our time, demanding innovative solutions. This chapter explores how data-driven smart governance can be a transformative tool in addressing climate change. By integrating technology, data, and participatory decision-making, smart governance allows for real-time insights and predictive modeling that enhance climate policies. Key technologies such as the Internet of Things (IoT), Artificial Intelligence (AI), and blockchain are essential in monitoring environmental data and optimizing resource use. The chapter discusses the challenges of data quality, privacy, and accessibility, and proposes a policy framework for effective climate governance. It also emphasizes the role of stakeholder engagement, international cooperation, and scalable solutions in shaping the future of climate action.

In 2017, Cape Town faced a severe drought. The incident would possibly culminate in that city's taps being turned off by March 2018. The likely day of the city's water facilities completely running dry and leaving more than three million residents without water was labelled ‘Day Zero’. Cape Town is part of the Western Province of South Africa which contains large stretches of vineyards which were highly affected by the drought. The South African wine industry, being one of the top major producers of wine in the world, stood to be severely affected. This chapter documents how the wine industry was impacted by this climate crisis and the adaptive mechanisms it implemented. Further, the incident provides a case example of how climate shocks may compel further futuristic thinking on the need to contemplate adaptive mechanisms by cities regarding managing future disruptive climate risks.

This chapter proposes a clearer and more holistic framework for understanding citizen engagement not only in decision-making but also in the implementation of environmental initiatives. Through a structured conceptual review of key theories, typologies, and related concepts, we highlight the overlooked role of citizens during the implementation phase, where authority-led action often dominates. We examine how this gap undermines inclusive governance and explore how implementation fits within the broader environmental project lifecycle. The chapter also analyzes related constructs such as co-production, co-design, co-creation and involvement clarifying their complementary roles in participatory implementation. Drawing from previous research, we underscore how sustained engagement fosters community ownership and long-term change. The integrative framework we propose refines the understanding of citizens' participation across project phases, advocating for a more inclusive and precise conceptualization to enhance future research and governance effectiveness.

The formation of sustainable and resilient cities has become a critical focus in modern urban governance. Rapid urbanization, infrastructure deficiencies, unplanned expansion, and social inequalities challenge the pursuit of sustainability in urban contexts. In response, smart city applications have emerged as effective instruments for addressing these issues. This chapter focuses on the Amsterdam Smart City

Preface

INTRODUCTION

The climate crisis represents one of the most pressing global challenges of the 21st century. Its effects are no longer abstract projections for a distant future; they are already reshaping lives, livelihoods, and landscapes across the globe. From rising sea levels and intensifying weather patterns to biodiversity loss and food insecurity, global warming impacts every aspect of human and environmental systems. It is within this urgent context that this edited volume, *Designing Effective Public Policies to Address Global Warming*, emerges as both timely and essential.

Our objective with this book is to explore the multifaceted nature of global warming and investigate how public policies—when designed thoughtfully and implemented effectively—can address its complex causes and far-reaching consequences. The volume brings together diverse perspectives from scholars across the globe, highlighting innovative strategies, empirical case studies, and theoretical insights. Each chapter offers a unique angle into how governments, communities, and institutions can develop policy interventions that are environmentally sound, socially just, and economically feasible.

By compiling research that spans climate justice, environmental economics, smart governance, sustainable urban design, and green finance, this book seeks to act as a bridge between policy theory and action. As editors, our hope is that this work not only informs but inspires. We aim to equip policymakers, researchers, practitioners, and students with the knowledge and tools necessary to participate meaningfully in the global effort to combat climate change.

CHAPTER OVERVIEW

Chapter 1 introduces the core framework for designing effective climate policies, with an emphasis on systemic approaches that integrate scientific evidence and social equity.

Chapter 2 explores co-housing in Austria as a model of participatory living that reduces carbon footprints and enhances social cohesion.

Chapter 3 focuses on the intersection of climate justice and public health among tribal communities in the Sundarbans, calling for the inclusion of marginalized voices in global solutions.

Chapter 4 examines the integration of climate action with social justice in policymaking, emphasizing equitable transitions and sustainable development goals.

Chapter 5 analyzes China's "Ecological Civilization" and its energy transition efforts, showcasing how national frameworks can influence global energy landscapes.

Chapter 6 investigates the fiscal and distributive implications of the EU's Carbon Border Adjustment Mechanism, especially for developing nations.

Chapter 7 explores how green finance mechanisms, such as bonds and sustainability-linked loans, can align with climate policy goals.

Chapter 8 shifts focus to local governance and community-led strategies, underscoring the importance of bottom-up climate resilience.

Chapter 9 presents a panel causality analysis of the relationship between environmental taxes and carbon footprints in EU member states.

Chapter 10 examines the complex dynamics between income inequality and environmental protection spending, advocating for socially equitable sustainability.

Chapter 11 discusses innovative tools used by local governments for effective climate action through political, cooperative, and technological lenses.

Chapter 12 highlights the transformative potential of smart governance in leveraging big data and digital tools for climate policy.

Chapter 13 documents the South African wine industry's response to Cape Town's 'Day Zero' crisis as a case study of adaptive climate governance.

Chapter 14 rethinks citizen participation in environmental projects, proposing a more integrated framework for inclusive governance.

Chapter 15 studies Amsterdam's smart city initiatives to evaluate the role of urban innovation in achieving climate goals.

Chapter 16 concludes with a deep dive into the relationship between urban poverty and climate change, offering policy insights for poverty alleviation.

CONCLUSION

As this book demonstrates, the climate crisis is not merely a scientific or environmental issue—it is a human issue, a justice issue, and a governance issue. The policies we design today will determine the trajectory of our societies and ecosystems for decades to come. The research presented here illustrates that effective climate policy requires not only technical expertise but also ethical considerations, cultural sensitivity, and inclusive engagement.

While each chapter addresses a distinct component of climate change, they are united by a common thread: the belief that well-crafted public policies can and must be a central part of the global climate response. The diversity of strategies presented—from carbon pricing and local governance to smart technologies and citizen engagement—shows the richness of innovation in this field. However, innovation must be matched with political will, cross-sectoral collaboration, and robust implementation mechanisms.

In closing, we express our gratitude to all contributing authors whose work enriches this volume. Their insights and dedication reflect the growing community of scholars committed to shaping a sustainable, just, and climate-resilient future. We hope this book will serve as a valuable resource for readers committed to understanding and shaping the public policies that will define our collective response to global warming.

Ömer Ugur
Gumushane University, Turkey

Kadir Caner Dogan
Gumushane University, Turkey

Chapter 1
Designing Effective Public Policies to Address Global Warming

Ömer Uğur
Gumushane University, Turkey

Kadir Caner Doğan
https://orcid.org/0000-0002-3476-8865
Gumushane University, Turkey

ABSTRACT

As one of the most urgent, multidimensional and deepening problems of our time, global warming represents one of the greatest challenges in human history. So why does this crisis require such an urgent response? How are rising temperatures, melting glaciers and increasing extreme weather events affecting our planet's ecosystems and human communities? Is climate change merely an environmental disaster, or is it an issue that deepens social inequalities, shakes economic systems and tests global cooperation? Moreover, what steps should governments, societies and individuals take to build an effective response to this threat? These questions reveal that global warming is not only a scientific reality, but also a complex challenge that requires the shared responsibility and solidarity of humanity.

INTRODUCTION

As one of the most urgent, multidimensional and deepening problems of our time, global warming represents one of the greatest challenges in human history. But why does this crisis require such an urgent response? How are rising temperatures, melting

DOI: 10.4018/979-8-3373-1310-8.ch001

glaciers and increasing extreme weather events affecting our planet's ecosystems and human communities? Is climate change merely an environmental disaster, or is it an issue that deepens social inequalities, shakes economic systems and tests global cooperation? Moreover, what steps should governments, societies and individuals take to build an effective response to this threat? These questions reveal that global warming is not only a scientific reality, but also a complex challenge that requires the shared responsibility and solidarity of humanity. The concentration of greenhouse gases released into the atmosphere since the Industrial Revolution has caused global average temperatures to rise by more than 1.1°C compared to pre-industrial times. The findings of the United Nations Intergovernmental Panel on Climate Change (IPCC) for 2023 indicate that if current emission trends continue, this increase could range between 2.4°C and 3.5°C by the end of the century. This would not only lead to the destruction of natural habitats, but also to dramatic declines in agricultural production, vulnerability in food security, mass climate migration and increased social unrest. However, global warming is not only an environmental challenge, but also a crisis that tests social justice with low-income countries disproportionately affected, economic sustainability with fossil fuel dependency, and global cooperation with a lack of international commitments. In this context, designing effective public policies against global warming should not be limited to reducing emissions, but should also strengthen adaptation strategies, promote technological innovation and put climate justice at the center. By blending scientific data with societal needs, public policies should both respond to current crises and build a livable world for future generations. This paper aims to provide a framework for the design of effective public policies by addressing the urgency and multidimensional impacts of global warming. Beginning with a section on "The Urgency of Climate Action", it proposes a framework for designing proactive, equitable and innovative public policies to address the climate crisis in light of scientific findings, socio-economic dynamics and policy imperatives.

THE URGENCY OF CLIMATE ACTION

Climate change is not only a threat that is expected to emerge in the future, but a concrete reality that is already being experienced and profoundly transforming the dynamics of our age with its multidimensional impacts. This transformation necessitates radical, preventive and systemic interventions that go beyond superficial regulations in public policies (Kaya, 2025). So why does this crisis require such an urgent intervention? Because the rapid approach to irreversible thresholds in the climate system is not only limited to the deterioration of ecological balances, but also has multidimensional consequences such as the erosion of social structures, increased economic vulnerabilities and deepening global injustices. Therefore, the climate crisis goes far beyond being a purely technical environmental issue and is a multidimensional existential crisis that requires a radical rethinking and collective will for transformation in areas ranging from politics to economy, social justice to global governance (Yılmaz Uğur, 2021; Kalkışım et. al., 2023).

Scientific data provides a clear and unambiguous picture of the scope, intensity and urgency of the climate crisis that leaves no room for debate. The IPCC's 2023 report found that the global average temperature rise exceeded 1.1°C compared to pre-industrial levels. However, recent observational data for 2024 show that this trend is not only continuing but also accelerating. According to the World Meteorological Organization's (WMO) Global Climate Outlook 2024, 2024 marked a turning point, with the global average temperature measuring 1.55 ± 0.13°C higher than in the pre-industrial period (1850-1900), making it the warmest year on record. If current policy approaches and greenhouse gas emission trends do not undergo a significant transformation, the temperature increase by the end of the century is likely to exceed the 2°C threshold, or even an increase between 3.3°C and 5.7°C in high emission scenarios, a scenario with high probability and serious consequences (Öner, 2023).

Viewing the climate crisis solely as an environmental problem weakens the capacity to produce a holistic solution by ignoring the economic, social and political dimensions of this multi-layered and intertwined phenomenon. Climate change has turned into a structural crisis that threatens not only ecosystems but also economic stability, social equality and the global security architecture. Rising temperatures and intensifying extreme weather events threaten agricultural production and water resources, weakening economic balances; this situation deepens vulnerabilities, especially in developing societies, accelerating social unrest and mass migration movements (Cengiz and Uğur, 2024). This process accelerates climate-induced migration and increases ethnic, cultural and class tensions in sending and receiving societies, setting the stage for a polarization dynamic that threatens social cohesion and political stability. On the other hand, inequalities in the transition to renewable energy jeopardize energy supply security, while the vulnerability of infrastructures to climate disasters necessitates radical transformations in the national security perspective. However, the intensification of resource scarcity and the potential for issues such as cross-border water sharing to escalate geopolitical tensions - as seen most recently in the Pakistan-India conflict - also make the climate crisis a conflict-generating factor at the level of international relations. For these reasons, policy approaches to the climate crisis should not only be reduced to reducing carbon emissions, but also include a systemic perspective that prioritizes social inclusion, secures just transition processes and strengthens economic resilience. Otherwise, interventions in the name of environmental sustainability can undermine both ecological and social goals by generating new forms of inequality, transforming the climate crisis from a global existential threat into a multifaceted source of instability.

The political urgency of climate action raises not only the effectiveness of political decision-making mechanisms, but also the legitimacy of states and the redefinition of social contracts. The direct and indirect impacts of the climate crisis push capacity limits in many areas from public administration to local governance, from regional cooperation to transnational arrangements, making managerial flexibility and institutional adaptability a strategic priority. Increasing disaster risk, agricultural and water-based conflicts, and uncertainties over energy security necessitate the prioritization of not only military but also climate-based threats in national security doctrines. This new security configuration requires a rethinking of climate-based risks beyond classical threat perceptions within the framework of foresight, prevention and response capacity. Moreover, the rise of social demands regarding the climate crisis creates a new ground of accountability in citizen-state relations, forcing governance forms to evolve into a more participatory, transparent and fair structure. In this context, postponing climate action or limiting it for political calculations may not only impose an ecological cost, but also bring serious political consequences such as governance crises, democratic backsliding and loss of social legitimacy. Therefore, responses to the climate crisis require determination and continuity at the level of political vision, ethical responsibility and collective will, beyond technical capacity.

In conclusion, the urgency of climate action is a critical imperative not only to prevent environmental destruction, but also to preserve global economic stability, ensure just transitions without deepening social inequalities, strengthen democratic governance and sustain international peace and security. In this context, any strategy to be developed against the climate crisis should be addressed in a holistic manner that is compatible with scientific data, socially inclusive and politically decisive; it should not be forgotten that delayed or superficial interventions can accelerate a multidimensional systemic collapse (Uğur and Aksoy, 2023; Efşan et. al, 2024).

FUNDAMENTALS OF CLIMATE POLICY DESIGN

Global warming is not only a complex and foreseeable threat by its very nature, but also one that involves deep uncertainties, has multifactorial cause-and-effect networks and is characterized by systemic risks. Faced with such a threat, it is clear that traditional public administration understanding and approaches are insufficient in planning, decision-making and implementation processes. Therefore, it is necessary to adopt a new understanding of public administration based on flexible and adaptive policy design processes against the complex and systemic threats posed by climate change. This new understanding should, above all, be designed in a flexible and dynamic structure capable of anticipating and adapting to the uncertainties inherent in climate change (Akın, 2006).

In this context, it should be based on participatory and pluralistic governance practices that encourage interaction between actors at different levels and scales, based on collective learning and institutional adaptation processes, moving away from traditional centralized and vertical hierarchical structures. Thus, decision-making processes for existing or potential problems related to climate change will not only be shaped by the monopoly of central authorities, but will also become an indispensable element of integrating the knowledge, experience and expectations of local governments, civil society organizations, private sector representatives and directly affected communities. Such a governance model is vital not only as a requirement of the democratic ideal, but also as the basis for more legitimate, responsive and viable policy-making processes.

The participatory governance mechanism to be built will not only build strong systems against current climate risks, but will also play a critical role in strengthening the ability to adapt to future uncertainties. The flexibility provided at the policy-making stage will enable learning processes and openness to innovative solutions in the face of the dynamic and unpredictable nature of climate change, supporting the continuous updating and development of policy making (Doğan and Uğur, 2020). Thus, instead of fixed and unidirectional approaches, realistic and feasible solutions that can be rapidly adapted to changing conditions, supported by feedback mechanisms and updated in the light of new data will emerge. Moreover, the lack of scientific knowledge on the thresholds in the climate system, which is one of the main challenges facing policymakers, can become more manageable in this way. Under uncertainty, the risk of policymakers either making wrong decisions or not taking action at all emerges as a significant threat. Such indecisions can lead to irreversible impacts on the climate system and narrow the window for intervention. However, the adoption of participatory and multi-level governance mechanisms not only lends social legitimacy to decision-making processes, but also makes it possible to generate conjuncture-specific, dynamic and up-to-date scientific data. Thus, decision-making processes become more evidence-based, predictable and strategic, and a more proactive and flexible policy architecture can be developed in the face of uncertainties.

Consequently, effective climate policy design in the context of systemic risks and uncertainty should be based on an adaptive and learning systems approach that balances normative objectives with scientific insights, and considers the adaptability of multi-actor and multi-level structures. Such a framework not only responds to environmental threats, but also underpins long-term social stability and institutional legitimacy.

TYPES OF CLIMATE POLICIES: TOOLS AND MECHANISMS

Policies to combat climate change are shaped by instruments and mechanisms that differ in line with each country's environmental conditions, level of economic development and social priorities. These policies are not only limited to technical intervention tools, but are also implemented within the framework of multi-actor and multi-scale strategies that support structural transformations. The main objectives of climate policies are to reduce greenhouse gas emissions from human activities, transform energy production and consumption systems in line with sustainable solutions and increase the resilience capacity of societies against the inevitable impacts of climate change. Policy instruments developed to achieve these goals have a multi-layered structure that includes economic, regulatory and voluntary-based approaches together (Kaymaz and Tut, 2020).

Carbon pricing stands out as one of the most widely used market-based instruments to combat climate change. This approach aims to internalize environmental externalities and imposes a financial cost on polluting activities by directly attributing an economic value to carbon emissions. In this context, carbon taxes and emissions trading systems (ETS) aim to direct the behavior of emitters towards more sustainable choices by determining the cost of environmentally damaging emissions through the market. In this framework, carbon taxes impose a fixed price on a certain amount of emissions, while emissions trading systems set a limit on the total level of emissions and create a market where emission permits can be bought and sold. By disadvantaging carbon-intensive forms of production, these two instruments aim to pave the way for the development of technologies that promote low-carbon production and clean energy investments. In other words, these instruments facilitate the achievement of environmental goals through economic instruments, while at the same time encouraging increased energy efficiency and the diffusion of sustainable technologies. Thus, by harnessing the power of market mechanisms, climate policies can be aligned with economic rationality and a more balanced relationship between environmental sustainability and economic competitiveness can be established (Uğur and Doğan, 2019).

In addition to economic instruments, regulatory policies also play a critical role in achieving climate goals. Emission standards impose a cap on greenhouse gas emissions in certain sectors, requiring technical compliance, and in some cases, the use or production of high-emission technologies may be banned altogether. For example, many countries are planning to phase out the sale of internal combustion engine vehicles in the coming years. Such direct interventions can be effective where market conditions are insufficient. However, regulatory policies should not only limit emissions, but also be complemented by incentive mechanisms to support the transition to low-carbon technologies. In this context, subsidies, tax breaks and public financial support to promote green innovation are critical in this process. State-funded research and development (R&D) activities, especially encouraging innovation in areas such as renewable energy, energy efficiency and energy storage technologies, increase the technical and economic viability of low-carbon solutions (Republic of Turkey Ministry of Environment, Urbanization and Climate Change, 2025). Such investments not only generate environmental benefits, but also have multifaceted impacts such as the creation of new jobs, the transformation of industrial policies and increased international competitiveness. Moreover, instruments such as cooperation models between the private and public sectors, venture capital support and public procurement guarantees are becoming increasingly important to accelerate the commercialization of these technologies. Thus, green transformation is becoming a process that is not only limited to environmental policies, but also central to economic development strategies.

Another important component of climate policies is public investments to adapt to changing climate conditions. In this context, practices such as developing infrastructure resilient to floods, droughts and extreme weather events, supporting climate-resilient methods in agricultural production and more efficient management of water resources stand out. In addition, community-based adaptation policies that encourage the participation of local communities in decision-making processes and aim to improve access to information and capacity building also create impacts that increase social resilience. As a result, an effective climate policy should not only focus on emission reductions; it should also support technological transformation, consider social justice and include long-term climate resilience goals. This multidimensional approach is indispensable for a sustainable and inclusive climate policy architecture.

APPROACHES TO ASSESSING THE EFFECTIVENESS OF CLIMATE POLICIES

The success of climate policies should be assessed not only by whether they are implemented or not, but also by the extent to which these policies create tangible, lasting and multidimensional impacts. Assessing policy effectiveness requires a multidimensional framework where economic, social and governance impacts are analyzed together, not only technical results. In this context, policy outcomes need to be monitored and evaluated holistically, using both quantitative and qualitative indicators (Uğur and Cengiz, 2023).

One of the most widely used quantitative indicators is the absolute reduction in greenhouse gas emissions and the change in per capita emission levels. However, these indicators alone are not sufficient. In addition to emission reduction, the extent to which these policies are compatible with economic sustainability should also be analyzed. In this context, it is important to evaluate the effects of emission policies on gross domestic product (GDP) together with economic indicators such as changes in employment levels, transformation in production structure and innovation capacity. In this way, the environmental benefits of climate policies as well as their effects on economic development processes can be made visible (Aydın, 2013). However, principles of social justice and equality are increasingly prominent criteria in policy effectiveness assessments. Market-based instruments, such as carbon pricing in particular, can create disproportionate burdens on low-income groups due to their impact on income distribution. Therefore, the impact of energy expenditures on household incomes should be analyzed together with income inequality indicators and whether these policies are supported by compensatory social mechanisms should be assessed. In addition, dimensions related to vulnerable groups such as gender, differences between rural and urban areas, and indigenous peoples' rights and lifestyles also necessitate fair measurement of the social impacts of climate policies. Finally, policy impact should be associated not only with outputs but also with the quality of decision-making and implementation processes. In this context, governance elements such as transparency, accountability, participation and institutional capacity are decisive in terms of both short- and long-term sustainability of effectiveness. Designing impact assessment processes to include these elements increases the legitimacy of climate policies and strengthens their implementation success (Doğan, 2021).

POLICY EFFECTIVENESS: CRITERIA AND EVALUATION

The Paris Agreement, which sets out the new path for international cooperation against global climate change, was adopted in December 2015 and entered into force in November 2016. The universal participation target of the Agreement was achieved in April 2019 when the number of parties reached 185. Following its adoption, the Agreement was made ready for implementation after approximately three years of negotiations, and its implementation rules were largely completed and adopted in December 2018. Although the Paris Agreement is an international agreement within the scope of the 1992 United Nations Framework Convention on Climate Change (UNFCCC), which establishes the general framework for cooperation against climate change, it has significantly differed from both the Convention and the Kyoto Protocol in terms of both the purpose of cooperation and the way in which the responsibilities of the parties are determined and regulated. The social and political impact of the Paris Agreement largely stems from its new understanding based on cooperation against global climate change. This understanding, which considers the fight against climate change and its effects as a matter of multi-actor and multi-level joint action within the framework of values such as human rights, climate justice, gender equality, intergenerational equality and ecosystem integrity, has found its expression particularly in the preamble of the Agreement and, to some extent, in its implementation provisions (Mazlum, 2019; Yılmaz Uğur, 2022).

The world's climate is changing due to human activities, and there is sufficient scientific evidence to show this change. Greenhouse gases released into the atmosphere are considered the most important cause of human-induced climate change. The main natural greenhouse gases include carbon dioxide (CO2), water vapor (H2O), nitrous oxide (N2O), methane (CH4) and ozone (O3), as well as man-made (artificial) chlorofluorocarbons (CFCs), hydrofluorocarbons (HFCs), perfluorocarbons (PFCs) and their various derivatives. Greenhouse gases that cause human-induced climate change and global warming generally originate from the burning of fossil fuels for energy and transformation, industrial activities, transportation, land use change, waste management and agricultural (non-energy; stubble burning, rice production, animal husbandry and fertilization, etc.) activities. In particular, the rate of greenhouse gases in the atmosphere has started to increase since the industrial revolution. For example, before the industrial revolution, the CO2 concentration in the atmosphere was 288 ppm (two hundred and eighty-eight parts per million by volume), while in 2023 it reached 419 ppm. The average global temperature in 2021 was recorded as approximately 1.11°C above pre-industrial (1850-1900) levels. Temperature increases trigger glacial melting, resulting in sea level rise. The global average sea level has risen by 21-24 cm since 1880 (Göksu and Somuncu, 2023).

There are numerous national and international agreements and protocols to combat climate change, and there are also important reports on solutions at the local level. For example, at the World Conference on Environment and Development adopted in 1994, the United Nations Framework Convention on Climate Change (UNFCCC) stated that the public should be informed and that society should be an active stakeholder in combating climate change. Although the Kyoto Protocol does not contain specific articles applicable at the local level, it has introduced provisions to increase awareness. The Paris Agreement aims to limit global warming to 1.5 °C and for local governments to play an active role in achieving this goal (Esen and Batuhan, 2024). Within the scope of combating climate change, the global transition to a low-carbon economy envisages a radical transformation that will change people's lifestyles, production and manufacturing methods. Therefore, efforts to combat and adapt to climate change should not be perceived solely as an environmental problem. In fact, this struggle can directly affect the growth strategies, energy policies, health and agriculture programs, water resource use, food security, transition to a low-carbon economy and sustainable development goals that countries will follow and can be decisive in their development. In order to achieve these goals, it is important for developed countries to fulfill their commitments regarding financing, technology transfer and capacity development (Ministry of Foreign Affairs of the Republic of Turkey, 2025).

The information gathered to date suggests that countries respond best to climate change by developing a portfolio of policy instruments that limit or reduce greenhouse gas emissions. These policies may include (Dunn and Flavin, 2002):

- Carbon/energy taxes
- Tradable emissions permits
- Removal of subsidies for carbon-intensive energy sources
- Subsidies and tax incentives for carbon-free sources
- Rebate systems
- Technology or performance standards
- Introduction of mixed energy requirements
- Product bans
- Voluntary agreements
- Investment in research and development

It is difficult to distinguish between these policies, but market-based instruments are likely to be more cost-effective. Energy efficiency standards are widely used and have been successfully implemented in many countries. Voluntary agreements with industry are often used as a precaution before more stringent requirements are introduced. Other measures include changing consumer and producer behavior through information campaigns, environmental labeling of products, green marketing, and various incentives. It is essential that government and private institutions increase their R&D activities to achieve technological developments that will further reduce costs (Dunn and Flavin, 2002).

INNOVATIONS IN CLIMATE GOVERNANCE

Climate change is a topic that is on the agenda of many organizations, governments and local governments at the global and local level and whose impact is increasing day by day (Kazancı and Tezer, 2021). Climate change is the most important agenda item when viewed from both a global governance perspective and as a global environmental problem. Essentially, climate change has created an international governance model driven by states. The inadequacies that emerged in this process, the problems related to international law, where the issue concerns not only states or neighboring states but also the international community, and the fact that climate change has opened a new dimension in human rights discussions, have revealed the need for a climate change governance model. Climate change governance is a newly developing field related to the activities of the private sector, including the state, public administration, business world, civil society and non-governmental organizations (Özer, 2017; Kalkışım, 2019).

The most important reason for this is that different levels can make different decisions in political systems where climate change policies are produced, and these decisions must be coordinated and interactive. Climate change governance takes into account the principles of accountability and strong administrative and institutional structures while combating the problems caused by climate change. The content of this concept covers a wide perspective from informal cooperation between institutions and actors to hierarchical regulations in legislation. The existence of many ecological, social and economic variables complicates climate change governance. The fact that climate change policies include many issues from energy to economy requires the participation of more than one actor. The fact that climate change governance is a multi-actor and multi-level formation brings some actors such as states, local governments and civil society organizations to the forefront (Ocaklı, 2021).

CONCLUSION: TOWARD A JUST AND SUSTAINABLE FUTURE

Peace, justice, truth, and compassion are central to most utopian and religious visions. For example, the concepts of Heaven and Nirvana each carry powerful connotations of justice, harmony, nonviolence, and unity. These aspirations are a religious way of saying that most people in most societies and cultures, when faced with a choice between order/chaos; peace/war; harmony/disharmony; structural stability/instability; equality/inequality; inclusion/exclusion; justice/injustice; tolerance/intolerance; waste/poverty, will always choose the former if possible (Clements, 2021).

The climate crisis, with its transformative impact not only on environmental systems but also on social structures, economic models and political institutions, creates an unprecedented testing ground in contemporary public policy literature. This crisis has the potential to deepen inequalities, erode institutional capacity limits and question existing development paradigms at both global and local scales. Therefore, combating climate change is not only a technical problem-solving process but also involves a multi-layered political process that requires redefining normative goals, governance forms and economic priorities. Therefore, this study demonstrates that effective climate actions should be shaped around environmental strategies based on scientific data, egalitarian approaches that observe social justice and strong governance structures that increase the capacity of institutions. Tools such as emission reduction, climate adaptation, promotion of green technologies and carbon pricing, which are at the intersection of these three axes, can only yield effective results with such a holistic approach. In this context, the success of climate policies should be measured not only by concrete environmental results but also by their effects on economic transformation processes and social cohesion; these effects should continue to be monitored and evaluated through continuous feedback mechanisms.

In addition, the success of climate policies is directly related to the level of democratization of governance forms, the inclusiveness of decision-making processes and the stability of long-term political visions, much more than the technical capacity of states. In particular, the transition to multi-level and multi-actor governance structures will not only increase the effectiveness of implementation. It will also help strengthen the social legitimacy of policies. In this context, the institutionalization of community-based adaptation policies, local knowledge systems and the participation of vulnerable groups will be decisive in ensuring that climate action serves a just and sustainable future. As a result, a just and sustainable future is a field of possibilities shaped by today's policy choices. Effective use of this field requires addressing the climate crisis not only as a matter of risk management, but also as a paradigm shift that necessitates the simultaneous transformation of social, economic and political systems. In other words, climate action is a comprehensive social transformation process based on collective responsibility, the search for normative justice and a new understanding of publicity, beyond environmental goals.

REFERENCES

Akın, G. (2006). Küresel Isınma, Nedenleri ve Sonuçları. *Ankara Üniversitesi Dil ve Tarih-Coğrafya Fakültesi Dergisi, 46*(2), 29–43. DOI: 10.1501/Dtcfder_0000000673

Aydın, D. (2013). Ülkeler Arası CO2 Emisyonun Azaltılması Yönde Etkinlik Çalışması, Muğla Sıtkı Koçman Üniversitesi Sosyal Bilimler Enstitüsü Dergisi, 30, Bahar, 120-135.

Cengiz, E., & Uğur, Ö. (Eds.) (2024). Doğal Afetler ve İklim Değişikliği, Nobel Basım Yayım, Ankara.

Çevre, T. C. Şehircilik ve İklim Değişikliği Bakanlığı (2025). İklim Değişikliği Azaltım Stratejisi Ve Eylem Planı (2024-2030), Ankara.

Clements, K. (2021). *Kriz Dönüşümüne ve Adil Bir Barışa Doğru, Çev.* Friedrich Naumann Vakfı, Berghof Handbook For Conflict Transformation.

Dışişleri Bakanlığı, T. C. (2025). İklim Değişikliğiyle Mücadelenin Önemi, https://www.mfa.gov.tr/iklim-degisikligiyle-mucadelenin-onemi.tr.mfa

Doğan, K. C. (2021). In Kavanoz, S. E., Uzun, A., Şengün, H., & Yayınları, A. (Eds.), *İklim Değişkliği ve Kent, Karmaşık Sorunlar ve Kent* (pp. 189–204).

Doğan, K. C., & Uğur, Ö. (Eds.). (2020). *From Efficiency Discussions to Democracy in Public Administration: A Theoretical Analysis.* Peter Lang.

Dunn, S., & Flavin, C. (2002). İklim Değişikliğini Gündemin Ön Sıralarına Taşımak, Tema Vakfı Yayınları, http://dogadostlari.free.fr/kuresel/ıklim%20degisikligini%20g%c3%bcndemin%20%c3%96n%20siralarina%20tasimak.pdf.

Efşan, N. Ç., Uğur, Ö. and Aksoy, M. (2024). Assesment in The Context of Climate Migration: Hurricane Katrina, Karadeniz Teknik Üniversitesi Sosyal Bilimler Enstitüsü Sosyal Bilimler Dergisi, 27 / Haziran, 119-143.

Esen, M., & Batuhan, T. (2024). İklim Değişikliğiyle Mücadelede Kent Yönetimleri. *EBYÜ İİBF Dergisi, 6*(2), 140–151.

Göksu, E., & Somuncu, M. (2023). İklim Değişikliğine Uyum İçin İş Birliğine Dayalı Yönetişimde Kent Konseylerinin Rolü ve Önemi. *Ankara Üniversitesi Çevrebilimleri Dergisi, 10*(2), 1–14.

Kalkışım, H. M. (2019). Yönetişim Kavram ve Teorisi Üzerine Bir Değerlendirme, E-Yönetişim, Eds. B. Parlak ve K.C. Doğan, Beta Yayıncılık, İstanbul, 19-32.

Kalkışım, H. M., Doğan, K. C., & Soylu, A. C. (2023). Türk Sendikalarının 1978 Ekonomik Krizi Karşısındaki Tutumları, Sosyal. *Beşerî ve İdari Bilimler Dergisi, 6*(5), 641–662.

Kaya, M. H. (2025). İklim Değişikliği ve Güvenlik: Bir Tehdit Analizi. *Nous Academy Journal, 4*, 36–53.

Kaymaz, C. and Tut, G. (2020). İklim Değişikliği Politikalarına Yönelik Bir Çerçeve Oluşturma Denemesi, Kent ve Çevre Araştırmaları Dergisi, 2(1), Haziran, 78-98.

Kazancı, G., & Tezer, A. (2021). İklim Değişikliğine Uyumda Mekânsal Planlama Ve Akıllı Yönetişim Çerçevesinde Türkiye. *Planlama, 31*(2), 302–320.

Mazlum, S. C. (2019). Küresel İklim Politikaları, İklim Değişikliği Eğitim Modülleri Serisi 2, WEglobal, Ankara.

Ocaklı, Ş. N. (2021). İklim Değişikliği Yönetişimi ve Türkiye Üzerine Bir Değerlendirme, III. Kent Konseyleri Sempozyumu Geleceğin Kentlerinde İşbirliği ve Dayanışma (Bildiri Kitabı), 18-20 Kasım, Balıkesir, 561-576.

Öner, Ş. (2023). İklim Değişikliği Sorununun Uluslararası Gelişmeler Eşliğinde Türkiye'nin Politika ve Kurumlarına Yansıması, Ombudsman Akademik, 18, Ocak-Haziran, 13-47.

Özer, Y. E. (2017). İklim Değişikliği Yönetişimindeki Aktörlerin Analizi ve Türkiye. *Uluslararası Yönetim İktisat ve İşletme Dergisi, 13*(4), 833–851.

Uğur, Ö., & Aksoy, M. (2023). İklim Değişikliğinin Göç Üzerindeki Etkisi, 2. *Bilsel International Truva Scientific Researches and Innovation Congress*, 16-17 December, 2023, Çanakkale /Türkiye.

Uğur, Ö., & Cengiz, E. (Eds.). (2023). *İklim Değişikliği Siyaseti*. Nobel Basım Yayım.

Uğur, Ö., & Doğan, K. C. (Eds.). (2019). *Contemporary Debates on Politics and Public Administration in the Postmodern Era*. Peter Lang. DOI: 10.3726/b16122

Yılmaz Uğur, C. (2021). In Dogan, K. C., Ugur, Ö., & de Lyon, L. (Eds.), *Struggling With Covid-19 Outbreak in The Framework of Global Public Goods, Public Administration and Public Finance Research* (pp. 292–311).

Yılmaz Uğur, C. (2022). The Polluter Pays Principle within the Framework of the European Union Emissions Trading System. *Abant Sosyal Bilimler Dergisi, 22*(2), 862–872. DOI: 10.11616/asbi.1102031

Chapter 2
Co-Housing in Austria as an Alternative Participatory Living Against Climate Change

Ozge Yalciner Ercoskun
https://orcid.org/0000-0003-2734-0374
Faculty of Architecture, Gazi University, Turkey

ABSTRACT

Climate change necessitates innovative approaches to sustainable urban development. Conventional housing models often exacerbate environmental and social challenges through resource inefficiency, high carbon emissions, and social fragmentation. This chapter investigates collaborative housing (co-housing) as a sustainable alternative, emphasizing shared spaces, participatory governance, and ecological design. Focusing on case studies in Vienna and in Lower Austria, it analyzes how energy-efficient construction, renewable energy systems, and sustainable mobility reduce ecological footprints while enhancing social cohesion. Additionally, the chapter examines Vienna's progressive housing policies that facilitate co-housing through targeted subsidies and incentives. While challenges such as finance and scalability remain, co-housing offers significant potential for climate change mitigation and urban resilience, with transferable insights for global sustainability efforts.

INTRODUCTION

Climate change, as a defining global challenge of the 21st century, demands innovative approaches to sustainable living and urban resilience. Traditional housing

DOI: 10.4018/979-8-3373-1310-8.ch002

models often perpetuate resource inefficiency, social isolation, and high carbon footprints, exacerbating environmental and societal challenges. In response, co-housing has emerged as a participatory living model that combines private dwellings with shared spaces and cooperative management, fostering sustainability, community cohesion, and ecological stewardship. This chapter explores the potential of green co-housing projects as an alternative living model to mitigate climate change impacts, with a focus on exemplary initiatives in or near Vienna, Austria.

Co-housing originated as a grassroots movement emphasizing shared resources and democratic governance, offering residents a balance between privacy and community engagement. This model has gained traction in recent decades, aligning with global efforts to transition towards low-carbon, socially inclusive, and resilient urban environments. In Europe, and particularly in Austria, co-housing projects have evolved to integrate green certification standards, which enhance their environmental performance and contribute to national and regional climate goals. Vienna, the city with the highest quality of life in the world, recognized for its innovative housing policies and commitment to sustainability, provides a fertile ground for analyzing the interplay between co-housing and climate change mitigation with the support of City of Vienna.

The chapter begins by contextualizing co-housing within the broader framework of sustainable urban development. It traces its roots across Europe. The discussion underscores the principles of co-housing including collaborative decision-making, shared facilities, and asset pooling as vital mechanisms for reducing environmental footprints while fostering social capital. The integration of green certification systems, such as Passive House standards, LEED, or BREEAM, further elevates the environmental credentials of co-housing projects, ensuring adherence to energy efficiency, renewable energy use, and ecological building materials.

The analysis then shifts to Vienna and its surrounding regions, examining several green co-housing projects as case studies. Vienna's progressive housing policies, underpinned by subsidies and incentives for cooperative and ecological living arrangements, have catalyzed the development of innovative co-housing communities. Prominent examples include:

1. **Wohnprojekt Wien**: Located in Vienna's 20th district, this project epitomizes the synergy between co-housing and environmental sustainability. Certified under stringent ecological standards, Wohnprojekt Wien incorporates energy-efficient building design, solar energy systems, rainwater harvesting, and urban gardening spaces. The project's participatory governance structure empowers residents to collectively manage resources and maintain ecological best practices, reducing their overall carbon footprint.

2. **Bikes and Rails**: Situated in Vienna, this project emphasizes sustainable mobility and urban living. By prioritizing bicycle-friendly infrastructure and integrating with public transit options, Bikes and Rails promotes low-carbon lifestyles among its residents. Shared facilities and energy-efficient designs further enhance its environmental impact.
3. **Gleis21**: Located in Vienna's Sonnwendviertel neighborhood, Gleis21 is a co-housing project that combines artistic expression, social inclusion, and ecological design. The building adheres to Passive House standards and incorporates renewable energy systems, shared spaces, and community-driven cultural initiatives, making it a hub for sustainable and creative urban living.
4. **Auenweide**: Situated near Vienna, in Lower Austria, Auenweide stands out for its integration with nature and emphasis on biodiversity. The project includes extensive green spaces, communal gardens, and ecological building practices that prioritize local materials and energy efficiency. Its participatory governance ensures that ecological values are embedded in daily living practices.

Beyond Vienna's city center, co-housing projects in suburban areas provide insights into the scalability of participatory living models. These projects often leverage the surrounding natural landscapes to enhance biodiversity, integrate local food systems, and create low-carbon mobility solutions, such as car-sharing programs and bike-friendly designs.

The chapter evaluates these projects against key climate change mitigation metrics, including energy consumption, carbon emissions, and waste reduction. It also assesses their contributions to social sustainability, such as fostering inclusive communities, reducing social isolation, and promoting participatory governance. By examining the synergies between ecological and social goals, the analysis provides a nuanced understanding of the co-housing model's potential to address multifaceted challenges posed by climate change.

Moreover, the chapter investigates the policy and institutional frameworks that have enabled the success of green-certified co-housing projects in Vienna. It highlights the role of subsidies, tax incentives, and public-private partnerships in facilitating the adoption of sustainable building practices and community-driven housing models. Vienna's integration of co-housing into its broader urban planning and climate strategies serves as a replicable model for Turkish cities aiming to promote low-carbon, participatory living arrangements.

The discussion extends to the limitations and challenges of co-housing as a climate mitigation strategy. Issues such as affordability, scalability, and the need for robust governance frameworks are critically examined. The chapter argues that while co-housing offers significant environmental and social benefits, its widespread adoption requires overcoming systemic barriers, including financial constraints, regulatory hurdles, and cultural resistance to shared living arrangements.

In conclusion, the chapter posits that co-housing, particularly when integrated with green certification standards, represents a viable and impactful alternative to traditional housing models in the face of climate change. By fostering collaborative consumption, resource efficiency, and community resilience, co-housing projects not only mitigate environmental impacts but also address social and economic inequalities. The case studies from Vienna exemplify how innovative policy interventions and grassroots initiatives can converge to create sustainable, inclusive, and climate-resilient urban futures.

The findings have broader implications for policymakers, urban planners, and community organizers seeking to promote participatory living models as part of comprehensive climate action strategies. The chapter calls for further research and cross-regional collaborations to scale up the co-housing model and explore its potential in diverse socio-economic and cultural contexts. Ultimately, the integration of co-housing into global climate agendas can catalyze a transformative shift towards sustainable and equitable living environments, contributing to the collective effort to combat climate change.

CONCEPTUAL FRAMEWORK

Collaborative housing refers to a variety of housing models where residents collectively participate in the design, development, and management of their living spaces. This concept encompasses a range of self-organized and participatory housing forms, such as co-housing, housing cooperatives, and Community Land Trusts (CLTs) (Lang et al., 2018; Czischke et al., 2020; Czischke, 2019). One of the defining characteristics of collaborative housing is collective self-organization, where residents actively engage in planning and managing their housing, often sharing spaces and resources to foster a sense of community (Czischke et al., 2020; Czischke, 2019). There are diverse models of collaborative housing, including resident-led cooperatives and self-build initiatives, each with distinct structures and objectives (Czischke et al., 2020; Griffith et al., 2022). Additionally, these housing models often feature unique property regimes, characterized by limited common property and self-governance, setting them apart from conventional private or public housing (Bossuyt, 2021).

The motivations behind collaborative housing and its benefits are multifaceted. One of the key advantages is the promotion of social inclusion and cohesion, as these housing models enhance social ties and encourage community engagement among residents (Czischke et al., 2020). Affordability and sustainability are also significant factors, as shared resources and spaces help reduce living costs while promoting environmentally responsible living practices (Brysch & Czischke, 2021; Czischke et al., 2020). Furthermore, collaborative housing fosters empowerment and control, allowing residents greater influence over their living environment, which can contribute to overall satisfaction and well-being (Thompson, 2018; Czischke, 2019).

Despite these benefits, collaborative housing faces certain challenges. Complex stakeholder relationships often arise, requiring effective collaboration between residents, housing providers, and local authorities to ensure successful implementation and management (Czischke, 2018; Szemző et al., 2019). The legal and institutional frameworks within which these projects operate also play a crucial role in their success, as they are influenced by the regulatory and economic environment, as well as the level of support from local authorities (Szemző et al., 2019).

Overall, collaborative housing represents a shift towards more participatory and community-oriented living arrangements. While it offers substantial benefits in terms of affordability, sustainability, and social cohesion, it also demands careful navigation of stakeholder relationships and supportive legal structures. As this model continues to gain traction worldwide, it evolves and adapts to diverse cultural and institutional contexts, demonstrating its potential to redefine urban living in the future.

BENEFITS AND CHALLENGES OF CO-HOUSING FOR SUSTAINABILITY

Co-housing communities are increasingly recognized as a viable model for promoting sustainable living. These communities aim to reduce environmental impact through shared resources and collaborative living arrangements. The concept of co-housing is intertwined with sustainability goals, addressing both ecological and social dimensions of urban development. Research indicates that co-housing communities often achieve lower ecological and carbon footprints compared to traditional housing models. A systematic review of quantitative studies highlights that intentional communities, including co-housing, demonstrate significant improvements in environmental sustainability, supporting claims of reduced ecological impact (Daly, 2017). These communities are designed to optimize resource use and minimize waste, contributing to broader sustainability objectives.

Reduced Energy Consumption: Co-housing communities often incorporate sustainable building practices that significantly lower energy consumption for heating, cooling, and domestic water heating, leading to reduced operational costs and environmental impact (Coimbra & Almeida, 2013).

Low Carbon Lifestyle: By sharing resources and facilities, such as gardens and workshops, co-housing promotes a low carbon lifestyle, reducing the overall ecological footprint of its residents (Wang et al., 2020; Daly, 2017). The governance of sustainable urban development through co-housing is explored in cities like Hamburg and Gothenburg. These cities have implemented programs to support self-build groups and co-housing as part of their urban restructuring efforts. The interaction between political institutions, civil society, and private actors is crucial in promoting co-housing as a sustainable urban development strategy (Scheller & Thörn, 2018).

Affordability: Co-housing can be more affordable than traditional housing due to shared resources and facilities, as well as the potential for tax benefits and reduced construction costs (Coimbra & Almeida, 2013; Hashim et al., 2021).

Cost Savings: Sustainable construction methods in co-housing lead to savings in energy and water consumption, which can decrease the payback period for investments in sustainable features (Coimbra & Almeida, 2013).

Co-housing is not only about environmental sustainability but also about creating socially and economically sustainable communities. Studies show that co-housing can foster vibrant community life, enhance social connections, and provide affordable living environments. However, the scale of these communities is often small, and there is a need for more quantitative evidence to substantiate these claims (Tummers, 2016). The integration of sustainable practices in cooperative housing can lead to substantial energy savings and improved environmental comfort (Coimbra & Almeida, 2013). Co-housing fosters strong social bonds and community engagement, which can enhance social well-being and reduce feelings of isolation (Wang et al., 2020; Thankamoniyan & Jagadisan, 2021). The collaborative nature of co-housing encourages shared responsibilities and decision-making, which can strengthen community ties and promote a sense of belonging (Thörn et al., 2019; Thankamoniyan & Jagadisan, 2021).

Co-housing represents a promising model for achieving sustainability in urban settings by reducing environmental impact and fostering social cohesion. However, challenges such as finance, governance, and integration into existing urban frameworks need to be addressed to realize its full potential. Further research and policy support are essential to expand the reach and impact of co-housing communities in promoting sustainable development. Co- housing offers several benefits for sustainability, encompassing social, economic, and environmental dimensions. It is recognized as a sustainable housing model that fosters community living and resource sharing, contributing to reduced ecological footprints and enhanced social interactions. Co-housing presents a viable model for sustainable living by integrating environmental, economic, and social benefits. It reduces ecological footprints, offers cost savings, and fosters community engagement, making it a promising approach to address contemporary urban and ecological challenges. However, its success depends on effective integration with urban policies and thoughtful design practices with the support of local governments.

KEY GOVERNANCE MODELS

Co-housing governance models for sustainability focus on collaborative and participatory approaches that integrate social, economic, and environmental goals. These models often involve interactions between civil society, political institutions, and private actors to promote sustainable urban development.

Self-Build and Co-Housing Groups: In cities like Gothenburg and Hamburg, self-build co-housing groups are supported as part of urban restructuring efforts. These groups aim to create sustainable housing alternatives, though challenges such as political and economic constraints persist (Scheller & Thörn, 2018).
Cooperative Housing Frameworks: In South Africa, a framework for sustainable housing cooperatives includes policy and legislation, support services, education, training, and governance. This framework is designed to guide government policy and cooperative governance (Wyk & Jimoh, 2015).
Instituonal Capital and Housing Alliances: In Vienna and Paris, housing alliances interact with institutional actors to promote social sustainability. These alliances focus on governance features and strategies to ensure housing affordability and socio-spatial equity (Paidakaki & Lang, 2021).
Ostrom's Design Principles: Swedish housing cooperatives are analyzed using Ostrom's principles for long-term sustainability. Adjustments in governance structures are suggested to enhance innovation and collective action (Vogel et al., 2016).
Public-Cooperative Policy Mechanisms: Cooperative housing is supported by public policies at various levels, which are crucial for expanding beyond niche markets. These policies focus on production, access, management, and maintenance to ensure long-term sustainability (Ferreri & Vidal, 2021).
Co-Housing Models: Across Europe, collaborative housing models, including cooperatives and community land trusts, emphasize social inclusion, affordability, and environmental sustainability. These models require collaboration among residents and external stakeholders (Czischke et al., 2020).

Governance models for co-housing emphasize collaboration, policy support, and integration of social, economic, and environmental objectives. Successful models often involve partnerships between civil society, government, and private sectors to address sustainability challenges and promote inclusive, affordable housing solutions.

Co-housing communities often adopt participatory governance models that align with principles of sustainability and resilience in the face of climate change. Central among these are consensus-based decision-making and sociocracy, both of which emphasize shared responsibility, transparency, and inclusivity. These models empower residents to actively engage in ecological decision-making, such as implementing renewable energy systems, energy-efficient retrofitting, and sustainable mobility options. Cooperative ownership structures further reinforce environmental stewardship by fostering collective investment in low-carbon infrastructure and shared resource management. Additionally, hybrid models that integrate legal condominium frameworks with community charters allow for formal accountability while preserving grassroots democratic practices. Overall, co-housing governance acts as a catalyst for climate action by embedding ecological values into daily living and institutional structures.

METHODOLOGY

The chapter builds on literature review, desktop research for project databases and online resources of intentional communities, co-housing projects and networks, videos and conference exchanges. For the Austrian cases, field trips and interviews with experts, planners, residents and other activists took place in 2023 and 2024. The selected cases are described following the typology of green co-housing projects.

Ten expert interviews were conducted between October and December 2023, using a semi-structured format that allowed for flexibility in questioning. Five interviews took place face-to-face, two were conducted online, and three were answered via email. The interviews aimed to capture diverse perspectives on urban-rural planning and sustainability, focusing on key themes such as cohousing, and participatory planning. The participants included experts from various fields, including environmental science, social ecology, media, sociocracy, community planning, and energy management. Their insights provided a broad understanding of governance structures, collaborative processes, and the role of networks in fostering sustainable communities.

A content analysis was conducted to identify recurring themes and insights from the interviews. Experts emphasized the importance of inclusive, community-driven processes, strong municipal networks, and clear governance structures in developing low-carbon communities. They also highlighted challenges such as bureaucratic inertia and limited public engagement to enhance participatory decision-making.

These findings offer valuable guidance for shaping future sustainable communities through participatory governance and strategic resource allocation.

AUSTRIAN CASE STUDIES

The Building Together and Living Initiative (Inigbw.org) is the platform which promotes co-housing projects and serves as a networking area for the ones interested in communal lifestyle. The database contains 128 records in Austria which open to new residential projects. The flat vacancy adds can be found in this database.

EU projects are made for co-housing practices for urban and rural parts of Austria. DEMOCHANGE was conducted on Alpine Regions which promotes intentional communities of rural parts for co-housing (demochange.at). Wankiewicz (2015) sees co-housing as an innovative potential transition from newly built secondary or primary detached houses to flexible living for disadvantaged groups in rural parts where a shrinking population and decline, with the re-use of empty housing stock. Co-hope is another EU project with 5 partners in Europe which will be ended in 2025 (CO-HOPE website). It explores interdisciplinary innovative approaches in collaborative living in the pandemic era. The solutions will contribute to a transition in the housing development to a more equal and sustainable community projects. The effect of crises such as Covid-19 pandemic or Ukraine-Russia War have shown the need of community projects in living, food and energy resilience (urbaninnovation.at).

Many co-housing projects use sociocracy for their organization and sustainability today. One of them is Wohnprojekt Wien. After the long hours and arguments, they selected sociocracy for having less stress, time consuming and deep democracy (Wohnprojekt.Wien.org). They organize themselves socioractically from the beginning a decade ago. In case of sociocracy, The solutions are found together by a consensus which no one has a serious objection. Many different ecological projects, solidarity activities are applied by small independent circles. These circles are led by two members and working on trust and wisdom. There is a management circle which coordinate tasks and organize cross-circle decisions. Important decisions are made in the larger circle by consensus with a shared vision. They have also built an association for sustainable living, their number is around 70 members who wants to seek alternative ways of living in solidarity and service to Earth. Some of the members are also workers of Sociocracy Center of Austria which is an active training, consulting and advice center for the organizations and neighborhoods (Soziokratiezentrum.org).

Wohnprojekt Wien is the first project built ten years ago in the 2nd district of Vienna. A number of companies have settled in the ground floor such as an organic food café and shop, an architectural office, physiotherapist, family consultancy and a shiatsu practitioner. There are repair rooms, bicycle rooms, cloth-book exchange rooms in the basement floor (Figure 1). A participatory planning process was conducted with the residents and the building was planned with these ideas. The building has 39 residential units which consist of 1 to 6 room flats. Each flat has a different plan which have their own balconies. There are also around 700 square meters of common areas: kitchen, children's playroom, roof garden, sauna, library, workshop, laundromat, event rooms and guest apartments. The central, open staircase serves as an interior access and communication space for the residents. The passive house standard has been reached: the roof has a 100 square meter photovoltaic system, a groundwater heat exchanger, controlled living space ventilation and mineral insulation. The facade is made of wood.

Figure 1. Wohnprojekt Wien

Wohnprojekt Wien, a pioneering co-housing initiative in Vienna's 2nd district, exemplifies the adoption of sociocracy for organizational sustainability. Established a decade ago, it prioritizes stress reduction, efficiency, and deep democracy through sociocratic principles. Decision-making involves consensus where objections are earnestly considered. Various ecological and solidarity projects are managed by small, autonomous circles overseen by two trusted members. A management circle coordinates tasks and facilitates cross-circle decisions. Key decisions are made in larger circles, aligning with a shared vision. Comprising about 70 members committed to alternative, Earth-centric living, Wohnprojekt Wien has fostered a community dedicated to solidarity and sustainability. Some members are also affiliated with the Sociocracy Center of Austria, an active consultancy and training hub for organizations and neighborhoods seeking sociocratic practices.

The Wohnprojekt Wien building accommodates diverse commercial enterprises on its ground floor, including an organic café, architectural office, and wellness services (Figure 2). A participatory planning process involving residents shaped the building's design, resulting in 39 unique residential units with individual balconies. The building features extensive communal spaces, such as kitchens, playrooms, and a roof garden, totaling around 700 square meters. Achieving passive house standards, it boasts eco-friendly features like a photovoltaic system, groundwater heat exchanger, and wood facade, exemplifying sustainable architecture and community living.

Figure 2. Wohnprojekt Wien terrace, garden, bicycle parking and studio

The Wohnprojekt Wien is the first co-housing project to implement sociocracy from its inception, serving as a model for subsequent initiatives. Celebrating their 10th anniversary along with the Sociocratic Center, they initially faced challenges as pioneers, navigating a system where community participation in building site contests was indirect. However, there's ongoing debate about social inclusivity in such projects, leading to political resistance. While the generous common areas may reduce personal space per inhabitant, they contribute to a lower carbon footprint. The community incorporates sustainable features like solar panels and a shared garden. They have also fostered connections with neighboring projects like "Hauswirtschaft," fostering synergy and community awareness through events like a yearly flea market.

Co-housing projects consist of private flats which can be owned or rented with community rooms and gardens. The private, semi-private and public spaces flow together (Hofer, 2017). The flats represent private, corners, atriums, staircases are semi-private and kitchen, seminar rooms, repair rooms, garages and permaculture gardens can serve as public spaces where people can interact. The architecture of these buildings is an area of expertise, they are often residential but can consist of home-office, café, playroom, shops, sports, meeting facilities as mixed use buildings. The decisions are made in a socioratic circular way, they use participatory planning, shared economy methods and some co-housing projects provide space for refugees in a solidarity project.

Co-housing projects in Vienna accumulate near the main train station- Hauptbahnhof, in order to find an affordable housing. Firstly, an association is built and this association buy the plot and finds architecture and construction companies to make a mixed-use building. Then members of the association become the renters of the flats and uses the common spaces of the building. The famous ones are Bikes and Rails in Sonnwendviertel and Gleis 21 buildings today which provide seminar halls for the public to hire. In the case of Bikes and Rails, public spaces include a bicycle workshop, a café and community rooms (Figure 3). The construction of the building was financed by the local investors which raised in the main capital and the City of Vienna granted the group with a subsidised loan designated for social housing projects (CooperativeCity, 2023).

Figure 3. Bikes and Rails: Co-housing

Gleis 21 seems like a village rather than a house, it is a hybrid building made of wood and concrete. There are private balconies on one side and the communal arcade on the other side. The attic includes common areas such as a kitchen, library, playroom and sauna inside and open spaces outside. A communal kitchen and children's playroom allow an open interaction. The library with a small office area is a private space and a sauna and relaxation room provide more intimate space. In the ground floor, an interactive book exchange shelf and bulletin board, a multifunctional hall, an artist studio and a café are located for the residents and public (Figure 4). Common spaces, such as a shared kitchen, a herb garden, a cultural space which is open to the public, reinforce the integrative effect of the neighborhood. The building holds more than five international awards on architecture such as new Bauhaus Award and klimaaktiv gold Award in 2022. Gleis 21 building is also managed by sociocracy. There are seven thematic working groups which all tasks are processed. The working groups are self-decided and management group consists of the leaders of the working groups and are elected in every two years. This double connection of all working groups with the management group ensures that the diversity of perspectives is incorporated into the decisions of the management group (Gleis21, 2023).

Figure 4. Gleis 21: Co-housing

Building construction sector is responsible for a third of all global greenhouse gas emissions. Cement production causes 7% of these emissions. A radical change is needed towards green planning and construction. The Auenweide is an ecological community settlement in the northwest of Vienna in Lower Austria (Figure 5). It is located in St.Adra-Wördern 30 minutes far away as an alternatively financed communal housing project 5 years ago. A large field next to the forest was on sale and Markus Spitzer realized his dream. He found the architectural firm named `einszueins` to make this project and he created an asset pool for the investment. He generated 650.000 Euros to buy the land and start the project (Auenweide website). The project is a mini village in the small town which accommodates 8 apartment buildings. The low- energy buildings were built is ecological timber. The land is located in a floodplain of Danube where ground water table is very high. They use heat pump for heating and cooling and the electricity comes from solar panels where all rooftops are full of. All residents can meet in the common kitchen with dining area, comfortable living area, children's playroom, youth lounge and a large exercise room, multi-functional hall and their guests can stay in the guest rooms in two circular community buildings. They have a large square in the middle where people can gather and children can play. Each residential unit is planned as a full apartment with bedrooms, kitchen, bathroom. The terraces and the community garden in the middle of the site used for exchange, play, inspiration and gathering.

A total of 46 residents are together in this project as being part of the community and renters of the buildings, their vision was an alternative way of communal living in the rural area. The owner is the association. The apartments belong to the association, while the residents are tenants. All living and non-living people who have money in the asset pool are investors who are secured by the trustee with a lien in the land register. Thus, the trustee represents the community of all investors. This saves them the high costs that would arise from every new entry in the land register if they had to submit a new land register application for every deposit or payment. hey did not want to live in a classical single-family house in an isolated way but they implemented the idea of ecological lifestyle against climate change. Their architect, M. Zilker emphasizes that the insulation is from straw which is a very ecological for CO_2 reduction. The straw is the material simply composts like organic waste. It is eco-tech material in a circular economy and straw can replace 90% of insulation materials in Austria. They plastered clay in some of the walls. Two floor apartments have been mirrored in the land and the typological diversity came from the different needs and sizes of the residents, they differ 35 m_2 to 120 m_2.

They requested to preserve the forest about 1200 square meters and they used dense building and underground car parking which is obligatory according to building regulations. The forest is used by the children for discovering wilderness as an adjacent area (UBM website).

One of the residents emphasizes that he lives here with his family because of the forest and floodplain and everything is fresh. He encouraged to find people to invest their money for this project. New people deposit their Money and old people can take their money out. They pay 10.50 Euros per square meter, part of which goes back into asset pool. They generate excess electricity from solar panels and sell it to the grid. They are organized in workgroups as a sociocratic way based on consensus or no serious objection principle (Fluter website).

Person Ma started this project 5 years ago when his daughter was 1.5 years old. He asked people in his environment who had a child and they wanted to live together especially for their children. The age group was between 30-45 years old with the same vision. People generally find their age category to understand and communicate easier and better.

The decision making in the large group and trying to get consensus needed a long time, so there should be a whole change in the decision making. When he visited famous eco-village Findhorn 10 years ago, he saw 200 people sitting together in a large community hall fighting each other. Sociocracy is a powerful tool in decision making, leading, doing and feedback. This cycle is important when something went wrong and feedback, you make a new decision and put into flow again. All the work is done by the small groups in sociocracy. If community house group decides something, they have the right to make all decisions and implement by themselves. If someone wants to join them, he can and does the work together. They have to stay connected with the whole group for feedback. If the kitchen group puts rules for using the kitchen and there is something wrong, they have to say that and they have to integrate new decisions. If the problem is so big, their whole group is 25 people so they can handle it in third round. They can split the decisions such as policy or operational decisions. Operational ones belong to daily life and not interested to big group. Policy decisions should be discussed together and make consent. Sociocracy is a tool to organize better. If the organization is well-structured and elect leaders and trust for the next 2 years, it is solved in a hierarchy inside.

It was started 4 years ago, they had €0 in the starting point and they found investors from their environment, family and friends and they had 6.5 million € which is the biggest pool in the world which is professional for the community responsibility. A responsible person has the only access to this account and every €10 they spent, he pays. Only this person has to spend the money, no one can reach the account. This is the highest security and transparency. They create another asset pool nowadays and find people to invest, because bank credits are higher.

The most difficult point in this project was to persuade the farmer to sell the plot. He always raised the price, they bought the plot about 1.8 million € in the end in the corona time. Every two months, the farmers saw the capital saved in the asset pool, they made campaign to raise the money in the asset pool and farmer gave the plot in the end but it was stressful for him.

Auenweide is the member of Inigbw, initiative for cohousing. They have good relationship with the mayors and pioneer in the region. Person Ma defines Auenweide as in the middle, not an isolated eco-village, still in the town and a community living in flats. They have their own income, they work outside but they feel like a community, growing children together, they attach their place and know their neighbors. This kind of co-housing is easier to connect to normal world than the eco-village. They just put the costs and they divide it by the square meters that they use. They made punch evening recently with the other neighbors as well. In eco-villages a lot of knowledge is learnt for transition but everything is shared and there are very few private space. They made a slight difference but closer to normal world. They share their cars and tools in the workshop. There are positive effects on sustainability, security, social interactions and childcare to sustain the investors.

They used wood and straw in the buildings and some metal columns. The main insulation material is straw. Some clay walls are seen in the kitchen and meeting room. Aluminum is used for the window/door framing. They have a big underground garage which they had separate 1.5 parking area per flat that is obligatory for the municipality. The garage is half empty. They will pass to e-car sharing system. They have a bike parking space as well. They have a large open space in the middle of the site where there are some elements for the children, sand pool and fountain. They have rented a large field for playground and they grow fruit trees as well. They have a wood just next to the site to have a forest scenery from the windows. They have guest rooms for accommodating short periods, a seminar room and party room (Figure 5). Their all rooftops are covered by PV panels. They use heat pump to heat the floors and they coo it to circulate in summer time. They think to apply to klimaaktiv national program or DGNB green building/settlement certificate system to get silver label for the site.

The mayor is conservative, they had a good relationship in the construction phase. It was difficult for them to explain what they were doing; he did not believe them. After five years, he expressed his good feelings and saw the difference in the opening ceremony. He could not support them he has a distance now because the community is close to left and green parties but he appreciated. After living here 1.5 years, the difference is great. However, he finished the design of his kitchen a few weeks ago, the construction absorbed his energy.

Children go to the schools in St.Andra Wördern, the town has many kinds of schools, 2 schools for primary and secondary level and 5 kindergartens. They have small gardens and grow some herbs there. All the food coming from a field from very near to the site, they make community supported agriculture to have organic food (Gartnerhof Distelfink). This farm provides food in 3 hectares for 120 families, they only use manpower and make permaculture, there is an incredible demand. The other fields making monoculture is very inefficient according to this. The potatoes come from another potato farmer; they buy the all food from professionals.

Figure 5. Auenweide photos from the field trip on 13 December 2023, the exterior, courtyard, the seminar room and the communal kitchen

The construction sector's significant environmental impact underscores the need for green planning and construction practices. Auenweide, an ecological community settlement northwest of Vienna, exemplifies this shift towards sustainability. Built with low-energy, eco-friendly materials like timber, and powered by solar energy and heat pumps, Auenweide offers a model for environmentally conscious communal living. Designed as a mini-village with communal spaces and gardens, it fosters a sense of community while minimizing its ecological footprint. The community operates through workgroups, adhering to sociocratic principles of consensus or lack of serious objection. This collaborative approach ensures that decisions are made collectively, fostering a sense of shared ownership and responsibility among residents (Fluter website).

In the Auenweide project, 46 residents form a vibrant community dedicated to alternative communal living in a rural setting. The project is owned by an association, with residents renting apartments owned by the association. All individuals who contribute to the asset pool are considered investors, securing their interests through a trustee who represents the community in legal matters, thus streamlining administrative processes and reducing costs associated with land registry entries. Rejecting the traditional isolated single-family house model, the community prioritizes ecological living practices to combat climate change (Figure 5). Architect M. Zilker highlights the use of straw insulation as a key eco-friendly measure, emphasizing its effectiveness in reducing CO_2 emissions and its ability to compost naturally like organic waste. This approach aligns with principles of circular economy, where materials are reused and recycled efficiently. Clay plastering is also employed in some walls, adding to the ecological ethos of the project. The architectural design reflects the diverse needs and preferences of residents, with apartments ranging from 35 to 120 square meters. Mirroring two-floor apartments on the land and incorporating typological diversity ensures that residents can find spaces tailored to their requirements, fostering a sense of inclusivity and community cohesion.

In the Auenweide project, residents prioritize environmental preservation by preserving a 1200 square meter forest area, providing a natural space for children to explore and connect with wilderness (UBM website). Compliance with building regulations necessitates dense construction and underground car parking, optimizing land use and meeting legal requirements. Residents are drawn to the project by its proximity to the forest and floodplain, enjoying the fresh environment it offers. To fund the project, residents invite individuals to invest money, with new investors able to deposit funds while existing investors retain the option to withdraw. Each resident pays 10.50 Euros per square meter, with a portion returned to the asset pool. The project generates excess electricity through solar panels, which is sold back to the grid, contributing to its financial sustainability.

Person Ma tells that the Inigbw cohousing initiative, exemplified by their Auenweide project, bridges eco-conscious living with mainstream urbanity. Residents enjoy a strong sense of community while maintaining individual spaces and sharing resources like cars and tools. Sustainable construction, including straw insulation and PV panels, complements communal amenities such as playgrounds and guest rooms. Plans for e-car sharing and green building certification-Klimaaktiv underscore their commitment to sustainability and social cohesion.

According to Person O, The Klimaaktiv Standard, akin to LEED or Breeam, is a national green building and settlement rating system, developed from 2014-2019 to align projects with global climate goals. Becoming an expert requires broad knowledge, and evaluations involve various groups and site visits. Both urban and rural projects face the same standards, but rural areas may struggle to meet criteria. Public involvement and future ownership structures are key considerations. While direct resilience measurement is not conducted, the standard evaluates concepts addressing resilience. Projects aim for climate neutrality, with awards given for exemplary initiatives listed on the Klimaaktiv website. Urban areas hold promise for eco-settlements, with collaboration extending to global networks like the Transition Towns Network or Global Ecovillage Network.

Investing in an asset pool offers the reassurance of supporting non-capitalist endeavors and fostering community, rather than enriching banks. It represents socially responsible investment in worthwhile causes. Investors enjoy the flexibility to withdraw their funds without penalties with a three-month notice period. Many investors in Auenweide's pool are friends or prospective residents, accessing non-public bonds. Their website, legally deemed a public bond under the Alternative Financing Act, provides comprehensive contractual and legal details. With a record-breaking €6.5 million collected, Auenweide covered all construction costs. This model, commonly adopted by associations for co-housing and eco-villages, facilitates project funding with reduced risk compared to private loans. It also allows foreign investors secure involvement through land registry guarantees (Auenweide website).

Person Ma from Auenweide outlines the journey of a project that began in 2019 with no initial funding, relying on investments from friends, family, and the community. They managed to accumulate €6.5 million, the largest professional pool for community responsibility worldwide. This money is managed by a single responsible person, ensuring transparency and security. However, as bank credits became less favorable, they decided to establish another asset pool and seek further investment.

Interest rates on asset pools are typically negligible or lower than bank rates. However, this model requires meticulous attention, including robust contracts and land registry entries. It demands a level of legal and financial expertise to manage effectively, especially in explaining unfamiliar financing methods. While more prevalent in German-speaking regions like Austria, wealth pools represent an innovative approach to sustainable business, transcending traditional capitalism.

Markus Distelberger, the developer of asset pools, underscores their transformative potential: "Asset pools revolutionize financial and property management, enabling sustainable business models beyond conventional capitalism. They bridge the gap between financial resources and meaningful projects, enriching society as a whole."

Person Ro has a consultancy firm which gives service on co-housing, architecture and urban development. He is the Austrian consortium leader of Co-Hope collaborative housing project funded by JPI Urban Europe. He has focus on co-housing on urban types. First types of these projects started in 1960-70s, it is not new. There are lots of people dealing with these projects. There are many projects in Germany to solve affordable housing. But Austria has subsidized housing policy for non-profit developers they are quite accessible for almost everyone. They have more space for self-determination they have much aspects of how they live and surroundings. They are part of bigger group as community who have a common goal with the vision of sustainability. Co-housing projects are a way to achieve high levels of sustainability than standard housing. Energy efficiency standards are quite high in Austria, energy was important in 1980s, but it is not important now. Building materials are important such as building wood or bricks, straw. Mobility is another topic such as car-sharing or bike-sharing. Food is another topic, almost every co-housing groups organize food-coops. Best examples in co-housing are Gleis 21 and Bikes and Rails in Wiener Sonnwendviertel, an area developed near central train station in 10th district. There are 6 co-housing in this area trying to achieve high standards of sustainability. Gleis 21 and Bikes and Rails projects are both built in wood in subsidized housing. The residents of these projects are members of an association, this association either owns the house or rents the house from a limited profit developer then association rents the single apartment to the residents, so everyone is at the same time tenant and landlord as part of the group, they work like a cooperative model.

The old projects had plenary meetings and majority decided, this was very ineffective. Most of the projects use sociocracy in 15 years in some forms. Not all of them uses this tool very strictly but they take aspects of it. It is effective that every member of the groups has to contribute and to co-decide in the working group structure but making meetings in strict sense actually takes lots of time. Wohnprojekt Wien in 2nd district Is the predecessor of Gleis 21. They are built by the same architect and the same limited profit developer. But Wohnprojekt is a concrete building with a wooden façade but the basement floor and ground floor, staircases of Gleis 21 are concrete and the rest is wood, mostly local wood in 2019 which was affordable. Construction costs rose up today and it is quite expensive now. There are not many urban designers in this area, there are not many big differences from standard spaces. The only difference is the size of the plots. The co-housing projects are smaller than other subsidized housing projects in Vienna. There are some architects specialized in these projects but there are not so many. There is an office in Wohnprojekt Wien who worked a lot in Vienna and Lower Austria in this subject. There 5-8 architects worked in other projects. There are common spaces used by residents themselves such as sauna, common kitchen, workshop, children playroom, many of them are on the rooftop, terrace or in the ground floor and basement floor. For example, Wohnprojekt Wien has a courtyard in the back side as green space and they have very few car spaces. They have also common spaces that are open to public in the neighborhood. Wohnprojekt Wien has a café, permaculture garden. Sometimes City of Vienna sells plots in urban development areas for co-housing and they decide on quality criteria, not the price. They want the groups to offer things to the neighborhood such as an event space or a garden. If the group offers some space like that, it is easier to get a plot.

Person Ro saw Auenweide co-housing from outside and he found quite interesting, they used alternative financing in a large amount- 4.5 million Euro which is unbelievable. He is also the national coordinator of Co-hope project. Its aim is to compare the situation of co-housing in several European countries- France, Sweden, Spain, Austria and Finland in the aspects of affordability, social integration and health. The first two is standard for co-housing in many research projects but health is newer. They also want to develop recommendations. They made workshops in living in those projects to collect recommendations and they would like to publish a guide book for policy recommendations at political level for future to emphasize the importance of these projects. There are good examples of co-housing in Germany, France. Barcelona and Zurich, Hamburg and Munich have good examples at city level. Nordic countries have many good projects in the past but they keep their traditions. They changed their housing policies recently, nowadays Sweden or Denmark changed their systems.

CONCLUSION

The question arises such as: Is the co-housing a resilient settlement model and does it enable the transition to low-carbon settlement? In Austrian case, the co-housing projects are the role models that are being mini-village in the city. All co-housing and ecovillage projects in Austria use Dragon Dreaming method for building their vision in the beginning and then use sociocracy for all their decision-making process. The goal of sociocracy is to decentralize decision-making, prioritizing the involvement of specialized working groups rather than central leadership in making as many decisions as possible. So, sociocracy is one of the important keywords for being and sustaining a low carbon community in a resilient settlement model. The main reason for the success of the Austrian examples of cohousing and eco-communities is that their members have been trained in sociocracy and have successfully resolved their issues using this Dragon Dreaming and sociocratic approach.

The other important keyword is to build an asset pool for the co-housing or eco-village investments. An asset pool aggregates funds from numerous individuals, with a portion being invested while some remains readily accessible. This structure facilitates smooth transitions between payments and new contributions. Austrian laws allow this alternative financing system. There are many good examples of asset pool such as Wohnprojekt Wien or Auenweide co-housing project.

Tummers (2016) analyzed many co-housing initiatives around Europe and he found some common features such as: their organizational unit fits a spatial location, they are collaborated on mutual living, they are seeking an alternative sustainable lifestyle and they live in a mixed income and mixed-use space which is self-managed.

There are some factors influencing the projects such as the role of the architects, planners, their locations, support from local governments or regulators for tenure. This is an affordable and alternative housing model and all fields of urban planning, urban design, strategic development and engineering and legal organization need to be combined. The co-housing model is produced for the lack of affordable response by conventional housing market. As a planner, it is an experimental field for testing the housing demand. It is also worth analyzing the spatial and social dynamic. Do intentional communities need new environments or transform attitude and communication? This question can contribute to the future understanding to multi-disciplines and international knowledge transfer for planning. There is a need for co-housing architects and planners specializing in social housing, as well as spatial scientists who can devise sustainable solutions.

Intentional communities in co-housing are the good examples for the low carbon living. For example, food is collectively ordered from organic wholesalers, or from local community-supported agriculture farms, or cultivated in allotments and horticultural areas, in place of individual food shopping visits and subsequently shared for preservation, storage, cooking, and consumption to reduce the need for packing and shipping. A network of low-carbon practices is created by the strong interdependence of one practice's performance on another's. For example, group cooking and food ordering are essential components of the collective dining practice. The low carbon experiences of intentional communities are not easily transferable to other rural and urban contexts since they are defined by certain structural constraints and members who share a desire to live a more sustainable life. Nonetheless, they have been viewed as "laboratories" or "models," providing knowledge and motivation for broader-scale societal change (Schafer et al. 2018). The cohousing system offers significant advantages for various demographics including the elderly, young people, and the broader population. Living in a community post-Covid-19 can extend life expectancy, enhance social connections, and provide affordable rental housing options. Moreover, communal production activities promote healthy nutrition, while environmentally friendly practices increase awareness about sustainability and low carbon communities.

In Austria, climate change mitigation and sustainable urban development are deeply intertwined, with co-housing emerging as a key strategy to reduce environmental impact while fostering social resilience. The country's commitment to sustainability is reflected in its ambitious climate goals, which include reducing greenhouse gas emissions, promoting renewable energy, and increasing energy efficiency in the built environment. Co-housing aligns with these objectives by integrating ecological building practices, such as Passive House standards, solar energy systems, and green roofs, which significantly lower energy consumption and carbon footprints.

The City of Vienna actively supports co-housing initiatives through a combination of financial subsidies, land allocation policies, and partnerships with non-profit developers. Co-housing projects, often structured as collective ownership models, benefit from general housing subsidies that help maintain affordability and accessibility for low- and middle-income households. A key mechanism is Vienna's "concept tendering" process, where municipally owned land is allocated to co-housing groups based on project quality rather than price, ensuring that land costs remain fixed and encouraging innovation and social diversity in urban development. These groups typically partner with Limited-Profit Housing Developers (LPDs), who manage construction and ensure cost-based, stable rents that are protected from market speculation. These illustrate Vienna's commitment to sustainable, community-driven urban living.

Lastly, Austria's emphasis on community-driven housing solutions supports participatory decision-making and resource-sharing, reducing waste and promoting low-carbon lifestyles. Vienna, a leader in sustainable housing policies, has actively supported co-housing initiatives through subsidies, land allocation, and urban planning strategies that encourage cooperative living. By blending environmental consciousness with social inclusivity, co-housing projects serve as models for climate-resilient urban living, demonstrating how collective action can drive meaningful ecological change.

ACKNOWLEDGEMENT

This research was supported by TÜBİTAK (The Scientific and Technological Research Council of Türkiye) under project number 1059B192201650. I would like to express my gratitude to Prof. Dr. Alois Humer for providing me with the opportunity to work at the Institute for Urban and Regional Research (ÖAW ISR) of the Austrian Academy of Sciences. I would also like to thank Maeve Hofer for her assistance with communication, fieldwork, and documentation in the early stages of the research.

REFERENCES

Auenweide. (2024). Auenweide website: https://www.auenweide.at/

Bo Brysch, S. L., & Czischke, D. (2021). Affordability through design: The role of building costs in collaborative housing. *Housing Studies*, *37*(10), 1800–1820. DOI: 10.1080/02673037.2021.2009778

Bossuyt, D. (2021). Who Owns Collaborative Housing? A Conceptual Typology of Property Regimes. Housing. *Theory and Society*, *39*(2), 200–216. DOI: 10.1080/14036096.2021.1888788

Carriere, Z., & Lesnikowski, A. (2024). The climate change adaptation readiness of co-operative housing in Nova Scotia, Canada. *Environmental Policy and Governance*, *34*(6), 724–739. DOI: 10.1002/eet.2116

CO-HOPE. https://jpi-urbaneurope.eu/project/co-hope-collaborative-housing-in-a-pandemic-era/, 11/12/2023

Coimbra, J., & Almeida, M. (2013). Challenges and benefits of building sustainable cooperative housing. *Building and Environment*, *62*, 9–17. DOI: 10.1016/j.buildenv.2013.01.003

Cooperative City. https://cooperativecity.org/2022/02/04/bikes-and-rails-co-housing-with-solidarity-and-an-environmental-mission/, 18/12/2023

Czischke, D. (2018). Collaborative housing and housing providers: Towards an analytical framework of multi-stakeholder collaboration in housing co-production. *International Journal of Housing Policy*, *18*(1), 55–81. DOI: 10.1080/19491247.2017.1331593

Czischke, D. (2019). Collaborative Housing: The Resurgence of Collectively Self-organised and Self-managed Housing in Europe, 5

Czischke, D., Carriou, C., & Lang, R. (2020). Collaborative Housing in Europe: Conceptualizing the Field. Housing. *Theory and Society*, *37*(1), 1–9. DOI: 10.1080/14036096.2020.1703611

Daly, M. (2017). Quantifying the environmental impact of ecovillages and co-housing communities: A systematic literature review. *Local Environment*, *22*(11), 1358–1377. DOI: 10.1080/13549839.2017.1348342

Ferreri, M., & Vidal, L. (2021). Public-cooperative policy mechanisms for housing commons. *International Journal of Housing Policy*, *22*(2), 149–173. DOI: 10.1080/19491247.2021.1877888

Fluter, (2024). Fluter website: https://fluter.de/wohnprojekte-oesterreich-zusammenleben

Griffith, E., Jepma, M., & Savini, F. (2022). Beyond collective property: A typology of collaborative housing in Europe. *International Journal of Housing Policy*, *24*(1), 121–141. DOI: 10.1080/19491247.2022.2123272

Hashim, M., Awaluddin, Z., R, A., Sarkum, S., Sholiha, A., & Aziz, A. (2021). Architectural Design in Rethinking the Future of Co-Housing in Malaysia. Journal of Design and Built Environment. DOI: 10.22452/jdbe.vol21no3.6

Hofer, A. (2017) Cohousing projects: The role of relationships and its dynamics, Masters thesis, J.Keppler University, Linz, Austria

Kamb, Z., Shi, L., Silva, S., & Spicer, J. (2022). Resident-Owned Resilience: Can Cooperative Land Ownership Enable Transformative Climate Adaptation for Manufactured Housing Communities?. *Housing Policy Debate*, 33, 1055 -. DOI: 10.1080/10511482.2021.2013284

Mellner, C., Niemi, M., Pöllänen, E., & Osika, W. (2021). *Enhancing social and individual sustainability in urban co-living*. International Journal of Housing Markets and Analysis., DOI: 10.1108/IJHMA-09-2020-0117

Paidakaki, A., & Lang, R. (2021). Uncovering Social Sustainability in Housing Systems through the Lens of Institutional Capital: A Study of Two Housing Alliances in Vienna, Austria. *Sustainability (Basel)*, *13*(17), 9726. DOI: 10.3390/su13179726

ADDITIONAL READING

Ruíz, M., & Mack-Vergara, Y. (2023). Resilient and Sustainable Housing Models against Climate Change: A Review. *Sustainability (Basel), 15*(18), 13544. DOI: 10.3390/su151813544

Schäfer, M., Hielscher, S., Haas, W., Hausknost, D., Leitner, M., Kunze, I., & Mandl, S. (2018). Facilitating low-carbon living? A comparison of intervention measures in different community-based initiatives. *Sustainability (Basel), 10*(4), 1047. DOI: 10.3390/su10041047

Scheller, D., & Thörn, H. (2018). Governing 'Sustainable Urban Development' Through Self-Build Groups and Co-Housing: The Cases of Hamburg and Gothenburg. *International Journal of Urban and Regional Research, 42*(5), 914–933. DOI: 10.1111/1468-2427.12652

Schmidt, K., & Walz, A. (2021). *Ecosystem-based adaptation to climate change through residential urban green structures: co-benefits to thermal comfort, biodiversity, carbon storage and social interaction.* One Ecosystem., DOI: 10.3897/oneeco.6.e65706

Soziokratiezentrum.org. (2023). https://soziokratiezentrum.org/ueber-soziokratie/ressourcen/

Szemző, H., Gerőházi, É., Droste, C., & Soetanto, D. (2019). Towards a Collaborative Housing Initiative: The Role of Local Authorities. *Built Environment, 45*(3), 398–415. DOI: 10.2148/benv.45.3.398

Thankamoniyan, N., & Jagadisan, S. (2021). *Enhancing Oneness through a Co-Housing Community*. Space and Culture., DOI: 10.20896/saci.v9i1.1126

Thompson, M. (2018). From Co-Ops to Community Land Trusts: Tracing the Historical Evolution and Policy Mobilities of Collaborative Housing Movements. Housing. *Theory and Society, 37*(1), 82–100. DOI: 10.1080/14036096.2018.1517822

Thörn, H., Larsen, H., Hagbert, P., & Wasshede, C. (2019). Constraints and possibilities for co-housing to address contemporary urban and ecological crises, 202-213. DOI: 10.4324/9780429450174-11

Tummers, L. (2016). The re-emergence of self-managed co-housing in Europe: A critical review of co-housing research. *Urban Studies (Edinburgh, Scotland), 53*(10), 2023–2040. DOI: 10.1177/0042098015586696

Tummers, L. (2017). Learning from co-housing initiatives: Between Passivhaus engineers and active inhabitants. *A+BE: Architecture and the Built Environment*, 1-282. DOI: 10.7480/ABE.2017.14

UBM, (2024). UBM website: https://www.ubm-development.com/magazin/auenweide-einszueins-architektur/, 13.01.2024

Vogel, J., Lind, H., & Lundqvist, P. (2016). Who is Governing the Commons: Studying Swedish Housing Cooperatives. Housing. *Theory and Society*, *33*(4), 424–444. DOI: 10.1080/14036096.2016.1186730

Wang, J., Hadjri, K., Bennett, S., & Morris, D. (2020). The role of cohousing in social communication and sustainable living environments. *WIT Transactions on the Built Environment*, *192*, 247–258. DOI: 10.2495/GD170201

Wien, W. (2023). Website: https://wohnprojekt.wien/leben.

Wyk, K., & Jimoh, R. (2015). Framework for the sustainability of housing co-operatives in South Africa. *Journal of Construction Project Management and Innovation*, *5*(1), 1100–1114.

KEY TERMS AND DEFINITIONS

Co-Housing Concept: A form of collaborative housing that combines private living spaces with shared communal areas.

Collaborative Housing: Various resident-led and participatory housing models. emphasize social inclusion, affordability, and environmental sustainability, and involve collaboration among residents and external stakeholders.

Community and Autonomy: Co-housing involves collaboration in daily activities like cooking and maintenance, fostering social bonds while respecting individual privacy and autonomy.

Sustainability and Urban Development: Co-housing is often seen as a sustainable housing form that addresses social, ecological, and economic challenges. It is contextualized within urban sustainability discourses, focusing on community building and autonomy within urban civil society and governance frameworks.

Design and Social Integration: The architectural design of co-housing aims to foster social integration and sustainable communities. It considers factors like demographic size, spatial configuration, and environmental psychology to enhance the social and economic benefits of co-housing.

Green-Certified Co-Housing: It refers to a collaboratively designed residential community that meets recognized environmental sustainability standards such as LEED, BREEAM, or Passive House. These communities emphasize energy efficiency, resource conservation, and eco-friendly construction while promoting shared spaces and collective living to reduce environmental impact and foster social connection.

Chapter 3
Climate Justice for the Sundarbans:
Integrating Local Tribal Health Rights With Global Policy Solutions

Oindrilla Ghosh
https://orcid.org/0000-0001-9609-0781
Christ University, India

Binod Kumar
https://orcid.org/0009-0002-9721-1367
Radha Govind University, India

ABSTRACT

Climate justice emphasizes the moral imperative to address the disproportionate impacts of climate change on marginalized communities while striving for global sustainability. As local communities often bear the brunt of environmental degradation, their needs must be prioritized to achieve equitable solutions. This concept acknowledges that climate issues intersect with social inequities, demanding actions that not only reduce carbon emissions but also address poverty, health, and access to resources. Effective climate justice strategies integrate diverse perspectives, fostering collaboration across borders and sectors. By aligning grassroots initiatives with global climate goals, we can create comprehensive solutions that enhance resilience and promote social justice. Ultimately, climate justice advocates for a holistic approach to sustainability, recognizing that local actions are fundamental to achieving global climate objectives.

DOI: 10.4018/979-8-3373-1310-8.ch003

INTRODUCTION

The Sundarbans, a UNESCO World Heritage Site and the largest mangrove ecosystem in the world, is a unique ecological treasure (Gómez-Baggethun et al., 2012) that spans across India and Bangladesh (Sahana et al., 2021). This vast delta region is home to a rich biodiversity, including the illustrious Bengal tiger, and functions as a critical buffer against climate-induced threats (Yohe and Tol, 2002) like rising sea levels and extreme weather events (Das & Hazra, 2020). However, the Sundarbans also face profound socio-economic challenges fueled by climate change (Cutter et al., 2003), which exacerbate existing inequalities and threaten the health rights of local communities (Kanjilal et al., 2010). Climate justice emerges as a vital framework for addressing these intersecting crises (Armitage, 2005), highlighting the need for equitable solutions that prioritize the health and wellbeing of marginalized populations (Ahsan & Warner, 2014).

Climate justice is rooted in the understanding that climate change disproportionately affects vulnerable communities those with the least resources and capacity (Adger et al., 2005) for adaptation are often on the frontlines of environmental degradation (Danda et al., 2019). In the Sundarbans, local populations rely on natural resources for their livelihoods, including fishing, agriculture, and forest products (Panda et al., 2016). Climate change, driven by global industrial practices, threatens their existence as intensified storms (Biesbroek et al., 2013), salinity intrusions, and unpredictable weather patterns disrupt traditional ways of life (Kanjilal et al., 2013). In this context, climate justice seeks to integrate local health rights recognizing the right to health as essential to survival into policy narratives (Smit and Pilifosova, 2003) that transcend local boundaries and engage with global climate politics (Savacool, 2018).

Integrating local health rights into global climate policy solutions is both a moral imperative and a practical necessity (Mistri, 2019). Health outcomes are directly influenced by environmental conditions, and the impacts of climate change manifest as increased morbidity and mortality from diseases (Berkes, 2009), mental health challenges, and food insecurity (Alam & Bhardwaj, 2020). For instance, extreme weather events can lead to malnutrition and waterborne diseases (Midgley, 1995), further compromising health systems already plagued by poverty and marginalization (Roy, 2018). By advocating for the health rights of Sundarbans communities within global climate frameworks, stakeholders aim to elevate local voices (Elliot, 1993), ensuring their needs and perspectives inform climate action plans (Rashad Al Mahamud et al., 2020).

Moreover, climate justice emphasizes accountability for historical injustices (Sovacool, 2011) particularly in the context of developed nations that have contributed disproportionately to greenhouse gas emissions (Saif et al., 2016). Global climate finance initiatives must recognize the unique vulnerabilities of regions (Moser and Ekstrom, 2010) like the Sundarbans and allocate resources that empower local populations to build resilience against climate impacts (Biswas, 2013). This involves not only addressing immediate health concerns (Pelling, 2003) but also investing in sustainable development practices that enhance livelihoods and safeguard the natural environment (Pramanik et al., 2021).

HISTORICAL EVOLUTION

The concept of climate justice has evolved significantly over the past few decades (Agrawal, 2008), emerging as a crucial discourse within the broader environmental justice movement (Chowdhury, 2019). In the Sundarbans, this evolution can be traced through a series of historical milestones that highlight the region's rich ecological heritage (Pelling & Manuel-Navarrete, 2011), changing climate dynamics, and the grassroots efforts of local communities advocating for their rights amid increasing environmental pressures (Adams et al., 2020).

Historically, the Sundarbans have been shaped by colonial forestry policies and land use practices (Mosher, 1979) that marginalized indigenous communities and disrupted traditional livelihoods (Molinari, 2017). British colonial rule in the 19th century introduced extensive deforestation to cultivate cash crops (Omer, 1979), fundamentally altering the landscape of the Sundarbans and exposing local communities to economic vulnerability (Satterthwaite & Mitlin, 2014). These historical injustices, rooted in colonial exploitation, established a legacy of marginalization (Paiva, 1977) that resonates today as communities confront the realities of climate change (Satterthwaite, Huq, Reid, Pelling, & Romero Lankao, 2007).

As awareness of climate change grew in the late 20th century (Healy, 2008), environmental activists began to connect global warming's impacts with social and economic inequalities (Soltesova, Brown, Dayal, & Dodman, 2014). The 1992 United Nations Earth Summit in Rio de Janeiro marked a pivotal moment in recognizing the link between environmental degradation and human rights (Ayers & Forsyth, 2009), laying the groundwork for climate justice advocates who would later highlight the unique vulnerabilities of regions like the Sundarbans (Cannon & Muller-Mahn, 2010). The call for climate justice gained momentum with the establishment of the Framework Convention on Climate Change (UNFCCC) and subsequent conferences (Peeters, 2012), emphasizing the need for equitable solutions in international climate agreements (Twigg, 2007).

In the Sundarbans, the early 2000s marked a significant shift, as local civil society organizations began mobilizing (Erickson, 2012) to address the adverse impacts of climate change on their communities (Pervin et al., 2013). Initiatives focused on community-based adaptations and resilience-building efforts, underscoring the need for local knowledge to inform climate action (Olhoff & Schaer, 2010). These grassroots movements not only challenged top-down bureaucratic approaches (Dominelli, 2010) but also sought to amplify the voices of marginalized groups (Anguelovski & Carmin, 2011), including women, indigenous peoples, and small farmers to ensure their health rights, livelihoods (Dubey et al., 2017), and cultural practices were prioritized in decision-making processes (Leck & Simon, 2012).

The impact of severe cyclones, such as Sidr in 2007 and Aila in 2009, further underscored the urgent need for climate justice in the Sundarbans (Roy, Hulme, and Jahan, 2013). These disasters exposed the vulnerabilities of local communities and prompted calls for both immediate relief and long-term strategies to enhance resilience (Anguelovski & Carmin, 2011). Civil society groups increasingly advocated for health rights to be integrated into climate adaptation plans (Mishra et al., 2021), emphasizing that access to healthcare, clean water, and nutritious food is critical for community survival (Tyler & Moench, 2012).

As global discussions around climate change intensified in the 2010s, the Sundarbans found its place on the international stage (Pelling, 2011). The region's struggles became emblematic of the broader battles for equity in climate justice discourse (Agrawal, Kononen, and Perrin, 2009). Initiatives such as the Paris Agreement in 2015 recognized the importance of integrating social justice into climate action, fostering international solidarity with vulnerable regions (Dodman and Mitlin, 2013). Nevertheless, the gap between policy rhetoric and actions remains a challenge, as local communities (Biswas & Nautiyal, 2020) continue to grapple with the immediate impacts of climate change and the slow pace of systemic change (Hurlimann & March, 2012).

THEORETICAL FRAMEWORKS

Climate justice is a burgeoning field of study that seeks to understand and address the disproportionate impacts of climate change on vulnerable populations (Few, Brown, and Tompkins, 2007). The Sundarbans, a unique mangrove ecosystem straddling India and Bangladesh, is one of the most affected regions (Rudra & Chattopadhyay, 2019) juxtaposing environmental degradation and socio-economic challenges, particularly for local tribal communities (Carmin et al., 2012). The theoretical frameworks for climate justice in the Sundarbans are rooted in various interdisciplinary approaches (Rudra & Chattopadhyay, 2018) that emphasize equity, rights, and the interconnectedness of local and global challenges (Roberts et al., 2012).

One prominent theoretical framework for understanding climate justice in the Sundarbans is the 'capabilities approach', conceptualized by Amartya Sen and Martha Nussbaum (Almansi et al., 2014). This framework posits that justice should be defined not merely in terms of resources (Satio et al., 2020) but in the actual capabilities or freedoms individuals have to lead a life they have reason to value (Syam, 2013). In the context of the Sundarbans, local tribal communities have historically faced systemic marginalization (Sarkar & Sanyal, 2019), leading to a lack of access to health care, education, and other essential resources (Brown, Dayal, & Rumbaitis del Rio, 2012). By applying the capabilities approach, researchers and activists can focus on enhancing the substantive freedoms of these communities (Hafiz Iqbal, 2020), enabling them to adapt to climate change impacts while also asserting their health rights (Corfee-Morlot et al., 2011).

Another vital framework is the 'political ecology' of climate change, which examines the socio-political dimensions of environmental issues (Dewulf, 2013). This framework recognizes that environmental crises are shaped by political, economic, and social structures, which often perpetuate inequities (Norris, Stevens, Pfefferbaum, Wyche, & Pfefferbaum, 2008). The Sundarbans' tribal communities are often caught in the crossfire of state policies, commercial interests, and global climate initiatives that may not adequately account for their needs (O'Brien, 2012). This highlights the importance of grassroots engagement, as local knowledge and leadership are essential for any climatic adaptation measures to be effective (Agrawal, 2010). Through political ecology, we can better understand how historical injustices, such as colonial exploitation (Roy, 2019) and contemporary neoliberal economic policies, contribute to the vulnerabilities faced by these communities (Folke et al., 2005).

Moreover, the 'Intersectionality' framework can be applied to analyze how overlapping identities (Mozumdar et al., 2018) such as ethnicity, gender, and class compound the vulnerabilities experienced by tribal communities in the Sundarbans (Smit and Wandel, 2006). For instance, women in these communities often face both climate-related challenges and gender-based discrimination, impacting their health and livelihood options (Berkes and Ross, 2013). By recognizing these intersecting issues, climate justice initiatives can be more targeted and effective (Dasgupta et al., 2021), ultimately contributing to a holistic approach to health rights and environmental sustainability (Adger et al., 2009).

Finally, the 'transnational advocacy network (TAN)' framework is crucial for understanding how local tribal health rights can be integrated with global policy solutions (Berkes, 2009). TANs, comprising various stakeholders such as NGOs, grassroots organizations, and international bodies, can amplify local voices in global discussions about climate change (Mertz et al., 2009). These networks can support advocacy efforts, help mobilize resources (Ahsan & Warner, 2014), and foster collaborative approaches to policy-making that respect local knowledge and rights (Eakin and Luers, 2006).

USE OF AI AND ML

The Sundarbans, a UNESCO World Heritage Site and one of the largest mangrove forests in the world (Smit and Skinner, 2002), is emblematic of the challenges posed by climate change, particularly to vulnerable communities like the tribes who inhabit the region (Gómez-Baggethun et al., 2012). The integration of Artificial Intelligence (AI) and Machine Learning (ML) into climate justice initiatives in the Sundarbans presents an innovative approach (Carmin et al., 2012) to addressing both local and global concerns regarding health rights and environmental sustainability (Almansi, Hardoy, & Pandiella, 2014).

AI and ML can enhance the ability to predict environmental changes and health impacts resulting from climate variability (DiGregorio, 2013). For instance, machine learning algorithms can analyze vast datasets to forecast extreme weather events (Dubey et al., 2017), such as cyclones or floods, which disproportionately affect local tribal populations (Mishra et al., 2021). By utilizing these predictive analytics, authorities and NGOs can issue timely warnings that enable communities to prepare better and minimize risks (Biswas & Nautiyal, 2020).

Moreover, AI can facilitate more efficient resource management (Rudra & Chattopadhyay, 2019). Drone technology powered by AI can conduct aerial surveys of the Sundarbans to monitor mangrove health and biodiversity (Rudra & Chattopadhyay, 2018). This data is crucial for conservation efforts, as the degradation of these ecosystems directly correlates with climate change (Satio et al., 2020) and, by extension, the health rights of local tribes that rely on these resources (Sarkar & Sanyal, 2019). Such AI-driven surveillance allows for the identification of ecological vulnerabilities (Hafiz Iqbal, 2020), thereby informing policies that align with the health and survival of these populations (Roy, 2019).

Machine Learning can also be instrumental in health monitoring (Mozumdar et al., 2018). By analyzing patterns in health data such as the incidence of climate-related diseases like malaria or dengue fever policymakers (Dasgupta et al., 2021) can make informed decisions about resource allocation, healthcare access, and the establishment of localized health interventions (Ahsan & Warner, 2014). This data-driven approach empowers local tribes by integrating their health rights (Smit and Skinner, 2002) into broader climate policies, enabling representation in decision-making processes (Gómez-Baggethun et al., 2012).

While the benefits of AI and ML are evident, their implementation must prioritize ethical considerations (Carmin et al., 2012). Efforts should be made to include local communities in the design and deployment of these technologies (Almansi, Hardoy, & Pandiella, 2014), ensuring that they are not just subjects of study but active participants in shaping the solutions that affect their lives (DiGregorio, 2013). Partnerships with local tribal leaders can foster a sense of ownership and help build trust, which is essential for the successful application of AI-driven measures (Sen, 2019).

At the global policy level, integrating AI and ML findings into international climate agreements and frameworks (Anwar et al., 2023) can help ensure that the rights and needs of local communities are reflected (Dey and Hazra, 2021). Collaborations between local tribal groups, scientists, policymakers (Jalais, 2007), and international organizations can set precedents for the inclusion of indigenous knowledge and health considerations in climate action plans (Rahman et al., 2021).

ROLE OF AR, VR, AND MR

Augmented Reality (AR), Virtual Reality (VR), and Mixed Reality (MR) technologies are increasingly recognized as powerful tools (Unogwu et al., 2024) for enhancing climate justice initiatives, particularly in complex environments like the Sundarbans (Szabo et al., 2016). Their ability to visualize complex data, immerse users in simulated experiences (Dasgupta et al., 2014), and facilitate interactions between virtual and physical environments can significantly enhance public awareness and engagement (Danda & Sriskanthan, 2011) with the pressing issues of health rights and climate change affecting local tribal communities (Ghosh, 2012).

One primary application of AR, VR, and MR technologies in the context of climate justice is education and awareness-raising (Bhowmick and Poddar, 2023). An immersive VR experience can transport users to the Sundarbans (Pitchaikani, 2020), allowing them to witness the impact of climate change on local ecosystems and communities firsthand (StC,D. E., 2008). By experiencing the devastating effects of rising sea levels, habitat loss, or extreme weather events (Mahadevia & Vikas, 2012), users including policymakers, activists, and the global public can cultivate empathy toward affected tribal communities (Afroz and Alam, 2013). This understanding is crucial for garnering support for policies that aim to protect their health rights and address climate injustices (Hiran et al., 2021).

Furthermore, AR can enhance the interaction between the physical and digital worlds (Doshi et al., 2024). For example, AR applications can enable community members to visualize environmental data in real time (Moore, 2023), such as changes in mangrove cover or air quality. This localized data can empower tribal communities to take active roles in conservation efforts (Mondal, Bowers, & Hossen, 2021) and hold authorities accountable for environmental degradation or neglect of health rights (Islam, Dey, Sanyal, & Bandyopadhyay, 2022). By facilitating informed dialogue, AR tools can bridge the gap between local knowledge and scientific expertise (Ray, 2020), enabling a more comprehensive approach to policy development (Begum, Lobry De Bruyn, Amirul Islam, & Kristiansen, 2022).

In research and planning, MR technologies can simulate potential future scenarios based on various climate models (Chanda & Akhand, 2023). Stakeholders can visualize and analyze different outcomes, allowing them to make more informed decisions regarding land use (Jamal, Ghosh, Hazarika, & Sen, 2022), resource management, and public health initiatives. Engaging local tribal populations in these simulations ensures that their unique knowledge of the landscape (Nicholls et al., 2020) and its socio-environmental dynamics are considered in the planning process, leading to more equitable and effective solutions (Mukhopadhyay, 2009).

Collaboration is essential for the success of AR, VR, and MR initiatives (Biswas, 2013). Engaging local communities, artists, scientists, and policymakers in co-designing these tools ensures they are relevant and culturally sensitive (Thakur et al., 2021). Training programs can empower tribal members to use these technologies, fostering local expertise and confidence while promoting collective action for climate justice (Pramanik et al., 2021).

Moreover, the integration of AR, VR, and MR into global climate discussions can elevate the voices of marginalized communities (Sarwar Hossain et al., 2016). By effectively showcasing the realities faced by local tribes in the Sundarbans, these technologies can help shift the narrative (Biswas & Nautiyal, 2020) from one of vulnerability to one that highlights resilience and agency. This shift can drive more substantial commitments to climate justice in national and international platforms (Adams et al., 2020).

CHALLENGES

While the integration of local tribal health rights with global climate policies in the Sundarbans presents several advantages, it is not without its challenges (Soltesova, Brown, Dayal, & Dodman, 2014). Understanding these obstacles is crucial for developing effective and equitable solutions (Few et al., 2007).

A primary challenge lies in the existing power dynamics between local communities and external stakeholders, such as government institutions and international entities (Roberts, 2008). Often, global policies are formulated without adequate input from indigenous populations, leading to solutions that may not be appropriate or effective in the local context (Roberts, 2013). This disconnect can exacerbate feelings of disenfranchisement and mistrust among tribal communities, who may perceive these policies as top-down impositions that overlook their unique needs and perspectives (Sivaramakrishnan, 2007).

Another obstacle stems from the complexity of tribal health rights, which may vary widely among different groups in the Sundarbans (Leck & Simon, 2012). Health needs, traditional practices, and societal structures may differ significantly among the various tribes (Kernaghan & da Silva, 2013), complicating the formulation of a unified policy that adequately addresses these diverse rights (Almansi, 2010). This fragmentation can hinder the development of shared strategies and may lead to unintentional marginalization (Almansi et al., 2014) within the very communities that policy solutions aim to support.

Climate change itself poses a formidable challenge, as the rapid and unpredictable nature of its impacts can outpace the development and implementation of effective health policies (Pelling & Manuel-Navarrete, 2011). For instance, health emergencies caused by climate events, such as floods or cyclones, can strain local healthcare resources (Honadle, 1999), making it difficult for tribal communities to access necessary services. This situation calls for an agile and responsive governance framework that is often lacking in bureaucratic systems (Adger, 2003).

Furthermore, issues of land rights and access to resources can complicate the picture (Uddin, 2019). Local tribes in the Sundarbans often rely heavily on their immediate environment for sustenance, and climate change threatens these resources (Sahana et al., 2021). Conflicting interests regarding land use, conservation efforts, and industrial development can disrupt traditional practices, displacing communities and undermining health rights (Uddin, 2019).

GOVERNANCE

Effective governance is crucial for achieving climate justice in the Sundarbans, particularly in integrating local tribal health rights with global policy solutions (Okoye et al., 2024). Good governance ideally involves transparency, inclusivity, accountability, and the capacity to respond to the unique challenges posed by climate change and public health (Aggarwal et al., 2024).

A critical aspect of governance in the Sundarbans involves recognizing and formalizing the participation of local tribes in decision-making (Pal and Poddar, 2024). Integrating indigenous voices into policy discussions ensures that their rights and health needs are adequately represented and prioritized (Tyagi et al., 2020). In practice, this means creating platforms for dialogue and participation, where tribal leaders can engage with local, regional, and global stakeholders (Unnikrishnan and Shankar, 2007). Such participatory governance fosters trust and helps build collaborative relationships, ensuring that indigenous perspectives are not sidelined (Polin & Alam, 2020).

Another important governance consideration is the alignment of policies across multiple sectors (Chanda & Akhand, 2023). Climate justice in the Sundarbans cannot be achieved through siloed approaches (Francesco & Rahman, 2020); rather, it requires a multisectoral framework that acknowledges the interconnections between environmental sustainability, health rights, and socioeconomic development (Ghosh, 2023). This can involve coordinated efforts among various governmental and non-governmental organizations (Mondal, Bowers, & Hossen, 2021), both locally and globally, to streamline resources and initiatives aimed at addressing climate-related health issues (Mitra, 2024).

Capacity-building is also vital for improving governance structures in the Sundarbans (Alam & Bhardwaj, 2020). This entails training local leaders and community members in areas such as advocacy, health policy, and climate resilience (Jamal, Ghosh, Hazarika, & Sen, 2022), enabling them to effectively navigate the complexities of governance (Moore, 2023). Strengthening local leadership and empowering communities will ultimately enable tribal populations to advocate more effectively for their rights and interests in the face of external pressures (Doshi et al., 2024).

Moreover, monitoring and evaluation mechanisms are essential for ensuring accountability and adapting policies in response to emerging challenges (Hiran et al., 2021). Governance frameworks should include feedback loops that allow local communities to convey their experiences and challenges with implemented policies (Pitchaikani, 2020). These insights are crucial for making informed adjustments that can lead to more effective health interventions and adaptation strategies (Bhowmick and Poddar, 2023).

International cooperation is another critical facet of governance in achieving climate justice in the Sundarbans (Unogwu et al., 2024). Global policy solutions must be informed by local knowledge, and cross-border collaborations between India and Bangladesh (Rahman et al., 2021) can enhance resource-sharing and collective action against climate impacts (Dey and Hazra, 2021). Organizations that work on climate justice can facilitate these dialogues and bridge the gap between local needs and international frameworks (Anwar et al., 2023).

Finally, ensuring a legal framework that protects local health rights is paramount (Sen, 2019). National and regional laws must be harmonized with international human rights standards to reflect the specific needs of indigenous tribes in the Sundarbans (Roy, 2019). This can include recognizing traditional health practices and ensuring access to healthcare irrespective of socio-economic status, reinforcing the concept that health is a fundamental human right (Hafiz Iqbal, 2020).

IMPLEMENTATION OF REGULATORY FRAMEWORKS

Implementing regulatory frameworks for climate justice in the Sundarbans requires a multi-faceted approach that integrates local tribal health rights with global policy solutions (Sarkar & Sanyal, 2019). Given the complexities of climate change impacts on this unique ecosystem and its communities (Satio et al., 2020), effective regulatory frameworks must encompass legal, institutional, and community-based initiatives that promote sustainability, resilience, and health rights (Rudra & Chattopadhyay, 2018).

Firstly, creating a legal framework that recognizes and enforces the rights of local tribal communities is imperative (Rudra & Chattopadhyay, 2019). Such frameworks should include provisions for land rights, access to resources, and participation in decision-making (Biswas & Nautiyal, 2020). National and local governments must align their policies with international human rights instruments (Mishra et al., 2021), such as the UN Declaration on the Rights of Indigenous Peoples (UNDRIP) and frameworks like the Paris Agreement, recognizing the specific needs and rights of tribal populations in their climate action plans (Dubey et al., 2017).

Furthermore, regulatory frameworks should promote collaborative governance, wherein local communities (Satio et al., 2020), government entities, NGOs, and stakeholders work together to develop and implement climate strategies (Dasgupta et al., 2021). This participatory approach enhances transparency, builds trust, and ensures that responses to climate change are more culturally relevant and effective (Anwar et al., 2023). By incorporating traditional knowledge into climate adaptation strategies, the unique insights of indigenous populations can inform sustainable resource management and ecosystem conservation (Dey and Hazra, 2021).

A vital aspect of implementation involves health rights and their integration into climate change frameworks (Unogwu et al., 2024). The Sundarbans face various health challenges exacerbated by climate change rising salinity, vector-borne diseases, and food insecurity, among others (Rahman et al., 2021). Regulatory frameworks must prioritize health assessments in climate planning, ensuring that health risks are mitigated, especially for vulnerable populations (Bhowmick and Poddar, 2023). This could involve establishing community health programs funded through climate adaptation financing (Pitchaikani, 2020), ensuring that local health practitioners are empowered to address health concerns arising from environmental changes (Mondal, Bowers, & Hossen, 2021).

Moreover, financial mechanisms play a crucial role in the implementation of regulatory frameworks (Islam, Dey, Sanyal, & Bandyopadhyay, 2022). Allocating funds from global climate finance initiatives, such as the Green Climate Fund (Ray, 2020), is essential for supporting local adaptation projects. These funds should be directly accessible to tribal communities (Begum, Lobry De Bruyn, Amirul Islam, & Kristiansen, 2022), ensuring they can implement projects that reflect their priorities and enhance their resilience against climate impacts (Chanda & Akhand, 2023).

Education and capacity-building programs are also crucial components of effective implementation (Jamal, Ghosh, Hazarika, & Sen, 2022). Enhancing local knowledge through training and awareness campaigns can empower communities to adapt to environmental changes while also promoting biodiversity conservation

(Nicholls et al., 2020). Such initiatives should focus on the interconnectedness of health, environment, and community well-being, fostering a holistic understanding of climate justice (Pramanik et al., 2021).

Lastly, robust monitoring and evaluation mechanisms must be established to assess the effectiveness of implemented strategies continually (Biswas & Nautiyal, 2020). These mechanisms should be inclusive, allowing local communities to provide feedback and submit data on health outcomes and environmental changes (Adams et al., 2020). This iterative process can ensure that policies remain adaptive, reflecting the dynamic nature of climate risks faced by the Sundarbans (Sahana et al., 2021).

CONCLUSION

The concept of climate justice in the Sundarbans encapsulates the urgent need to integrate local tribal health rights into the broader spectrum of global climate policy (Uddin, 2019). The Sundarbans, a UNESCO World Heritage Site, faces an existential threat from climate change, including rising sea levels, increasing salinity, and extreme weather events (Okoye et al., 2024). These environmental transformations adversely impact the diverse and vulnerable populations residing in the region, particularly the indigenous tribal communities (Aggarwal et al., 2024). These communities have historically inhabited the region and possess a profound connection to its ecosystems (Tyagi et al., 2020), making them uniquely exposed to the repercussions of climate change.

The interplay of local health rights and climate justice underscores a vital paradigm shift necessary for addressing climate challenges (Polin & Alam, 2020). This integration can establish pathways toward equity, ensuring that responses to climate change do not just focus on environmental sustainability but also prioritize human wellbeing (Francesco & Rahman, 2020). Additionally, acknowledging the rights of indigenous peoples as essential stakeholders in climate discourse empowers them to hold authorities accountable and advocate for their needs (Mitra, 2024).

SUGGESTIONS

To promote climate justice in the Sundarbans while integrating local tribal health rights with global policies, several actionable suggestions emerge:

Health Impact Assessments: Include health impact assessments in climate adaptation and mitigation policies (Alam & Bhardwaj, 2020). These assessments should specifically focus on the implications for local tribal populations and their unique health challenges.

Culturally Relevant Health Programs: Design and implement health programs that respect and incorporate local tribal culture, traditional knowledge, and practices (Tyagi et al., 2020). Such programs must be participatory, ensuring local voices are heard in health and climate interventions.

Capacity Building: Invest in capacity-building initiatives that educate local communities about their rights, climate change, and adaptation strategies (Pal and Poddar, 2024). Such education can empower them to advocate for their health and environmental rights more effectively.

International Collaboration: Foster international partnerships that allow local tribal representatives to present their needs and rights directly at global climate negotiations (Sahana et al., 2021). These partnerships can advocate for the inclusion of indigenous rights within international climate agreements.

Research and Data Collection: Facilitate research and data collection on the health impacts of climate change specifically affecting tribal communities (Uddin, 2019). This data should inform local and global policy frameworks, ensuring they are evidence-based and pertinent.

Policy Coherence: Ensure coherence among various policies - environmental, health, and social - at all levels (Sahana et al., 2021). A unified approach helps in creating policies that address the multifaceted challenges posed by climate change.

FUTURE RESEARCH DIRECTIONS

Future research on climate justice in the Sundarbans should explore the following key directions to strengthen the integration of local tribal health rights with global policy solutions:

Longitudinal Studies: Conduct longitudinal studies to evaluate the long-term health impacts of climate change on tribal populations (Adams et al., 2020). Understanding these trends over time can inform better policy frameworks.

Ethno-ecological Research: Investigate the traditional ecological knowledge of tribal communities to identify sustainable practices that can contribute to biodiversity conservation and climate resilience (Biswas & Nautiyal, 2020).

Health and Climate Nexus: Explore the intersection of health and climate change at local levels, particularly how weather events and environmental trends influence tribal health outcomes, including mental health (Pramanik et al., 2021).

Impact of Policy Interventions: Assess the effectiveness of existing climate policies in addressing the health rights of local communities (Thakur et al., 2021). Evaluating what works and what doesn't will provide insights for crafting improved policies.
Participatory Action Research: Engage tribal communities in participatory action research to co-create knowledge about their needs and aspirations related to climate and health, fostering mutual understanding with policymakers (Nicholls et al., 2020).
Comparative Studies: Conduct comparative studies with other regions facing similar challenges to derive lessons from diverse experiences in integrating community health rights within climate justice frameworks (Ray, 2020).
Technological Innovations: Investigate how emerging technologies can aid in improving health outcomes and enhancing climate resilience for tribal communities (Hiran et al., 2021), ensuring that these technologies are accessible and culturally appropriate. Thus, by advancing research in these directions, a more robust and inclusive framework for climate justice can be developed (Doshi et al., 2024), ensuring that local tribal health rights are adequately represented in global policy discussions and that marginalized voices are empowered in addressing climate change (Unogwu et al., 2024).

REFERENCES

Adams, H., Adger, W. N., Ahmad, S., Ahmed, A., Begum, D., Matthews, Z., Rahman, M. M., Nilsen, K., Gurney, G. G., & Streatfield, P. K. (2020). Multi-dimensional well-being associated with economic dependence on ecosystem services in deltaic social-ecological systems of Bangladesh. *Regional Environmental Change*, *20*(42), 2–16. DOI: 10.1007/s10113-020-01620-x

Adger, W. N. (2003). *Social aspects of adaptive capacity*. Climate Change, Adaptive Capacity and Development, 29-49. DOI: 10.1142/9781860945816_0003

Adger, W. N., Dessai, S., Goulden, M., Hulme, M., Lorenzoni, I., Nelson, D. R., Naess, L. O., Wolf, J., & Wreford, A. (2009). Are there social limits to adaptation to climate change? *Climatic Change*, *93*(3-4), 335–354. DOI: 10.1007/s10584-008-9520-z

Adger, W. N., Hughes, T. P., Folke, C., Carpenter, S. R., & Rockström, J. (2009). Social-ecological resilience to coastal disasters. *Science*, *309*(5737), 1036–1039. DOI: 10.1126/science.1112122 PMID: 16099974

Aggarwal, K., Doshi, R., Mijwil, M. M., Hiran, K. K., Gök, M., & Bala, I. (2024). A Simple Way to Comprehend the Difference and the Significance of Artificial Intelligence in Agriculture. *Intelligent Techniques for Predictive Data Analytics*, 189-201.

Agrawal, A. (2008). *The role of local institutions in adaptation to climate change.* Social Dimensions of Climate Change: Equity and Vulnerability in a Warming World, 173-198. https://hdl.handle.net/10986/28274

Ahsan, M. N., & Warner, J. (2014). The socioeconomic vulnerability index: A pragmatic approach for assessing climate change led risks-A case study in the south-western coastal Bangladesh. *International Journal of Disaster Risk Reduction*, *8*, 32–49. DOI: 10.1016/j.ijdrr.2013.12.009

Al Mahmud, R., Titumir, M., Afrin, T., Islam, M. S., Titumir, R. A. M., Afrin, T., Islam, M. S., Onneshan, U., & Dhaka, B. (2020). *Traditional Knowledge, Institutions and Human Sociality in Sustainable Use and Conservation of Biodiversity of the Sundarbans of Bangladesh.* Managing Socio-ecological Production Landscapes and Seascapes for Sustainable Communities in Asia. Springer (pp. 67–91).

Alam, A. (2012). Some Indian bryophytes known for their biologically active compounds. *International Journal of Applied Biology and Pharmaceutical Technology*, *3*(2), 239–246.

Alam, M., & Bhardwaj, S. (2020). *Temporal variation in glacier's area and identification of glacial lakes in Sikkim*. DOI: 10.1007/978-981-15-2097-6_8

Ali, A. M. (2006). Rice to shrimp: Land use land cover changes and soil degradation in Southwestern Bangladesh. *Land Use Policy*, *23*(4), 421–435. DOI: 10.1016/j.landusepol.2005.02.001

Ali, B. (2020). Community participation and quality of life in naturebased tourism: Exploring the antecedents and moderators. *Journal of Hospitality & Tourism Research (Washington, D.C.)*, *20*(10), 1–32. DOI: 10.1177/1096348020980094

Almansi, F. (2010). *The relationship between disaster risks and urban planning in Argentina,* Case study prepared for ISDR Global Assessment Report 2011.

Almansi, F., & Hardoy, J. (2013). *Climate change impact in coastal areas of the Rio de la Plata River: Actions for prevention and adaptation*. DOI: 10.4324/9781315658735

Almansi, F., Hardoy, J., & Pandiella, G. (2014). Impacto del cambio climático en ciudades costeras del estuario del Rio de la Plata. *Medio Ambiente y Urbanización*, *79*, 217–237. DOI: 10.1080/17565529.2014.918868

Anguelovski, I., & Carmin, J. (2011). Something borrowed, everything new – current opinion: Innovation and institutionalisation in urban climate governance. *Current Opinion in Environmental Sustainability*, *3*(3), 169–175. DOI: 10.1016/j.cosust.2010.12.017

Anurag Danda, A., Ghosh, N., Bandyopadhyay, J., & Hazra, S. (2019). Managed retreat: Adaptation to climate change in the Sundarbans ecoregion in the Bengal Delta. *Journal of the Indian Ocean Region*, *15*(3), 317–335. DOI: 10.1080/19480881.2019.1652974

Anwar, G., Arianto, W., Simarmata, M., Bertham, Y. H., Yuwana, Y., Romeida, A., Ruslan, D. (2023). Floristic Composition and Species Richness of Mangrove Ecosystems in Pantai Panjang–Pulau Baai Nature Park, Bengkulu Province, Indonesia. *International Journal of Advancement in Life Sciences Research*, 6(3), 10-18. 10-18. DOI: 10.31632/ijalsr.2023.v06i03

Armitage, D. (2005). Adaptive capacity and community-based natural resource management. *Environmental Management*, *35*(6), 703–715. DOI: 10.1007/s00267-004-0076-z PMID: 15940398

Ascoli, F. D. (1921). *A revenue history of the Sundarbans from 1870 to 1920*. Bengal Secretariat Book Depot, Calcutta. https://cslrepository.nvli.in//handle/123456789/7362

Ayeb-Karlsson, S., van der Geest, K., Ahmed, I., Huq, S., & Warner, K. (2016). A people-centred perspective on climate change, environmental stress, and livelihood resilience in Bangladesh. *Sustainability Science, 11*(4), 679–694. DOI: 10.1007/s11625-016-0379-z PMID: 30174739

Ayers, J. (2011). Resolving the adaptation paradox: Exploring the potential for deliberative adaption policy-making in Bangladesh. *Global Environmental Politics, 11*(1), 62–88. DOI: 10.1162/GLEP_a_00043

Ayers, J., & Forsyth, T. (2009). Community-based adaptation to climate change, strengthening resilience through development. *Environment, 51*(4), 23–31. DOI: 10.3200/ENV.51.4.22-31

Aziz, M. A., Tollington, S., Barlow, A., Goodrich, J., Shamsuddoha, M., Islam, M. A., & Groombridge, J. J. (2017). Investigating patterns of tiger and prey poaching in the Bangladesh Sundarbans: Implications for improved management. *Global Ecology and Conservation, 9*, 70–81. DOI: 10.1016/j.gecco.2016.12.001

Begum, F., Lobry De Bruyn, L., Amirul Islam, M., & Kristiansen, P. (2022). *The Role of Women's Participation in Co-Management of the Sundarban Mangrove Forest of Bangladesh*. University of New England. Retrieved from https://hdl.handle.net/1959.11/55652

Berger, R. M., & Kelly, J. J. (1993). Social Work in the Ecological Crisis. *Social Work, 38*, 521–526. https://www.jstor.org/stable/23717148

Berkes, F. (2009). Evolution of co-management: Role of knowledge generation, bridging organizations and social learning. *Journal of Environmental Management, 90*(5), 1692–1702. DOI: 10.1016/j.jenvman.2008.12.001 PMID: 19110363

Berkes, F. (2009). Indigenous ways of knowing and the study of environmental change. *Journal of the Royal Society of New Zealand, 39*(4), 151–156. DOI: 10.1080/03014220909510568

Berkes, F., & Ross, H. (2013). Community resilience: Toward an integrated approach. *Society & Natural Resources, 26*(1), 5–20. DOI: 10.1080/08941920.2012.736605

Besthorn, F. H., & McMillen, D. P. (2002). '*The Oppression of Women and Nature: Ecofeminism as a Framework for Social Justice Oriented Social Work*', Families and Society. *The Journal of Contemporary Human Service, 83*, 221–232. DOI: 10.1606/1044-3894.20

Beveridge, H. (1876). Favorite Share Flag Were the Sundarbans inhabited in ancient times. *Journal of the Asiatic Society of Bengal, 45*. https://archive.org/details/biostor-281250

Bhowmik, D., & Poddar, S. (2023). Impacts of Flood in India from 1953 to 2020. [AMT]. *Advancement in Management and Technology*, *4*(2), 12–33.

Biesbroek, G. R., Klostermann, J. E. M., Termeer, C. J. A. M., & Kabat, P. (2013). On the nature of barriers to climate change adaptation. *Regional Environmental Change*, *13*(5), 1119–1129. DOI: 10.1007/s10113-013-0421-y

Biswas, B. (2013). "The God of Death Takes Half of Our Children": Health of Women and Children in the Sundarbans Islands. *Journal of Health Care for the Poor and Underserved*, 24(2), 730–740. . 2013.0069DOI: 10.1353/hpu

Biswas, S., & Nautiyal, S. (2020). An assessment of socio-economic vulnerability at the household level: A study on villages of the Indian Sundarbans. *Environment, Development and Sustainability*, *23*(7), 11120–11137. DOI: 10.1007/s10668-020-01085-2

Breil, M., & Johnson, K. (2012). *Conceptualizing urban adaptation to climate change*. Review of Environment, Energy and Economics (Re3). Available at https://ssrn.com/abstract=2125728

Brown, A., Dayal, A., & Rumbaitis del Rio, C. (2012). From practice to theory: Emerging lessons from Asia for building urban climate change resilience. *Environment and Urbanization*, *24*(2), 531–556. DOI: 10.1177/0956247812456490

Bulkeley, H., & Tuts, R. (2013). Understanding urban vulnerability, adaptation and resilience in the context of climate change. *Local Environment*, *18*(6), 646–662. DOI: 10.1080/13549839.2013.788479

Cannon, T., & Mueller-Mahn, D. (2010). Vulnerability, resilience and development discourses in context of climate change. *Natural Hazards*, *55*(3), 621–635. DOI: 10.1007/s11069-010-9499-4

Carmin, J., Anguelovski, I., & Roberts, D. (2012). Urban climate adaptation in the global South: Planning in an emerging policy domain. *Journal of Planning Education and Research*, *32*(1), 18–32. DOI: 10.1177/0739456X11430951

Chakrabarti, R. (2009). *Local People and the Global Tiger: An Environmental History of the Sundarbans*. Global Environment, Vol. 2, (2009),(No.3), pp. 72-95. https://www.jstor.org/stable/43201488

Chanda, A., & Akhand, A. (2023, August). Challenges towards the Sustainability and Enhancement of the Indian Sundarban Mangrove's Blue Carbon Stock. *Life (Basel, Switzerland)*, *13*(8), 1–19. DOI: 10.3390/life13081787 PMID: 37629645

Chattopadhyay, H. (1999). *The Mystery of the Sunderbans* (First ed.). A Mukherjee & Co. https://archive.org/details/the-mystery-of-the-sundarbans/page/18/mode/2up

Chowdhury, A. N. (2019). Environment and wellbeing: Eco-psychiatry in Sundarban Delta, India. *International Journal of Management and Social Sciences, 37*, 2. DOI: 10.21013/jmss.v14.n2sp.p5

Corfee-Morlot, J., Cochran, I., Hallegatte, S., & Teasdale, P. (2011). Multilevel risk governance and urban adaptation policy. *Climatic Change, 104*(1), 169–197. DOI: 10.1007/s10584-010-9980-9

Cox, D., & Pawar, M. (2006). *International Social Work: Issues, Strategies and Programs*. SAGE., DOI: 10.4135/9781544308685

Cutter, S. L., Boruff, B. J., & Shirley, W. L. (2003). Social vulnerability to environmental hazards. *Social Science Quarterly, 84*(2), 242–261. DOI: 10.1111/1540-6237.8402002

Danda, A. A. (2019, November). *Environmental Security in the Sundarban in the Current Climate Change Era Strengthening India-Bangladesh Cooperation*. Observer Research Foundation, ORF Occasional Paper No. 220.

Das, S., & Hazra, S. (2020). Trapped or resettled: Coastal communities in the Sundarbans Delta. *Climate Crisis and Local Communities, 64*, 15–17.

Dasgupta, S., Wheeler, D., & Ghosh, S. (2021). Fishing in Salty Waters: Poverty, Occupational Saline Exposure, and Women's Health in the Indian Sundarban. *Journal of Management and Sustainability, 12*(1), 1. DOI: 10.5539/jms.v12n1p1

Dewulf, A. (2013). Contrasting frames in policy debates on climate change adaptation. *Wiley Interdisciplinary Reviews: Climate Change, 4*(4), 321–330. DOI: 10.1002/wcc.227

Dey, M. K., & Hazra, A. K. (2021). Abundance of microarthropods population in different sites of Midnapore east coast of West Bengal, India. *International Journal of Advancement in Life Sciences Research*, 13-25. DOI: DOI: 10.31632/ijalsr.2021.v04i03.003

DiGregorio, M. (2013a). Planning for climate resilience: Learning from grassroots, Presentation at *CBA 7 conference*, Dhaka, April 2013. DOI: 10.4324/9781315658735

Dodman, D., & Mitlin, D. (2013). Challenges for community-based adaptation: Discovering the potential for transformation. *Journal of International Development, 25*(5), 640–659. DOI: 10.1002/jid.1772

Dodman, D., & Satterthwaite, D. (2008). Institutional capacity, climate change adaptation and the urban poor. *IDS Bulletin*, *39*(4), 67–74. DOI: 10.1111/j.1759-5436.2008.tb00478.x

Dominelli, L. (2010). '*Globalization, Contemporary Challenges and Social Work Practice'*. *International Social Work*, *53*(5), 599–612. DOI: 10.1177/0020872810371201

Dominelli, L. (2011). '*Climate Change: Social Workers' Roles and Contributions to Policy Debates and Interventions*'. *International Journal of Social Welfare*, *20*(4), 430–438. DOI: 10.1111/j.1468-2397.2011.00795.x

Doshi, R., Dadhich, M., Poddar, S., & Hiran, K. K. (2024). *Integrating generative AI in education to achieve sustainable development goals.*

Dubey, S., Trivedi, R., Chand, B., Mandal, B., & Rout, S. (2017). Farmers' perception of climate change, impacts on freshwater aquaculture and adaptation strategies in climatic change hotspots: A case of the Indian Sundarban delta. *Environmental Development*, *21*, 38–51. DOI: 10.1016/j.envdev.2016.12.002

Eakin, H., & Lemos, M. C. (2006). Adaptation and the state: Latin America and the challenge of capacity-building under globalization. *Global Environmental Change*, *16*(1), 7–18. DOI: 10.1016/j.gloenvcha.2005.10.004

Eakin, H., & Luers, A. L. (2006). Assessing the vulnerability of social-environmental systems. *Annual Review of Environment and Resources*, *31*(1), 365–394. DOI: 10.1146/annurev.energy.30.050504.144352

Elliot, D. (2012). *Social Development and Social Work.* In Healy, L. M., & Link, R. J. (Eds.), *Handbook of International Social Work* (pp. 102–108). Oxford University Press., DOI: 10.1093/acprof:oso/9780195333619.003.0014

Elliott, D. (1993). Social Work and Social Development: Towards an Integrative Model for Social Work Practice. *International Social Work*, *36*(1), 21–36. DOI: 10.1177/002087289303600103

Erickson, C. L. (2011). *Environmental Degradation and Preservation.* In Healy, L. M., & Link, R. J. (Eds.), *Handbook of International Social Work* (pp. 184–189). Oxford University Press., DOI: 10.1093/acprof:oso/9780195333619.003.0027

Eriksen, S. H., Brown, K., & Kelly, P. M. (2011). The dynamics of vulnerability: Locating coping strategies in Kenya and Tanzania. *The Geographical Journal*, *177*(4), 287–305. DOI: 10.1111/j.1475-4959.2005.00174.x

Eriksen, S. H., Nightingale, A. J., & Eakin, H. (2015). Reframing adaptation: The political nature of climate change adaptation. *Global Environmental Change*, *21*(2), 670–680. DOI: 10.1016/j.gloenvcha.2015.09.014

Few, R., Brown, K., & Tompkins, E. L. (2007). Public participation and climate change adaptation: Avoiding the illusion of inclusion. *Climate Policy*, *7*(1), 45–59. DOI: 10.1080/14693062.2007.9685637

Folke, C., Hahn, T., Olsson, P., & Norberg, J. (2005). Adaptive governance of social-ecological systems. *Annual Review of Environment and Resources*, *30*(1), 441–473. DOI: 10.1146/annurev.energy.30.050504.144511

Forsyth, T. (2013). Community-based adaptation: A review of past and future challenges. *Wiley Interdisciplinary Reviews: Climate Change*, *4*(5), 439–446. DOI: 10.1002/wcc.231

Francesco, E. D., & Rahman, M. (2020, December). *Natural, social and economic regeneration in the Sundarbans of Bangladesh*. Circular Conversations. Retrieved from https://www.circularconversations.com/earth-heroes/bangladesh-environment-and-development-society

Germain, C. B. (1978). General-systems Theory and Ego Psychology: An Ecological Perspective. *The Social Service Review*, *52*(4), 535–550. https://www.jstor.org/stable/30015668. DOI: 10.1086/643676

Ghosh, A. (2012). *Living with changing climate–impact, vulnerability and adaptation challenges in Indian Sundarbans*. Centre for Science and Environment.

Ghosh, A. (2019). *The Tides of Life*. Indira Gandhi National Centre for the Arts.

Ghosh, O. (2023). Environmental hazards and disasters - A response towards mitigating disaster management. *AIP Conference Proceedings*, *2763*(1), 050008. DOI: 10.1063/5.0158892

Gómez-Baggethun, E., Corbera, E., & Reyes-García, V. (2012). Traditional ecological knowledge and global environmental change: Research findings and policy implications. *Ecology and Society*, *18*(4), 72. DOI: 10.5751/ES-06288-180472 PMID: 26097492

Government of West Bengal. (2020). (*rep.*). State Action Plan on Climate Change.

Gray, M. (2006). The Progress of Social Development in South Africa. *International Journal of Social Welfare*, *15*(s1, Suppl. 1), S53–S64. DOI: 10.1111/j.1468-2397.2006.00445.x

Gray, M., & Coates, J. R. (2012). Environmental Ethics for Social Work: Social Work's Responsibility to the Non-human World. *International Journal of Social Welfare*, *21*(3), 239–247. DOI: 10.1111/j.1468-2397.2011.00852.x

Gray, M., Mazibuko, F., & O'Brien, F. (1996). Social Work Education for Social Development. *Journal of Social Development in Africa*, *11*(1), 33–42. https://api.semanticscholar.org/CorpusID:221926574

Grothmann, T., & Patt, A. (2005). Adaptive capacity and human cognition: The process of individual adaptation to climate change. *Global Environmental Change*, *15*(3), 199–213. DOI: 10.1016/j.gloenvcha.2005.01.002

Gutierrez, K. S., & LePrevost, C. E. (2016). Climate justice in rural southeastern United States: A review of climate change impacts and effects on human health. *International Journal of Environmental Research and Public Health*, *13*(2), 189. DOI: 10.3390/ijerph13020189 PMID: 26848673

Hafiz Iqbal, M. (2020). Valuing ecosystem services of Sundarbans mangrove forest for improved conservation: Approach of randomized conjoint experiment. *Forestry Economics Review*, *2*(1), 2631–3030. DOI: 10.1108/FER-04-2020-0008

Healy, L. M. (2008). *International Social Work: Professional Action in an Interdependent World* (2nd ed.). Oxford University Press., Available at https://global.oup.com/ushe/product/international-social-work-9780190922252?cc=in&lang=

Hiran, K. K., Khazanchi, D., Vyas, A. K., & Padmanaban, S. (Eds.). (2021). *Machine learning for sustainable development* (Vol. 9). Walter de Gruyter GmbH & Co KG., DOI: 10.1515/9783110702514

Honadle, G. (1999). *How Context Matters: Linking Environmental Policy to People and Place*. Kumarian Press., https://api.semanticscholar.org/CorpusID:152578133

Hurlimann, A. C., & March, A. P. (2012). The role of spatial planning in adapting to climate change. *Wiley Interdisciplinary Reviews: Climate Change*, *3*(5), 477–488. DOI: 10.1002/wcc.183

Islam, S. T., Dey, A., Sanyal, N., & Bandyopadhyay, S. (2022). *Troubles in The Sundarbans: A study of social and ecological issues in Hingalganj Block, North 24 Parganas West Bengal*. ActionAid International.

Jabeen, H., Johnson, C., & Allen, A. (2010). Built-in resilience: Learning from grassroots coping strategies for climate variability. *Environment and Urbanization*, *22*(2), 415–431. DOI: 10.1177/0956247810379937

Jahan, R. (1995). *The elusive agenda: Mainstreaming women in development*. London: ZED Books. https://www.jstor.org/stable/41260001

Jalais, A. (2007). The Sundarbans: Whose World Heritage Site? *Conservation & Society*, *5*(3), 335–342. https://www.jstor.org/stable/26392892

Jamal, S., Ghosh, A., Hazarika, R., & Sen, A. (2022, July). Livelihood, conflict and tourism: An assessment of livelihood impact in Sundarbans, West Bengal. *International Journal of Geoheritage and Parks*, *10*(1), 383–399. DOI: 10.1016/j.ijgeop.2022.07.004

Kanjilal, B., Bose, S., Patra, N., Barman, D., Ghosh, U., Mandal, A., Vardrevu, L., & Sengupta, P. (2013). *How Healthy are the Children of Indian Sundarbans?* http://r4d.dfid.gov.uk/pdf/outputs/FutureHealth_RPC/SHWforweb.pdf

Kanjilal, B., Mazumdar, P., Mukherjee, M., Mondal, S., Barman, D., Singh, S., & Mandal, A. (2010). Health care in the Sundarbans (India): Challenges and plan for a better future. *Future Health Systems Research Programme* (Issues 1–91). DOI: 10.6133/apjcn.012016.01

Kanjilal, B., Mazumdar, P. G., Mukherjee, M., Mondal, S., Barman, D., & Singh, S. M. (2010, January). *Health care in the Sundarbans (India): Challenges. Future Health Systems Research Programme*, 21-30. Retrieved from https://assets.publishing.service.gov.uk/media/57a08b24ed915d3cfd000b68/sundarbans.pdf

Kemp, S. P., Whittaker, J. K., & Tracy, E. M. (1997) *Person–Environment Practice: The Social Ecology of Interpersonal Helping*. New York: Aldine De Gruyter. https://api.semanticscholar.org/CorpusID:141687227

Kernaghan, S., & da Silva, J. (2013). Initiating and sustaining action: Experiences building resilience to climate change in Asian cities. *Urban Climate*, *7*, 47–63. DOI: 10.1016/j.uclim.2013.10.008

Khan, S. A. R., Umar, M., Yu, Z., & Nawaz, M. T. (2023). A Recent Digitalization in Recycling Industry Attaining Ecological Sustainability: A Comprehensive Outlook and Future Trend. *Environmental Science and Pollution Research International*, *30*(47), 103760–103775. DOI: 10.1007/s11356-023-29537-y PMID: 37695483

Leck, H., & Simon, D. (2012). Fostering multiscalar collaboration and co-operation for effective governance of climate change adaptation. *Urban Studies (Edinburgh, Scotland)*, *50*(6), 1221–1238. DOI: 10.1177/0042098012461675

Leichenko, R. (2011). Climate change and urban resilience. *Current Opinion in Environmental Sustainability*, *3*(3), 164–168. DOI: 10.1016/j.cosust.2010.12.014

Mahadevia, K., & Vikas, M. (2012). Climate change—Impact on the Sunderbans: A case study. *International Scientific Journal-Environmental Science*, *2*(1), 7–15.

Mertz, O., Mbow, C., Reenberg, A., & Diouf, A. (2009). Farmers' perceptions of climate change and agricultural adaptation strategies in rural Sahel. *Environmental Management*, *43*(5), 804–816. DOI: 10.1007/s00267-008-9197-0 PMID: 18810526

Midgley, J. (1994). Defining Social Development: Historical Trends and Conceptual Formulations. *Social Development Issues*, *16*(3), 3–19. DOI: 10.4135/9781446221839.n2

Midgley, J. (1995). *Social Development: The Developmental Perspective in Social Welfare*. SAGE., DOI: 10.4135/9781446221839.n2

Miller, S. E., Hayward, R. A., & Shaw, T. V. (2012). Environmental Shifts for Social Work: A Principles Approach. *International Journal of Social Welfare*, *21*(3), 270–277. DOI: 10.1111/j.1468-2397.2011.00848.x

Mimi, R., & Haynes, L. (2018). *The role of livelihood activities and income portfolios in shaping women's perceptions of wellbeing and adaptation choices, in the Indian Bengal Delta Statement of Originality*.

Mishra, M., Acharyya, T., Santos, C. A. G., da Silva, R. M., Kar, D., Mustafa Kamal, A. H., & Raulo, S. (2021). Geo-ecological impact assessment of severe cyclonic storm Amphan on Sundarban mangrove forest using geospatial technology. *Estuarine, Coastal and Shelf Science*, *260*, 107486. DOI: 10.1016/j.ecss.2021.107486

Mistri, A. (2019). Is the migration from Indian Sundarban an environmental migration? Investigating through Sustainable Livelihood Approach (SLA). *Asian Profile*, *47*(3), 195–218.

Mitlin, D., & Satterthwaite, D. (2012). Addressing poverty and inequality new forms of urban governance in Asia. *Environment and Urbanization*, *24*(2), 395–401. DOI: 10.1177/0956247812458062

Mitra, R. (2024). *Nature-based Solutions can Help Sundarbans Survive. DownToEarth*. Retrieved from https://www.downtoearth.org.in/news/climate-change/nature-based-solutions-can-help-sundarbans -survive-95230

Mizanur Rahman, M., Rakib Hossain, M., & Nazrul Islam, M. (2021). Climate Change Impact on Sundarbans: Challenges for Mitigation Strategies. *Climate Change in Bangladesh: A Cross-Disciplinary Framework*, 47-64.

Molinari, N. (2017). Intensifying Insecurities: The impact of climate change on vulnerability to human trafficking in the Indian Sundarbans. *Anti Trafficking Review, 363*(8), 1–11. DOI: 10.14197/atr.20121784

Mondal, M. K., Bhuiyan, S. I., & Franco, D. T. (2001). Soil salinity reduction and prediction of salt dynamics in the coastal rice lands of Bangladesh. *Agricultural Water Management, 47*(1), 9–23. DOI: 10.1016/S0378-3774(00)00098-6

Mondal, T., Bowers, W. W., & Hossen, M. S. (2021, March 25). Sustainable Management of Sundarbans: Stakeholder Attitudes Towards Participatory Management and Conservation of Mangrove Forests. *Journal of Sustainable Development, 14*(3), 23–40. DOI: 10.5539/jsd.v14n3p23

Moore, J. (2023, March 6). *PayBitoPro saves Sundarbans' child education.* Retrieved from PayBitoPro: https://www.paybito.com/lost-in -the -mangroves -paybito -steps -up -to -save -child -education -in -sundarbans/

Moser, S. C., & Ekstrom, J. A. (2010). A framework to diagnose barriers to climate change adaptation. *Proceedings of the National Academy of Sciences of the United States of America, 107*(51), 22026–22031. DOI: 10.1073/pnas.1007887107 PMID: 21135232

Mozumdar, M., Shamsuzzaman, M., Un-Nabi, R., & Harun-Al-Rashid, A. (2018). Socio-economic characteristics and fishing operation activities of the artisanal fishers in the Sundarbans Mangrove Forest, Bangladesh. *Turkish Journal of Fisheries and Aquatic Sciences, 18*(6), 788–799. DOI: 10.4194/1303-2712-v18_6_05

Mukhopadhyay, A. (2009). *Cyclone Aila and the Sundarbans: An Enquiry into the Disaster and Politics of Aid and Relief.* In Research Report on CRG Annual Winter Course on Forced Migration.

Nicholls, R. J., Adger, W. N., Hutton, C. W., & Hanson, S. E. (2020). *Deltas in the Anthropocene.* Palgrave Macmillan Springer Nature., DOI: 10.1007/978-3-030-23517-8

Norgaard, K. M. (2017, September). The Sociological Imagination in a Time of Climate Change. *Global and Planetary Change, Elsevier.* DOI: DOI: 10.1016/j.gloplacha.2017.09.018

Norris, F., Stevens, S., Pfefferbaum, B., Wyche, K., & Pfefferbaum, R. (2008). Community resilience as metaphor, theory, set of capacities, and strategy for disaster readiness. *American Journal of Community Psychology, 41*(1–2), 127–150. DOI: 10.1007/s10464-007-9156-6 PMID: 18157631

Nyong, A., Adesina, F., & Elasha, B. O. (2007). The value of indigenous knowledge in climate change mitigation and adaptation strategies in the African Sahel. *Mitigation and Adaptation Strategies for Global Change, 12*(5), 787–797. DOI: 10.1007/s11027-007-9099-0

O'Brien, K. (2012). Global environmental change II: From adaptation to deliberate transformation. *Progress in Human Geography, 36*(5), 667–676. DOI: 10.1177/0309132511425767

O'Brien, K., Eriksen, S., Nygaard, L. P., & Schjolden, A. (2007). Why different interpretations of vulnerability matter in climate change discourses. *Climate Policy, 7*(1), 73–88. DOI: 10.1080/14693062.2007.9685639

Okoye, K., Nganji, J. T., Hiran, K. K., & Hosseini, S. (2024, May). Impact and implications of AI methods and tools for the future of education. [). Frontiers Media SA.]. *Frontiers in Education, 9*, 1434052. DOI: 10.3389/feduc.2024.1434052

Olhoff, A., & Schaer, C. (2010). *Screening tools and guidelines to support the mainstreaming of climate change adaptation into development assistance a stocktaking report.* New York: UNDP. http://www.undp.org/climatechange/library.shtml

Omer, S. (1979). *'Social Development'. International Social Work, 22*(3), 11–26. DOI: 10.1177/002087287902200303

Pal, P., & Poddar, S. (2024). Applications of Smart Farming Sensors: A Way Forward. *Data Science for Agricultural Innovation and Productivity*, 175.

Panda, S., Sadhu, C., Pramanik, G., Pahari, S., & Hossain, J. (2016). Concerning public health situation of under-nutrition in children and anemia in women in Indian Sundarbans delta: A community based cross sectional investigation. *BMC Nutrition, 2*(1), 1–10. DOI: 10.1186/s40795-016-0105-3

Peeters, J. (2012). The Place of Social Work in Sustainable Development: Towards Ecosocial Practice. *International Journal of Social Welfare, 21*(3), 287–298. DOI: 10.1111/j.1468-2397.2011.00856.x

Pelling, M. (2003). *The vulnerability of cities: Natural disasters and social resilience.*

Pelling, M. (2011). *Adaptation to climate change: From resilience to transformation.* Routledge., DOI: 10.4324/9780203889046

Pelling, M., & Manuel-Navarrete, D. (2011). From resilience to transformation: The adaptive cycle in two Mexican urban centers. *Ecology and Society, 16*(2), 11. DOI: 10.5751/ES-04038-160211

Pervin, M., Sultana, S., Phirum, A., Camara, I., Nzau, M., Phonnasane, V., & Anderson, S. (2013). *A framework for mainstreaming climate resilience into development planning*. IIED Working Paper, IIED, London https://www.iied.org/10050iied

Pitchaikani, J. S. (2020, March). Integrated coastal zone management practices for Sundarbans, India. *Indian Journal of Geo-Marine Sciences*, *49*(03), 352–356. https://core.ac.uk/download/pdf/326041666.pdf

Polin, F., & Alam, D. (2020, March 13). *Enhancing Community Participation in Forest Management: Case Study of the Sundarbans, Bangladesh*. South-West Region (Khulna Division). Retrieved from https://www.researchgate.net/publication/339916369_Enhancing_Community_Participation_in_Forest_Management_Case_Study_of_the_Sundarbans_Bangladesh

Pramanik, M., Szabo, S., Pal, I., Udmale, P., Pongsiri, M., & Chilton, S. (2021). Population health risks in multi-hazard environments: Action needed in the Cyclone Amphan and COVID-19 - hit Sundarbans region, India. *Climate and Development*, *14*(2), 99–104. DOI: 10.1080/17565529.2021.1889948

Rahman, M., Giedraitis, V. R., Lieberman, L. S., Akhtar, T., & Taminskienė, V. (2013). Shrimp cultivation with water salinity in Bangladesh: The implications of an ecological model. *Universal Journal of Public Health*, *1*(3), 131–142. DOI: 10.13189/ujph.2013.010313

Ray, P. S. (2020). Socio-Economic History in Sundarbans. *Journal of Critical Reviews*, *7*(2), 1030–1034.

Reid, H., Alam, M., Berger, R., Cannon, T., Huq, S., & Milligan, A. (2009). *Community-based adaptation to climate change: An overview*. Participatory Learning and Action: Community Based Adaptation to Climate Change, 60, 11–33. Available at https://www.researchgate.net/publication/272791313_Community-based_adaptation_to_climate_change_An_overview

Ribot, J. C., Najam, A., & Watson, G. (2009). *Climate Variation, Vulnerability and Sustainable Development in the Semi-arid Tropics*. In Schipper, E. L. F., & Burton, I. (Eds.), *The Earthscan Reader on Adaptation to Climate Change* (pp. 117–160). Earthscan., Available at https://www.researchgate.net/publication/291177311_Climate_variation_vulnerability_and_sustainable_development_in_the_semi-arid_tropics

Roberts, D. (2008). Thinking globally, acting locally institutionalising climate change at the the local government level in Durban, South Africa. *Environment and Urbanization*, *20*(2), 521–537. DOI: 10.1177/0956247808096126

Roberts, D. (2013). *Busting the Myths – An African reality check on Community Based Adaptation in Durban, South Africa.* Presentation at *CBA 7 conference,* Dhaka, April 2013. DOI: 10.1177/0956247808096126

Rogers, P. P., Jalal, K. F., & Boyd, J. A. (2007). *An Introduction to Sustainable Development.* Harvard University Press., DOI: 10.4324/9781849770477

Roy, D. (2020). 'On the Horns of a Dilemma'! Climate change, forest conservation and the marginal people in Indian Sundarbans. *Forum for Development Studies,* 47(2), 307–326. . 1786452DOI: 10.1080/08039410.2020

Roy, M., Hulme, D., & Jahan, F. (2013). Contrasting adaptation responses by squatters and low-income tenants in Khulna, Bangladesh. *Environment and Urbanization, 25*(1), 157–176. DOI: 10.1177/0956247813477362

Roy, S. (2018). Livelihood resilience of the indigenous Munda community in the Bangladesh Sundarbans Forest. *Handbook Of Climate Change Resilience*, 10(1), 1–21. DOI: 10.1007/978-3-319-71025-9_10-1

Roy, S. (2019). *Women's Use of the Sundarbans Forest Resources*. Springer Briefs in Environment, Security. *Development and Peace*, *29*, 59–71. DOI: 10.1007/978-981-13-6776-2_4

Rudra, A., & Chattopadhyay, A. (2018). *Environmental vulnerabilities and risks in population: A comparison of two Sundarban Delta.* Population Association of America, May, 1–14. https://www.researchgate.net/publication/326201083

Rudra, A., & Chattopadhyay, A. (2019). Environmental change of coastal Sundarbans: Impact on livelihood and standard of living status of indigenous people. *Environmental Quality Management*, *29*(2), 77–84. DOI: 10.1002/tqem.21667

Sahana, M., Rehman, S., Ahmed, R., & Sajjad, H. (2021). Analyzing climate variability and its effects in Sundarban Biosphere Reserve, India: Reaffirmation from local communities. *Environment, Development and Sustainability*, *23*(2), 2465–2492. DOI: 10.1007/s10668-020-00682-5

Saif, S., Russell, A. M., Nodie, S. I., Inskip, C., Lahann, P., Barlow, A., Greenwood Barlow, C., Islam, M. A., & MacMillan, D. C. (2016). Local usage of tiger parts and its role in tiger killing in the Bangladesh Sundarbans. *Human Dimensions of Wildlife*, *21*(2), 95–110. DOI: 10.1080/10871209.2015.1107786

Saito, O., Subramanian, S., Hashimoto, S., & Takeuchi, K. (2020). *Managing socio-ecological production landscapes and seascapes for sustainable communities in Asia mapping and navigating stakeholders, policy and action.* Springer https://www.springer.com/series/11884

Sarkar, S., & Sanyal, P. (2019). Community perception towards tourism development in the Indian Sundarbans: Case study of Jharkhali Island. *International Journal of Engineering Research & Technology (Ahmedabad)*, *8*(12), 526–540. www.ijert.org

Sarwar Hossain, M., Eigenbrod, F., Amoako Johnson, F., & Dearing, J. A. (2016). Unravelling the interrelationships between ecosystem services and human wellbeing in the Bangladesh delta. *International Journal of Sustainable Development and World Ecology*, *24*(2), 120–134. DOI: 10.1080/13504509.2016.1182087

Satterthwaite, D., Huq, S., Reid, H., Pelling, M., & Romero Lankao, P. (2007). *Adapting to climate change in urban areas: the possibilities and constraints in low- and middle-income nations*. Human Settlements Working Paper, IIED, London. https://www.iied.org/10549iied

Satterthwaite, D., & Mitlin, D. (2014). *Reducing urban poverty in the global south*. Routledge., DOI: 10.4324/9780203104316

Schipper, E. L., Ayers, J., Reid, H., Huq, S., & Rahman, A. (Eds.), *Community-based adaptation to climate change: scaling it up* (pp. 214–225). Routledge., DOI: 10.4324/9780203105061

Sen, H. S. (2019). *The Sundarbans: A disaster-prone ecoregion*. Berlin: Springer. 10, 978-3 https://link.springer.com/content/pdf/10.1007/978-3-030-00680-8.pdf

Sharma, D., & Tomar, S. (2010). Mainstreaming climate change adaptation in Indian cities. *Environment and Urbanization*, *22*(2), 451–465. DOI: 10.1177/0956247810377390

Sharma, R., Webster, P., & Bhattacharyya, S. (2014). Factors affecting the performance of community health workers in India: A multi-stakeholder perspective. *Global Health Action*, *7*(1), 25352. DOI: 10.3402/gha.v7.25352 PMID: 25319596

Sivaramakrishnan, K. C. (2007). *Democracy in urban India*. Urban Age. https://urbanage.lsecities.net/essays/democracy-in-urban-india

Smit, B., & Pilifosova, O. (2003). *From adaptation to adaptive capacity and vulnerability reduction*. In Climate Change, Adaptive Capacity and Development (pp. 9-28). Imperial College Press. DOI: 10.1142/9781860945816_0002

Smit, B., & Skinner, M. W. (2002). Adaptation options in agriculture to climate change: A typology. *Mitigation and Adaptation Strategies for Global Change*, *7*(1), 85–114. DOI: 10.1023/A:1015862228270

Smit, B., & Wandel, J. (2006). Adaptation, adaptive capacity and vulnerability. *Global Environmental Change*, *16*(3), 282–292. DOI: 10.1016/j.gloenvcha.2006.03.008

Soltesova, K., Brown, A., Dayal, A., & Dodman, D. (2014). *Community participation in urban adaptation to climate change: Potential and limits for CBA approaches*. Community-Based Adaptation to Climate Change., DOI: 10.4324/9780203105061

Sovacool, B. K. (2011). An international comparison of four polycentric approaches to climate and energy governance. *Energy Policy, 39*(6), 3832–3844. DOI: 10.1016/j.enpol.2011.04.014

Sovacool, B. K. (2018). Bamboo beating bandits: Conflict, inequality, and vulnerability in the political ecology of climate change adaptation in Bangladesh. *World Development, 102*, 183–194. DOI: 10.1016/j.worlddev.2017.10.014

StC. (2008). D. E. *Convention on the Conservation of Migratory Species of Wild Animals*.

Sundarbans National park. (1987). *UNESCO World Heritage Centre*. https://whc.unesco.org/en/list/452

Syam, D. (2013). *National engagement and replication in Indonesia*. Presentation at the CBA7 conference, Dhaka, April 2013. DOI: 10.1177/0956247808096126

Szabo, S., Hossain, M. S., Adger, W. N., Matthews, Z., Ahmed, S., Lázár, A. N., & Ahmad, S. (2016). Soil salinity, household wealth and food insecurity in tropical deltas: Evidence from south-westcoast of Bangladesh. *Sustainability Science, 11*(3), 411–421. DOI: 10.1007/s11625-015-0337-1 PMID: 30174734

Thakur, S., Maity, D., Mondal, I., Basumatary, G., Ghosh, P. B., Das, P., & De, T. K. (2021). Assessment of changes in land use, land cover, and land surface temperature in the mangrove forest of Sundarbans, northeast coast of India. *Environment, Development and Sustainability, 23*(2), 1917–1943. DOI: 10.1007/s10668-020-00656-7

Tripodi, T., & Potocky-Tripodi, M. (2007). *International Social Work Research: Issues and Prospects*. Oxford University Press., DOI: 10.1093/acprof:oso/9780195187250.001.0001

Twigg, J. (2007). *Characteristics of a disaster-resilient community, a guidance note*. Available at https://www.researchgate.net/publication/315799865_Characteristics_of_a_Disaster-resilient_Community_A_Guidance_Note_Characteristics_of_a_Disaster-resilient_Community_A_Guidance_Note

Tyagi, S. K. S., Mukherjee, A., Pokhrel, S. R., & Hiran, K. K. (2020). An intelligent and optimal resource allocation approach in sensor networks for smart agri-IoT. *IEEE Sensors Journal, 21*(16), 17439–17446. DOI: 10.1109/JSEN.2020.3020889

Tyler, S., & Moench, M. (2012). A framework for urban climate resilience. *Climate and Development*, *4*(4), 311–326. DOI: 10.1080/17565529.2012.745389

Uddin, S. M. (2019). Religion, Nature, and Life in the Sundarbans. *Asian Ethnology*, *78*(2), 289–310. https://www.jstor.org/stable/26845329

Unogwu, O. J., Doshi, R., Hiran, K. K., Mijwil, M. M., Catherine, A. T., & Abotaleb, M. (2024). *Exploring the Implications of Emerging Artificial Intelligence Technologies at Edge Computing in Higher Education*. In Integrating Generative AI in Education to Achieve Sustainable Development Goals (pp. 169-182). IGI Global. DOI: 10.4018/979-8-3693-2440-0.ch009

Yohe, G., & Tol, R. S. J. (2002). Indicators for social and economic coping capacity moving toward a working definition of adaptive capacity. *Global Environmental Change*, *12*(1), 25–40. DOI: 10.1016/S0959-3780(01)00026-7

ADDITIONAL READING

Andtfolk, M., Nyholm, L., Eide, H., Rauhala, A., & Fagerström, L. (2022). Attitudes Toward the Use of Humanoid Robots in Healthcare—A Cross-Sectional Study. *AI & Society*, *37*(4), 1739–1748. DOI: 10.1007/s00146-021-01271-4

Asiedu-Berkoe, F., Chandi, G. M., Bandoh, D. A., Atsu, B. K., Lokossou, V. K., Antara, S. N., Sarkodie, B., & Kenu, E. (2022). State of Public Health Emergency Preparedness and Response Capacity of Ghana. *Journal of Interventional Epidemiology and Public Health*, *5*(4), 1–16. DOI: 10.37432/jieph.supp.2022.5.4.05.4

Chauhan, P., Bali, A., & Kaur, S. (2024). Breaking Barriers for Accessible Health Programs: The Role of Telemedicine in a Global Healthcare Transformation. In *Transformative Approaches to Patient Literacy and Healthcare Innovation*. IGI Global., DOI: 10.4018/979-8-3693-3661-8.ch014

KEY TERMS AND DEFINITIONS

Climate: Adaptation: Climate adaptation refers to strategies and measures implemented to help communities and ecosystems adjust to the adverse effects of climate change.

Climate Justice: Environmental justice is a social movement and legal principle that advocates for fair treatment and meaningful involvement of all individuals, particularly marginalized communities, in environmental decision-making.

Health Equity: Health equity refers to the principle that everyone should have a fair opportunity to attain their highest level of health, regardless of social or economic status.

Indigenous Rights: Indigenous rights refer to the collective and individual entitlements of Indigenous peoples, recognizing their unique cultural identities, traditional lands, and ways of life.

Chapter 4
Climate Action Meets Social Justice:
Designing Public Policies That Address Global Warming and Promote Sustainable Development

Naboshree Bhattacharya
https://orcid.org/0000-0001-5572-4098
Amity University, Ranchi, India

Sweta Sharan
Magadh Mahila College, India

ABSTRACT

Global warming represents an urgent crisis requiring immediate action. According to the IPCC's latest report, human-induced climate change is causing widespread disruption to nature and affecting billions worldwide, with vulnerable populations bearing the heaviest burden. While multiple feasible solutions exist to reduce emissions and adapt to climate change, immediate action is crucial for securing a liveable future. The challenge of addressing climate change is intrinsically linked to social justice and sustainable development. Marginalized communities, often having the least resources to adapt, face disproportionate impacts from rising sea levels, extreme weather events, and food insecurity. Effective climate policies must therefore embrace climate justice principles while ensuring a just transition for workers in carbon-intensive industries. Success requires integrating climate action with broader sustainable development goals, including poverty alleviation, food security, and gender equality. By pursuing these synergies, policymakers can maximize benefits.

DOI: 10.4018/979-8-3373-1310-8.ch004

1. INTRODUCTION

The convergence of climate change mitigation and social justice represents one of the most pressing challenges facing humanity in the 21st century. Recent evidence indicates that the climate crisis has reached unprecedented levels of urgency, with the Intergovernmental Panel on Climate Change (IPCC, 2023) warning that the window for limiting global temperature rise to 1.5°C above pre-industrial levels is rapidly closing. This situation is further emphasized by NASA and NOAA's (2024) confirmation that 2023 was the warmest year on record, with global temperatures averaging 1.18°C above the pre-industrial baseline, highlighting the critical nature of immediate action. The severity of the climate crisis is underscored by multiple indicators that have surpassed previous projections. The World Meteorological Organization (2024) reports that atmospheric CO2 concentrations have reached 421.7 ppm, the highest level in 2 million years. Ocean heat content has achieved record levels, with significant implications for marine ecosystems and coastal communities (von Schuckmann et al., 2024). These changes have manifested in increasingly frequent and severe extreme weather events, which affected over 185 million people in 2023 alone, with economic losses exceeding $290 billion (Stroeve & Notz, 2023).

The integration of social justice considerations into climate policy has become increasingly crucial as evidence mounts regarding the disproportionate impact of climate change on vulnerable populations. The World Bank (2024) reports that low-income communities face 5-6 times higher exposure to climate risks compared to affluent populations. This disparity is particularly evident in indigenous communities, which bear disproportionate climate impacts despite contributing minimally to emissions (United Nations Development Programme, 2024). Gender disparities in climate vulnerability persist, with women facing higher risks in developing regions (UN Women, 2024).Recent research has demonstrated that successful climate policies must address both environmental sustainability and social equity to achieve lasting impact.

The United Nations Development Programme's Human Development Report (2024) emphasizes that community-based adaptation programs show 40% higher success rates when incorporating local knowledge and participatory approaches. This finding aligns with emerging evidence that integrated approaches combining climate action with sustainable development goals demonstrate superior outcomes in both environmental and social dimensions. The urgency of addressing these interlinked challenges requires a comprehensive policy framework that acknowledges both the environmental and social dimensions of climate change.

As global temperatures continue to rise and climate impacts intensify, the need for policies that promote both climate resilience and social equity becomes increasingly apparent. The evidence suggests that successful climate action must be rooted in principles of social justice and sustainable development to ensure effective and equitable outcomes for all communities. Looking forward, research priorities should focus on quantifying the co-benefits of integrated climate and social justice policies, developing robust metrics for measuring equity outcomes, and identifying best practices for community engagement in policy development. These efforts will be crucial in designing and implementing effective policies that address both the environmental and social dimensions of the climate crisis while promoting sustainable development goals.

2. UNDERSTANDING CLIMATE JUSTICE

2.1 Definition and Principles of Climate Justice

Climate justice is a conceptual framework that recognizes the disproportionate impacts of climate change on marginalized and vulnerable communities, and advocates for equitable solutions that address these disparities (Shue, 2018). This principle is rooted in the understanding that climate change is not only an environmental issue but also a social justice concern, as its effects exacerbate existing inequalities and disproportionately burden those who have contributed the least to the problem (Klinsky et al., 2017).

The core tenets of climate justice are grounded in the principles of environmental justice, which emerged in the 1980s to address the unequal distribution of environmental burdens and benefits across different communities (Bullard, 1990). Climate justice extends this concept to the global scale, recognizing that the impacts of climate change are not evenly distributed across nations, regions, or socioeconomic groups (Schlosberg & Collins, 2014).

Several key principles underpin the climate justice framework:

i. Equitable distribution of burdens and benefits: Climate policies should ensure that the costs and benefits of climate action are distributed fairly, with particular attention to protecting and empowering marginalized communities (Shue, 2018).
ii. Participatory decision-making: Affected communities, especially those most vulnerable, should have a meaningful voice in the development and implementation of climate policies (Schlosberg & Collins, 2014).
iii. Recognition of historical responsibilities: The disproportionate contribution of industrialized nations to greenhouse gas emissions should be acknowl-

edged, and they should bear a greater responsibility for mitigation and adaptation efforts (Klinsky et al., 2017).

iv. Intergenerational equity: Climate policies should consider the long-term implications for future generations and ensure that their interests are protected (Shue, 2018).
v. Intersectionality: Climate justice must address the intersecting forms of oppression and marginalization, such as race, gender, class, and disability, that shape vulnerability to climate change (Kaijser & Kronsell, 2014).

By incorporating these principles, climate justice aims to ensure that climate policies do not exacerbate existing inequalities but rather promote equitable and inclusive solutions that benefit all members of society, especially those most vulnerable to the impacts of climate change.

2.2 Disproportionate Impacts of Climate Change on Marginalized and Vulnerable Communities

Numerous studies have documented the disproportionate impacts of climate change on marginalized and vulnerable communities, both within and across nations. These impacts are often compounded by pre-existing socioeconomic and environmental inequalities, creating a vicious cycle of vulnerability (IPCC, 2022).

Socioeconomic Factors:

i. Low-income communities often lack the resources and infrastructure to adapt to climate impacts, such as extreme weather events, rising sea levels, and food insecurity (Hallegatte et al., 2016).
ii. Indigenous communities, whose livelihoods and cultural practices are deeply tied to the land and natural resources, face threats to their traditional ways of life due to climate-induced environmental changes (Ford et al., 2020).

Geographic Vulnerability:

i. Coastal and island communities are at heightened risk from sea-level rise, storm surges, and coastal erosion, threatening their homes, livelihoods, and cultural heritage (IPCC, 2022).
ii. Many developing nations in tropical regions are particularly susceptible to the impacts of climate change, such as droughts, floods, and heat waves, due to their geographic location and limited adaptive capacity (Diffenbaugh & Burke, 2019).

Gender Inequality:

Women, especially in developing countries, often bear a disproportionate burden of the impacts of climate change due to existing gender inequalities, limited access to resources, and their traditional roles in securing food, water, and energy for their households (UN Women, 2022).

Intersectional Vulnerabilities:

The impacts of climate change are exacerbated for individuals and communities experiencing multiple and intersecting forms of marginalization, such as race, ethnicity, disability, and age (Kaijser & Kronsell, 2014).

These disproportionate impacts not only threaten the livelihoods, health, and well-being of vulnerable communities but also undermine their ability to adapt and build resilience to the changing climate, perpetuating a cycle of vulnerability and inequality (Pelling & Garschagen, 2019).

2.3 The Need for Equitable Climate Policies that Protect and Empower Marginalized Communities

Addressing the disproportionate impacts of climate change on marginalized and vulnerable communities requires a fundamental shift in how climate policies are developed and implemented. Traditional approaches that prioritize technical solutions and economic efficiency often fail to consider the social and distributional consequences, potentially exacerbating existing inequalities (Klinsky et al., 2017).

To achieve climate justice, climate policies must be designed and implemented through an equity lens, ensuring that the needs and perspectives of marginalized communities are central to the decision-making process (Schlosberg & Collins, 2014). This requires:

i. Inclusive and participatory processes: Meaningful engagement and consultation with affected communities, particularly those most vulnerable, to ensure their voices and traditional knowledge are incorporated into policy development and implementation (Naess, 2013).
ii. Targeted adaptation and mitigation strategies: Climate policies should prioritize adaptation measures that directly address the specific vulnerabilities and needs of marginalized communities, while also promoting inclusive and sustainable mitigation efforts (Pelling & Garschagen, 2019).

iii. Capacity building and resource allocation: Investing in building the adaptive capacity and resilience of vulnerable communities through access to resources, technology transfer, and capacity-building initiatives (Hallegatte et al., 2016).
iv. Addressing structural inequalities: Recognizing and addressing the underlying social, economic, and political factors that contribute to vulnerability and marginalization, such as poverty, discrimination, and lack of access to decision-making processes (Klinsky et al., 2017).
v. Monitoring and evaluation: Establishing mechanisms for monitoring and evaluating the distributional impacts of climate policies, and making necessary adjustments to ensure equitable outcomes (Pelling & Garschagen, 2019).

By integrating these principles into climate policy design and implementation, nations can work towards achieving climate justice and ensuring that the burdens and benefits of climate action are distributed equitably, empowering and protecting the most vulnerable communities.

3. INTEGRATING CLIMATE ACTION WITH SUSTAINABLE DEVELOPMENT GOALS

3.1. Overview of the Sustainable Development Goals (SDGs)

The Sustainable Development Goals (SDGs) are a comprehensive and ambitious set of 17 global goals adopted by the United Nations in 2015 as part of the 2030 Agenda for Sustainable Development (United Nations, 2015). These goals aim to address the most pressing challenges facing humanity and the planet, ranging from poverty, inequality, and climate change to environmental degradation, peace, and justice.

The 17 SDGs are interconnected and indivisible, recognizing that progress in one area is often linked to progress in others. This interconnectedness is a fundamental principle of the 2030 Agenda, which emphasizes the need for a holistic and integrated approach to sustainable development (Nilsson et al., 2016).

i. **No Poverty**: Eradicating poverty in all its forms remains one of the greatest challenges facing humanity. This goal aims to ensure social protection systems, access to economic resources, and building resilience for the poor and vulnerable.
ii. **Zero Hunger**: Ending hunger, achieving food security, improving nutrition, and promoting sustainable agriculture are critical objectives under this goal.
iii. **Good Health and Well-being**: Ensuring healthy lives and promoting well-being for all at all ages is essential for sustainable development. This goal addresses

various health issues, including maternal and child health, communicable and non-communicable diseases, and universal health coverage.

iv. **Quality Education**: Inclusive and equitable quality education and promoting lifelong learning opportunities for all are key to empowering individuals and communities.

v. **Gender Equality**: Achieving gender equality and empowering all women and girls is a fundamental human right and a necessary foundation for a peaceful, prosperous, and sustainable world.

vi. **Clean Water and Sanitation**: Ensuring availability and sustainable management of water and sanitation for all is crucial for human health, environmental sustainability, and economic development.

vii. **Affordable and Clean Energy**: Ensuring access to affordable, reliable, sustainable, and modern energy for all is essential for addressing climate change and promoting sustainable development.

viii. **Decent Work and Economic Growth**: Promoting sustained, inclusive, and sustainable economic growth, full and productive employment, and decent work for all is a key driver of sustainable development.

ix. **Industry, Innovation, and Infrastructure**: Building resilient infrastructure, promoting inclusive and sustainable industrialization, and fostering innovation are vital for economic growth and sustainable development.

x. **Reduced Inequalities**: Reducing inequalities within and among countries is essential for achieving sustainable development and promoting social cohesion.

xi. **Sustainable Cities and Communities**: Making cities and human settlements inclusive, safe, resilient, and sustainable is crucial for promoting sustainable development and addressing the challenges of rapid urbanization.

xii. **Responsible Consumption and Production**: Ensuring sustainable consumption and production patterns is essential for managing the world's natural resources and addressing environmental degradation.

xiii. **Climate Action**: Taking urgent action to combat climate change and its impacts is a critical challenge for humanity and a prerequisite for sustainable development.

xiv. **Life Below Water**: Conserving and sustainably using the oceans, seas, and marine resources for sustainable development is essential for the health of the planet and human well-being.

xv. **Life on Land**: Protecting, restoring, and promoting sustainable use of terrestrial ecosystems, sustainably managing forests, combating desertification, halting and reversing land degradation, and halting biodiversity loss are crucial for sustainable development.

xvi. **Peace, Justice, and Strong Institutions**: Promoting peaceful and inclusive societies, providing access to justice for all, and building effective, accountable, and inclusive institutions at all levels are essential for sustainable development.

xvii. **Partnerships for the Goals**: Revitalizing the global partnership for sustainable development and enhancing the means of implementation are critical for achieving the SDGs.

The SDGs provide a comprehensive framework for addressing the complex and interrelated challenges facing humanity and the planet, emphasizing the need for a holistic and integrated approach to sustainable development (United Nations, 2015).

3.2. Synergies between Climate Action and Sustainable Development

Climate action, as outlined in SDG 13, is intrinsically linked to the achievement of many other Sustainable Development Goals. Addressing climate change is not only an environmental imperative but also a critical enabler for sustainable development across various sectors and dimensions. The synergies between climate action and sustainable development are multifaceted and far-reaching (Stafford-Smith et al., 2017).

i. **Poverty Reduction and Economic Growth (SDGs 1 and 8):** Climate change disproportionately impacts the poor and vulnerable, threatening their livelihoods, food security, and access to water and other essential resources (IPCC, 2022). Effective climate action can help mitigate these impacts and create opportunities for sustainable economic growth through green jobs, climate-resilient infrastructure, and the development of low-carbon technologies (Hallegatte et al., 2016). For example, investing in renewable energy can create employment opportunities while reducing greenhouse gas emissions and promoting energy access (IEA, 2020).

ii. **Health and Well-being (SDG 3):** Climate change poses significant risks to human health, including increased incidence of vector-borne diseases, heat-related illnesses, malnutrition, and respiratory problems due to air pollution (WHO, 2021). Climate action, such as reducing greenhouse gas emissions, promoting sustainable agriculture, and improving air quality, can help reduce these health risks and promote healthier environments. Additionally, climate-resilient healthcare systems can better withstand the impacts of extreme weather events.

iii. **Food Security and Water Availability (SDGs 2 and 6):** Climate change threatens agricultural productivity and water availability, particularly in vulnerable regions (FAO, 2019). Climate-smart agriculture practices, such as drought-resistant crops, efficient irrigation systems, and sustainable land management, can help ensure food security and access to clean water. Additionally, reducing greenhouse gas emissions from agriculture and promoting sustainable water management practices can contribute to climate action while supporting these goals.

iv. **Sustainable Cities and Communities (SDG 11):** Urbanization and climate change are closely linked, with cities being major contributors to greenhouse gas emissions and also being highly vulnerable to climate impacts such as heat waves, flooding, and sea-level rise (UN-Habitat, 2020). Climate action, including sustainable urban planning, resilient infrastructure, and green spaces, can promote sustainable cities and communities while also reducing emissions and enhancing climate resilience.

v. **Biodiversity and Ecosystem Protection (SDGs 14 and 15):** Climate change is a significant threat to biodiversity and ecosystem services (IPBES, 2019). Climate action, including conservation efforts, sustainable land and marine management practices, and reducing deforestation and habitat degradation, can help protect and restore these vital resources. Healthy ecosystems, in turn, can contribute to climate change mitigation and adaptation through carbon sequestration and enhancing resilience to climate impacts.

vi. Energy Access and Sustainability (SDG 7): Transitioning to clean and renewable energy sources is a critical component of climate action and can also promote energy access, security, and sustainability (IEA, 2020). Investing in renewable energy technologies and improving energy efficiency can reduce greenhouse gas emissions while providing affordable and reliable energy to communities, supporting economic development and improved living standards.

By recognizing and leveraging these synergies, climate action can serve as a catalyst for achieving multiple Sustainable Development Goals, creating co-benefits across various sectors and dimensions of sustainable development (Stafford-Smith et al., 2017).

3.3 Maximizing Co-benefits and Minimizing Trade-offs through Policy Integration

While climate action and sustainable development are inherently linked, their effective integration requires careful policy design and implementation to maximize co-benefits and minimize potential trade-offs. Achieving this integration requires a holistic and coordinated approach across various sectors and levels of governance (Nilsson et al., 2018).

i. **Policy Coherence:** Ensuring policy coherence across different sectors and levels of governance is crucial for aligning climate action with sustainable development objectives (OECD, 2019). This involves coordination and collaboration among various stakeholders, including governments, civil society, the private sector, and international organizations. Policy coherence can be achieved through mechanisms such as inter-ministerial coordination, integrated policymaking processes, and harmonized regulatory frameworks (Mukherjee & Bali, 2019).
ii. **Prioritizing Co-Benefits:** Climate policies should be designed to prioritize co-benefits across multiple SDGs, such as job creation, improved air quality, enhanced energy access, and better health outcomes (Ürge-Vorsatz et al., 2014). By highlighting these co-benefits, policymakers can build broader support for climate action and promote sustainable development. For example, investing in renewable energy can create jobs, reduce air pollution, and contribute to climate change mitigation (IEA, 2020).
iii. **Addressing Trade-Offs:** In some cases, climate action may conflict with other development priorities, such as economic growth or energy access. Identifying and addressing these trade-offs through inclusive stakeholder engagement, impact assessments, and targeted interventions is essential (Nilsson et al., 2018). For instance, the transition to a low-carbon economy may initially result in job losses in certain sectors, requiring targeted policies to support affected communities and facilitate a just transition (ILO, 2018).
iv. **Capacity Building and Technology Transfer:** Ensuring access to resources, capacity building, and technology transfer, particularly for developing countries, is crucial for enabling effective integration of climate action and sustainable development (UNFCCC, 2015). This includes financial support, technical assistance, and knowledge sharing to build the necessary institutional and human capacities for implementing integrated policies and adopting sustainable technologies.
v. **Monitoring and Evaluation:** Establishing robust monitoring and evaluation frameworks to track progress, identify synergies and trade-offs, and inform adaptive policy-making is essential for effective integration (Allen et al., 2018). This involves developing appropriate indicators, collecting and analyzing data, and using this information to adjust policies and strategies as needed. Participatory monitoring and evaluation processes that involve relevant stakeholders can also enhance transparency and accountability.
vi. **Multi-Stakeholder Collaboration**: Effective integration of climate action and sustainable development requires collaboration among various stakeholders, including governments, civil society organizations, the private sector, academia, and local communities (Pattberg & Widerberg, 2016). This collaboration can take the form of multi-stakeholder partnerships, knowledge-sharing platforms,

and participatory decision-making processes, ensuring that diverse perspectives and interests are represented.

vii. **Localized Approaches:** While global frameworks and policies are essential, effective integration of climate action and sustainable development often requires localized approaches that consider the unique contexts, priorities, and capacities of different regions and communities (Oteros-Rozas et al., 2015). This may involve decentralized decision-making, community-based initiatives, and the incorporation of traditional knowledge and practices.

By adopting an integrated approach that maximizes co-benefits and minimizes trade-offs, policymakers can leverage the synergies between climate action and sustainable development, creating a virtuous cycle of progress towards a more sustainable and resilient future (Stafford-Smith et al., 2017).

4. POLICY INSTRUMENTS AND APPROACHES FOR JUST AND SUSTAINABLE CLIMATE ACTION

4.1 Carbon Pricing Mechanisms

Carbon pricing mechanisms, such as carbon taxes and emissions trading systems, are widely recognized as effective policy instruments for reducing greenhouse gas emissions and promoting a transition towards a low-carbon economy. However, the design and implementation of these mechanisms must carefully consider social justice implications to ensure equitable outcomes.

A. Design Considerations for Social Justice:
   a. Addressing potential regressive impacts on low-income households through revenue recycling or targeted compensation measures (Cronin et al., 2017).
   b. Ensuring a just transition for workers and communities affected by the economic restructuring associated with carbon pricing (Robins et al., 2018).
   c. Incorporating principles of procedural justice by involving affected stakeholders in the design and decision-making processes (Maor et al., 2013).
   d. Considering the distributional impacts across different sectors, regions, and demographic groups to mitigate disproportionate burdens (Boyce, 2018).
   e. Aligning carbon pricing policies with broader sustainable development goals, such as poverty reduction and access to clean energy (Campiglio et al., 2017).

B. Examples and Case Studies:

a. The European Union's Emissions Trading System (EU ETS) has implemented measures to address potential regressive impacts, such as allocating a portion of revenues to support low-income households and energy efficiency programs (Sartor, 2013).
b. British Columbia's carbon tax in Canada includes revenue-neutral recycling, with revenues used to reduce personal and corporate income taxes, providing a progressive redistribution of the tax burden (Beck et al., 2015).
c. California's cap-and-trade program allocates a portion of revenues to disadvantaged communities for investments in renewable energy, energy efficiency, and sustainable transportation projects (Cushing et al., 2018).

4.2 Renewable Energy Policies

Transitioning to renewable energy sources is crucial for mitigating climate change and promoting sustainable development. However, renewable energy policies must prioritize energy access, affordability, and community ownership to ensure just and equitable outcomes.

A. Prioritizing Energy Access, Affordability, and Community Ownership:
   a. Implementing policies and incentives to expand access to affordable renewable energy, particularly in underserved and marginalized communities (Yadav et al., 2019).
   b. Promoting community-owned and decentralized renewable energy projects to foster local economic development and empowerment (Haf et al., 2019).
   c. Addressing potential energy affordability concerns through targeted subsidies, energy efficiency programs, and progressive pricing structures (Mundaca et al., 2019).
   d. Ensuring meaningful participation and decision-making power for local communities in the planning and implementation of renewable energy projects (Yenneti & Day, 2015).
B. Examples and Case Studies:
   a. Bangladesh's Infrastructure Development Company Limited (IDCOL) has facilitated the installation of over 4 million solar home systems in rural areas, providing affordable and clean energy access (Khandker et al., 2014).
   b. Germany's Energiewende (Energy Transition) policy framework has enabled the growth of community-owned renewable energy projects, with over 50% of renewable energy capacity owned by citizens and cooperatives (Yildiz et al., 2015).
   c. South Africa's Renewable Energy Independent Power Producer Procurement Program (REIPPPP) has included socio-economic development require-

ments, such as local job creation and community ownership, in its bidding process for renewable energy projects (Wlokas et al., 2017).

4.3 Sustainable Land Use Policies

Sustainable land use policies play a crucial role in addressing climate change, biodiversity conservation, and the rights of indigenous communities. Effective policies in this area must support indigenous rights and promote biodiversity conservation while ensuring equitable outcomes.

A. Supporting Indigenous Rights and Biodiversity Conservation:
   a. Recognizing and protecting the land rights and traditional knowledge of indigenous communities in land use policies and decision-making processes (Brondizio & Le Tourneau, 2016).
   b. Promoting sustainable land management practices that align with indigenous traditions and support biodiversity conservation (Garnett et al., 2018).
   c. Ensuring free, prior, and informed consent (FPIC) from indigenous communities for any land use decisions or projects that may affect their territories and livelihoods (Hanna & Vanclay, 2013).
   d. Providing financial and technical support for indigenous-led conservation and sustainable land management initiatives (Ding et al., 2016).

B. Examples and Case Studies:
   a. The Brazilian Amazon Fund has provided financial resources to support indigenous communities in their efforts to protect and sustainably manage their territories, contributing to biodiversity conservation and climate change mitigation (Duchelle et al., 2014).
   b. The Community Conserved Areas (CCAs) approach in India has recognized and supported the traditional conservation practices of indigenous and local communities, leading to the protection of significant biodiversity hotspots (Pathak-Broome et al., 2017).
   c. The Indigenous Peoples' and Community Conserved Territories and Areas (ICCAs) initiative, supported by the International Union for Conservation of Nature (IUCN), has promoted the recognition and support of indigenous and community-led conservation efforts globally (Corrigan & Granziera, 2018).

4.4 Green Industrial Policies

Green industrial policies aim to promote sustainable economic growth, create decent jobs, and spur innovation in low-carbon technologies and industries. These policies must prioritize social justice considerations to ensure a just transition and equitable outcomes.

A. Creating Decent Jobs and Spurring Innovation:
   a. Investing in research, development, and deployment of clean technologies and sustainable industries to create new job opportunities (Pollin, 2015).
   b. Providing workforce training and reskilling programs to facilitate the transition of workers from carbon-intensive industries to green jobs (ILO, 2018).
   c. Promoting decent work conditions, fair wages, and worker rights in the green economy through labor standards and regulations (Olsen, 2009).
   d. Fostering innovation and entrepreneurship in sustainable technologies and business models through incentives, public-private partnerships, and supportive regulatory frameworks (Kemp & Never, 2017).
B. Examples and Case Studies:
   a. South Korea's Green New Deal has focused on creating jobs in renewable energy, electric vehicles, and energy-efficient buildings, while providing support for worker retraining and transition assistance (Kim et al., 2020).
   b. The European Union's Circular Economy Action Plan aims to promote sustainable production and consumption patterns, create new business opportunities, and generate green jobs through initiatives such as eco-design, waste reduction, and recycling (European Commission, 2020).
   c. Costa Rica's National Bioeconomy Strategy has leveraged the country's rich biodiversity to develop sustainable industries, create green jobs, and promote innovation in areas such as biotechnology, ecotourism, and sustainable agriculture (Rodríguez et al., 2019).

4.5 Adaptation and Resilience Strategies

Adaptation and resilience strategies are essential for addressing the unavoidable impacts of climate change and building resilience in vulnerable communities. These strategies must prioritize the needs of marginalized communities and promote social, economic, and ecological resilience.

A. Prioritizing the Needs of Vulnerable Communities:
   a. Conducting participatory vulnerability assessments to identify the specific needs and priorities of marginalized communities (Oppenheimer et al., 2019).
   b. Ensuring that adaptation and resilience strategies are designed and implemented with the active participation and leadership of vulnerable communities (Ayers et al., 2014).

c. Allocating adequate resources and funding to support community-based adaptation initiatives and address the specific vulnerabilities of marginalized groups (Soanes et al., 2017).
d. Promoting capacity-building and knowledge-sharing to empower vulnerable communities in adapting to climate change impacts (Tschakert et al., 2017).

B. Building Social, Economic, and Ecological Resilience:
a. Integrating climate change adaptation into broader sustainable development strategies to address the interconnected social, economic, and environmental dimensions of resilience (Dilling et al., 2019).
b. Promoting ecosystem-based adaptation approaches that leverage the protective and regulatory services provided by healthy ecosystems (Seddon et al., 2020).
c. Investing in climate-resilient infrastructure, housing, and basic services to enhance the adaptive capacity of vulnerable communities (Hallegatte et al., 2019).
d. Fostering diversified and sustainable livelihoods to reduce economic vulnerability and enhance resilience to climate shocks and stresses (Suckall et al., 2015).

C. Examples and Case Studies:
a. The Least Developed Countries Fund (LDCF) and the Special Climate Change Fund (SCCF), managed by the Global Environment Facility (GEF), have supported adaptation projects in vulnerable countries, with a focus on community-based approaches and addressing the needs of marginalized groups (GEF, 2018).
b. The Ecosystem-based Adaptation (EbA) approach has been implemented in various countries, such as Nepal, where community-based initiatives have focused on restoring and conserving ecosystems to enhance resilience to climate impacts like floods and landslides (Ayers et al., 2017).
c. The Productive Safety Net Program (PSNP) in Ethiopia has combined social protection measures with climate-smart agriculture and livelihood diversification strategies to enhance the resilience of vulnerable rural communities to climate shocks and food insecurity (Berhane et al., 2014).

5. OVERCOMING BARRIERS TO JUST AND SUSTAINABLE CLIMATE ACTION

5.1 Lack of Urgency and Political Will

One of the significant barriers to effective climate action is the lack of urgency and political will among some governments and decision-makers (Giddens, 2015). Despite the overwhelming scientific evidence and growing public concern, some leaders and policymakers have been slow to prioritize climate change mitigation and adaptation measures. This lack of urgency can stem from various factors, including short-term political interests, economic considerations, and the influence of vested interests in fossil fuel industries (Lockwood, 2018).

To overcome this barrier, it is crucial to raise awareness about the severe consequences of inaction and the urgent need for transformative change. This can be achieved through effective communication strategies that highlight the scientific evidence, the economic and social costs of climate change, and the potential benefits of transitioning to a sustainable, low-carbon economy (Moser, 2016). Additionally, strengthening democratic processes, promoting transparency, and empowering citizens to hold their leaders accountable can help build political will and pressure for ambitious climate action (Dryzek & Pickering, 2019).

Overcoming the lack of urgency and political will in climate action requires a multifaceted approach, and several real-world applications demonstrate how these barriers can be addressed effectively. Public awareness campaigns and grassroots movements have played a pivotal role in creating a sense of urgency and pressuring political leaders to act. For example, the Fridays for Future movement, initiated by Greta Thunberg in 2018, rapidly mobilized millions of young people worldwide to advocate for stronger climate policies, significantly elevating climate change as a political priority in many countries (de Moor et al., 2020; Wahlström et al., 2022). Similarly, Extinction Rebellion's disruptive protests in the United Kingdom were instrumental in prompting the UK government to declare a climate emergency in 2019, illustrating how direct action can force political acknowledgment and response (Extinction Rebellion, 2019; Saunders et al., 2020).

On a local scale, movements such as Karachi Bachao Tehreek in Pakistan and Save Aru in Indonesia have demonstrated the power of community-led activism in raising awareness and compelling authorities to reconsider environmentally harmful projects. Karachi Bachao Tehreek, for instance, has highlighted the intersection of climate emergencies and urban inequalities, focusing on housing rights as a component of climate justice (Qazi, 2022). The Save Aru movement, led by local women activists, successfully protested against a government-approved commercial project, leading to its cancellation and showcasing the effectiveness of grassroots mobilization in environmental justice (Ganobal, 2019).

In addition to activism, innovative governance models and policy frameworks have also proven effective. Scotland's Citizens' Assembly, for example, placed ordinary citizens at the center of climate policy discussions, fostering broad-based support and overcoming political inertia by making the policymaking process more inclusive and participatory (Scottish Government, 2021). At the national level, the adoption of comprehensive climate change framework legislation, as advocated by the World Bank, has helped establish clear targets, accountability mechanisms, and cross-sectoral coordination, thereby sustaining political commitment to climate action even across changing administrations (World Bank, 2020). These examples collectively show that combining grassroots mobilization, public engagement, and robust policy frameworks can effectively overcome the barriers of insufficient urgency and political will, paving the way for just and sustainable climate action.

5.2 Rejection of Climate Change by Some Actors

Another significant barrier is the rejection of climate change science and the denial of its anthropogenic causes by some actors, including certain political groups, industries, and individuals (Dunlap & McCright, 2015). This rejection can be fueled by various factors, such as ideological beliefs, economic interests, or misinformation campaigns aimed at sowing doubt and confusion (Oreskes & Conway, 2010).

Overcoming this barrier requires a multifaceted approach that combines scientific education, effective communication, and addressing the underlying drivers of climate change denial. Promoting scientific literacy and critical thinking skills can help counter misinformation and enable individuals to make informed decisions based on evidence (Somerville & Hassol, 2011). Additionally, addressing the economic and social concerns of communities and industries affected by the transition to a low-carbon economy can help reduce resistance and build broader support for climate action (Newell & Mulvaney, 2013).

Some examples are as follows:

a. **Industry-Led Misinformation**: Fossil fuel companies such as ExxonMobil have funded advertising, digital campaigns, and think tanks to cast doubt on climate science and delay policy action, often framing climate change as uncertain or merely a "risk" rather than a reality (U.S. House Committee on Oversight and Reform, 2022; Union of Concerned Scientists, 2015).

b. **Astroturfing and Front Groups**: Industry actors have created fake grassroots organizations to simulate public opposition to climate policies, misleading both the public and policymakers (Union of Concerned Scientists, 2015).
c. **Political Lobbying**: Major polluters spend millions on lobbying to block or weaken climate legislation; for example, ExxonMobil spent over $6.8 million on lobbying in 2024 alone (OpenSecrets, 2025).
d. **Social Media Misinformation**: Climate denial and delay narratives are widely spread on platforms like Facebook and Twitter, amplifying skepticism and confusion about climate science, especially among youth (Williams et al., 2015; Pearce et al., 2019).
e. **Media Manipulation**: Companies sponsor pseudo-scientific reports and use advertorials to influence public opinion and downplay the urgency of climate action (Brulle, 2014).

These coordinated efforts by some actors have significantly shaped public perception, delayed policy implementation, and hindered effective climate action (Brulle, 2014; U.S. House Committee on Oversight and Reform, 2022).

5.3 Need for Global Institutions to Address Climate-Related Security Threats

Climate change poses significant security threats, including the potential for conflict over scarce resources, mass migration due to environmental degradation, and the destabilization of fragile regions (Gemenne et al., 2014). Addressing these threats requires a coordinated global effort and robust international institutions capable of facilitating cooperation, conflict resolution, and collective action.

Strengthening existing international frameworks, such as the United Nations Framework Convention on Climate Change (UNFCCC) and the Paris Agreement, is crucial for promoting global cooperation and accountability (Falkner, 2016). Additionally, enhancing the capacity of international organizations and institutions to address climate-related security threats, such as resource scarcity, food insecurity, and environmental migration, is essential for maintaining global stability and peace (Werrell & Femia, 2015).

Furthermore, integrating climate change considerations into existing security frameworks and strategies can help anticipate and mitigate potential conflicts and instability. This may involve developing early warning systems, promoting regional cooperation, and fostering dialogue and conflict resolution mechanisms in regions particularly vulnerable to climate-related security threats (Busby, 2018).

Overcoming these barriers requires a concerted effort from governments, civil society, the private sector, and international organizations. By raising awareness, promoting scientific education, addressing underlying economic and social concerns, strengthening international cooperation, and integrating climate change into security frameworks, we can overcome the obstacles and pave the way for just and sustainable climate action.

6. CONCLUSION

6.1 Designing Just and Sustainable Climate Policies Is A Moral Imperative and an Economic and Social Opportunity

Addressing climate change through a lens of justice and sustainability is not only a moral obligation but also an opportunity to create a more equitable, prosperous, and resilient society. The impacts of climate change disproportionately affect the most vulnerable populations, exacerbating existing inequalities and threatening hard-won development gains (IPCC, 2022). By prioritizing justice and sustainability in climate policies, we can ensure that the transition to a low-carbon economy does not leave anyone behind and that the benefits are shared equitably.

Furthermore, investing in just and sustainable climate solutions can unlock significant economic and social opportunities. The transition to a green economy has the potential to create millions of new jobs, spur innovation, and drive economic growth (ILO, 2018). Sustainable practices, such as renewable energy, energy efficiency, and sustainable agriculture, can also improve resource efficiency, reduce costs, and enhance long-term competitiveness (OECD, 2017).

6.2 The Benefits of a Just and Sustainable Transition for People and the Planet

A just and sustainable transition to a low-carbon economy can yield numerous benefits for both people and the planet. By addressing climate change while promoting social equity and environmental sustainability, we can create a virtuous cycle of progress that improves the well-being of current and future generations.

For people, a just transition means ensuring that the costs and benefits of climate action are distributed fairly, with particular attention to vulnerable and marginalized communities (Newell & Mulvaney, 2013). This includes creating decent jobs, providing access to affordable and clean energy, ensuring food and water security, and promoting resilience to climate impacts. Additionally, a just transition can foster social cohesion, reduce inequalities, and empower communities to participate in decision-making processes that affect their lives.

For the planet, a sustainable transition means protecting and restoring ecosystems, conserving biodiversity, and promoting the efficient use of natural resources (UNEP, 2019). Sustainable practices, such as renewable energy, sustainable agriculture, and circular economy approaches, can reduce greenhouse gas emissions, minimize environmental degradation, and enhance the resilience of ecosystems to climate change impacts.

6.3 The Need for Bold Leadership, Innovative Thinking, and Challenging Entrenched Interests and Power Structures

Achieving a just and sustainable transition to a low-carbon economy requires bold leadership, innovative thinking, and a willingness to challenge entrenched interests and power structures (Gills & Morgan, 2020). Policymakers must be willing to confront the influence of fossil fuel industries, vested economic interests, and those who prioritize short-term gains over long-term sustainability.

Innovative thinking is needed to develop and implement policies that address the complex and interconnected challenges of climate change, social justice, and environmental sustainability. This may involve exploring new governance models, fostering cross-sectoral collaboration, and embracing transformative approaches that challenge traditional paradigms (Meadowcroft et al., 2019).

6.4 A Call to Action for Policymakers to Embrace a Just and Sustainable Approach to Climate Action

In light of the urgent need for action and the potential benefits of a just and sustainable transition, policymakers must embrace a comprehensive and integrated approach to climate action. This approach should prioritize social equity, environmental sustainability, and economic opportunities, while addressing the interconnected challenges of climate change, poverty, inequality, and environmental degradation.

Policymakers should:

i. Develop and implement policies that address the root causes of climate change while promoting social justice and environmental sustainability (Leichenko & Silva, 2014). This includes adopting a holistic and integrated approach that considers the synergies and trade-offs between climate action and sustainable development goals.
ii. Foster inclusive and participatory decision-making processes that involve diverse stakeholders, including marginalized communities, civil society, and the private sector (Schlosberg & Collins, 2014). Ensuring that the voices

and perspectives of those most affected by climate change and the transition to a low-carbon economy are heard and incorporated into policy decisions is crucial for achieving just and sustainable outcomes.

iii. Invest in capacity-building, technology transfer, and financial support to enable developing countries and vulnerable communities to participate in and benefit from the transition to a low-carbon economy (UNFCCC, 2015). Addressing the uneven distribution of resources and capabilities is essential for promoting a just and equitable transition.
iv. Promote international cooperation, knowledge-sharing, and collective action to address the global challenges of climate change and sustainable development (Bäckstrand & Lövbrand, 2019). Climate change is a global issue that requires coordinated and collaborative efforts across borders and sectors.
v. Challenge entrenched interests and power structures that impede progress towards a just and sustainable future, and embrace transformative approaches that challenge traditional paradigms (Meadowcroft et al., 2019). Vested interests and established systems may resist change, but bold and innovative solutions are necessary to address the scale and complexity of the challenges we face.

By embracing a just and sustainable approach to climate action, policymakers can create a more equitable, prosperous, and resilient future for all. The time to act is now, and the benefits of a just and sustainable transition are too great to ignore.

REFERENCES

Allen, C., Metternicht, G., & Wiedmann, T. (2018). Initial progress in implementing the Sustainable Development Goals (SDGs): A review of evidence from countries. *Sustainability Science, 13*(5), 1453–1467. DOI: 10.1007/s11625-018-0572-3

Ayers, J., Huq, S., Forer, R., Forsyth, T., Vaughan, C., & Visman, E. (2014). Ensuring climate resilient poverty alleviation. In Ensor, J., Berger, R., & Huq, S. (Eds.), *Community-based adaptation to climate change: Emerging lessons* (pp. 145–168). Practical Action Publishing., DOI: 10.3362/9781780448749.007

Ayers, J., Kaur, N., & Anderson, S. (2017). Negotiating climate resilience in Nepal. In Filho, W. L. (Ed.), *Climate change adaptation in Pacific countries* (pp. 115–132). Springer., DOI: 10.1007/978-3-319-50094-2_7

Bäckstrand, K., & Lövbrand, E. (2019). The road to Paris: Negotiating a stronger climate agreement. *Journal of Environmental Policy and Planning, 21*(5), 519–532. DOI: 10.1080/1523908X.2019.1603774

Beck, M., Rivers, N., Wigle, R., & Yonezawa, H. (2015). Carbon tax and revenue recycling: Impacts on households in British Columbia. *Resource and Energy Economics, 41*, 40–69. DOI: 10.1016/j.reseneeco.2015.04.005

Berhane, G., Gilligan, D. O., Hoddinott, J., Kumar, N., & Taffesse, A. S. (2014). Can social protection work in Africa? The impact of Ethiopia's Productive Safety Net Programme. *Economic Development and Cultural Change, 63*(1), 1–26. DOI: 10.1086/677753

Boyce, J. K. (2018). Carbon pricing: Effectiveness and equity. *Ecological Economics, 150*, 52–61. DOI: 10.1016/j.ecolecon.2018.03.030

Brondizio, E. S., & Le Tourneau, F. M. (2016). Environmental governance for all. *Science, 352*(6291), 1272–1273. DOI: 10.1126/science.aaf5122 PMID: 27284179

Brulle, R. J. (2014). Institutionalizing delay: Foundation funding and the creation of U.S. climate change counter-movement organizations. *Climatic Change, 122*(4), 681–694. DOI: 10.1007/s10584-013-1018-7

Bullard, R. D. (1990). *Dumping in Dixie: Race, class, and environmental quality*. Westview Press.

Busby, J. (2018). Taking stock: The field of climate and security revisited. Social Science Research Council. https://www.ssrc.org/publications/view/taking-stock-the-field-of-climate-and-security-revisited/

Campiglio, E., Godin, A., Kemp-Benedict, E., & Matikainen, S. (2017). The tightening links between financial systems and the low-carbon transition. In *Economic Policies since the Global Financial Crisis* (pp. 313–356). Springer., DOI: 10.1007/978-3-319-60459-6_8

Corrigan, C., & Granziera, A. (2018). Governance of protected areas from impact evaluations to celebrating success. In Worboys, G. L., Lockwood, M., Kothari, A., Feary, S., & Pulsford, I. (Eds.), *Protected area governance and management* (pp. 771–788). ANU Press., DOI: 10.22459/PAGM.04.2018.04

Cronin, J. A., Fullerton, D., & Sexton, S. (2017). Vertical and horizontal redistribution from a carbon tax and rebate. NBER Working Paper No. 23250. National Bureau of Economic Research. DOI: 10.3386/w23250

Cushing, L., Blaustein-Rejto, D., Wander, M., Pastor, M., Sadd, J., Hyun, A., & Morello-Frosch, R. (2018). Carbon trading, co-pollutants, and environmental equity: Evidence from California's cap-and-trade program (2011–2015). *PLoS Medicine, 15*(7), e1002604. DOI: 10.1371/journal.pmed.1002604 PMID: 29990353

de Moor, J., Uba, K., Wahlström, M., Wennerhag, M., & De Vydt, M. (2020). Protest for a future: Composition, mobilization and motives of the participants in Fridays For Future climate protests on 15 March, 2019 in 13 European cities. *Sustainability, 12*(6), 2191. DOI: 10.3390/su12062191

Diffenbaugh, N. S., & Burke, M. (2019). Global warming has increased global economic inequality. *Proceedings of the National Academy of Sciences of the United States of America, 116*(20), 9808–9813. DOI: 10.1073/pnas.1816020116 PMID: 31010922

Dilling, L., Daly, M. E., Travis, W. R., Wilhelmi, O. V., & Klein, R. A. (2019). The dynamics of vulnerability: Why adapting to climate variability will not always prepare us for climate change. *Wiley Interdisciplinary Reviews: Climate Change, 10*(1), e571. DOI: 10.1002/wcc.571

Ding, H., Veit, P. G., Blackman, A., Gray, E., Reytar, K., Altamirano, J. C., & Hodgdon, B. (2016). Climate benefits, tenure costs: The economic case for securing indigenous land rights in the Amazon. World Resources Institute. https://www.wri.org/research/climate-benefits-tenure-costs

Dryzek, J. S., & Pickering, J. (2019). *The politics of the Anthropocene*. Oxford University Press.

Duchelle, A. E., Crompton, T., Gebara, M. F., Guerra, R., Melo, T., Larson, A., & Sills, E. (2014). Linking forest tenure reform, environmental compliance, and incentives: Lessons from REDD+ initiatives in the Brazilian Amazon. *World Development, 55*, 53–67. DOI: 10.1016/j.worlddev.2013.01.014

Dunlap, R. E., & McCright, A. M. (2015). Challenging climate change: The denial countermovement. In Dunlap, R. E., & Brulle, R. J. (Eds.), *Climate change and society: Sociological perspectives* (pp. 300–332). Oxford University Press. DOI: 10.1093/acprof:oso/9780199356102.003.0010

European Commission. (2020). A new Circular Economy Action Plan for a cleaner and more competitive Europe. https://eur-lex.europa.eu/legal-content/EN/TXT/?qid=1583933814386&uri=COM:2020:98:FIN

Extinction Rebellion. (2019). The truth behind the UK Parliament's climate emergency declaration. https://rebellion.earth/2019/05/02/the-truth-behind-the-uk-parliaments-climate-emergency-declaration

Falkner, R. (2016). The Paris Agreement and the new logic of international climate politics. *International Affairs, 92*(5), 1107–1125. DOI: 10.1111/1468-2346.12708

FAO. (2019). The State of Food and Agriculture 2019. Moving forward on food loss and waste reduction. https://www.fao.org/3/ca6030en/ca6030en.pdf

Ford, J. D., Maillet, M., Pouliot, V., Meredith, T., & Cavanaugh, A. (2020). Adaptation and Indigenous peoples in the United Nations framework convention on climate change. *Climatic Change, 164*(3), 1–14. DOI: 10.1007/s10584-020-02849-6

Ganobal, M. (2019). Save Aru: Grassroots resistance and the cancellation of a commercial project in Indonesia. [Case study].

Garnett, S. T., Burgess, N. D., Fa, J. E., Fernández-Llamazares, Á., Molnár, Z., Robinson, C. J., Watson, J. E. M., Zander, K. K., Austin, B., Brondizio, E. S., Collier, N. F., Duncan, T., Ellis, E., Geyle, H., Jackson, M. V., Jonas, H., Malmer, P., McGowan, B., Sivongxay, A., & Leiper, I. (2018). A spatial overview of the global importance of Indigenous lands for conservation. *Nature Sustainability, 1*(7), 369–374. DOI: 10.1038/s41893-018-0100-6

GEF. (2018). Least Developed Countries Fund and Special Climate Change Fund. Global Environment Facility. https://www.thegef.org/topics/least-developed-countries-fund-ldcf

Gemenne, F., Barnett, J., Adger, W. N., & Dabelko, G. D. (2014). Climate and security: Evidence, emerging risks, and a new agenda. *Climatic Change, 123*(1), 1–9. DOI: 10.1007/s10584-014-1074-7

Giddens, A. (2015). *The politics of climate change*. Polity Press.

Gills, B. K., & Morgan, J. (2020). Global climate emergency: After COP24, climate science, urgency, and the threat to humanity. *Globalizations*, *17*(6), 885–900. DOI: 10.1080/14747731.2019.1669915

Haf, S., Robins, N., & Huq, S. (2019). Defining a policy agenda to mobilize climate investment through a just transition. In Robins, N., Muguet, S., & Marquant, A. (Eds.), *The Routledge handbook of climate change and finance* (pp. 193–208). Routledge., DOI: 10.4324/9781351127318-13

Hallegatte, S., Bangalore, M., Bonzanigo, L., Fay, M., Kane, T., Narloch, U., Vogt-Schilb, A. (2016). Shock waves: Managing the impacts of climate change on poverty. The World Bank. https://openknowledge.worldbank.org/handle/10986/22787

Hallegatte, S., Rentschler, J., & Rozenberg, J. (2019). *Lifelines: The resilient infrastructure opportunity*. World Bank., DOI: 10.1596/978-1-4648-1430-3

Hanna, P., & Vanclay, F. (2013). Human rights, Indigenous peoples and the concept of Free, Prior and Informed Consent. *Impact Assessment and Project Appraisal*, *31*(2), 146–157. DOI: 10.1080/14615517.2013.780373

IEA. (2020). Sustainable Recovery: World Energy Outlook Special Report. International Energy Agency. https://www.iea.org/reports/sustainable-recovery

ILO. (2018). World Employment and Social Outlook 2018: Greening with jobs. International Labour Organization. https://www.ilo.org/weso-greening/documents/WESO_Greening_EN_web2.pdf

IPBES. (2019). Global assessment report on biodiversity and ecosystem services of the Intergovernmental Science-Policy Platform on Biodiversity and Ecosystem Services. IPBES Secretariat. https://ipbes.net/global-assessment

IPCC. (2018). *Global Warming of 1.5°C: An IPCC Special Report*. Intergovernmental Panel on Climate Change.

IPCC. (2022). Climate Change 2022: Impacts, Adaptation and Vulnerability. Cambridge University Press. https://www.ipcc.ch/report/ar6/wg2/

Kaijser, A., & Kronsell, A. (2014). Climate change through the lens of intersectionality. *Environmental Politics*, *23*(3), 417–433. DOI: 10.1080/09644016.2013.835203

Kemp, R., & Never, B. (2017). Green transition, industrial policy, and economic development. *Oxford Review of Economic Policy*, *33*(1), 66–84. DOI: 10.1093/oxrep/grw037

Khandker, S. R., Samad, H. A., Ali, R., & Barnes, D. F. (2014). Who benefits most from rural electrification? Evidence in India. *The Energy Journal (Cambridge, Mass.)*, *35*(2), 75–96. DOI: 10.5547/01956574.35.2.4

Kim, J., Chung, J., & Suh, K. (2020). Green new deal policy and wind energy utilization in the Republic of Korea. *Sustainability*, *12*(18), 7599. DOI: 10.3390/su12187599

Klinsky, S., Roberts, T., Huq, S., Okereke, C., Newell, P., Dauvergne, P., O'Brien, K., Schroeder, H., Tschakert, P., Clapp, J., Keck, M., Biermann, F., Liverman, D., Gupta, J., Rahman, A., Messner, D., Pellow, D., & Bauer, S. (2017). Why equity is fundamental in climate change policy research. *Global Environmental Change*, *44*, 170–173. DOI: 10.1016/j.gloenvcha.2016.08.002

Leichenko, R., & Silva, J. A. (2014). Climate change and poverty: Vulnerability, impacts, and alleviation strategies. *Wiley Interdisciplinary Reviews: Climate Change*, *5*(4), 539–556. DOI: 10.1002/wcc.287

Lockwood, M. (2018). Right-wing populism and the climate change agenda: Exploring the linkages. *Environmental Politics*, *27*(4), 712–732. DOI: 10.1080/09644016.2018.1458411

Maor, M., Gilad, S., & Ben-Nun Bloom, P. (2013). Organizational reputation, regulatory talk, and strategic silence. *Journal of Public Administration: Research and Theory*, *23*(3), 627–649. DOI: 10.1093/jopart/mus047

Meadowcroft, J., Banister, D., Holden, E., Langhelle, O., Linnerud, K., & Gilpin, G. (2019). *What next for sustainable development? Our common future at thirty*. Edward Elgar Publishing. DOI: 10.4337/9781788975209

Moser, S. C. (2016). Reflections on climate change communication research and practice in the second decade of the 21st century: What more is there to say? *Wiley Interdisciplinary Reviews: Climate Change*, *7*(3), 345–369. DOI: 10.1002/wcc.403

Mukherjee, V., & Bali, A. S. (2019). Policy coherence for sustainable development in India: An analysis of sustainable development goals. NITI Aayog. https://niti.gov.in/sites/default/files/2019-07/SDG-Policy-Coherence-in-India.pdf

Naess, L. O. (2013). The role of local knowledge in adaptation to climate change. *Wiley Interdisciplinary Reviews: Climate Change*, *4*(2), 99–106. DOI: 10.1002/wcc.204

NASA, & NOAA. (2024). Annual Global Climate Report 2023. U.S. Government Printing Office.

Nilsson, M., Chisholm, E., Griggs, D., Howden-Chapman, P., McCollum, D., Messerli, P., Neumann, B., Stevance, A.-S., Visbeck, M., & Stafford-Smith, M. (2018). Mapping interactions between the sustainable development goals: Lessons learned and ways forward. *Sustainability Science*, *13*(6), 1489–1503. DOI: 10.1007/s11625-018-0604-z PMID: 30546483

Nilsson, M., Griggs, D., & Visbeck, M. (2016). Policy: Map the interactions between sustainable development goals. *Nature*, *534*(7607), 320–322. DOI: 10.1038/534320a PMID: 27306173

Olsen, L. (2009). The employment effects of climate change and climate change responses: A role for international labour standards? Global Union Research Network Discussion Paper No. 12. https://www.gurn.info/en/discussion-papers/no12-may09-the-employment-effects-of-climate-change-and-climate-change-responses-a-role-for-international-labour-standards

OpenSecrets. (2025). Exxon Mobil: Lobbying profile. https://www.opensecrets.org/federal-lobbying/clients/summary?cycle=2024&id=D000000129

Oppenheimer, M., Glavovic, B. C., Hinkel, J., van de Wal, R., Magnan, A. K., Abd-Elgawad, A., Sebesvari, Z. (2019). Sea level rise and implications for low-lying islands, coasts and communities. In H.-O. Pörtner, D. C. Roberts, V. Masson-Delmotte, P. Zhai, M. Tignor, E. Poloczanska, ... A. Okem (Eds.), IPCC Special Report on the Ocean and Cryosphere in a Changing Climate. https://www.ipcc.

Pearce, W., Holmberg, K., Hellsten, I., & Nerlich, B. (2019). Climate change on Twitter: Topics, communities and conversations about the 2013 IPCC Working Group 1 report. *PLoS One*, *9*(4), e94785. DOI: 10.1371/journal.pone.0094785 PMID: 24718388

Qazi, A. L. (2022). *Karachi Bachao Tehreek: Housing rights, climate justice, and urban activism in Pakistan*. Peabody College, Vanderbilt University.

Saunders, C., Doherty, B., Hayes, G., & McCarthy, J. (2020). The impact of Extinction Rebellion in the UK: Disruptive protest and climate emergency declarations. *Environmental Politics*, *29*(5), 856–875. DOI: 10.1080/09644016.2020.1786331

Scottish Government. (2021). Scotland's Climate Assembly: Research findings. https://www.gov.scot/publications/scotlands-climate-assembly-research-findings

Somerville, R. C., & Hassol, S. J. (2011). Communicating the science of climate change. *Physics Today*, *64*(10), 48–53. DOI: 10.1063/PT.3.1296

Union of Concerned Scientists. (2015). The climate deception dossiers: Internal fossil fuel industry memos reveal decades of corporate disinformation. https://www.ucsusa.org/resources/climate-deception-dossiers

U.S. House Committee on Oversight and Reform. (2022). Fueling the climate crisis: Exposing big oil's disinformation campaign to prevent climate action. https://oversight.house.gov/legislation/hearings/fueling-the-climate-crisis-exposing-big-oil-s-disinformation-campaign-to

Wahlström, M., Kocyba, P., De Vydt, M., & de Moor, J. (2022). *Youth activism and the Fridays for Future movement: Mobilizing for climate justice*. Routledge.

Werrell, C. E., & Femia, F. (Eds.). (2015). Climate change as threat multiplier: Understanding the broader nature of the risk. The Center for Climate and Security. https://climateandsecurity.org/climate-change-as-threat-multiplier/

Williams, H. T. P., McMurray, J. R., Kurz, T., & Lambert, F. H. (2015). Network analysis reveals open forums and echo chambers in social media discussions of climate change. *Global Environmental Change*, *32*, 126–138. DOI: 10.1016/j.gloenvcha.2015.03.006

World Bank. (2020). Reference guide to climate change framework legislation. https://openknowledge.worldbank.org/handle/10986/34593

Chapter 5
China's Ecological Civilization:
An Effective Instrument for Solving Energy Challenges and Climate Change Complications

Elizabeth Yoneva
https://orcid.org/0000-0003-3911-063X
University of National and World Economy - Sofia, Bulgaria

ABSTRACT

China faces its own complications in adapting to the new realities in the energy sector. The distinct strategic vision of the country was boosted by an on-target assortment of practical tools for energy readjustment. The motivation to transform the national energy mix triggered the efforts to join the global trend for advancement in green energy. In the light of these issues, the paper explores the principal dimensions of the integration of the green pillar in the energy strategy and policy of PRC, in order to evaluate their repercussions in the domestic and the international realm. The research showcases the essence of the promoted concept for "ecological civilization", defined by many experts as a form of sustainable development with Chinese characteristics, and its role in the rise of the country as a renewable energy superpower. The study is oriented towards outlining the main dimensions of the metamorphoses in the Asian country's energy policy and presenting its potential for the use of renewable energy to generate green growth.

DOI: 10.4018/979-8-3373-1310-8.ch005

INTRODUCTION

The global energy landscape has been marked in recent decades by major changes, induced by sector-specific processes with their distinctive endogenous and exogenous dynamics. This industry, which is vital to the functioning of all countries and societies, had to go through a number of struggles in relation to the challenges generated by the reduction of the fossil fuel reserves, provoking an escalation of competition between buyers of energy resources on the global markets. The transformations were additionally intensified by the intervention of the newly emerging "resource hungry" countries, especially in Asia, where their economic performance remodeled the design of several supply chains. As a result of the inconstant conditions and the altering order, achieving and safeguarding energy security has been elevated to the status of a fundamental foreign policy goal. In this perplexing situation, discussions on the risks of militarization of energy policy have come to the top of the political agenda. The ghost of mushrooming conflicts, and in this context of energy wars, was far from being the only one hovering in the public domain in terms of predicted future disasters.

Furthermore, the threatening levels of the negative impact of the ecological footprint of human activities, including the extraction and consumption of energy resources, emerged as no less alarming with regard to the loss of natural capital and the shrinking opportunities for its restoration. These concerns brought about increasing calls for transformations in the public perceptions with regard of the practices, according to which the use of energy resources is managed, both for the needs of the economy and in our daily activities. In this fashion, the signals for the accelerating destruction of the biosphere have stimulated the urge for a shift towards new energy paradigm on international level. The search for alternatives to fossil fuels placed special emphasis on renewable energy and green technologies. On the one hand, focusing public attention on them was logical in the context of the issues of environmental protection, limiting climate change and achieving sustainable development. On the other hand, however, the necessity and existence of an international consensus on the imperative for their launch were questioned. Gradually, they succeeded in overcoming the label of fashionable trend and proved their qualities as widely embraced tools for solving existing problems.

The unfolding debate on the search for a panacea in the energy sector had its reasonable supplement. It was also necessary to find mechanisms to overcome the damages caused by the financial and economic crisis of 2007-2008. The European Union used the situation, in order to promote the green technologies and to generate profits on this base. In due time, the potential of this instrument has been recognized by other actors on the international scene. The picture of investment activity for renewables has progressively started to change, adding new agents in the competition equation. China's success in this area cannot be described as anything other than impressive. However, the world had to withstand soon another disaster. The global economic recovery never fully unfolded because of the outbreak of COVID-19 in 2020. Nonetheless, the pandemic crisis was also seen as a source of chances for an energy model rearrangement. It served as a new test confirming the resilience of the renewable energy vector. The calls for turning towards "sustainable recovery" (IEA, 2020) were largely resounding. However, transforming the foundation and the structure of the current energy configuration represents also a challenge to political will, and setting new trajectories for its development depends on the specific drivers for change in the different countries and regions.

Given its status as the top world polluter, China also faces its own complications in adapting to the new realities in the energy sector. The distinct strategic vision of the country was boosted by an on-target assortment of practical tools for energy readjustment. The motivation to transform the national energy mix triggered the efforts to join the global trend for advancement in green energy. PRC's performance altered not only the internal setup, but it had as well its spillover effects externally on the international scenery. Remarkably, the Asian power has not only entered the competition, but has also taken the lead in it. Nevertheless, these developments are activating simultaneously social, economic and geopolitical implications, which are in need of even more detailed analyses. China's energy reforms and achievements have set, and will continue to set, plentiful parameters in the scientific discussions in the field of exploring the accomplishments in renewable energy, and this is precisely the reason for the orientation of the current research work towards this topic. The paper explores the principal dimensions of the integration of the green pillar in the energy strategy and policy of the Asian giant, in order to evaluate their repercussions in the domestic and the international realm. The research showcases the essence of the promoted concept for "ecological civilization", defined by many experts as a form of sustainable development with Chinese characteristics, and its role in the rise of the country as a renewable energy superpower.

In the light of these issues, the study is oriented towards outlining the main dimensions of the metamorphoses in the Asian country's energy policy and presenting its potential for the use of renewable energy to generate green growth after the end of the pandemic crisis. In view of the current energy dilemmas, the aim of the study is to discuss the transformations not only through the theoretical lens, but also per using practical optics with the help of distinguishing cases. On methodological ground, it has primarily to deal with specific theoretical challenges. In the scientific literature on energy issues, there are different definitions about unsimilar in content terms such as renewable energy, green energy, clean energy, alternative energy, sustainable energy, etc. Green energy is a subcategory of renewable energy that does not generate harmful emissions. For example, biomass belongs to renewable energy, but not to green energy. Accordingly, all types of green and renewable energy are included in the clean energy category, but nuclear energy is similarly incorporated in it. Sustainable energy does not cover sources that are doomed to depletion and whose recovery therefore requires a certain period of time or human intervention. Solar and wind energy, for instance, are sustainable, but biomass is not. Alternative energy, by definition, has no destructive impact on the environment - for this reason, hydropower and biomass are not an element of it. The various international institutions, organizations and companies, operating in the energy sector, apply different methodologies when presenting statistics on renewable energy. The current study is based on the use of the prestigious annual report "Statistical Review of World Energy" of the Energy Institute (specifically the 2024 edition with data for 2023), where hydropower data is moved out of the renewable energy category.

PLANS AND ACTIONS TO OVERCOME CHINA'S ENERGY INSECURITY

Described as an emblematic example of one of the emerging economies within Asia, China is often seen through the lens of its rivalry with Western countries in energy resource markets. However, in view of the serious challenges it faces with regard to its energy security, PRC is inevitably focused on expanding the geographical scope of its operational field for sourcing energy commodities, thus clashing in certain areas with EU and US interests. In 2010, China has become the world's second largest economy and the world's biggest energy consumer. The insufficient domestic resources to power its growing industrial and commercial strength have inescapably commanded a trend towards increasing import dependence to secure the necessary energy supplies. These developments were also reflected in the foreign policy of the Asian colossus, which had as its new goal the unfolding of explicit cooperation with resource-abundant countries in Africa, Central Asia and Latin America. China's neoteric approach to securing the necessary volumes of raw materials was based on negotiating long-term exploitation rights agreements, acquiring stakes in energy companies and projects, and investing in various types of energy infrastructure in different parts of the world.

Until these external partnerships are developed and strengthened, however, China will continue to rely on coal as the foundation of its energy sector. The existence of substantial national reserves of this type of hydrocarbons has put it in the position of being the largest consumer of this resource globally, but also helped in overtaking the US as the largest polluter. The widespread use of coal could not but contribute to the harmful large-scale greenhouse gas emissions with their devastating effect on the environment. The plans to gradually limit the coal consumption were tailored towards replacing it with natural gas. Nevertheless, this is not an optimal option either, as limited domestic reserves also imply the need to import massive volumes. Renewables have been given more of a supporting role in the efforts to achieve energy diversification. However, the transformations in the global energy sector have further enhanced the attractiveness of green technologies, thus playing their part in urging China to make the correcting arrangements. The different crises since the start of the current decade, expressly with their energy repercussions, served as valuable catalysts for the country's solidification as a leader in this field.

With the start of the pandemics in 2020, the global primary energy consumption has declined by 4.5%, demonstrating a record collapse in the post-World War II period. In contrast to countries such as the US, India and Russia, which were the major contributors to the drop, China has found itself at the opposite pole, ranking among the few countries that have not only registered an increase, but also achieved the most significant growth in energy consumption at 2.1%. The Asian giant's economy has demonstrated its excellent ability to recover quickly from the pandemic shock with the help of stringent measures and impressive discipline. At the same time, the issue of renewable energy progress has proven to be particularly relevant in the context of the pandemic crisis. The unfolding recession has led to an energy slump mainly in oil and coal, but green energy has shown remarkable resilience, growing against the background of the general decline in energy consumption. China's role in these achievements is evident - for 2020, the country served as the largest contributor to renewable energy growth, followed by the US, Japan, the UK, India and Germany. Renewable energy consumption grew by 15%, accounting for 25% of global renewable energy demand and 36% of the global growth (BP, 2021a). Solar energy grew by 15.8%, wind by 14% and other renewable types by 16.2%. In wind power alone, PRC added 72 gigawatts of new capacity - more than the combined total of the previous three years (BP, 2021b). The concerns that the recession will cause a more serious collapse and threaten investment activity in renewables proved to be excessive.

In 2022, the repercussions from the war in Ukraine drew the spotlight towards the disruptions in the energy markets. For the EU member states, the imperative for reorganizing their energy relations was a pressing one, contributing to the redrawing of the supply routes not only for the continent, but also on a global scale. In these turbulent times, China managed to take advantage of the situation. The formula of "normal trade cooperation" with Russia was applied beneficially, contributing to the purchase of the vacant energy resources and their resell. Resultantly, the Asian leader was described as the winner in the crisis.

In 2023, China continued its prosperous energy quest. Its primary energy consumption grew by 6.5% (above the world average of 2.0%). Its share in global primary energy consumption is 27.6% (in front of the US with 15.2% and EU with

9.1%) (Energy Institute, 2024a). The country is also cementing its position of the top emitter of carbon dioxide with an increase of 6.1%, thereby having a 31.9% share of global emissions (followed by the US with 13.2%, India with 8.0%, and the EU with 7.2%). China's primary energy consumption in 2023 was structured in the following manner: coal accounted for 53.85%, prior to oil – 19.17%, renewables – 9.45%, natural gas – 8.53%, hydro – 6.72%, and nuclear – 2.28%. In the electricity generation also dominates coal – 60.84%, followed by renewables – 17.64%, hydro – 12.96%, nuclear – 4.6%, natural gas – 3.1%, and oil – 0.1%. With regard to electricity, China is the leader in installed capacity for grid-scale battery energy storage systems (BESS) – 27.1 gigawatts in 2023 with the astonishing growth by 249.1%. Nearly 50% of the global BESS capacity additions were installed in the country. The top position with a share of 48.6% of the world's total BESS capacity puts PRC far ahead of the US (28.3%) and the EU (6.0%).

The oil consumption of the country was marked by a growth of 10.7% in 2023, resulting in the use of 16.658 million barrels per day. When tracking the ratio, it can be found that the share of 4.4% of the global production of the "black gold" is far below the 16.1% share of global consumption (which is equivalent to second place, following the US 19.6% and in front of the EU 10.6%). Furthermore, China's oil imports grew by 13.0%, thus reaching 13.717 million barrels per day or 20.1% of global oil imports with the leading position in front of the EU (18.8%) and the US (12.5%). It is noteworthy that for the first time ever the Asian leader became the largest oil refining market by capacity (with 18,484 kb/d), exceeding the US one (18,429 kb/d). Obviously, the oil import dependency of the country remains at high levels. Supplies from areas such as the Middle East and Africa involve the use of sea routes, which pose risks in terms of their possible blockage. This pushes Beijing to look at tracing overland transport corridors for oil and gas imports, with the hydrocarbon-rich Central Asian states offering attractive opportunities in this direction.

With regard to natural gas, the modest reserves of the "blue fuel" define the share of only 5.8% in world production. In 2023, the national consumption with its rise of 7.2% reached 404.8 billion cubic meters (bcm) or 10.1% of the world total. This is the third highest share – just behind the US (22.1%) and Russia (11.3%), but already in front of the EU 8.0% due to the decreasing consumption in Europe. On the import side, the growth of 9.4% (12.4% for LNG and 5.0% for pipeline imports) resulted in a share of 17.0% of the global total. The general growth in LNG demand is triggered primarily by the Asia Pacific region with China, India, and other non-OECD countries in the region being the major contributors to this trend. In 2020, China surpassed Japan as the world's largest LNG importer, but it lost this place in the next years. However, according to the data for 2023, the Asian giant regained the head position with a share of 17.8% of the global LNG imports, followed by

Japan (16.4%) and South Korea (11.0%). The country imported 97.8 bcm of LNG with major trading partners being Australia (33.0 bcm), Qatar (22.9), Russia (11.0), Malaysia (9.7), Indonesia (5.5), and the US (4.3). The pipeline imports of 61.3 bcm came from Turkmenistan (30.5 bcm), Russia (21.3), Kazakhstan (4.6), Myanmar (3.7), and Uzbekistan (1.2). Although Russian Federation pipeline deliveries to Europe experienced an overall decline, its exports to China increased by 45%.

PRC's addiction to coal is largely debated and often criticized in the framework of the international community. The large reserves of this type of fossil fuel promote logically its use as a domestically available resource. Nevertheless, the negative impact of its consumption on the environment sparked concerns about the pollution and its repercussions on people's health. The tools to induce a shift in the energy model proved successful, but the process was delayed by stumbling into a number of unexpected obstacles. In 2021, China had to deal with a destructive energy crisis, similar to that in other regions. The power shortages and the blackouts disturbed several provinces and sectors of the national economy. The measures for rationing electricity forced the industrial users to limit their operating days. As the recovery after the start of the pandemics was based mainly on construction and heavy industry, the need for coal expanded (Myllyvirta, 2021a). The increasing prices of the resources urged the government to intervene, resulting in increasing imports, restarting of old coal mines and expanding the production of operating ones. Obviously, the struggles with minimizing the negative effects of the energy crisis undermined the plans for decarbonization (Inman, 2021). The registered surge in the permits for the construction of new coal-fired power plants is in balance with the lack of statement, when China will quit with building new coal facilities. The 106GW of new coal projects permitted is more than four times 2021's 23GW score (Hannam, 2023).

In 2023, global coal production reached its highest ever level, with China alone being responsible for just over half of total global production (51.8%). Its growth by 3.3% in national production was supplemented by 4.7% rise in consumption. The Asian leader is by far the largest consumer of coal, beating its own record set in 2022 and accounting now for 56.0% of the world's total. This contrasts markedly with the performance in Western countries where the efforts to reduce coal use produced decline by 17.3% in the US and 20.4% in the EU. Nevertheless, with the decrease of coal's share in the national energy mix (to the above-mentioned 53% compared to 72% in 2009), the country proceeds successfully with its commitment to promote the green vector of development.

In the department of nuclear energy, PRC continues to lead the way in creating new capacity. Since 2000, it has built around 60% of all new nuclear capacity additions. Notably, China commissioned the world's first onshore demonstration of a small modular reactor (SMR). In late 2023, the commercial operations at the facility started with featuring the latest Generation IV technology and incorporating inherent safety systems. The nuclear energy production in the country was marked by a growth of 4.1%, resulting in a share of 15.9% of the global output – a second place in the rankings to the leader US (29.8%) and in front of France (12.4%). The switch of positions between China and France became reality in 2020, when the Asian giant became the operator of the second largest nuclear energy fleet. Even in the context of the situation after the Fukushima accident, China was one of the few countries to confirm its plans to build new nuclear power plants. Emerging as one of the supporters of nuclear energy, the Asian country has set the goal of achieving a closed cycle in terms of nuclear fuel. Alongside this, it has achieved almost complete self-sufficiency in reactor design and construction, adapting and improving on Western technology. Beijing's policy in this sector goes even further, focusing on the export of nuclear technology with global dimensions.

The achievements in the field of hydropower are also extremely significant (Bin, 2021). Since the start of the century, China has completed emblematic projects such as the construction of the Three Gorges hydropower plant on the Yangtze River, the world's largest in terms of installed capacity (22 500 megawatts as of 2012). With its commissioning, the plant was hailed as the "eighth wonder of the world" and the "new Great Wall of China" (Bettwy, 2007). In June 2021, the first two generators of the second largest hydropower plant globally, the Baihetan (16 000 megawatts), started their operations (Stanway, 2021). Currently, China is the leader in hydroelectricity generation. Despite the decline in this direction by 5.6% in 2023, which was in sync with the global reduction by 1.9%, the country has a share of 28.9% of the world's total, putting it in front of the following powers Brazil (10.1%) and Canada 8.6%.

There is a clear national orientation towards the construction of pumped storage hydropower plants. In August 2024, the construction of the Fengning plant was finished with the start of its twelfth reversible turbine unit. The facility is located in Hebei province and with its total installed capacity of 3.6 GW ranks as the largest of its kind in the world, surpassing the Bath County (Virginia) project in the U.S. In 2023, China added more than 6.2 GW of pumped storage hydropower capacity. Furthermore, the Asian country is aiming at expanding its pumped storage capacity to 80 GW by 2027 and to 120 GW by 2030 (IHA, 2024).

The Asian country works also on remarkable energy projects for constructing hybrid energy plants. In 2023, the Lianghekou Hydropower Station, located on the Yalong River, was connected to the nearby Kela Solar Power Station in Sichuan Province, in order to combine the output from the two facilities. The aim is to reduce the negative impact of daily and seasonal fluctuations, and so to achieve a stable supply of power. The world's largest hybrid energy plant will operate with the latest information technologies, such as highly computerized operation platform with self-learning AI (Wei, 2022). The expectations for the results of the project are oriented towards a reduction of carbon dioxide emissions by over 1.6 million tons, an equivalent of burning 600,000 tons of coal.

In 2022, another important project was finished and the country obtained its first hybrid power station for both solar and tidal energy. The facility was built in Wenling City of east China's Zhejiang Province. The use of two green energy sources in a complementary mode will provide more stability, as solar energy supply is available during the day and tidal waves could replace it during the night. With an installed capacity of 100 megawatts, the power plant will cover the annual electricity demand of about 30,000 urban households. It will save around 28,716 tons of coal, thus reducing the carbon dioxide emissions by 76,638 tons annually (Xun, 2022). Specific vanguard technologies are used for ensuring the efficient operation of the facility such as a digital platform with an early warning model for rapid remote diagnosis of the condition of the equipment and the use of drones for routine patrol and maintenance of the station.

China solidifies its status as the biggest power generator from renewables and remains the main driver of solar and wind capacity growth. In 2023, its renewable energy generation increased by 21.5%, leading to a share of 35.1% of the world's total, in contrast to the shrinking shares for the EU (18.7%) and the US (15.5%) (Energy Institute, 2024b). The share of the Asia-Pacific reached 49.6%, demonstrating that the center of gravity for renewable energy has already shifted to East. Globally, wind energy achieved a record year for new build. Nearly 66% of the capacity additions came from China and its total installed capacity is now equal to North America and Europe combined. The installed wind turbine capacity of PRC marked an increase by 20.7%, reaching 441 895 megawatts, which accounts for 43.4% of the global total against the background of the reducing shares for Europe (26.4%) and the US (14.5%). The installed photovoltaic (PV) power and concentrated solar power (CSP) capacity grew by 55.2% to 609 921 megawatts, obtaining a share of 43.0% (compared to Europe's 21.1%, and the US's one of 9.8%). PRC was responsible for around a quarter of the rise of the solar capacity additions.

In December 2024, another key project in this sector was finished, when the Aksay Huidong CSP-PV plant became operational. It is located in the desert zone of Jiuquan in China's Gansu province. With a capacity of 750 MW (including a 110 MW CSP (concentrated solar power) tower and a 640 MW solar plant), it is the largest hybrid solar facility in the country. By integrating the two types of energy, the plant could ensure stable energy supply, as the electricity generation of photovoltaic panels during the day could be replaced by nighttime release for up to eight hours from the CSP system thanks to its ability to store solar energy in molten salt (Shaw, 2024). The 200-meter-high absorption tower is surrounded by 11,960 pentagonal heliostats. The plant is part of the solar energy initiative for the Gobi desert and other arid regions.

Definitely, the segment of solar thermal power is developing progressively in the Asian country. In 2024, PRC brought into operation the world's first dual tower CSP plant. It is located near Guazhou County in northwest China's Gansu Province. Its innovative 30,000 mirrors form two overlapping concentric circles to maximize reflection. They are equipped with ultra-white glass that reflects 94% of solar energy onto the receiver. They are also adaptable due to their automatic sun-tracking capabilities throughout the day. The cutting-edge design and setup of the facility should enhance power generation efficiency and storage. The twin endothermic towers are 200 meters tall and their configuration reduces the need for greater installed capacity, needed by the conventional options. Beneath them lie the large storage tanks containing molten salt. As it has a high boiling point than water - 600°C instead of 100°C, the liquid can store significantly more heat. Therefore, the solar energy collected during the daytime is converted into thermal energy and stored in the molten salt. Acting like a thermal battery, the towers can release it during the night to ensure uninterrupted power supply. The dual tower CSP plant is expected to have an annual output of 1.8 billion kilowatt hours. Its operation will reduce the carbon dioxide emissions by 1.53 million tons annually (Gong, 2024). The facility takes the title of the world's largest solar thermal power plant from Morocco's Noor Complex, operating in the Sahara Desert and capable of generating 510 megawatts of electricity. The Chinese plant shares several similarities with the Ivanpah Solar Electric Generating System, built in 2014 in the Mojave Desert in the United States. With a capacity of 392 megawatts, it has three towers, but each of them is surrounded with its own field of heliostats. Unlike the American model, the dual CSP facility in Gansu uses a common turbine for both towers and mirrors in overlapping concentric circles for optimizing efficiency of power generation (Suness, 2024).

Notably, the Asian giant is also the largest global producer and exporter of clean energy technologies. However, it is not a remarkable player in the sector of biofuels, as it has a share of only 3.8% in the global production. PRC has the world's largest car fleet and it is the second largest gasoline market (just behind the US) and third largest diesel market (behind the US and EU), both of them marked by expanding consumption. The use of biodiesel in the country is focused predominantly on electrical power generation, fishing vessels, and farm equipment, and only about one-third of the demand comes from the on-road transport. The modest interest is rooted in the low quality of the biodiesel produced. China's fuel ethanol production is over 80% grain-based (i.e., corn, wheat, and rice) and only 10% cassava or sugarcane-based with regard to the limited supplies of sugarcane and record high molasses prices in the last years (IEA Bioenergy, 2023). The biofuel industry is undergoing a transformation towards advanced types such as cellulosic bioethanol as well as coal and industrial flue gas-based synthetic ethanol. However, most of the plants export their production to the EU with regard to the lucrative tax policies.

Outstandingly, the country is a prominent factor with regard to critical minerals, which have a crucial role in the process of energy transition. Their consumption (especially the one of lithium, cobalt, nickel, and copper) is expected to bloom with regard to the needs of the renewable energy sector. The global demand of critical minerals is projected to quadruple by 2040. China is responsible for around 20% of the world's lithium production and for nearly 74% of the graphite supply. It is both the world's leading producer of refined cobalt and the largest consumer of it, as nearly 87% of the national consumption are used in the lithium-ion battery industry. The Chinese dominance across critical minerals supply chains generates concerns about the risks of such concentration and respective dependence. PRC is also the leader in the domain of refining - it refines 68% of nickel globally, 40% of copper, 59% of lithium, and 73% of cobalt (Castillo & Purdy, 2022). The country also holds 78% of the world's cell manufacturing capacity for EV batteries. It is a prominent player in the downstream activities, as in the upstream for critical minerals dominate countries such as Australia and Chile (for lithium extraction), Democratic Republic of the Congo (in cobalt extraction), Indonesia (in nickel extraction), etc. Moreover, China puts special efforts in securing new deals for critical minerals globally through investments in mining projects and companies and extended supply agreements.

Recently, the supremacy in the Asian country in the sector intensified the debate about energy security challenges. In December 2024, in response to Donald Trump's decision to introduce 10% tariffs on Chinese goods, PRC reacted with a ban on exports to the United States of critical raw materials. Especially valuable are the rare earths - as experts underline, "gallium is used in semiconductors, germanium in both semiconductors and infrared technologies, antimony in the production of everything from bullets to missiles, while graphite is critical for electric batteries. China dominates the world market in all of these materials, producing roughly 60% of the world's germanium, 80% of global gallium production and 78% of antimony (Rockwell, 2025). These tensions add further volatility to global trade. Obviously, the geopolitics of critical minerals will play a crucial role in the future shaping of the minerals supply chains. From current perspective, the global energy transition would be not possible without the input of the Asian giant.

CHINA AS THE GLOBAL LEADER IN THE RENEWABLE ENERGY SECTOR

The developments in PRC's energy mix show that expanding of the green vector is especially impressive. With regard to the use of renewables, lessons have been drawn from the experience, collected in the EU and the US. Europe has made considerable efforts over the last decade to establish itself as a world leader in environmental measures, although it has also come under criticism in this respect for launching alternative technologies in order to secure a source of economic recovery in the 2008 crisis. Despite the initial perception of renewables as a costly luxury, a number of countries around the world have moved towards their implementation. Furthermore, experts have disproved the perception that they are only suitable for highly developed countries (Johnston, 2016).

At the same time, theoretical inventions have also contributed to viewing renewable energy in a positive perspective. Emerging in the 1980s, ecological modernization theory reconciled capitalism, industrialization, economic growth and technological development with environmental sustainability, declaring them compatible. It opened up the prospect of using green energy to solve ecological problems, since through the adoption of new technologies and practices companies can generate profits. The strategic industry theory is another appropriate tool for promoting the use of renewable energy. Focusing on the public good, national prosperity and well-being, it relies on a regime of public policies, actions and investments in key sectors, where the market and private business appear unable to produce the desired results on their own (Dent, 2015). This approach is particularly appropriate for renewable energy, as this industry relies on state support to provide incentives for the start-up and early stages of development, innovation and commercialization of this type of technology. Considering the specificities of the Chinese economic model, it is easy to explain the favoritism of this theoretical scheme in the country as well as in other Asian states. It is also particularly attractive in other strategic sectors, where the government intervenes directly, relying on state-owned enterprises (either by creating new ones or by using existing ones) and establishing mechanisms for cooperation with the business.

These theoretical insights helped China to pave the way for the launch of its green energy segment as a strategic industry and as an essential element of ecological modernization. Undeterred by claims that renewables are only applicable at a local level and are not a suitable tool to power a large-scale economy, PRC accelerated the use of this type of technology and cultivated a vision and practical platform for its progress. In the national macro development plans were included clear targets for achievement in this direction, supplemented by actions to establish and improve the legal framework for supporting the expansion of green energy. In 2006, the first Renewable Energy Act began its operation, introducing the main regulations on pricing, taxation and investment activities in the industry (Schuman, 2012). A subsidy system was installed with regard to renewable energy pricing and a special fund was set up to support its development. In 2007, targets were also set for the share of renewable energy in total energy consumption of 10% by 2010 and 15% by 2020.

Special emphasis was placed on the development of wind energy. Since 2003, the government has introduced new rules regarding concessions (ESCAP, 2012), with specific requirements for foreign investors that all wind turbines should be assembled in the country, and that 70% of them should be manufactured locally. In 2008, China also had its first wind generator with fully independent intellectual property (Xinyu, 2011). Despite EU accusations towards Beijing that the authorities are pursuing a policy of energy protectionism (Johnson, 2009), the measures are clearly having an effect. In 2009, the Asian giant overtook the US and obtained the lead in renewable energy investment. Logically, the 70% domestic participation requirement was abolished. The following year, China became the leader in wind power generation. The efforts to build its own renewable energy industry proved successful and contributed to the creation of 4 million jobs. The other components of renewable energy had their expansion as well. China is reaching dominant positions in solar and tidal energy technologies. It is also a leader in the use of geothermal energy, although this area is developing more slowly due to the availability of limited high-temperature resources.

Currently, the Asian country has the status of the world's largest producer of wind and solar energy, as well as of renewable energy in general. It is also the largest producer and exporter of green energy technologies. Undoubtedly, this champion position of China in the renewable energy sector is achieved and maintained thanks to a continuous optimization process of energy policies. Its catalysts, however, appear to be specific arguments, both internal and external. National ambitions to form an "ecological civilization" correspond to the need to reduce pollution levels and lower the risks of socio-economic instability. At the same time, there are also impulses from the international arena to take measures to limit emissions and minimize the use of fossil fuels. For the Asian giant, diversification is not proving to be appropriate solely on environmental grounds. There is also an explicit geopolitical logic behind it. Renewable energies are a suitable instrument for reducing energy dependence on unstable regions such as the Middle East or North Africa, as well as on problematic chokepoints in resource transport routes. Security of supply also has an important role to play in China's efforts to build naval capacity to help overcome the threats posed by the so-called Malacca Strait dilemma (Lanteigne, 2008). It is of strategic importance to China, as nearly 80% of the country's oil supplies pass through it. Similar complications are present in the other corridor for transporting energy resources through the Indian Ocean. Since the sea routes are vulnerable to blockade by external factors (Ji, 2007), reducing the Asian country's energy dependence will generate geopolitical benefits as it will reduce the need for militarization for energy security purposes.

A distinctive feature of the Chinese model of renewable energy development is the creation of incentives to build own industry for this type of technology. Starting with the import of components and equipment, the government is moving towards implementing a long-term plan to build domestic capacity in the sector. National markets are strengthened in the initial phases of deployment through protectionist measures, in order to block the strong external competition from solid and experienced companies. The specific local participation requirements were only removed once the innovations implemented had sufficiently hardened the position of the Chinese renewable energy industry. The Western criticism of Beijing's energy nationalism is understandable in view of the Asian country's success in shortening its distance from industrialized countries. The application of the 'negative list' helps to prevent foreign investment in a number of strategic sectors and ignites various trade disputes (Chiu, 2017). However, once the measures have played their role, the country opened up the renewable energy sector.

An interesting element of the Chinese model for the promotion of green energy can also be found in the architecture of energy policies, which is based on a three-level system (NREL, 2004) where the central government is responsible for shaping the first two layers, providing the general guidelines for the advancement of the sector and then developing plans, setting targets and proposing regulations and standardization. Local authorities at provincial, prefecture and county level set the practical and managerial modus operandi for renewable energy, building their own policies to support its progress. They can implement appropriate measures, such as subsidies or tax incentives.

In elucidating the specifics of China's approach to renewable energy, the peculiar relationship between government and business should be emphasized, as it shapes the complex and multi-layered configuration of the green industry. Woven into its foundation is a web of cooperation and competition between different types of state-owned enterprises and a large number of private companies. The former logically play a more significant role in terms of the expansion of green technologies - their nature should also be taken into account, whether they are powerful corporations administered by the central government or smaller companies owned by local authorities. Despite their close ties to the ruling elite, both types enjoy wide operational autonomy and compete with each other. This ability distinguishes them from being classified as inert or paralyzed political instruments. The unique business approach to renewable energy development in China, combining public and private entrepreneurship, is also based on intense intra-sectoral partnerships. The hard-to-demarcate lines between public and private activities in this industry clearly construct a model without parallel, and it remains a matter of debate whether it can become a pattern to be copied in other countries.

In view of the international calls to seize the moment so that new energy technologies can be mobilized as a source of “green growth” after the last energy crises, China's potential to lead the way in energy transition is extremely promising. With regard to its targets for the current decade, the Asian giant has set the ambition to increase the share of non-fossil energy in overall primary energy consumption to

20% by 2030. Renewable energy (including hydro) is planned to reach at least 35% in electricity generation by the end of the same period (REW, 2018). China's targets of peaking carbon dioxide emissions before 2030 and reaching carbon neutrality before 2060 (China's MidCentury Long-Term Low Greenhouse Gas Emission Development Strategy, 2021) are also remarkable, especially in view of the 30-year interval between them, against a target of 43 years for the US and 71 years for the EU.

Currently China is described as the world's renewable powerhouse. In 2025, PRC is projected to meet its goal for 2030 to reach 1,200 gigawatts of renewables capacity. By keeping the present pace, the country could have around as 1,000 gigawatts of solar power alone by the end of 2026. The impressive achievements of its renewable energy sector are considered crucial for reducing prices worldwide and for driving down the cost barrier for renewable systems in the developing countries. The perspectives for further advancement of renewable energy in the Asian leader are promising, as the plans for future expansion include guidelines to intensify the solar exploitation of the vast desert regions in the western part of the country. The high solar radiation there is not the only advantage, because the cheap and largely available land is also an upper hand. As experts underline, "China has 2.6m square kilometers of desertified land, accounting for a quarter of its landmass or roughly four times the size of France" (Myllyvirta & Zhang, 2022). For that reason, the use of sandy and rocky deserts is especially suitable for building of clean energy bases. As an alternative, such facilities could be constructed in other wastelands – for example such left behind by the coal mining industry. Accordingly, PRC aims at implementing more than 200 projects for building of energy bases in the desert zones, so to enlarge its renewables capacity to about 3.9 terawatts by 2030. The largest ones will be situated in Inner Mongolia and Gansu. The solar bases will combine their outputs with these of the massive wind farms operating in northern China. As a connection between them, high-speed transmission lines should be constructed, leading to the markets in the eastern arrays.

The establishment of internal energy links appears to be of cardinal importance in view of the shaped belts comprising different provinces in the country, distinguished by their specificities in the energy dimension. Economically important production centers such as Guangdong, Zhejiang, Jiangsu and Shandong are located along the eastern and southern coasts. As major drivers of GDP, they are characterized by large-scale electricity consumption and high carbon dioxide emissions. On the other hand, China's coal deposits are mainly concentrated in the north of the country in areas such as Shanxi, Inner Mongolia and Shaanxi. The three provinces form the so-called "coal triangle" as they produce around 70% of PRC's coal. Additionally, north-eastern provinces such as Liaoning, Jilin and Heilongjiang, which once acted as important industrial bases, are now experiencing economic collapse. They constitute the so-called "rust belt", but are emerging as the most important agricultural areas in the country. The north-western provinces, for their part, are characterized by the development of heavy, chemical and refining industries, which inevitably condemns them to problems such as high levels of pollution and large-scale energy demand (described with the term "dual high"). A solution to these has been found there partly on the basis of the deployment of most of the China's wind and solar power, in order to exploit the vast deserts. The home of PRC's hydropower industry is based in the south-western provinces, situated east of the Tibetan plateau, where most of the water resources are concentrated. However, traditional hydropower centers such as Sichuan province have also faced energy crises in recent years in view of the hot and dry periods.

In order to overcome these differences in the resource capacity of the provinces, since the beginning of the twenty-first century, the Asian leader has launched major power transfer programs along the West-East axis so that electricity can be brought from the coal- and water-rich western regions to the eastern areas. Resources are also beginning to be shared between provinces in other zones, such as natural gas transfer along the West-East axis, coal transfer along the North-South axis and water transfer along the South-North axis (You 2022).

Besides the desert projects, the Asian leader is focused on improving rural grid transmission. There is a growing need for allowing village collectives to invest in distributed renewable power. In June 2021, the National Energy Administration (NEA) launched The Whole County PV program. It is based on the following requirements for percentage of rooftops to be covered by PV: 1) government office buildings – 50%, 2) public buildings such as schools and hospitals – 40%; 3) industrial and commercial plants- 30%; 4) rural residential buildings – 20%. This pilot scheme should promote the adoption of distributed rooftop solar by installing photovoltaics for ¼ of the population. The emergence of a rural model is welcomed as a radical move, as it will provide customers with the opportunity to draw energy down and sell it back to the grid. As the rural population is considered neglected in the process because of the favoring of large, centralized plants, the new course will finally include it in the green revolution. According to the experts, by the end of 2022, "676 counties had signed up for the scheme, and more than 51 gigawatts of new distributed solar photovoltaic was installed, nearly half of it on rural rooftops. In total, by the end of 2022, China had built roughly 157 gigawatts of distributed photovoltaic capacity, more than double that of the U.S. (Hilton, 2024). The end of the traditional pattern for scale-up of PV was enforced by the phasing out of subsidized feed-in tariffs for central PV plants in 2019, stimulating the rise of distributed solar – "residential rooftop solar installations reached 21 GW in 2021, double the previous year, and roughly 40% of annual installations. The majority of PV installations in both 2021 and 2022 were classified as distributed PV, a substantial change from pre-2020, when virtually all PV was at large centralized plants" (Hove, 2023).

Nevertheless, PRC is facing several challenges in integrating renewable power into its giant grid due to its variable nature. More reliable sources as coal are still preferred, in order to ensure easier matching between supply and demand. The approach of developing regional "renewable energy zones" for large-scale electricity generation introduced also the instruments to "firm" or stabilize the supply of power from them by using a mix of pumped hydro and battery storage. Additionally, China has oriented towards power systems reforms, in order to find solutions to these fluctuations and the general energy complications. The National Development and Reform Commission (NDRC) presented plans for building a unified national power market by 2030 by merging its six regional grids into one nationwide electricity market. It will help the country to position itself in a more optimal way as a green energy leader. Undoubtedly, power systems modifications represent an imperative for PRC with regard to the pace of its clean energy transition, which is roughly the equivalent of installing five large-scale nuclear power plants worth of renewables every week (Purtill, 2024). The cheap and safe style of solar energy forced the country to embrace it, as building solar farms is speedy than constructing nuclear reactors. Therefore, China has downgraded its plans for nuclear energy expansion by choosing solar as the basis for electricity generation advancement. In 2023, the installed 1GW of nuclear compared to 300GW of solar and wind shows clearly the new direction.

The clean energy sector is not only the main pillar of Chinese energy transition. It also serves as a crucial instrument for ensuring wider economic stability and advancement. In 2023, it provided a larger share of economic growth than any other sector, accounting for 40% of the expansion of GDP. Clean energy obtained a share of 9% of PRC's GDP (compared to 7.2% in 2022). The boom in investment in the sector acted as contributor for all of the growth in investment in the national economy against the background of shrinking spending in other areas, such as the real estate. According to the analyses, "China's $890 billion investment in clean-energy sectors is almost as large as total global investments in fossil fuel supply in 2023 – and similar to the GDP of Switzerland or Turkey" (Myllyvirta & Qin, 2024).

In October 2024, the government presented the "Guiding Opinions on Vigorously Implementing the Renewable Energy Substitution Initiative". It is perceived as a new plan, deigned to accelerating renewable energy advancement. According to expert opinions, unlike previous ones "focused primarily on capacity installation, the new strategy emphasizes maximizing renewable energy utilization through infrastructure upgrades and sectoral integration across industries, including transportation, agriculture, and manufacturing" (Interesse, 2024). Therefore, it represents a herald of a systemic shift towards a multifaceted approach for creating a more integrated and efficient energy model. The building of advanced power infrastructure will include construction of flexible DC transmission and smart grids, optimization of cross-regional transmission channels, improvement of distribution networks, etc. Moreover, the infrastructure development is associated with enhancing charging and refueling stations, and arranging the use of advanced technologies like AI and IoT to harmonizing the energy systems. Driving innovation should be not underestimated as well, because an indispensable component of the strategic vision lies in pushing pilot projects for emerging technologies, such as floating offshore wind farms and green hydrogen production.

On the industry level, the new plan highlights the need for transition towards regions abundant in renewable resources and for electrification, including one based on the development of integrated energy systems in industrial parks for expediting the direct supply of green electricity. These moves are crucial for key industries such as steel, petrochemicals, and textiles. Furthermore, specific action in agriculture and rural areas will grease the wheels of the renewable energy progress. Urban planning should also incorporate the renewable course and ensure that the new buildings meet the green standards by promoting energy-efficient appliances and heating systems, as well as installation of solar panels on rooftops. In the transportation sector, the plan sets the goals for arranging renewable energy corridors, establishing multi-functional charging stations at crucial transport hubs, speeding the electrification of public transport, developing electric heavy-duty vehicles and maritime vessels, providing sustainable aviation fuels, etc.

Providing regulatory and financial support is vital for the success of the renewable energy prosperity. New legal frameworks should be forged, combined with an improved financial apparatus for backing the investment process in green energy. In this regard, the role of fiscal tools should not be undervalued, as they could provide the funds for infrastructure upgrades and the development of innovative technologies. The opportunities for the business in this realm are especially vast. Additionally, the guidelines envisage a special focus on public participation, so in the center of attention are the actions towards educating and engaging the society in adopting the sustainable mode and cultivating green mindset and culture. The dimension of international cooperation is also not left behind – the worldwide spread of Chinese technologies is highlighted, inclusive of the expanding energy partnerships with the countries within the Belt and Road Initiative (BRI).

THE INCORPORATION OF RENEWABLE ENERGY IN THE BROADER STRATEGIC FRAMEWORK

China's ambitions on the environmental and energy front were unfolded on a distinguishing ideological fundament. The concept for "ecological civilization" was developed as a synthesis of economic, political, educational and other societal transformations. Based on the idea of reforms, which should be both long-term and systemic in orientation, it was integrated into the national development plans, in order to upgrade the notion of progress. The term "ecological civilization" was proposed in 1984 by environmental experts in the Soviet Union. In 1987, the agricultural economist Qianji Ye proceeded with the theoretical attempts on Chinese ground. At the end of XX century, the shift in PRC from pure developmentalism towards a new eco-modus stimulated the narrative of "environmental culture". Starting from 2007, the term "ecological civilization" was put in the spotlight (Chun, 2015) as a final goal of environmental and societal reform, in order to form an energy and resource efficient, environmentally friendly structure of industries, pattern of growth, and mode of consumption. It was embraced by the Chinese Communist Party (CCP), which included the goal of achieving this ideal in its 2012 constitution and five-year plan. "Ecological civilization" was further raised as one of the five development goals and incorporated into the grand design of the "Chinese Dream". The glorious journey of the concept continued with the inclusion into the constitution of PRC with the 2018 amendments and the advancement in 2022, when it became the core developmental goal of the CCP.

China's "ecological civilization" stays closer to the Western model of "sustainable development", but tries to add some specific elements as well. As experts highlight, its implementation "requires not only a fundamental transformation of modes of production, and the development model, but also a fundamental transformation of worldview, values, and lifestyle" (Wang et al., 2014). Such perception underscores the moral dimension, as it views the environmental degradation as a problem of values. This element is seen as the heart of the concept, as "globally, however, framing environmental issues as a moral problem is a unique approach to addressing the environmental crisis" (Oswald, 2016). The efforts to promote the concept expanded beyond the borders of China, in order to form an international movement. In April 2014, a sub-committee on ecological civilization was established under the framework of the United Nations Alliance of Civilizations and the International Ecological Safety Collaborative Organization.

In the light of the efforts to transform its energy depiction, China also rebranded its "Belt and Road" Initiative. With regard to the Western critiques, that it stimulates the dependence on fossil fuels in the participating countries, PRC changed the normative basis with a set of document including such as "Guidance on Promoting Green Belt and Road" (2017), "The Belt and Road Ecological and Environmental Cooperation Plan" (2017), and "Vision and Actions on Energy Cooperation in Jointly Building Silk Road Economic Belt and 21st-Century Maritime Silk Road" (2017). The new green BRI approach was defined by the ambition to "share the ecological civilization philosophy and achieve sustainable development". Additionally, it aims at building "a favorable pattern of well-grounded cooperation on eco-environmental protection".

Furthermore, president Xi Jinping announced at a meeting of the United Nations General Assembly in 2021, that his country is quitting funding the construction of new coal-fired power plants abroad (Myllyvirta, 2021b). The investments in the BRI countries were oriented towards new energy projects, promoting more sustainable practices. Alongside the interest in renewable energy endeavors, a special emphasis was put on ventures in the hydropower sector.

Such projects are especially alluring for areas with low capacity in critical infrastructure. For vulnerable zones in the energy domain, finding a helping hand is extremely important, because mobilizing support for large-scale projects is not always a successful undertaking. Consequently, such energy partnership with China is described as a "win-win" solution based on mutually beneficial economic interaction. Nevertheless, these relations are subject to critical evaluations as well with regard to the approach of the Asian giant, which is based not on actual FDI, but on provided loans. This makes PRC more a "lending-power" than an "investment-power" (Fenkart 2021). Such loans generate a donor-recipient connection, introducing dependence for the small economies. However, labels such as "debt-trap diplomacy" and "checkbook diplomacy" are partially denounced with regard to the quicker implementation of the projects due to the reduced bureaucratic hurdles as a result of their rooting in interstate agreements (Novaković 2020).

Beyond the efforts to promote the "Green Silk Road" with the incorporation of the ecological civilization narrative, in 2012, the government presented the strategic concept of "Building a beautiful China", focused at achieving harmonious coexistence between humanity and nature, effective ecological protection, and sustainable resource use. The initiative set specific goals for different sectors, but the general aspiration was to advance green development, conserve the ecosystem, and achieve carbon neutrality. Embracing sustainability was patronized with regard to the business that was urged to align with eco standards, invest in green projects, and adopt environmentally friendly practices. This type of actions should be not based only on domestic efforts – foreign involvement is welcomed as well. In 2017, the "Beautiful China" initiative was upgraded by overall requirements, in order to provide consistency with the concept of ecological civilization.

The next phase in its evolution started in January 2024, when the government introduced the guidelines for promoting the furtherance of the initiative with distinctive goals. For example, they refer to reaching a 45% proportion of new energy vehicles among all new automobiles by 2030 and the elimination of old diesel locomotives, while ensuring the sustained growth of port container transport. By 2035, the peak and declining carbon emissions will be guaranteed by the spread of green production methods and lifestyles. As specific targets are set, including "railway freight turnover reaching about 25%, attaining global advanced efficiency in energy and water resource utilization, achieving full coverage of zero-waste city construction, increasing national forest coverage to 26%, and elevating the soil and water conservation rate to 75% to foster a virtuous cycle in ecosystems (Wu, 2024). Moving towards 2050, China wants to achieve a far-reaching elevate of its ecological civilization by expansive decarburization in important areas and deep greening of economy and society, that should culminate in the building of a "Beautiful China". Generally, the initiative not only favors the holistic approach for providing solutions to environmental challenges – it also campaigns for establishing a multi-participation action system for involving all stakeholders.

CONCLUSION

The outlined evolution of the Chinese energy sector shows in a clear way the transformations that have taken place in this important area, which helped to overcome the perception that the country's energy security should be linked to the use of coal. The progress made in the field of renewable energy demonstrates that this segment is now cemented as critical to solving a number of existing problems. In parallel, it is helping to generate economic growth. China's remarkable success in the realm of green energy since the beginning of the current decade reinforce its position as a leader, and the implemented large-scale energy projects testify to serious innovative potential as well. Furthermore, the planned energy endeavors are not only extremely ambitious, but they will also contribute to elevating the country's active role in global efforts to achieve an energy transition.

Looking at the development prospects for the sector, 2025 is expected to serve as a turning point in this respect, as it is the last year of the implementation of the fourteenth five-year plan. The forthcoming adoption of the next one, whose implementation should start in 2026, will show whether the Asian leader will move towards reducing the absolute levels of coal consumption. At the same time, the launch of the following climate plan for the next 10 years will also test the ambitions in this direction. However, the biggest questions remain in the domain of the future geopolitical context within which the identified energy targets will have to be achieved.

For the time being, however, as a clear trend is emerging the active investment of PRC in green energy projects in partner countries within the context of the "One Belt, One Road" Initiative. Greater interest appears to be concentrated in the South, where a number of developing states are keen to benefit from China's experience in this area. This targeted cooperation will on the one hand help to achieve local energy priorities, but on the other, it is expected to form a broader framework of interaction, including in the field of critical minerals. Undoubtedly, China's leadership position in green energy also implies the promotion of a certain model that can be copied by other countries. However, their willingness to adopt it will also be subject to the innovative solutions that the Asian leader can offer in this field.

ACKNOWLEDGEMENTS

This paper is based on the broader research within the project, "China's One Belt – One Road" Initiative in the Black Sea Region – Analysis of the Opportunities and Risks for Bulgaria" (КП-06–Н55/15), financed by National Research Fund – Bulgaria.

REFERENCES

Bettwy, M. (2007). NASA Satellites Watch as China Constructs Giant Dam. – NASA, 14 June 2007, available at: https://www.nasa.gov/centers/goddard/news/topstory/2007/dam_construct.html

Bin, D. (2021, March). Discussion on the development direction of hydropower in China. -. *Clean Energy*, *5*(1), 10–18. DOI: 10.1093/ce/zkaa025

Bioenergy, I. E. A. (2023). Biofuels Production and Development in China. - IEA Bioenergy T39 – Biofuel News, Issue 63, October 2023, available at: https://www.ieabioenergy.com/wp-content/uploads/2023/12/IEA-BIOENERGY-T39-BIOFUEL-NEWS-63-1.pdf

BP. (2021a). China's energy market in 2020. Statistical Review of World Energy – 2021. London.

BP. (2021b). Statistical Review of World Energy 2021, 70th edition. London, available at: https://www.bp.com/content/dam/bp/business-sites/en/global/corporate/pdfs/energy-economics/statistical-review/bp-stats-review-2021-full-report.pdf

Castillo, R., & Purdy, C. (2022). China's Role in Supplying Critical Minerals for the Global Energy Transition: What Could the Future Hold? – Brookings, 1 August 2022, available at: https://www.brookings.edu/articles/chinas-role-in-supplying-critical-minerals-for-the-global-energy-transition-what-could-the-future-hold/

China's Mid-Century Long-Term Low Greenhouse Gas Emission Development Strategy. 28 October 2021, available at: https://unfccc.int/sites/default/files/resource/China%E2%80%99s%20Mid-Century%20Long-Term%20Low%20Greenhouse%20Gas%20Emission%20Development%20Strategy.pdf

Chiu, D. (2017). The East Is Green: China's Global Leadership in Renewable Energy. - New Perspectives in Foreign Policy, Issue 13, Summer 2017, Center for Strategic and International Studies, available at: https://www.csis.org/east-green-chinas-global-leadership-renewable-energy

Chun, Z. (2015). China's New Blueprint for an "Ecological Civilization". – The Diplomat, 30 September 2015, available at: https://thediplomat.com/2015/09/chinas-new-blueprint-for-an-ecological-civilization/

Dent, C. (2015). China's renewable energy development: policy, industry and business perspectives. - Asia Pacific Business Review, Volume 21, Issue 1: Managing China's Energy Sector: Between the Market and the State, pp. 26-43.

Energy Institute. (2024a), Statistical Review of World Energy, 73rd edition, available at: https://www.energyinst.org/__data/assets/pdf_file/0006/1542714/684_EI_Stat_Review_V16_DIGITAL.pdf

Energy Institute. (2024b), Statistical Review of World Energy – all data, 1965–2023, available at: https://www.energyinst.org/__data/assets/excel_doc/0020/1540550/EI-Stats-Review-All-Data.xlsx

ESCAP. (2012). Finding a green engine for economic growth: China's renewable energy policies", Low Carbon Green Growth Roadmap for Asia and the Pacific, Economic and Social Commission for Asia and the Pacific, Bangkok, available at: https://www.unescap.org/sites/default/files/14.%20CS-China-Renewable-Energy-Policies.pdf

Fenkart, S. (2021). China's Influence in the Western Balkans: Partnership or Confrontation? – Institute of New Europe, 29 November 2021, available at: https://ine.org.pl/en/chinas-influence-in-the-western-balkans-partnership-or-confrontation/

Gong, Z. (2024). China Innovates Dual-tower Solar Thermal Plant for Better Efficiency. – CGTN, 16 July 2024, available at: https://news.cgtn.com/news/2024-07-16/China-innovates-dual-tower-solar-thermal-plant-for-better-efficiency-1vhVgy6Jp2E/p.html#:~:text=The%20two%20towers%20are%20each%20200%20meters%20tall,the%20concentrated%20sunlight%20heats%20water%20to%20generate%20electricity

Guidance on Promoting Green Belt and Road. (2017). – Belt and Road Portal, 8 May 2017, available at: https://eng.yidaiyilu.gov.cn/p/12479.html

Hannam, P. (2023). Confusion Surrounds China's Energy Policies as GDP and Climate Goals Clash. – The Guardian, 28 February 2023, available at: https://www.theguardian.com/world/2023/feb/28/china-energy-policies-gdp-climate-goals-clash?_hsenc=p2ANqtz-815FZ3Nq9xcpSsG5O2AfYsdZTVvCD-43lGMIR_eeIuQ-ZKHLyBJyEsnz4oO7V4rH9nj6dy

Hilton, I. (2024). How China Became the World's Leader on Renewable Energy. - Yale School of the Environment, 13 March 2024, available at: https://e360.yale.edu/features/china-renewable-energy

Hove, A. (2023). Synergies between China's Whole County PV Program and Rural Heating Electrification. – The Oxford Institute for Energy Studies, May 2023, available at: https://www.oxfordenergy.org/wpcms/wp-content/uploads/2023/05/CE6-Synergies-between-Chinas-Whole-County-PV-program-and-rural-heating-electrification.pdf

IHA. (2024). China's Fengning Station: World's Largest Pumped Hydro Power Plant Sets New Global Benchmark. - International Hydropower Association, 16 August 2024, available at: https://www.hydropower.org/news/chinas-fengning-station-worlds-largest-pumped-hydro-power-plant-sets-new-global-benchmark

Inman, P. (2021). How bad is China's energy crisis? – The Guardian, 29 September 2021, available at: https://www.theguardian.com/world/2021/sep/29/how-bad-is-chinas-energy-crisis

Interesse, G. (2024). China's New Renewable Energy Plan: Key Insights for Businesses. - China Briefing, 26 November 2024, available at: https://www.china-briefing.com/news/chinas-new-renewable-energy-plan-key-insights-for-businesses/

International Energy Agency. (2020). Sustainable Recovery. Paris, available at: https://www.iea.org/reports/sustainable-recovery

Ji, Y. (2007). Dealing with the Malacca Dilemma: China's effort to protect its energy supply. – Strategic Analysis, Vol. 31/3/2007, Institute for Defense Studies and Analyses, available at: www.idsa.in

Johnson, K. (2009). Protectionist Breezes: Wind-Power Companies Cry Foul on China. - Wall Street Journal, 28 May 2009, available at: http://europe.wsj.com

Johnston, I. (2016). Developing world invests more in renewable energy than rich countries for first time, new study says. - The Independent, 1 June 2016, available at: https://www.independent.co.uk/climate-change/news/renewable-energy-investment-developed-world-developing-world-ren21-report-a7058436.html

Lanteigne, M. (2008). China's maritime security and the "Malacca dilemma". – Asian Security, Vol. 4, Issue 2/2008, pp. 143–161.

Myllyvirta, L. (2021a, October). The Real Reasons Behind China's Energy Crisis. –. *Foreign Policy*, *7*, https://foreignpolicy.com/2021/10/07/china-energy-crisis-electricity-coal-pricing-renewables

Myllyvirta, L. (2021b, September). Why Xi's Coal Pledge Is a Big Deal. –. *Foreign Policy*, *28*, https://foreignpolicy.com/2021/09/28/china-xi-jinping-coal-pledge-unga-clean-energy developing-countries/

Myllyvirta, L., & Qin, Q. (2024). Clean Energy Was Top Driver of China's Economic Growth in 2023. – Carbon Brief, 25 January 2024, available at: https://www.carbonbrief.org/analysis-clean-energy-was-top-driver-of-chinas-economic-growth-in-2023/

Myllyvirta, L., & Zhang, X. (2022). What Do China's Gigantic Wind and Solar Bases Mean for Its Climate Goals? – Carbon Brief, 3 May 2022, available at: https://www.carbonbrief.org/analysis-what-do-chinas-gigantic-wind-and-solar-bases-mean-for-its-climate-goals/

Novaković, I. (2020). China in the Western Balkans Energy Sector: Between Development and Ecological Concerns. - European Western Balkans, 2 November 2020, available at: https://europeanwesternbalkans.com/2020/11/02/china-in-the-western-balkans-energy-sector-between-development-and-ecological-concerns/

NREL. (2004). Renewable Energy in China. - National Renewable Energy Laboratory, U.S. Department of Energy - Office of Energy Efficiency and Renewable Energy, April 2004, Colorado, available at: www.nrel.gov

Oswald, J. (2016). China Turns to Ecology in Search of "Civilisation". – Asian Studies Association of Australia, 03 August 2016, available at: https://asaa.asn.au/china-turns-to-ecology-in-search-of-civilisation/

Purtill, J. (2024). China Is Installing the Wind and Solar Equivalent of Five Large Nuclear Power Stations per Week. – ABC News, 15 July 2024, available at: https://www.abc.net.au/news/science/2024-07-16/chinas-renewable-energy-boom-breaks-records/104086640

REW. (2018). China Sets New Renewables Target of 35 Percent by 2030. - Renewable Energy World, 26 September 2018, available at: https://www.renewableenergyworld.com/baseload/china-sets-new-renewables-target-of-35-percent-by-2030/#gref

Rockwell, K. (2025). How China is Weaponising Its Dominance in Critical Minerals Trade. – East Asia Forum, 19 February 2025, available at: https://eastasiaforum.org/2025/02/19/how-china-is-weaponising-its-dominance-in-critical-minerals-trade/

Sara, S., (2012). China's Renewable Energy Law and its impact on renewable power in China: Progress, challenges and recommendations for improving implementation. -. *Energy Policy*, *51*(December), 89–109.

Shaw, V. (2024). SDIC Gansu Commissions 750 MW CSP-PV Plant in China. – PV Magazine, 18 December 2024, available at: https://www.pv-magazine.com/2024/12/18/sdic-gansu-commissions-750-mw-csp-pv-plant-in-china/

Stanway, D. (2021). China starts Baihetan hydro project, biggest since Three Gorges. – Reuters, 28 June 2021, available at: https://www.reuters.com/business/energy/chinas-giant-baihetan-hydro-plant-begins-generating-power-cctv-2021-06-28/

Suness. (2024). China's Dual Tower CSP Innovation. – Suness Blog, 22 July 2024, available at: https://www.suness.com/Blog/china-dual-tower-csp-innovation.html

The Belt and Road Ecological and Environmental Cooperation Plan. (2017). – Belt and Road Portal, 14 May 2017, available at: https://eng.yidaiyilu.gov.cn/p/13392.html

Vision and Actions on Energy Cooperation in Jointly Building Silk Road Economic Belt and 21st-Century Maritime Silk Road. (2017). – Belt and Road Portal, 25 December 2017, available at: http://obor.nea.gov.cn/detail/669.html

Wang, Z., (2014). The Ecological Civilization Debate in China: The Role of Ecological Marxism and Constructive Postmodernism—Beyond the Predicament of Legislation. - Monthly Review, 01 November 2014, available at: https://monthlyreview.org/2014/11/01/the-ecological-civilization-debate-in-china/#en73

Wei, L. (2023). China Set to Open World's Largest Hybrid Energy Plant in 2023. – CGTN, 1 October 2022, available at: https://news.cgtn.com/news/2022-10-01/China-set-to-open-world-s-largest-hybrid-energy-plant-in-2023-1dLLxnhKDbW/index.html

Wu, Y. (2024). Embracing Sustainability: How Businesses Can Contribute to the 'Beautiful China' Initiative. – China Briefing, 30 January 2024, available at: https://www.china-briefing.com/news/embracing-sustainability-how-businesses-can-contribute-to-the-beautiful-china-initiative/

Xinyu, G., Bo, J., Bin, L., Kai, Y., Hongguang, Z., & Boyuan, F. (2011). Study on Renewable Energy Development and Policy in China. -. *Energy Procedia, 5/2011*, 1284–1290. DOI: 10.1016/j.egypro.2011.03.224

Xun, L. (2022). China's First Solar-tidal Photovoltaic Power Plant Fully Operational. – CGTN, 02 June 2022, available at: https://news.cgtn.com/news/2022-05-30/China-s-first-solar-tidal-photovoltaic-power-plant-fully-operational-1asqPv0xC2k/index.html

You, X. (2022). Why China's Provinces Are So Important for Action on Climate Change. – Carbon Brief, 10 October 2022, available at: https://www.carbonbrief.org/explainer-why-chinas-provinces-are-so-important-for-action-on-climate-change/

Chapter 6
Carbon Pricing in the Fight Against Climate Change:
The Impact of the Carbon Border Adjustment Mechanism on Public Revenues and Revenue Redistribution

Candan Yilmaz Uğur
https://orcid.org/0000-0002-7915-8040
Gümüshane University, Turkey

ABSTRACT

This study focuses on the fiscal aspect of the Carbon Border Adjustment Mechanism (CBAM) introduced by the European Union (EU) and aims to examine the use of revenues from the mechanism for developed and developing countries. The CBAM aims to prevent carbon leakage and promote environmental sustainability through carbon pricing, while at the same time generating significant public revenue for the EU. The envisaged use of these revenues in strategic areas such as green transformation, sustainable infrastructure projects and similar is an important step in contributing to the EU's environmental goals. However, given the risk that underdeveloped and developing countries are disproportionately affected by CBAM, the need for effective use of these revenues in these countries becomes even more evident. In this context, the effective use of CBAM revenues in less developed and developing countries is considered as an important tool to support not only environmental sustainability but also the economic and social transformation processes of these countries.

DOI: 10.4018/979-8-3373-1310-8.ch006

1. INTRODUCTION

Climate change is an undisputed reality of the twenty-first century. Global greenhouse gas emissions are increasing day by day, triggering disasters such as droughts, floods, forest fires and hurricanes, resulting in thousands of deaths and millions of displaced people worldwide (Leal-Arcas et. al., 2022). The impacts of climate change are not confined to the borders of a particular country or region, but are a global threat. Therefore, it is not possible for any state to overcome this problem through its own efforts alone (Uğur, 2019; Uğur & Cengiz, 2022). Considering that the problem will not be solved on its own and its effects will know no borders, the need for a coordinated and multilateral intervention at the international level in the fight against climate change emerges (Soylu, 2022).

In this framework, the European Union (EU) has assumed a leading role in the fight against climate change on a global scale. Since the 1990s, the EU has developed various strategies for environmental sustainability and reduction of greenhouse gas emissions. One of these strategies is the European Green Deal (EGC), which guides the green transformation process to fulfill the requirements of the Paris Climate Agreement. Announced on 11 December 2019 by European Commission President Ursula von der Leyen, the ECC is defined as the EU's new growth strategy that aims to make Europe the world's first climate-neutral continent with net zero greenhouse gas emissions by 2050 (Ministry of Foreign Affairs Directorate for EU Affairs, 2024). This strategy also contributes to ensuring the unity of voice of all member states in terms of the EU's global climate change leadership, energy security, international competition and economic developments (Uğur & Efşan, 2024).

However, the reluctance of other countries to reduce carbon emissions and green transformation is complicating the EU's efforts. This creates a risk known as carbon leakage, which creates a disadvantageous picture for EU countries. This leakage manifests itself as a shift of some of the production activities of carbon-intensive industries to countries where climate policies are more flexible and imports partially replace local production (Beken & Cebeci, 2024). The fact that production techniques used in less regulated countries are often carbon-intensive undermines the European Commission's efforts to achieve climate neutrality by 2050 (Bellora & Fontagne, 2023). Accordingly, the EU announced the Carbon Border Regulation Mechanism in 2021 to prevent carbon leakage. The Border Carbon Adjustment Mechanism (CBAM) can essentially be defined as an import tax based on carbon emissions generated during the production process of goods. This mechanism requires the reporting of emissions for every product imported into the EU. Countries planning to export goods to the EU will be obliged to purchase carbon certificates corresponding to the EU's current carbon price. The carbon price will be set by the EU Emissions Trading System (ETS), not by countries, and will be aligned with the carbon price in the ETS market. In the first phase, CBAM will cover iron and steel, cement, fertilizer, aluminum and electricity generation, where the risk of carbon leakage is high, and other sectors will be included in CBAM in the following stages (Newman, 2022).

This study focuses on CBAM and the distribution of the revenues generated/ to be generated from CBAM to EU countries and less developed and developing countries. In the first part of the three-part study, combating climate change is defined as a market failure and carbon pricing and emissions trading systems are analyzed in a theoretical framework as a solution to this failure. In the second part, the carbon pricing strategy of the European Union (EU), which has assumed a global leadership role in the fight against climate change, is discussed and in this context, CBAM, which was put into practice with the European Green Deal, is discussed at the theoretical level. The third section discusses the revenues generated from CBAM and policy recommendations for their redistribution, and concludes the study with conclusions and recommendations. In a period of intense debates on CBAM, the study is expected to contribute to the national and international literature by addressing the issue with the dimensions of revenue impact and revenue redistribution. Moreover, the study can provide valuable clues for policymakers and relevant institutions in shaping current and future policies in the fight against climate change.

2. CARBON PRICING AS A STRATEGY TO ADDRESS MARKET FAILURES: CARBON TAX AND EMISSIONS TRADING SYSTEM

Combating climate change is a global public good that raises the question of collective action. Every country benefits from the reduction of carbon emissions in the fight against climate change, but countries do not share a common denominator (Bellora & Fontagne, 2023). In this context, carbon pricing policy has been adopted to reduce carbon emissions under the leadership of the EU. Carbon pricing, which was first implemented in Finland in 1990, has been adopted by 75 countries, mainly EU countries, as of 2024 (Chu et. al., 2024).

Carbon pricing aims to correct market failures caused by negative externalities resulting from carbon emissions from fossil fuel use. In theory, producers set the price and quantity of the goods they produce at the level at which the quantity of goods supplied meets the quantity demanded. When a firm adopts carbon-intensive production, the resulting carbon emissions are not taken into account in production costs and hence in the supply curve. In other words, the firm produces more than the optimum amount, creating an additional cost for society. Carbon pricing aims to correct this by using a market-based solution. With carbon pricing, the costs incurred by the firm in the production process increase, the supply curve shifts to the right and lower quantities of goods are produced. In this context, it can be ensured that the firm produces the optimal amount of goods, including social benefits. Carbon pricing also encourages firms to reduce their carbon-intensive production in order to reduce their costs, leading to an increase in the use of resources such as renewable energy for production (Newman, 2022).

The carbon pricing system includes carbon taxes and emissions trading systems (ETS) (World Bank, 2022). Carbon taxes aim to reduce carbon emissions by taxing goods that produce carbon emissions, such as fossil fuels or energy. In the ETS, the system works by producing tradable carbon allowances and trading them on a special trading platform (Uyar & Bilgici, 2021, pp. 53). While each firm has the right to use the emission allowances allocated to it, firms that emit more emissions can purchase additional allowances through the market, while firms that emit less emissions can sell their excess allowances or accumulate them for future use (Green, 2021, pp. 1). The total volume of these mechanisms is regulated to be equivalent to the emission cap (World Bank, 2022, pp. 13).

Carbon pricing instruments are being adopted by more and more countries on a global scale, and their effectiveness varies according to their mode of implementation and the sectors they cover. In this framework, emission trading systems, which aim to limit carbon emissions through market-based mechanisms, have become an important policy instrument, especially in large economies. The most advanced and comprehensive example of these systems is the ETS established by the EU.

3. DEVELOPMENT OF THE EUROPEAN UNION EMISSIONS TRADING SYSTEM

The European Union Emissions Trading System (EU ETS) has been operational since 2005 and stands out as the largest and most comprehensive market-based mechanism implemented with the aim of reducing the carbon emissions of member states. The system was developed to encourage the reduction of greenhouse gas emissions through economic efficiency and cost effectiveness (Bartkowiak, 2023, pp. 1).

The EU ETS consists of different phases, each defined by specific rules and targets. Currently, three main phases have been completed and the fourth phase is underway (Salvagnin et. al., 2024, pp. 1). The first phase of the EU ETS, covering the years 2005-2007, was designed as a pilot program. The aim of this phase was not only to establish a market price for CO_2, but also to build administrative capacity at the European Commission, national governments and industry level ahead of the first commitment period of the Kyoto Protocol (2008-2012) (Sato et. al., 2022, pp. 5). In the first phase, only CO_2 emissions from electricity producers and energy-intensive industries were covered (Tandon & Merle, 2024, pp. 13). Member States were left free to determine the emission allowances to be allocated for each company within their borders within the framework of their "national allocation plans". By combining these national allowance plans, the total CO_2 emission limit set in the first phase was set at 2.096 million tons of CO_2 ($MtCO_2$). Within this total cap, almost all of the allocated emission permits (2,100 million) were distributed free of charge. Furthermore, the fine imposed on companies in case of non-compliance is set at 40 Euros per ton of CO_2. In the second phase of the EU ETS covering 2008-2012, the CO_2 cap was set at 2,049 million tons and emission allowances were reduced by 6.5% compared to 2005. Iceland, Norway and Liechtenstein were included in the system and the system was expanded to cover a total of 11,500 installations. 10% of the emission permits, all of which were given free of charge in the first phase, were auctioned. The penalty for non-compliance was increased to EUR 100 per ton. Furthermore, the aviation sector was included in the EU ETS on January 1, 2012 (Usta, 2024, pp. 4). In the third phase of the ETS, it was aimed to gradually reduce the emission limits and a Market Stability Reserve (MSR) was established in this direction. In order to limit the negative effects of excess emission allowances accumulated in the market, the MSR aimed to stabilize price fluctuations by transferring excess allowances to reserves. At this stage, it was implemented through a combination of both free allocation of allowances and their sale through auctions. Among the sectors covered by the ETS, electricity/heat (82 percent), paper and pulp (78 percent), minerals and metals (75 percent) have the largest share, while the chemicals sector (42 percent) has a lower share. In addition, new sectors were included in the system and 300 million permits were reserved for new participants. Banking of permits was also allowed, meaning that permits remaining at the end of the second phase could not be used for the third phase, but were canceled and an equivalent amount of new permits were created. In addition, measurement, reporting and verification (MRV) processes were implemented as a compliance requirement each year; businesses had to verify their emission reports through accredited verifiers and submit their permits by April 30 each year (Tandon & Merle, 2024, pp. 15). The fourth phase started in 2022 and aimed to reduce greenhouse gas emissions by at least 40% by 2030 compared to 1990 levels. In line with this target, the Linear Reduction Factor (LRF) was increased to 2.2% per year, thus accelerating the reduction in the emission limit. In addition, the proportion of emission allowances allocated through auctioning was increased and free allowance allocation in certain sectors started to be phased out. This change particularly affected sectors that benefited from free allowances in the third phase, but free allowance allocation continued for sectors of strategic importance and high carbon dependence (European Commission, 2023).

The institutional and technical experience gained with the implementation of the EU ETS has paved the way for a more holistic and effective structuring of climate policies.

Figure 1. ETS and carbon tax uptake by countries' income group over time (2005-2024)

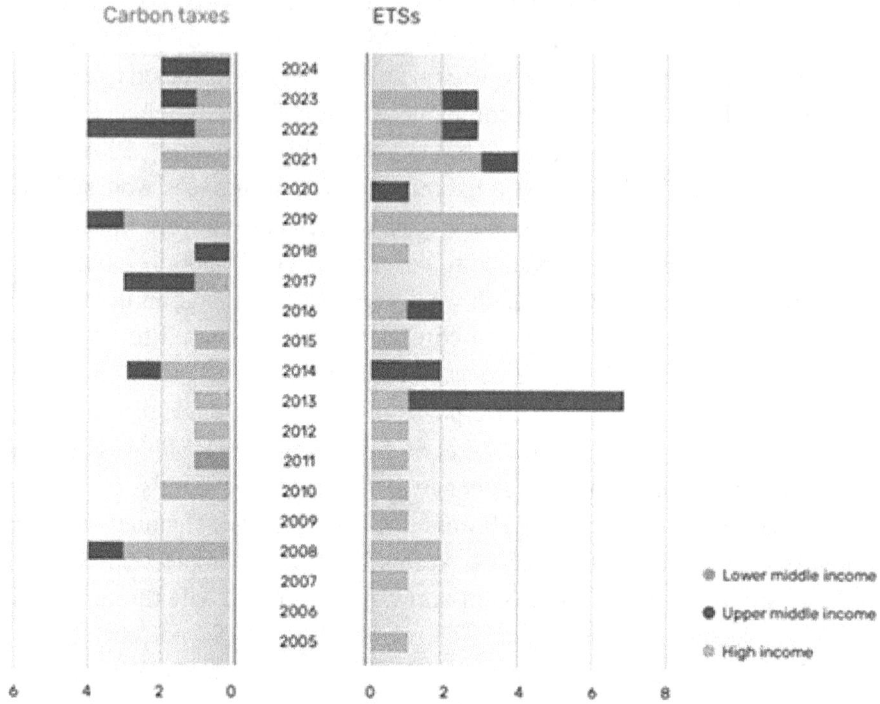

Source: *World Bank, 2024, pp. 19.*

Figure 1 shows the number of carbon taxes and ETSs enacted each year since 2005 and the rate of implementation of carbon pricing in low-, middle- and high-income countries over the last decade (World Bank, 2024, pp. 19). For example, Indonesia launched an ETS covering coal-fired power plants at the beginning of 2023. Turkey, on the other hand, announced at the 2023 United Nations Climate Change Conference (COP28) that an ETS pilot covering the energy and industry sectors will be launched in October 2024 and market operations are planned to be operational in early 2025. India has laid the institutional foundations by adopting the legal framework to establish a national carbon market, including an ETS, in 2022. India's intensity-based ETS model aims to increase energy efficiency in emission-intensive sectors and has the potential to transform into a holistic carbon market in the long run. Similar developments are taking place in Latin America: Brazil has submitted legislation to the National Congress to establish a pilot system over the next decade, while Argentina is in the process of evaluating the establishment of an ETS for the energy sector under a EU-funded project. Countries such as Chile and Colombia are also working towards the establishment of carbon pricing mechanisms at the national level. These developments reveal that carbon markets have become widespread on a global scale and have assumed an increasingly decisive role in countries' climate policies.

When examining the World Bank database shown in Figure 1, it can be seen that nearly one-third of high-income countries have a carbon pricing mechanism, but even among upper-middle-income countries, the adoption of carbon pricing is limited to 16% (Cornago & Berg, 2024, pp. 6). In total, 26 countries have not yet adopted any carbon pricing mechanism. This situation is largely attributed to the limited institutional capacity, low administrative readiness, inadequate technical infrastructure, and divergent economic priorities of these countries. However, with global climate policies gaining momentum, awareness of carbon pricing is increasing in these countries, and policy development processes are becoming more visible. Indeed, the level of adoption of carbon pricing mechanisms varies significantly between countries, and these differences are directly related to factors such as economic structure, energy production profile, institutional development, and political will. Developing and transition economies face various challenges in the implementation phase, but some countries have taken concrete steps towards integrating into carbon markets by implementing structural reforms in recent years. This trend shows that carbon pricing is not limited to developed economies, but is instead becoming a policy tool adopted on a global scale. However, the limitations of existing systems in achieving long-term emission reduction targets, particularly in the context of carbon leakage risk and sectoral inequalities, have necessitated the development of new and complementary policy tools. In this context, the European Union Emissions Trading System (EU ETS) has been revised and made more comprehensive and effective within the framework of the European Green Deal; at the same time, the Carbon Border Adjustment Mechanism (CBAM) has been introduced as a complementary tool to balance the external effects of this system. The CBAM stands out as a strategic policy tool aimed at internalising carbon costs in global trade in line with the EU's climate goals, thereby strengthening the economic dimension of climate policy.

4. EMISSIONS TRADING SYSTEM AND BORDER CARBON REGULATION MECHANISM UNDER THE GREEN DEAL

The European Green Deal (EGC) is a comprehensive climate policy blueprint created by the European Commission that includes a set of strategies and legislative proposals to achieve the EU’s goal of zero carbon emissions by 2050. It sets out the EU's roadmap for tackling climate change, while also aiming to ensure a “just and inclusive transition” and create jobs by supporting industry and the green economy (Becker et.al., 2020, pp. 2). The European Green Deal (EGC) is designed as a comprehensive structural transformation process in line with the EU's sustainable development goals, and reducing carbon emissions has been identified as one of the main priorities in this framework. This transformation aims not only to limit environmental impacts, but also to ensure that the transition to a low-carbon economy provides multidimensional benefits such as creating new jobs in the labor market and supporting economic growth.

Accordingly, the EGC envisages the extension of the scope of the ETS, making carbon pricing more effective and introducing new regulations to prevent carbon leakage. Indeed, through carbon pricing, the ETS affects producers' costs within the EU, shaping their competitiveness both within the EU and in international markets. Depending on the characteristics of the products produced (e.g. carbon intensity and trade intensity), the degree to which producers are affected by the ETS varies. While the impact of carbon pricing creates direct cost increases for businesses operating in non-displacement sectors (e.g. road transport), EU producers in trade-intensive sectors (e.g. iron and steel, cement) tend to shift their production facilities to countries where carbon regulations do not exist. This raises two main problems for the EU. First, shifting production abroad may lead to a loss of production and employment. Secondly, if the production facilities of sectors subject to carbon regulations are relocated to other countries, these sectors may no longer be exempt from the obligation to reduce their emissions and may continue to maintain their emission levels at the same level (Uyduranoğlu et. al., 2023, pp. 261 - 262). Shifting production and investments to countries with more flexible climate policies or substituting imports with higher carbon intensity for EU products brings the risk of carbon leakage (Özsoy, 2022, pp. 6). This situation not only negatively affects the economy of the European Union, but also hinders the EU's efforts to achieve global climate goals, making it difficult for the European Commission to achieve its goal of achieving climate neutrality by 2050 (Bellora & Fontagne, 2023, pp. 1). The CBAM developed in this context aims to prevent products imported from countries outside the EU with high carbon intensity from gaining a competitive advantage in the EU domestic market. Thus, CBAM is considered as a complementary element to the functionality of the ETS and contributes to the expansion of carbon pricing on a global scale.

Figure 2. Timeline for implementation of CBAM

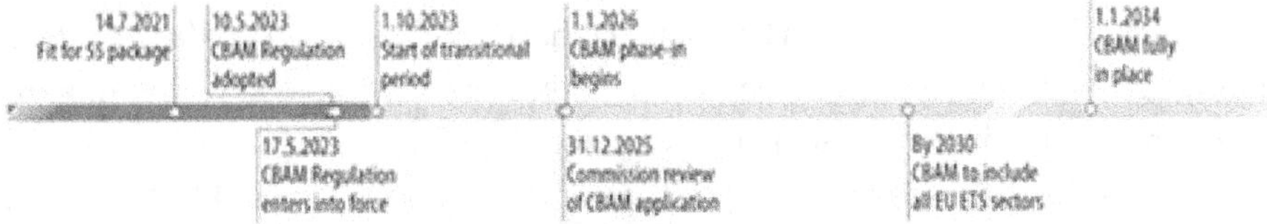

Source: *European Parliamentary Research Service, 2023: 1.*

As can be seen in Figure 2, the EU aims to strengthen its role in the fight against global climate change by submitting a comprehensive proposal for a CBAM in July 2021. The mechanism aims to integrate the EU's efforts to price and reduce carbon emissions within the EU with external trade, imposing an additional financial burden on imports of carbon-intensive products. It envisages establishing a tax mechanism in parallel with the carbon emissions trading system within the EU, especially for goods from countries that do not apply carbon pricing or strict emission restrictions (Huang et. al., 2022, pp. 1). The transition period for this carbon border tax is set to be between 2023 and 2025. These importers are required to report on the quantity of imported goods, direct and indirect carbon emissions, and the carbon price (if any) to be paid in the country of origin (Park et. al., 2023, pp. 2). This phase allows both companies within the EU and exporting countries to make the necessary preparations to comply with the requirements of the mechanism before the full implementation of the carbon regulation mechanism at the border. The 2023-2025 transition period also plays a critical role in overseeing the mechanism's implementation processes, testing its effectiveness and addressing potential challenges (Gupta et. al., 2024, pp. 8-9). As of January 1, 2026, producers importing into the EU will be required to obtain CBAM certificates to offset carbon emissions associated with the production of products that are not subject to an equivalent carbon pricing system in their country of origin. This regulation aims to ensure a level playing field between producers within the EU and producers importing into the EU by creating a carbon pricing system integrated with foreign trade. The price of certificates will be set in line with the carbon prices set under the EU ETS, ensuring that EU companies and non-EU importers compete on a level playing field in terms of carbon emissions. It is also envisaged that free emission allowances under the EU ETS will be phased out with the implementation of the CBAM. Starting from 2026, free emission allowances granted to producers within the EU will be phased out gradually until 2034. This will take place in parallel with the introduction of the CBAM, replacing free permits with regulations on the carbon emissions of imported products. In this phased transition, the rate of reduction of free emission allowances will be slow at the beginning and will gradually increase towards the end (European Parliamentary Research Service, 2023, pp. 1).

Under the EU ETS, the Carbon Cap-and-Trade Mechanism will gradually replace free allowances from 2026 in some industrial sectors such as cement, aluminum, fertilizer, hydrogen, iron and steel. These sectors represent around 54% of the total free allowances in 2021-2025. From 2026, importers of goods in the CBAM sectors into the EU are required to purchase and deliver certificates for embedded emissions (EU Commission, 2024, pp. 12).

Through the carbon regulation mechanism at the border, the EU monitors the carbon emissions emitted in the production processes of imported products and imposes a certain carbon price on these products through carbon taxes and trade policies imposed on third countries. This system functions as an important tool for maintaining environmental standards within the EU and supporting green transformation goals. However, there are significant concerns that the CBAM could deepen economic disparities between developed and developing countries.

Figure 3. Top 25 countries by total quantity declared of CBAM goods to the EU

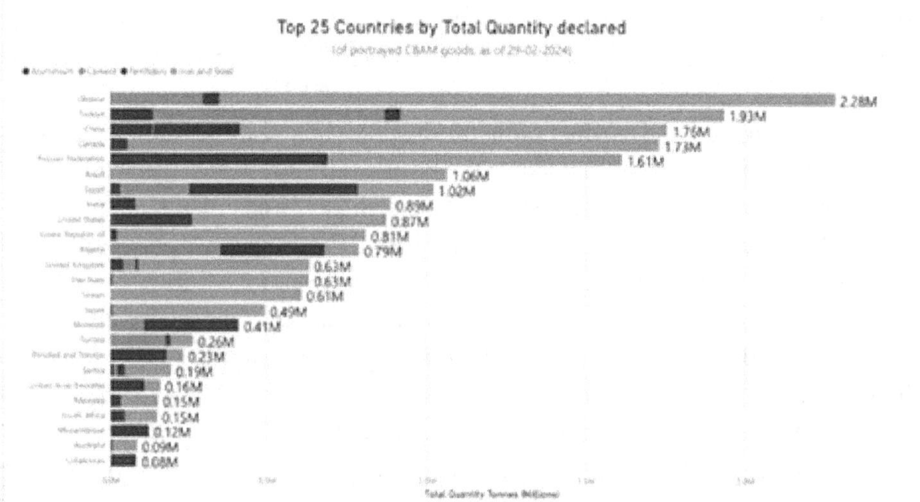

Source: *European Commission, 2025.*

Indeed, Figure 3 shows that in terms of the total value of exports of goods subject to the CBAM to the EU, the most affected countries are Ukraine, China, Turkey, Russia, India, the United States, and the United Kingdom (Cornago & Berg, 2024, pp. 4). China, which produces about 28% of global emissions and is the world's largest carbon emitter, holds an important place among the countries that may be affected under the CBAM. China is responsible for a large share of steel imports into the EU and stands out as one of the sectors targeted in the first phase of the mechanism. In addition, China's plastics sector has also been identified as one of the areas that will be affected by the CBAM (Leal-Arcas et. al., 2022, pp. 226 - 229). From Turkey's point of view, the total exports to the EU in 2023 is 9.8 billion USD, including 233 million USD in cement products, 117 million USD in fertilizer products, 6.3 billion USD in iron and steel products, 3 billion USD in aluminum products, 149 million USD in electricity and 8 thousand USD in hydrogen, which constitutes 9.37% of total exports to the EU. As of 2026, chemicals, polymers, plastics, glass, ceramic products and many intermediate inputs used as inputs may also be subject to the implementation of the CBAM, which will have a negative impact on the exports of the sectors included in the scope of the EU CBAM as of 2026 and on competitiveness in the EU market (Ministry of Trade, 2024, pp. 1). In this context, countries such as China and Turkey have stated that they have adopted a carbon pricing policy, thus aiming to adapt to the border carbon regulation mechanism (Leal-Arcas et. al., 2022, pp. 226 - 229).

Figure 4. CBAM exposure depends both on carbon emissions intensity and on EU export dependence

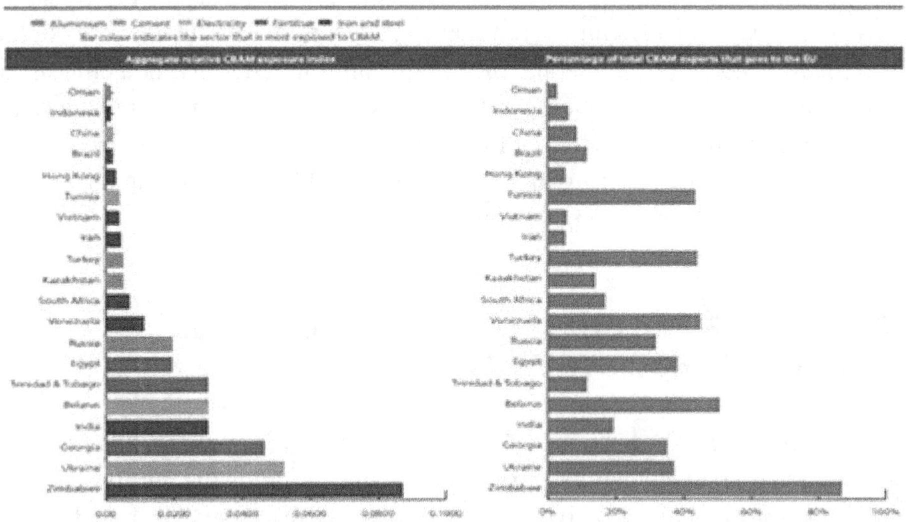

Source: *Cornago & Berg, 2024, pp. 6.*

On the other hand, an analysis of the leading exporters of goods covered by the CBAM to the EU shows that economies with carbon-intensive sectors, such as Brazil, India, Ukraine and Vietnam, may need technical and financial support from the EU in this transformation process. These countries face structural challenges in the process of aligning their heavy industries with net zero targets. However, as shown in Figure 4, low-income countries such as Mozambique and Zimbabwe are in an important position in terms of the weight of CBAM products in their economic structure, despite their limited share in total EU imports. For example, Zimbabwe's trade relations with the European Union are highly dependent. More than 80% of the country's exports under CBAM - most of which are iron and steel - are destined for the EU. Therefore, the potential impacts of CBAM on such economies may not be fully understood by looking at export volumes alone. However, when technical indicators such as carbon intensity and the share of exports to the EU in the country's total exports are assessed, a more holistic picture of why countries like Zimbabwe face structural and economic risks under CBAM can be provided (Cornago & Berg, 2024, pp. 4).

The main challenges that less developed countries face in adapting to the CBAM include limitations not only in their economic, but also in their social and industrial structures. If these countries are to make payments based on carbon emissions for carbon-intensive goods that they export to the EU, the unit prices of these goods are expected to increase. This would weaken their competitive advantage and increase the economic impact. However, due to the technological and economic weaknesses of these countries, their efforts to decarbonize their carbon-intensive industries will involve huge costs and difficulties (Bayat et. al., 2025, pp. 2239). In this context, the EU's CBAM is not only a policy instrument that aims to reduce carbon emissions, but also aims to generate economic revenue based on carbon intensity in foreign trade. Indeed, the implementation of this mechanism makes it possible to generate economic resources by ensuring that carbon pricing obligations within the EU are reflected on products imported from outside the EU.

5. IMPACTS OF THE CBAM ON PUBLIC REVENUES AND REVENUE REDISTRIBUTION

The Carbon Border Adjustment Mechanism is a regulatory system that provides for the carbon pricing obligations imposed on producers within the European Union to also apply to carbon-intensive products imported from outside the EU. This mechanism not only encourages the reduction of carbon emissions but also creates an economic resource through revenues generated based on carbon intensity in international trade. Indeed, Figure 5 shows the revenues generated from the European Union Emissions Trading System.

Figure 5. 2013 - 2022 European Union countries emissions trading system revenues

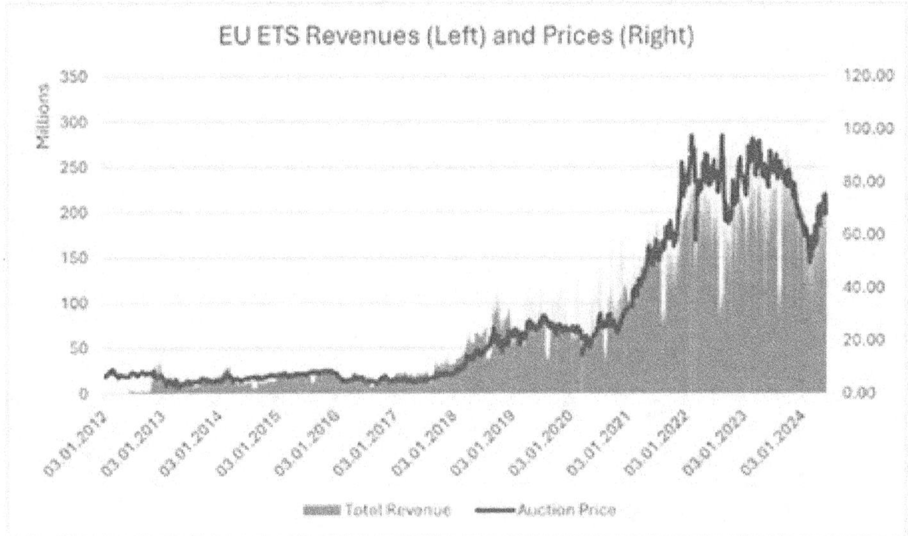

Source: *Mazzarano, 2024, pp. 1.*

Figure 5 compares auction revenues reported by member states between 2013 and 2022 and changes in carbon prices within the framework of the European Union Emissions Trading System (EU ETS). 2017 was the first year in which these revenues were reported annually. While EU member states reported approximately EUR 5 billion in EU ETS revenues in 2017, this figure exceeded EUR 29 billion by 2022, representing an increase of approximately 83% in EU ETS revenues over five years.

Figure 6. ETS revenues and reported utilization in EU Member States for 2023 (Euro)

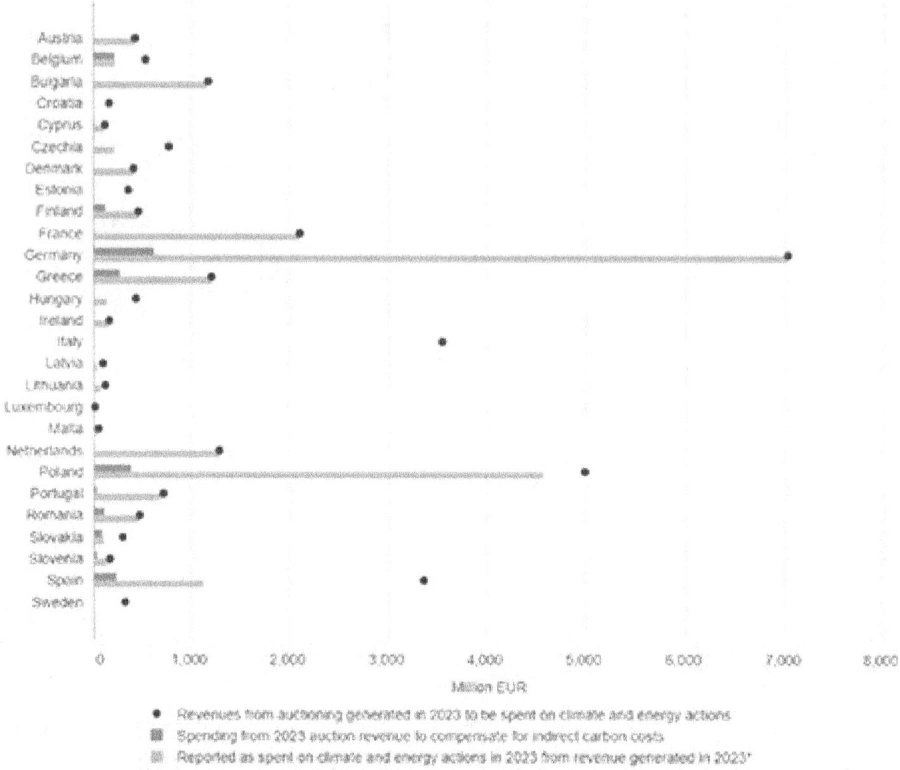

Source: *European Environment Acency, 2024*

Figure 6 shows the ETS revenues and reported utilization of EU member states for 2023. In this context, Germany, Poland and Spain were the three Member States with the highest revenues in 2023 with EUR 6.8, 4.9 and 3.2 billion respectively. In 2023, total auction revenues amounted to EUR 43.6 billion (EU Commission, 2024, pp. 22).

Figure 7. ETS revenues and reported utilization for the EU-27 between 2013-2022 (EUR billion)

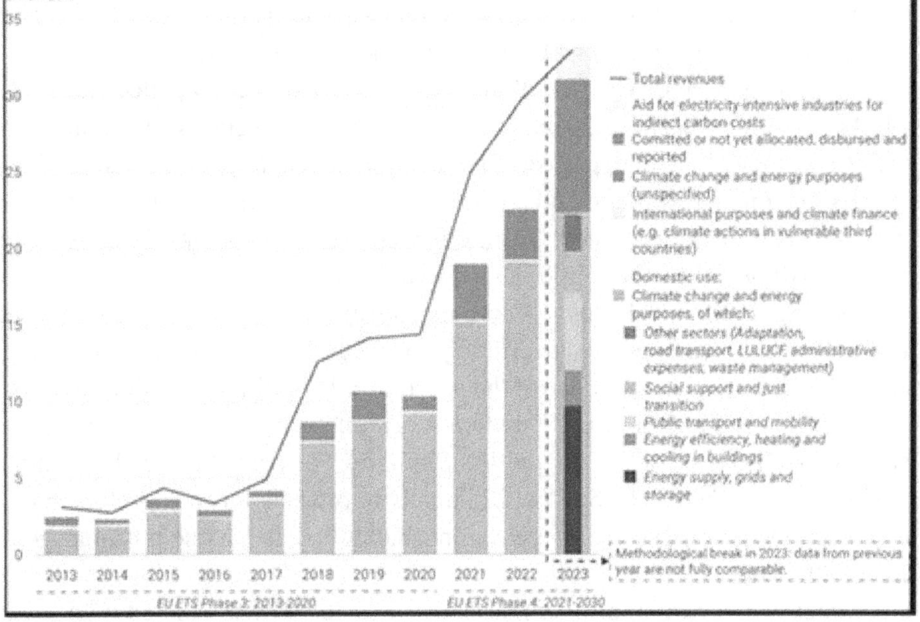

Source: *European Environment Agency, 2024.*

As shown in Figure 7, by 2023, the total auction revenue generated under the EU ETS reached €43.6 billion, of which €33 billion was transferred directly to Member States. The remaining amount was allocated to various EU-level funds, notably the Innovation Fund and the Modernization Fund. In the period 2013-2022, around 75% of auction revenues generated across the EU-27 were allocated or spent on climate and energy-related objectives. This is well above the minimum target of 50% set by EU legislation (European Environment Agency, 2024, pp. 1).

The way Member States use ETS revenues varies in line with national priorities and institutional structures. In Germany, for example, all revenues from national carbon pricing are transferred to the Energy and Climate Fund (ECF) and used for energy transition targets. In 2021, around 40% of this fund was used to alleviate the financial burden on households and businesses by reducing the renewable energy tax; in 2022, this proportion dropped to 32%, with the remaining revenues directed to subsidy programs (Edenhofer, 2021, pp. 5). Denmark, on the other hand, reported allocating all revenues from carbon pricing as grants for research and development projects through the Energy Technology and Demonstration Program (EUDP) (Marten & Dender, 2019, pp. 21).

ETS revenues are used not only for national environmental objectives but also to contribute to international development goals. In particular, various projects are supported to transform carbon-intensive transport systems in underdeveloped and developing countries and to promote greener transport infrastructure. In this context, Italy has established two different funds to promote the transition to low-emission modes of transportation. The first of these, PON METRO (2016), allocated a total of 100 million euros between 2016 and 2018 for cities to integrate sustainable transport systems; with this fund, projects for the improvement of public transport lines, the construction of new metro and railway lines and the strengthening of transport infrastructure were implemented. On the other hand, the temporary "Green Mobility Program" (Buono Mobilità Verde) aimed to promote the use of sustainable urban transport (e.g. electric vehicles, e-bikes, e-scooters) through 2020 and envisaged the reimbursement of expenses incurred by individual users through an online registration system. Within the scope of the program, €215 million was allocated from ETS revenues, and a total of €121 million was reimbursed to nearly half a million beneficiaries (Haase et. al., 2022, pp. 43). Similarly, Portugal allocated €1.1 billion in revenues from the EU ETSa in 2022, largely to projects for the development of green transport infrastructure, thereby contributing to sustainable transport goals (Sabio, 2023, pp. 1).

Revenues from carbon pricing mechanisms can also be integrated with social policy instruments to reduce the energy consumption costs of low-income households that are exposed to the negative impacts of these policies. In this context, revenues are used through subsidy programs for energy efficiency investments, thereby supporting carbon reduction targets and ensuring social justice. For example, France uses ETS revenues to finance the National Housing Agency, which operates to improve energy efficiency in buildings. These resources are specifically structured to facilitate the participation of low-income groups in the energy transition process (CPLC, 2016, pp. 3).

On the other hand, the European Commission's impact assessment clearly states the need for targeted policies to reallocate revenues from the CBAM, which aims to prevent carbon leakage, to support developing and least developed countries. The Commission recognizes the need to provide financial support in this context. However, at this stage, there is no binding or explicit arrangement that the revenues generated under the CBAM will be directed directly to these countries (Pirlot, 2022, pp. 41 - 42).

In this context, he suggests using the proceeds from CBAM as financial assistance to less developed and developing countries (Pirlot, 2022; Keane et. al., 2021). Indeed, developing countries such as India, Vietnam, Brazil, Ukraine and Turkey may find it difficult to cope with this shock without EU support. The main challenge that developing countries face in adapting to the CCS is the lack of capacity to change their industrial systems and socioeconomic structures. With the European Union's introduction of CCS, the competitive advantage of these countries could be negatively affected by increased product prices. Moreover, technological and economic weaknesses in these countries make decarbonization of carbon-intensive industries difficult and costly (Bayat et. al., 2025, pp. 2239). In this perspective, the revenue from CBAM may be used to facilitate the adoption of clean technologies in sectors that will be negatively affected by CBAM in less developed and developing countries (Brandi, 2021). As a matter of fact, Perdana and Vielle (2022) stated in their study that if the revenue from CBAM is used to subsidize renewable electricity sources such as solar and wind in less developed countries, global emissions will decrease by 79 million tons of CO2 and carbon leakage rate will decrease by 7.6%. The study also found that by 2040, solar and photovoltaic electricity generation will increase by 168 TWh, while electricity from wind power will increase by 107 TWh. Significant increases in renewable electricity generation will mostly occur in India (+121 TWh) and Africa (+107 TWh). By directing the revenue from CBAM to sectors where the transition to clean technologies can be most effective, the adoption of low-carbon solutions in targeted industries is encouraged, thereby increasing both environmental sustainability and economic competitiveness (Perdana & Vielle, 2022, pp. 7). On the other side, the CBAM creates additional costs, especially for firms operating in carbon-intensive sectors, which may put them at a competitive disadvantage (Kosonen & Nicodeme, 2009, pp. 8). However, revenues from this mechanism can be used to fund various financial instruments to support firms in sectors adversely affected by CBAM (e.g. iron and steel, cement) to transition to low-carbon production processes. These instruments include production and investment loans, R&D incentives and support for energy efficiency investments (CPLC, 2016, pp. 4). In addition to this approach, another approach that is gaining increasing traction concerns the repayment of CBAM revenues based on emissions intensity performance.

Such policies aim to reward additional emissions intensity reductions, either by making per output rebates more generous for cleaner firms (intensity-based output rebates) or by refunding a portion of emissions payments (intensity-based emissions rebates). Recent economic research suggests that intensity-based rebates can protect against leakage and competition losses, while at the same time lowering the carbon price needed to achieve a given level of emission reductions. This effect can improve the enforceability of policies, reduce carbon price inequalities and promote a faster clean energy transition (Böhringer et. al., 2022, pp. 28).

On the other hand, less developed countries, such as Mozambique and Zimbabwe, may need special solutions and support (Cornago & Berg, 2024, pp. 6). In this context, the possibility of transferring the revenue from the Border Carbon Regulation Mechanism to the general budgets of underdeveloped and developing countries may also come to the fore (Perdana & Vielle, 2022, pp. 6). This revenue could be used by these countries for various strategic purposes. For instance, it could be used to increase employment through investments, promote economic competitiveness, or finance significant investments that are under-resourced in budgeting processes. In addition, countries with high public debt stocks can also use CBAM revenues to pay down debt or close existing budget deficits. Reducing debt stocks can reduce debt servicing costs, mitigate perceived risks to borrowing, and ultimately lower the cost of borrowing (CPLC, 2016, pp. 4). However, the diversion of CBAM revenues to the general budget may also have some negative consequences. First, there could be deviations in the use of these revenues for the purpose of reducing carbon emissions. If revenues are diverted to the general budget, the lack of transparency on their use could pose another problem by making it difficult to trace them within the budget (Navigant and The Generation Foundation, 2018, pp. 29).

6. CONCLUSION

Carbon pricing stands out as a central policy instrument for the EU to achieve its environmental sustainability goals and is considered as one of the main economic mechanisms aimed at reducing greenhouse gas emissions. However, the implementation of this policy has brought along important controversies such as the risk of carbon leakage and the need to protect international competition. In this framework, the EU introduced the CBAM to both offset carbon costs in its domestic market and contribute to emission reductions on a global scale. CBAM aims to achieve carbon cost equivalence between EU producers and importers by introducing regulations based on the carbon content of imported products. The mechanism is also designed with the multi-layered objectives of supporting the EU's emission reduction targets, maintaining competitiveness in carbon-intensive sectors and encouraging trading partners to adopt lower-carbon production processes.

However, CBAM not only contributes to environmental targets, but also generates significant public revenues for the European Union based on the carbon intensity of foreign trade. It is envisaged that these revenues will be largely directed to strategic areas such as green transformation, sustainable transportation and infrastructure in EU countries. However, considering the social and economic inequalities created by CBAM at the global level, the transfer of these revenues to underdeveloped and developing countries is also on the agenda. This is because the faster decarbonization of high-income countries by setting more ambitious emission targets causes underdeveloped and developing countries to be disproportionately exposed to CBAM, which has the potential to disrupt their capital accumulation processes and development efforts. In this context, reallocating CBAM revenues to these countries through investments in decarbonized technologies or direct financial support would have positive impacts not only in terms of environmental, but also economic and social transformation.

For the establishment of a fair and inclusive global climate regime, it is essential to redistribute resources from mechanisms such as CBAM in a way that strengthens international cooperation and solidarity. To this end, carbon pricing should be considered not only as an environmental policy but also as a strategic tool for sustainable and equitable development. However, in the current situation, the inadequacy of institutional and financial structures to channel CBAM revenues to support the decarbonization processes of less developed and developing countries seriously hinders the equitable development of the climate regime on a global scale.

This study, which focuses on the distribution of CBAM revenues, has some limitations. First of all, the study was conducted within a theoretical framework. Accordingly, it is important for future studies to be supported by empirical analyses using more comprehensive and representative data sets in order to reveal the potential effects of the redistribution of these public revenues, especially on less developed and developing countries. On the other hand, this study focuses only on the financial aspects of the mechanism, leaving out the international competition dynamics and legal dimensions of CBAM. Therefore, it is considered that research that addresses the multilayered and multidimensional effects of CBAM with a holistic approach will make important contributions to the literature and policy-making processes.

REFERENCES

Bartkowiak, K. M. (2023). The European Union emission trading system and its role for green budgeting development — The case of EU member states. *Current Opinion in Environmental Sustainability, 65*, 101390. DOI: 10.1016/j.cosust.2023.101390

Bayat, T., Ozsalman, E., & Derindağ, O. F. (2025). Evaluating export vulnerability through import demand elasticity in carbon border adjustment contexts: A focus on Türkiye. *Environmental Science and Pollution Research International, 32*(5), 2238–2250. DOI: 10.1007/s11356-024-35861-8 PMID: 39762524

Becker, W., Norlen, H., Dijkstra, L., & Athanasoglou, S. (2020). Wrapping up the Europe 2020 strategy: A multidimensional indicator analysis. *Environmental and Sustainability Indicators, 8*, 100075. DOI: 10.1016/j.indic.2020.100075

Beken, H. G., & Cebeci, A. (2024). Çatışmalı bir süreç olarak sınırda karbon düzenleme mekanizması uygulaması. *Paradigma: İktisadi ve İdari Araştırmalar Dergisi, 13*(2), 115–128.

Bellora, C., & Fontagne, L. (2023). EU in search of a carbon border adjustment mechanism. *Energy Economics, 123*, 106673. DOI: 10.1016/j.eneco.2023.106673

Böhringer, C., Fischer, C., Rosendahl, K. E., & Rutherford, T. F. (2022). Potential impacts and challenges of border carbon adjustments. *Nature Climate Change, 12*(1), 22–29. DOI: 10.1038/s41558-021-01250-z

Brandi, C. (2021). Priorities for a development-friendly EU carbon border adjustment (CBAM): Briefing Paper 20/2021. https://hdl.handle.net/10419/242609

Chu, L., Do, T. N., Le, T. H. L., Ho, Q. A., & Dang, K. (2024). Carbon border adjustment mechanism, carbon pricing, and within-sector shifts: A partial equilibrium approach to Vietnam's steel sector. *Energy Policy, 193*, 114293. DOI: 10.1016/j.enpol.2024.114293

Cornago, E., & Berg, A. (2024). Learning from CBAM's transitional phase: Early impacts on trade and climate efforts. https://www.cer.eu/publications/archive/policy-brief/2024/learning-cbams-transitional-impacts-trade

CPLC (Carbon Pricing Leadership Coalition). (2016). What Are the Options for Using Carbon Pricing Revenues? https://thedocs.worldbank.org/en/doc/668851474296920877-0020022016/original/CPLCWhataretheOptionsforUsingCarbonPricingRevenues092016.pdf

European Commission. (2023). About the EU ETS, https://climate.ec.europa.eu/eu-action/eu-emissions-trading-system-eu-ets/about-eu-ets_en

European Commission. (2025). Application User Manual. CBAM Declarant Portal ver 1.4.1.1, https://taxation-customs.ec.europa.eu/document/download/47322ae4-27c9-49f1-bdb7-2a1fef73d647_en?filename=CBAM-UMN-UI%20manual%20for%20Declarants-Release%201.4.1.1-v1.81.pdf (Accessed 20 April 2025).

European Environment Acency. (2024). Use of auctioning revenues generated under the EU Emissions Trading System, https://www.eea.europa.eu/en/analysis/indicators/use-of-auctioning-revenues-generated (Accessed 12 February 2025).

European Parliamentary Research Service. (2023). Carbon border adjustment mechanism, https://www.europarl.europa.eu/RegData/etudes/ATAG/2023/754626/EPRS_ATA%282023%29754626_EN.pdf (Accessed 25 February 2025).

Green, J. F. (2021). Does carbon pricing reduce emissions? A review of ex-post analyses. *Environmental Research Letters, 16*(4), 1–17. DOI: 10.1088/1748-9326/abdae9 PMID: 33897808

Gupta, A., Pandey, R., & Sapatneka, S. (2024). Potential implications of the EU's carbon border adjustment mechanism. NIPFP Working Paper Series.

Haase, I., Velten, E. K., Branner, H., & Reyneri, A. (2022). *The use of auctioning revenues from the EU ETS for climate action – An analysis based on eight case studies. Technical report.* Ecologic Institute.

Huang, T., Liu, Z., & Zhao, T. (2022). Evolutionary Game Analysis of Responding to the EU's Carbon Border Adjustment Mechanism. *Energies, 15*(2), 427. DOI: 10.3390/en15020427

Keane, J., Mendez-Parra, M., Pettinotti, L., & Colenbrander, S. (2021). Carbon markets and standards. Policy Brief. London, www.odi.org/publications/carbon-markets-andstandards

Kosenen, K., & Nicodime, G. (2009). The role of fiscal instruments in environmental policy. CESifo Working Paper, No. 2719.

Leal-Arcas, R. Faktaufon, M., & Kyprianou, A. (2022). A legal exploration of the European Union's carbon border adjustment mechanism. *European Energy and Environmental Law Review*, 223 – 240.

Mazzarano, M. (2024). Emission trading systems, their fiscal progressivity and its limits. https://fsr.eui.eu/emission-trading-systems-their-fiscal-progressivity-and-its-limits/

Navigant ve The Generation Foundation. (2018). Raising the acceptability and effectiveness of carbon pricing: The crucial role of carbon revenue recycling, Carbon Pricing Unlocked Partnership. https://guidehouse.com/-/media/www/site/downloads/energy/2018/cpu2018carbonrevenuerecycling.pdf

Newman, N. S. (2022). The European Union opean Union's CBAM: Is ıt effective economic climate policy? *Pepperdine Policy Review*, 14 (3), https://digitalcommons.pepperdine.edu/ppr/vol14/iss1/3

Özsoy, E. C. (2022). Sınırda karbon düzenleme mekanizması: Türkiye için yeni bir risk mi fırsat mı? VI. Anadolu International Conference on Economics, 13-15 Mayıs, Eskişehir, Türkiye.

Park, C. Y., Yamamoto, Y., & Doong, M. A. L. (2023). European Union carbon border adjustment mechanism: Economic impact and implications for Asia. *ADB Briefs*, (276). DOI: 10.22617/BRF230561-2

Perdana, S., & Vielle, M. (2022). Making the EU carbon border adjustment mechanism acceptable and climate friendly for least developed countries. *Energy Policy, 170*, 113245. DOI: 10.1016/j.enpol.2022.113245

Pirlot, A. (2022). Carbon border adjustment measures: A straightforward multi-purpose climate change instrument? *Journal of Environmental Law, 34*(1), 25–52. DOI: 10.1093/jel/eqab028

Sabio, N. R. (2023). How EU ETS in Portugal supports cyclists and helps birds to fly again. Carbon Market Watch, https://carbonmarketwatch.org/2023/10/09/how-eu-ets-in-portugal-supports-cyclists-and-helps-birds-to-fly-again/

Salvagnin, C., Glielmo, A., Giulis, M. E., & Mira, A. (2024). Investigating the price determinants of the European Emission Trading System: A non-parametric approach. *Quantitative Finance, 24*(10), 1529–1544. DOI: 10.1080/14697688.2024.2407895

Sato, M., Rafaty, R., Calel, R., & Grubb, M. (2022). Allocation, allocation, allocation! The political economyof the development of the European Union EmissionsTrading System. *Wiley Interdisciplinary Reviews: Climate Change, 13*(5), 1–19. DOI: 10.1002/wcc.796

Soylu, A. C. (2022). Sürdürülebilir kalkınma ve gıda güvenliği ilişkisi. *Paradigma: İktisadi ve İdari Araştırmalar Dergisi, 11*(2), 100–111.

Tandon, S., & Merle, K. (2024). Evaluatıng the impact of cbam on developıng countries: The EU's new approach to multilateralism and trade. Policy Study, https://feps-europe.eu/wp-content/uploads/2024/11/Impact-of-CBAM.pdf

Uğur, Ö. (2019). Climate change in global politics in the postmodern eera: A new challenge and responses. Ömer Uğur ve Kadir Caner Doğan (Ed). *Contemporary Debates on Politics and Public Administration in the Postmodern Era*, 83-99.

Uğur, Ö., & Cengiz, E. (2022). İklim değişikliği siyaseti ve ekolojik diplomasi: Küresel sorunlara karşı küresel mücadele, *1 St International Conference on Scientific and Academic Research*, 318 – 322, Konya.

Uğur, Ö., & Efşan, N. Ç. (2024). Avrupa Birliği İklim Değişikliği Politikası: Politik Fırsatlar İçin İtici ve Destekleyici Faktörler. *Uluslararası Ekonomi İşletme ve Politika Dergisi*, *8*(2), 305–327. DOI: 10.29216/ueip.1474284

Usta, A. E. (2024). Sınırda karbon düzenleme mekanizması: Türkiye'yi neler bekliyor?. İktisadi Kalkınma Vakfı, No: 277.

Uyar, B., & Bilgici, Y. (2021). Karbon fiyatlandırmasında emisyon ticaret sistemi ve önemi. *Kırklareli Üniversitesi İktisadi ve İdari Bilimler Fakültesi Dergisi*, *10*(1), 47–72.

Uyduranoğlu, A., Gevrek, Z. E., & Acar, S. (2023). Avrupa Yeşil Mutabakatı'na uyum çerçevesinde Türkiye'deki işletmelerin emisyon ticaretine verdikleri desteği etkileyen faktörler. *Ankara Avrupa Çalışmaları Dergisi*, *22*(1), 257–289. DOI: 10.32450/aacd.1327099

World Bank. (2022). State and Trends of Carbon Pricing. https://documents1.worldbank.org/curated/en/099045006072224607/pdf/P1780300092e910590acb201757ecd54322.pdf

World Bank. (2024). State and Trends of Carbon Pricing 2024, https://documents1.worldbank.org/curated/en/099081624122529330/pdf/P50228315fd8d1050186341ea02e1c107bc.pdf

Chapter 7
Aligning Green Finance With Climate Governance:
Strategies for Mitigating Global Warming

Ananya Pandey
https://orcid.org/0000-0002-2419-6314
Christ University, Bangalore, India

ABSTRACT

Green finance plays a pivotal role in aligning financial systems with climate policy objectives to mitigate global warming. This chapter examines strategies that integrate green finance with regulatory frameworks, ensuring that capital flows support climate resilience and sustainability. It explores mechanisms such as green bonds, sustainability-linked loans, carbon pricing, and public-private partnerships to mobilize investments toward low-carbon technologies. Additionally, the chapter highlights the role of financial institutions in promoting climate disclosure and risk assessment while addressing challenges like greenwashing and policy misalignment. Case studies illustrate successful implementations of green finance policies in different jurisdictions, offering insights into best practices and regulatory advancements. By fostering collaboration between governments, financial markets, and international organizations, green finance can accelerate the transition toward a net-zero economy.

INTRODUCTION

Global warming remains one of the most pressing challenges of the 21st century, posing severe risks to ecosystems, economies, and human societies. The steady increase in global temperatures is largely driven by anthropogenic activities,

DOI: 10.4018/979-8-3373-1310-8.ch007

primarily the burning of fossil fuels, deforestation, and industrial processes that release greenhouse gases (GHGs) such as carbon dioxide (CO2), methane (CH4), and nitrous oxide (N2O) into the atmosphere (Abbas et al., 2022). According to the Intergovernmental Panel on Climate Change (IPCC), global surface temperatures have risen by approximately 1.1°C above pre-industrial levels, with projections indicating a continued rise unless significant mitigation measures are implemented (Forster et al. 2023).

The consequences of global warming manifest in more frequent and severe climate-related disasters, including hurricanes, wildfires, droughts, and heatwaves. Additionally, rising sea levels threaten coastal communities, and biodiversity loss accelerates due to shifting climate conditions. Beyond environmental impacts, climate change exacerbates socio-economic inequalities, disproportionately affecting vulnerable populations in developing nations (Ciscar et al., 2011). These challenges necessitate immediate and comprehensive action, integrating policy measures, technological advancements, and financial strategies to mitigate and adapt to climate change. Addressing global warming requires a systemic transition from fossil fuel-dependent economies to sustainable, low-carbon models (Dwivedi et al., 2022). While technological innovations and regulatory frameworks play crucial roles, the availability of adequate financial resources is fundamental to achieving this transition. Green finance emerges as a key instrument in mobilizing capital towards sustainable projects, reducing carbon footprints, and aligning economic growth with climate objectives (Fu et al., 2023).

Green finance encompasses a broad spectrum of financial instruments and investment strategies designed to support environmentally sustainable projects. It includes green bonds, climate funds, sustainability-linked loans, carbon pricing mechanisms, and impact investing, all of which contribute to reducing GHG emissions and promoting sustainable economic development (Mohd & Kaushal, 2018). The fundamental objective of green finance is to channel financial resources towards activities that foster resilience to climate change and accelerate the transition to a low-carbon economy. One of the most significant aspects of green finance is its ability to fund renewable energy projects, such as solar, wind, and hydroelectric power. Transitioning from fossil fuels to renewable energy sources is imperative for mitigating climate change, as energy production accounts for a substantial proportion of global carbon emissions. Investment in clean energy not only reduces emissions but also enhances energy security, creates job opportunities, and fosters economic stability (Dai & Xiong, 2023).

Additionally, green finance plays a crucial role in supporting climate-resilient infrastructure. Infrastructure such as flood defenses, sustainable transport systems, and energy-efficient buildings are essential for adapting to the impacts of climate change while simultaneously reducing carbon footprints. Financial institutions and governments increasingly recognize the need for green financing mechanisms to support the development of resilient cities that can withstand climate-induced risks (Yadav et al., 2023). Carbon pricing, another component of green finance, serves as an economic tool to internalize the environmental costs of carbon emissions. Mechanisms such as carbon taxes and cap-and-trade systems incentivize companies to reduce their emissions by imposing financial costs on high-carbon activities. These market-based approaches drive industries towards cleaner production processes and innovation in low-carbon technologies (Pan et al., 2024). Countries that have implemented carbon pricing mechanisms, such as the European Union Emissions Trading System (EU ETS), have demonstrated significant reductions in emissions while maintaining economic growth (Jung & Song, 2023).

Moreover, green finance facilitates the growth of sustainable agriculture and forestry practices. Agriculture is a major contributor to GHG emissions due to deforestation, soil degradation, and methane emissions from livestock. Investments in sustainable agricultural techniques, such as precision farming, agroforestry, and regenerative agriculture, can significantly mitigate emissions while enhancing food security (Raihan et al., 2023). Similarly, afforestation and reforestation projects financed through green bonds or climate funds contribute to carbon sequestration, preserving biodiversity, and restoring ecosystems. Despite its potential, green finance faces several challenges, including regulatory inconsistencies, greenwashing risks, and financing gaps in developing nations. The absence of standardized criteria for what qualifies as a "green" investment can lead to misallocation of resources and undermine investor confidence (Reyer et al., 2009). Additionally, many emerging economies struggle to access sufficient green funding due to weak financial institutions and high perceived investment risks. Addressing these barriers requires coordinated efforts among governments, financial institutions, and international organizations to establish robust regulatory frameworks, transparency measures, and risk-mitigation mechanisms (Finance, 2017).

Therefore, green finance represents a critical pillar in global efforts to combat climate change. By mobilizing financial resources towards sustainable projects, green finance not only reduces carbon emissions but also fosters economic resilience and environmental sustainability. As climate policies evolve, integrating green financial mechanisms with regulatory frameworks and technological innovations will be essential to achieving long-term climate goals. Governments, private investors, and multilateral institutions must work collaboratively to scale up green finance and ensure that financial flows align with climate mitigation objectives, ultimately paving the way for a more sustainable and climate-resilient future.

INTERSECTION OF GREEN FINANCE AND CLIMATE POLICY

Green finance refers to financial activities that support sustainable environmental outcomes, particularly those aligned with climate action, conservation, and energy transition. It encompasses a broad spectrum of financial instruments, policies, and mechanisms that channel investments toward environmentally friendly projects (Bhattacharya, 2022). The concept of green finance has evolved alongside the global recognition of climate change as an economic and financial risk, necessitating the redirection of capital flows toward sustainable development. The key components of green finance include green bonds, green loans, sustainable and responsible investing (SRI), environmental, social, and governance (ESG) criteria, carbon markets, and blended finance (Qing et al., 2024). Green bonds, for instance, are fixed-income instruments specifically earmarked to finance projects with environmental benefits, such as renewable energy, energy efficiency, and sustainable agriculture. Green loans, similarly, are debt instruments provided to entities that demonstrate clear environmental benefits from their use of funds (Bhutta et al., 2022). The integration of ESG factors into investment decisions has gained prominence as investors recognize the material risks and opportunities associated with climate change. ESG investing ensures that financial decision-making incorporates sustainability considerations, enhancing long-term value creation (Mupa et al., 2024).

Another crucial pillar of green finance is carbon markets, which enable the trading of carbon credits and emissions allowances. Market-based mechanisms, such as the European Union Emissions Trading System (EU ETS), have played a critical role in setting a price on carbon, thereby incentivizing emissions reductions and financing climate-friendly projects (Bayer & Aklin, 2020). Additionally, blended finance, which combines public and private sector capital, has emerged as a strategic tool to de-risk green investments and catalyze financial flows into climate-related projects in developing economies. This approach helps bridge the investment gap by leveraging concessional funding from development finance institutions to attract commercial investments (Solwa, 2019). Green finance is driven by both policy frameworks and voluntary initiatives. Regulatory measures, such as the EU Taxonomy for Sustainable Activities and the Task Force on Climate-related Financial Disclosures (TCFD), have introduced guidelines for defining and reporting sustainable investments (Eccles & Krzus, 2019). Meanwhile, voluntary initiatives, such as the Principles for Responsible Investment (PRI) and Green Bond Principles (GBP), provide best practices for incorporating sustainability into financial activities. These initiatives aim to prevent greenwashing, improve transparency, and foster accountability in sustainable finance (Garcia et al., 2023).

Despite its progress, green finance faces several challenges, including the lack of standardized definitions, regulatory fragmentation, and the risk of greenwashing.

Many financial institutions struggle to distinguish truly sustainable investments from those merely marketed as "green" (Beerbaum, 2021). Additionally, capital allocation disparities between developed and developing economies hinder the equitable distribution of green finance resources. Addressing these challenges requires coordinated global efforts to strengthen regulatory frameworks, improve data transparency, and enhance financial incentives for sustainable investments (Yadav et al., 2024).

Global Climate Policies with Focus on SDG-13 (Climate Action)

Climate policies provide the overarching framework that guides green finance by setting emission reduction targets, enforcing regulatory standards, and establishing financial incentives for sustainability. Three major global climate policy frameworks—the Paris Agreement, the European Green Deal, and the United Nations Sustainable Development Goals (SDGs)—have significantly influenced the trajectory of green finance (Yadav et al., 2023).

The Paris Agreement, adopted in 2015 under the United Nations Framework Convention on Climate Change (UNFCCC), is a landmark accord aimed at limiting global temperature rise to well below 2°C above pre-industrial levels, with efforts to restrict it to 1.5°C (Denchak, 2021). The agreement emphasizes the role of financial flows in achieving climate objectives, as outlined in Article 2.1(c), which calls for "making finance flows consistent with a pathway towards low greenhouse gas emissions and climate-resilient development" (D'Orazia & dirks, 2022). This provision underscores the need for mobilizing capital to finance the transition to a low-carbon economy. Nationally Determined Contributions (NDCs), submitted by countries under the Paris Agreement, outline national strategies for reducing emissions and adapting to climate change, with many including financial mechanisms such as green taxation, climate funds, and carbon markets (Prasad et al., 2022).

The European Green Deal, introduced in 2019, represents the European Union's ambitious roadmap to achieving climate neutrality by 2050. It integrates sustainability into all economic and financial activities, ensuring that financial markets contribute to the transition (Almeida et al., 2023). The EU Taxonomy for Sustainable Activities, a classification system for sustainable economic activities, plays a pivotal role in directing investments toward projects aligned with climate goals. Additionally, the Sustainable Finance Disclosure Regulation (SFDR) requires financial institutions to report on the sustainability impacts of their investments, enhancing transparency and accountability (Andrioli, 2023). The EU Green Bond Standard (EUGBS) sets rigorous criteria for green bonds, ensuring credibility and preventing greenwashing. The European Green Deal also promotes carbon pricing, green public procurement, and sustainable corporate governance, further embedding climate considerations into financial decision-making (Pana, 2021).

The United Nations Sustainable Development Goals (SDGs), particularly SDG 13: Climate Action, underscore the role of finance in addressing climate change. SDG 13 calls for integrating climate measures into national policies, improving climate finance mobilization, and strengthening resilience against climate impacts. Financial institutions have increasingly aligned their investment strategies with the SDGs, recognizing the long-term economic benefits of sustainable investments (Ziolo et al., 2021). The UN's Global Compact and Principles for Responsible Banking encourage financial actors to integrate sustainability into their core operations. However, achieving SDG 13 requires scaling up climate finance to developing economies, where adaptation needs are greatest but financial resources are often limited (Ferri & Acosta, 2019). The alignment of green finance with global climate policies is crucial for achieving systemic change. Policymakers must ensure that financial regulations incentivize low-carbon investments, while financial institutions must commit to sustainable finance principles. Collaborative efforts between governments, investors, and international organizations are essential for closing the climate finance gap and accelerating the transition to a net-zero economy (Monesterolo et al., 2024).

Support of Financial Instruments to Policy Objectives

Financial instruments serve as the operational mechanisms through which climate policies translate into tangible economic actions. Various instruments, including green bonds, carbon pricing, climate funds, and sustainable lending, enable governments and businesses to mobilize capital for climate action, aligning financial flows with environmental priorities (Bradford & Fraser, 2008). Green bonds, one of the most widely used instruments, provide dedicated funding for climate-related projects. Governments and corporations issue these bonds to finance initiatives such as renewable energy, energy-efficient infrastructure, and sustainable transportation. The growth of the green bond market, surpassing $2 trillion in cumulative issuance by 2023, highlights investor demand for sustainable investment opportunities. Regulatory frameworks, such as the EU Green Bond Standard, ensure that proceeds from green bonds are transparently allocated to projects with verifiable climate benefits (Bhutta et al., 2022).

Carbon pricing mechanisms, including carbon taxes and cap-and-trade systems, create financial incentives for emissions reductions. A carbon tax sets a fixed price on carbon emissions, encouraging businesses to reduce their carbon footprint, while a cap-and-trade system imposes a limit on emissions and allows firms to trade emissions allowances (Pan et al., 2024). The EU Emissions Trading System (EU ETS), the largest carbon market globally, has significantly influenced corporate investment in low-carbon technologies. Revenue generated from carbon pricing is often reinvested into climate adaptation and mitigation initiatives, ensuring a circular flow of financial resources. Climate funds, such as the Green Climate Fund (GCF) and the Climate Investment Funds (CIF), play a crucial role in financing climate action, particularly in developing economies (Sarkar & Dash, 2011). These funds provide concessional financing for renewable energy projects, climate resilience programs, and low-carbon development strategies. Blended finance approaches, combining public and private sector funding, help de-risk climate investments and attract institutional capital to emerging markets (Brown, 2022).

Sustainable lending and ESG-linked loans further support policy objectives by incentivizing corporate sustainability. Financial institutions increasingly offer sustainability-linked loans (SLLs), where interest rates are tied to the borrower's sustainability performance, creating a direct financial incentive for businesses to enhance their ESG commitments (Auzepy et al., 2023). The integration of financial instruments with climate policies is fundamental to achieving global climate goals. Governments, regulators, and financial institutions must work together to strengthen sustainable finance mechanisms, ensuring that capital flows are directed toward climate-resilient and low-carbon solutions. By aligning financial instruments with policy frameworks, the global financial system can play a transformative role in mitigating global warming and fostering a sustainable future (Yilan et al., 2022).

STRATEGIES FOR ALIGNING GREEN FINANCE WITH CLIMATE ACTION

Public-Private Sector Collaboration

Role of Government and Financial Institutions

Green finance is a critical tool for addressing climate change, yet its effectiveness depends on a well-coordinated approach between governments, financial institutions, and corporations. Governments play a central role by establishing regulatory frameworks, setting emission targets, and creating economic incentives to drive sustainable investments (Fu et al., 2023). Regulatory policies such as the European Green Deal, the U.S. Inflation Reduction Act, and China's Green Finance Guidelines demonstrate how national governments can shape the green financial landscape. These policies provide the foundation for market confidence by setting clear guidelines and establishing accountability mechanisms for sustainable finance initiatives (Yadav et al., 2024).

Financial institutions, including banks, investment firms, and pension funds, are key intermediaries in mobilizing capital toward sustainable projects. These institutions integrate Environmental, Social, and Governance (ESG) criteria into their investment decisions, directing capital to projects that align with climate goals (Lehner, 2016). They develop and promote financial instruments such as green bonds, sustainability-linked loans, and climate-focused mutual funds that facilitate the transition toward a low-carbon economy. The financial sector also plays an essential role in risk assessment by integrating climate risk disclosures and stress testing into decision-making frameworks, as seen with the Task Force on Climate-related Financial Disclosures (TCFD) (Clark et al., 2018).

Corporations, particularly in high-emission industries, must align their strategies with climate policies by investing in clean energy, sustainable supply chains, and emissions-reduction technologies. Corporate sustainability commitments, such as net-zero pledges and science-based targets, help reinforce the effectiveness of green finance initiatives (Erb et al., 2022). By issuing corporate green bonds and participating in carbon markets, businesses contribute to scaling green finance while ensuring regulatory compliance. A strong public-private synergy, underpinned by transparency, accountability, and shared incentives, is essential to bridging the financing gap and ensuring the effective deployment of green finance strategies (Nicol et al., 2018).

Policy Incentives

To accelerate the transition to a sustainable financial ecosystem, governments and regulatory bodies implement policy incentives that encourage green investments. Tax benefits are among the most widely used tools, allowing businesses and investors to offset the costs of sustainable projects (Fu et al., 2023). Incentives such as investment tax credits (ITCs) and production tax credits (PTCs) reduce the financial burden associated with renewable energy projects, energy efficiency improvements, and sustainable infrastructure development. For instance, the U.S. government provides significant tax credits for solar and wind energy projects, fostering large-scale private-sector participation in the green economy (Wiser et al., 2015).

Carbon pricing, including carbon taxes and cap-and-trade systems, serves as a market-driven approach to internalizing the environmental costs of greenhouse gas emissions. Carbon taxes directly impose a fee on emissions, creating a financial disincentive for polluting activities and encouraging businesses to adopt cleaner technologies (Stavins, 2019). Cap-and-trade systems, on the other hand, establish emission limits while allowing companies to buy and sell allowances, thus promoting cost-effective emissions reductions. The European Union's Emissions Trading System (EU ETS) is a leading example of how carbon markets can drive substantial emissions reductions in energy-intensive industries (Carl & Fedor, 2016).

Subsidies for renewable energy projects, sustainable agriculture, and clean transportation further bolster the green finance landscape. Direct subsidies and government-backed grants lower the initial investment risks associated with sustainable ventures, making them more attractive to private investors (Yang et al., 2024). Additionally, blended finance mechanisms—where public funds are used to de-risk private-sector investments—have gained prominence as an effective tool for scaling climate-related investments in emerging markets. Through a combination of tax benefits, carbon pricing mechanisms, and targeted subsidies, governments can align financial incentives with climate policy objectives, ensuring a steady flow of capital toward sustainable projects (Brown, 2022).

Sustainable Investment Mechanism

Sustainable investment mechanisms are essential for channeling capital into environmentally responsible projects. Among these mechanisms, green bonds have emerged as a critical financial instrument, allowing governments, corporations, and financial institutions to raise funds specifically for climate-friendly initiatives (Alamgir & Cheng, 2023). These bonds finance projects such as renewable energy infrastructure, energy efficiency upgrades, and climate adaptation measures. The global green bond market has experienced exponential growth, with issuance exceeding $500 billion annually, reflecting strong investor interest in sustainable finance (Shcherbakova, 2023).

Sustainability-linked loans (SLLs) offer another innovative approach by tying borrowing costs to the achievement of sustainability performance targets. Unlike green bonds, which finance specific projects, SLLs incentivize companies to improve their overall sustainability performance (Emose, 2021). Companies that meet predefined ESG targets benefit from lower interest rates, whereas those that fail to do so face financial penalties. This mechanism has gained traction among multinational corporations seeking to integrate sustainability into their financial strategy.

ESG investments have also reshaped the financial sector, influencing investment decisions and capital allocation. Institutional investors increasingly prioritize ESG criteria when selecting assets, leading to a significant shift in capital flows. ESG-focused exchange-traded funds (ETFs) and mutual funds provide retail and institutional investors with opportunities to support companies committed to environmental sustainability (Chaudhry et al., 2023). As financial markets evolve, the integration of ESG principles into mainstream investment practices will be crucial for aligning capital markets with climate policy goals.

Role of Financial Market

Financial markets play a pivotal role in scaling green finance by facilitating liquidity, price discovery, and risk management for sustainable investments. Stock exchanges worldwide have introduced sustainability indices that track the performance of ESG-compliant companies, providing investors with transparency and benchmarks for sustainable investing (Lannoo & Thomadakis, 2020). The rise of green financial products, including climate-focused hedge funds and carbon credit trading platforms, further supports the growth of the green finance ecosystem. Moreover, regulatory measures such as mandatory ESG disclosures and standardized reporting frameworks enhance investor confidence and prevent greenwashing (Naeem et al., 2023). The development of global taxonomies, such as the EU Sustainable Finance Taxonomy, helps classify sustainable economic activities, ensuring consistency and credibility in green finance markets. By fostering innovation and regulatory clarity, financial markets can significantly accelerate the transition to a low-carbon economy (Hilbrich et al., 2024).

Carbon Prices and Market Based Solutions

Carbon pricing mechanisms are essential for aligning economic activities with climate policy. Cap-and-trade systems establish a market-driven approach to emissions reductions by setting a cap on total emissions and allowing companies to trade allowances (Boyce, 2018). This system provides flexibility for businesses while ensuring that emissions remain within the established limit. The EU ETS is the largest and most developed carbon market, successfully reducing emissions in energy-intensive industries by promoting cost-effective decarbonization (Re et al., 2020).

Carbon taxes, in contrast, impose a fixed cost per ton of carbon emitted, providing a straightforward financial incentive to reduce emissions. Countries such as Sweden and Canada have implemented carbon tax policies that have proven effective in lowering emissions while generating government revenue for climate adaptation programs (Haites, 2018). The choice between cap-and-trade and carbon taxes depends on economic and political contexts, with hybrid approaches often yielding the best results (Pan et al., 2024).

Both carbon pricing mechanisms have demonstrated effectiveness in reducing emissions, but their success depends on policy design, enforcement, and market conditions. Studies indicate that well-designed carbon pricing policies encourage innovation in clean technologies, drive energy efficiency improvements, and shift investment toward low-carbon solutions (Muth, 2023). However, to maximize impact, these policies must be complemented by broader regulatory frameworks, sector-specific incentives, and public-private collaboration.

Technological Advancements in Green Finance

Technological innovations are reshaping green finance by enhancing transparency, efficiency, and scalability. Blockchain technology enables secure and transparent tracking of sustainable investments, reducing the risk of greenwashing. Smart contracts facilitate automated compliance with sustainability commitments, ensuring that funds are allocated as intended (Nwariaku et al., 2024). Artificial intelligence (AI) enhances risk assessment by analyzing large datasets to predict climate-related financial risks. AI-driven tools help investors evaluate ESG performance, detect fraudulent sustainability claims, and optimize green investment portfolios. The Internet of Things (IoT) further supports green finance by providing real-time data on emissions, resource consumption, and environmental impact, enabling more accurate sustainability assessments (Lim, 2024). The integration of digital platforms in sustainable finance has democratized access to green investments. Crowdfunding platforms and decentralized finance (DeFi) solutions offer innovative channels for financing climate projects, increasing participation from retail investors. As technology continues to evolve, digital solutions will play an increasingly vital role in advancing sustainable finance and aligning financial markets with climate policy objectives (Kashif et al., 2024).

POTENTIAL CHALLENGES AND BARRIERS

One of the primary challenges in aligning green finance with climate policy is the inconsistency in regulatory frameworks across different jurisdictions. While international agreements, such as the Paris Agreement and the European Green Deal, set overarching climate goals, national and regional policies often diverge in implementation (Chin et al., 2024). This lack of harmonization creates uncertainty for investors and financial institutions, limiting the scalability of green finance initiatives. For instance, variations in carbon pricing mechanisms—ranging from carbon taxes to cap-and-trade systems—result in disparate financial incentives, reducing the effectiveness of market-driven climate solutions (Vorster et al., 2011). Some countries have robust green finance taxonomies, such as the European Union's Sustainable Finance Taxonomy, which classifies sustainable economic activities, whereas others lack clear definitions, leading to confusion and potential market fragmentation.

Moreover, policy stability is crucial for attracting long-term green investments. Sudden changes in government priorities, regulatory reversals, or weak enforcement of climate-related financial disclosures can deter investors. In many developing economies, the absence of clear legal frameworks for sustainable finance leads to hesitancy among banks and private investors, who fear regulatory uncertainties or potential policy shifts (Abbas, 2020). Furthermore, financial institutions often struggle to align their lending and investment practices with climate goals due to conflicting mandates. Central banks and financial regulators, tasked with ensuring financial stability, sometimes prioritize short-term economic growth over long-term sustainability objectives, creating a misalignment between climate policy aspirations and financial system realities (de Galhau, 2019). Addressing these inconsistencies requires stronger international coordination, transparent regulatory frameworks, and mechanisms to ensure policy continuity to enhance investor confidence and mobilize capital effectively (Colaert, 2022).

Greenwashing Risks and Lack of Standardization

Greenwashing has emerged as a significant barrier in the integration of green finance with climate policy. The absence of universally accepted definitions and standardized criteria for sustainable investments allows financial institutions and corporations to misrepresent their environmental impact (Free et al., 2024). Greenwashing refers to the practice of portraying financial products, corporate activities, or investment portfolios as environmentally friendly when they do not meet genuine sustainability standards. This misrepresentation erodes investor trust, misallocates capital, and undermines the credibility of green finance initiatives (Griese et al., 2017).

A key driver of greenwashing is the lack of standardized sustainability reporting and disclosure requirements. While frameworks such as the Task Force on Climate-related Financial Disclosures (TCFD) and the Global Reporting Initiative (GRI) aim to improve transparency, their adoption remains uneven across industries and geographies. Some companies selectively disclose favorable sustainability metrics while omitting critical environmental risks, making it difficult for investors to assess the true sustainability impact of their investments (Hassan, 2024). Additionally, the proliferation of ESG (Environmental, Social, and Governance) rating agencies, each using different methodologies, has resulted in inconsistent assessments of corporate sustainability performance. A company may receive a high ESG rating from one agency and a low rating from another, creating confusion in the financial markets (Lin et al., 2023).

To combat greenwashing, regulatory bodies must enforce stricter sustainability disclosure requirements, ensuring that companies and financial institutions provide accurate, comparable, and verifiable information (Marcatajo, 2023) Establishing a globally recognized green finance taxonomy, similar to the EU Taxonomy, can help define what qualifies as a sustainable investment and reduce the risk of misleading claims (Pfaff et al., 2023). Additionally, independent audits and third-party verification of green financial products can enhance transparency and ensure that capital is genuinely directed toward climate-friendly projects. Without these measures, the risk of greenwashing will continue to hinder the effectiveness of green finance in supporting climate policy objectives (Zirek & Unsal, 2023).

Financing Gaps in Emerging Economies

Despite the growing momentum in green finance, a significant financing gap persists in emerging economies, limiting their ability to transition to low-carbon and climate-resilient development pathways (Esposito & Mastromatteo, 2019). According to estimates from the International Energy Agency (IEA) and the United Nations Environment Program (UNEP), trillions of dollars in annual investment are required to meet global climate targets, yet developing nations receive only a fraction of the needed financing (Kircher, 2019). Several structural barriers contribute to this gap, including limited access to affordable capital, weak financial markets, and high perceived risks associated with climate-related investments.

One major issue is the cost of capital, which remains disproportionately high in emerging economies compared to developed nations. Many green projects, such as renewable energy infrastructure or sustainable agriculture initiatives, require substantial upfront investment and long payback periods (Hu et al., 2018). However, due to economic instability, currency fluctuations, and lack of creditworthiness, financial institutions often charge higher interest rates or impose stricter lending conditions on borrowers in developing countries. This discourages private sector participation and slows the expansion of green finance (Hawkins & turner, 2000). Additionally, limited availability of financial instruments tailored for climate adaptation and mitigation—such as green bonds, blended finance models, or risk-sharing mechanisms—further constraint investment flows to these regions (Brown, 2022).

Moreover, underdeveloped regulatory and institutional frameworks in some emerging economies make it difficult to attract international green investment. A lack of clear policies on carbon pricing, renewable energy incentives, or environmental regulations creates uncertainty, making investors hesitant to commit large-scale funding (Amer & Daim, 2010). Bureaucratic inefficiencies, corruption risks, and weak enforcement of environmental laws further exacerbate the challenge. International financial institutions, such as the World Bank and the Green Climate Fund (GCF), play a crucial role in addressing these barriers by providing concessional financing, technical assistance, and de-risking mechanisms to mobilize private capital (Anjanappa, 2024).

To bridge the financing gap, a multi-faceted approach is needed, including strengthening local financial markets, implementing robust regulatory frameworks, and fostering innovative financing mechanisms. Governments in emerging economies should develop targeted policies to incentivize private investment in green projects, such as tax breaks, loan guarantees, and public-private partnerships (Shrimali & Jena, 2023). Additionally, enhancing climate risk assessment tools and integrating sustainability criteria into national financial policies can help attract more investment in climate-friendly sectors. Without adequate financial resources, emerging economies will struggle to implement effective climate policies, highlighting the urgent need for coordinated global efforts to ensure equitable access to green finance (Mungai et al., 2022).

POLICY RECOMMENDATION AND FUTURE DIRECTION

Addressing climate change requires coordinated efforts across nations, financial institutions, and regulatory bodies. The fragmented nature of climate policies and green finance initiatives creates inconsistencies that hinder the full potential of sustainable finance. To strengthen international cooperation, global financial institutions such as the International Monetary Fund (IMF), World Bank, and United Nations Framework Convention on Climate Change (UNFCCC) must play a more proactive role in harmonizing green finance standards. Establishing a unified taxonomy for sustainable investments, similar to the European Union's Green Taxonomy, would provide clear definitions for green assets and prevent greenwashing (Buchner et al., 2013). Furthermore, international climate finance agreements should focus on equitable distribution of funds, ensuring that developing economies receive adequate support to transition toward sustainability. Countries should also align financial regulations with the Paris Agreement goals by adopting mandatory climate-related financial disclosures, as recommended by the Task Force on Climate-related Financial Disclosures (TCFD). Bilateral and multilateral partnerships, such as the Just Energy Transition Partnerships (JETPs), should be expanded to facilitate technology transfer, capacity building, and financial assistance (Romano et al., 2017). Additionally, central banks and financial supervisors need to integrate climate risk into their macroprudential frameworks, enhancing the resilience of financial systems against climate-induced disruptions. The success of international cooperation depends on transparent governance structures, strong legal frameworks, and sustained political commitment to climate finance (Olovsson, 2018).

Strategic Directives for Financial Regulators and Investors

Financial regulators and investors play a crucial role in aligning green finance with climate policy. Regulators should implement stringent disclosure requirements to improve transparency in sustainable investments. Standardizing environmental, social, and governance (ESG) reporting across jurisdictions would ensure comparability and credibility of green finance instruments (Steuer & Tröger, 2022). Moreover, financial authorities should introduce incentives such as tax benefits and concessional financing to promote investment in renewable energy, low-carbon technologies, and climate adaptation projects. Central banks should incorporate climate-related risks into stress testing and monetary policy frameworks to mitigate systemic risks (Shirai, 2023). For investors, integrating climate risks into portfolio management strategies is essential for long-term financial stability. Institutional investors, including pension funds and sovereign wealth funds, should prioritize climate-resilient assets and divest from carbon-intensive industries. The adoption of sustainable finance principles, such as the United Nations' Principles for Responsible Investment (PRI), can guide investors in aligning their portfolios with climate goals. Public-private partnerships (PPPs) should also be leveraged to mobilize capital for large-scale green infrastructure projects (Clark et al., 2018). To further enhance market confidence, the establishment of an independent verification system for green financial products can prevent greenwashing and ensure adherence to sustainability criteria. A combination of regulatory enforcement, market incentives, and investor accountability is necessary to drive systemic change in financial markets (Wang & Zhang, 2023).

Scaling up Financial Instruments for Climate Resilience

Developing and scaling financial instruments tailored to climate resilience is imperative for accelerating the transition to a low-carbon and climate-resilient economy. Green bonds, sustainability-linked loans, and climate resilience funds have emerged as key financial mechanisms to mobilize capital for climate adaptation and mitigation projects (Meyer & Schwarze, 2019). To increase their effectiveness, governments should provide guarantees and risk-sharing mechanisms that attract private-sector investments. Enhancing the liquidity and accessibility of green bonds through regulatory support and financial incentives can encourage broader market participation (Anjanappa, 2024). Additionally, the development of innovative financial products, such as catastrophe bonds and climate insurance schemes, can help vulnerable economies mitigate climate-related risks (Steuer & Tröger, 2022). Financial institutions should integrate climate resilience metrics into credit assessments and investment evaluations to ensure funding is directed toward sustainable projects (BCBS, 2021). Digital finance solutions, including blockchain and artificial intelligence (AI), can further enhance the transparency and efficiency of green finance transactions (Loukoianova et al., 2024). Additionally, integrating climate risk considerations into sovereign credit ratings would incentivize countries to adopt sustainable policies. Multilateral development banks (MDBs) should expand concessional financing and technical assistance to help developing countries implement climate-resilient infrastructure projects (Vishnoi, 2022). Aligning financial instruments with science-based climate targets and risk-adjusted returns can ensure a sustainable financial ecosystem that supports both economic development and environmental sustainability. By fostering innovation, regulatory alignment, and strategic financial interventions, green finance can be scaled up to drive global climate resilience (Lu et al., 2024).

CONCLUSION

Achieving a sustainable financial ecosystem requires a systemic transformation in global finance, integrating sustainability at its core. Financial institutions, governments, and international organizations must collaborate to align investment flows with long-term environmental objectives (Agarwal et al., 2024). A critical step in this transition is the integration of Environmental, Social, and Governance (ESG) criteria into financial decision-making, ensuring that capital allocation supports low-carbon and climate-resilient projects. Governments must enforce stringent sustainability reporting standards, reducing greenwashing and ensuring transparency (Shao et al., 2020). Additionally, financial regulators should mandate climate risk assessments to ensure that financial institutions account for environmental risks in their lending and investment strategies. By embedding sustainability within financial regulations and policies, the global financial system can shift towards a more resilient and responsible model that actively mitigates climate change (Dikau & Volz, 2021).

Innovative financial instruments will play a key role in advancing green finance and scaling climate-related investments. Green bonds, sustainability-linked loans, and impact investment funds provide mechanisms for channeling capital into environmentally sustainable projects. Additionally, the development of carbon markets, enhanced through blockchain technology, can improve transparency and efficiency in carbon credit trading (Surayya, 2012). Digital finance solutions, including AI-driven risk assessments and IoT-enabled tracking of sustainable assets, can enhance financial decision-making. Strengthening the regulatory frameworks around these instruments will further ensure their credibility and effectiveness. By fostering financial innovations that prioritize sustainability, the financial sector can play a transformative role in achieving global climate goals while maintaining economic stability and growth.

A robust and inclusive financial ecosystem must also address the financing gap for developing economies, ensuring equitable access to green finance. Many emerging economies face structural barriers to attracting sustainable investments due to high capital costs, policy uncertainty, and limited financial infrastructure. Governments and international financial institutions should provide blended finance solutions, combining public funds with private investments to de-risk sustainable projects. Additionally, global financial institutions should support capacity-building initiatives, equipping local financial actors with the knowledge and tools to implement green finance strategies. Strengthening cooperation between developed and developing nations can ensure a just transition, where all countries have the resources to finance their climate commitments. By fostering financial inclusivity and leveraging international partnerships, green finance can drive a truly global response to climate change, ensuring that sustainability becomes a cornerstone of the financial system.

REFERENCES

Abbas, J. (2020). Impact of total quality management on corporate green performance through the mediating role of corporate social responsibility. *Journal of Cleaner Production, 242*, 118458. DOI: 10.1016/j.jclepro.2019.118458

Abbass, K., Qasim, M. Z., Song, H., Murshed, M., Mahmood, H., & Younis, I. (2022). A review of the global climate change impacts, adaptation, and sustainable mitigation measures. *Environmental Science and Pollution Research International, 29*(28), 42539–42559. DOI: 10.1007/s11356-022-19718-6 PMID: 35378646

Agrawal, R., Agrawal, S., Samadhiya, A., Kumar, A., Luthra, S., & Jain, V. (2024). Adoption of green finance and green innovation for achieving circularity: An exploratory review and future directions. *Geoscience Frontiers, 15*(4), 101669. DOI: 10.1016/j.gsf.2023.101669

Alamgir, M., & Cheng, M. C. (2023). Do green bonds play a role in achieving sustainability? *Sustainability (Basel), 15*(13), 10177. DOI: 10.3390/su151310177

Almeida, D. V., Kolinjivadi, V., Ferrando, T., Roy, B., Herrera, H., Gonçalves, M. V., & Van Hecken, G. (2023). The "greening" of empire: The European Green Deal as the EU first agenda. *Political Geography, 105*, 102925. DOI: 10.1016/j.polgeo.2023.102925

Amer, M., & Daim, T. U. (2010). Application of technology roadmaps for the renewable energy sector. *Technological Forecasting and Social Change, 77*(8), 1355–1370. DOI: 10.1016/j.techfore.2010.05.002

Andrioli, W. (2023). EU's ESG Regulatory Framework: a Study of SFDR.

Anjanappa, J. (2024). Role of Private Sector in Driving the Green Bond Market in India. *Available at SSRN* 4959091. DOI: 10.2139/ssrn.4959091

Auzepy, A., Bannier, C. E., & Martin, F. (2023). Are sustainability-linked loans designed to effectively incentivize corporate sustainability? A framework for review. *Financial Management, 52*(4), 643–675. DOI: 10.1111/fima.12437

Bayer, P., & Aklin, M. (2020). The European Union emissions trading system reduced CO2 emissions despite low prices. *Proceedings of the National Academy of Sciences of the United States of America, 117*(16), 8804–8812. DOI: 10.1073/pnas.1918128117 PMID: 32253304

BCBS, A. (2021). Climate-related financial risks–Measurement methodologies.

Beerbaum, D. (2021). Green Quadriga?-EU-Taxonomy, TCFD, Non-Financial-Reporting Directive and EBA ESG Pillar III/IFRS Foundation. *IFRS Foundation (April 11, 2021).*

Bhattacharyya, R. (2022). Green finance for energy transition, climate action and sustainable development: Overview of concepts, applications, implementation and challenges. *Green Finance, 4*(1), 1–35. DOI: 10.3934/GF.2022001

Bhutta, U. S., Tariq, A., Farrukh, M., Raza, A., & Iqbal, M. K. (2022). Green bonds for sustainable development: Review of literature on development and impact of green bonds. *Technological Forecasting and Social Change, 175*, 121378. DOI: 10.1016/j.techfore.2021.121378

Boyce, J. K. (2018). Carbon pricing: Effectiveness and equity. *Ecological Economics, 150*, 52–61. DOI: 10.1016/j.ecolecon.2018.03.030

Bradford, J., & Fraser, E. D. (2008). Local authorities, climate change and small and medium enterprises: Identifying effective policy instruments to reduce energy use and carbon emissions. *Corporate Social Responsibility and Environmental Management, 15*(3), 156–172. DOI: 10.1002/csr.151

Brown, L. H. (2022). Addressing the adaptation finance gap: scaling up investment in climate change adaptation using blended finance solutions.

Buchner, B., Hervé Mignucci, M., Trabacchi, C., Wilkinson, J., Stadelmann, M., Boyd, R., & Micale, V. (2013). Global Landscape of Climate Finance.

Carl, J., & Fedor, D. (2016). Tracking global carbon revenues: A survey of carbon taxes versus cap-and-trade in the real world. *Energy Policy, 96*, 50–77. DOI: 10.1016/j.enpol.2016.05.023

Chaudhry, S. M., Chen, X. H., Ahmed, R., & Nasir, M. A. (2023). Risk modelling of ESG (environmental, social, and governance), healthcare, and financial sectors. *Risk Analysis*. PMID: 37480163

Chin, N., Pinko, N., Olutoke, J., Eschalier, C., Zamarioli Santos, L., & Bendahou, S. (2024). Approaches to Meeting the Paris Agreement Goals. Options for Public Development Banks-June 2024.

Ciscar, J. C., Iglesias, A., Feyen, L., Szabó, L., Van Regemorter, D., Amelung, B., Nicholls, R., Watkiss, P., Christensen, O. B., Dankers, R., Garrote, L., Goodess, C. M., Hunt, A., Moreno, A., Richards, J., & Soria, A. (2011). Physical and economic consequences of climate change in Europe. *Proceedings of the National Academy of Sciences of the United States of America, 108*(7), 2678–2683. DOI: 10.1073/pnas.1011612108 PMID: 21282624

Clark, R., Reed, J., & Sunderland, T. (2018). Bridging funding gaps for climate and sustainable development: Pitfalls, progress and potential of private finance. *Land Use Policy*, *71*, 335–346. DOI: 10.1016/j.landusepol.2017.12.013

D'Orazio, P., & Dirks, M. W. (2022). Exploring the effects of climate-related financial policies on carbon emissions in G20 countries: A panel quantile regression approach. *Environmental Science and Pollution Research International*, *29*(5), 7678–7702. DOI: 10.1007/s11356-021-15655-y PMID: 34476706

Dai, L., & Xiong, Y. (2023). Does the green finance development and renewable energy affect the economic recovery in Asian economies. *Renewable Energy*, *216*, 118922. DOI: 10.1016/j.renene.2023.118922

De Galhau, V. (2019, June). Climate change: central banks are taking action. In *FSR* (p. 7).

Denchak, M. (2021). *Paris climate agreement: Everything you need to know*. National Resource Defend Council.

Dikau, S., & Volz, U. (2021). Central bank mandates, sustainability objectives and the promotion of green finance. *Ecological Economics*, *184*, 107022. DOI: 10.1016/j.ecolecon.2021.107022

Dwivedi, Y. K., Hughes, L., Kar, A. K., Baabdullah, A. M., Grover, P., Abbas, R., Andreini, D., Abumoghli, I., Barlette, Y., Bunker, D., Chandra Kruse, L., Constantiou, I., Davison, R. M., De', R., Dubey, R., Fenby-Taylor, H., Gupta, B., He, W., Kodama, M., & Wade, M. (2022). Climate change and COP26: Are digital technologies and information management part of the problem or the solution? An editorial reflection and call to action. *International Journal of Information Management*, *63*, 102456. DOI: 10.1016/j.ijinfomgt.2021.102456

Eccles, R. G., & Krzus, M. P. (2019). Implementing the task force on climate-related financial disclosures recommendations: An assessment of corporate readiness. *Schmalenbach Business Review*, *71*(2), 287–293. DOI: 10.1007/s41464-018-0060-4

Emose, G. (2021). Sustainability bond for the Pacific feasibility study.

Erb, T., Perciasepe, B., Radulovic, V., & Niland, M. (2022). Corporate climate commitments: The trend towards net zero. In *Handbook of climate change mitigation and adaptation* (pp. 2985–3018). Springer International Publishing. DOI: 10.1007/978-3-030-72579-2_146

Esposito, L., & Mastromatteo, G. (2019). Sustainable finance, the good, the bad and the ugly: A critical assessment of the EU institutional framework for the green transition. In *Sustainable finance, the good, the bad and the ugly: a critical assessment of the EU institutional framework for the green transition* (Vol. 2019, No. 4, pp. 1-42). Vita e Pensiero.

Ferri, G., & Acosta, B. A. (2019). Sustainable finance for sustainable development. *Center for Relationship Banking and Economics Working Paper Series, 30.*

Finance, S. U. P. (2017). National Green Banks in Developing Countries.

Forster, P. M., Smith, C. J., Walsh, T., Lamb, W. F., Lamboll, R., Hauser, M., Ribes, A., Rosen, D., Gillett, N., Palmer, M. D., Rogelj, J., von Schuckmann, K., Seneviratne, S. I., Trewin, B., Zhang, X., Allen, M., Andrew, R., Birt, A., Borger, A., & Zhai, P. (2023). Indicators of Global Climate Change 2022: Annual update of large-scale indicators of the state of the climate system and human influence. *Earth System Science Data, 15*(6), 2295–2327. DOI: 10.5194/essd-15-2295-2023

Free, C., Jones, S., & Tremblay, M. S. (2024). Greenwashing and sustainability assurance: A review and call for future research. *Journal of Accounting Literature*. DOI: 10.1108/JAL-11-2023-0201

Fu, C., Lu, L., & Pirabi, M. (2023). Advancing green finance: A review of sustainable development. *Digital Economy and Sustainable Development, 1*(1), 20. DOI: 10.1007/s44265-023-00020-3

García, C. J., Herrero, B., Miralles-Quirós, J. L., & del Mar Mirallles-Quirós, M. (2023). Exploring the determinants of corporate green bond issuance and its environmental implication: The role of corporate board. *Technological Forecasting and Social Change, 189*, 122379. DOI: 10.1016/j.techfore.2023.122379

Griese, K. M., Werner, K., & Hogg, J. (2017). Avoiding greenwashing in event marketing: An exploration of concepts, literature and methods. *Journal of Management and Sustainability*, *7*(4), 1. DOI: 10.5539/jms.v7n4p1

Haites, E. (2018). Carbon taxes and greenhouse gas emissions trading systems: What have we learned? *Climate Policy*, *18*(8), 955–966. DOI: 10.1080/14693062.2018.1492897

Hassan, S. M. (2024). Greenwashing in ESG: Identifying and Addressing False Claims of Sustainability. *Journal of Business and Strategic Management*, *9*(8), 90–105. DOI: 10.47941/jbsm.2390

Hawkins, J., & Turner, P. (2000). Managing foreign debt and liquidity risks in emerging economies: An overview. *BIS Policy Papers*, *8*, 3–60.

Hilbrich, S., Berensmann, K., Artmann, G., Ashman, S., Herbold, T., Lötters-Viehof, S., & Steenkamp, L. A. (2024). Implementing sustainability taxonomies to redirect capital flows: The case of South Africa. *Climate Policy*, 1–15.

Hu, J., Harmsen, R., Crijns-Graus, W., & Worrell, E. (2018). Barriers to investment in utility-scale variable renewable electricity (VRE) generation projects. *Renewable Energy*, *121*, 730–744. DOI: 10.1016/j.renene.2018.01.092

Jung, H., & Song, C. K. (2023). Effects of emission trading scheme (ETS) on change rate of carbon emission. *Scientific Reports*, *13*(1), 912. DOI: 10.1038/s41598-023-28154-6 PMID: 36650382

Kashif, M., Pinglu, C., Ullah, S., & Zaman, M. (2024). Evaluating the influence of financial technology (FinTech) on sustainable finance: A comprehensive global analysis. *Financial Markets and Portfolio Management*, *38*(1), 123–155. DOI: 10.1007/s11408-023-00439-w

Kircher, M. (2019). Bioeconomy: Markets, implications, and investment opportunities. *Economies*, *7*(3), 73. DOI: 10.3390/economies7030073

Lannoo, K., & Thomadakis, A. (2020). Derivatives in sustainable finance. *CEPS-ECMI Study. Brussels: Centre for European Policy Studies, 3*.

Lehner, O. M. (2016). *Routledge handbook of social and sustainable finance*. Routledge. DOI: 10.4324/9781315772578

Lim, T. (2024). Environmental, social, and governance (ESG) and artificial intelligence in finance: State-of-the-art and research takeaways. *Artificial Intelligence Review*, *57*(4), 76. DOI: 10.1007/s10462-024-10708-3

Lin, Y., Lu, Z., & Wang, Y. (2023). The impact of environmental, social, and governance (ESG) practices on investment efficiency in China: Does digital transformation matter? *Research in International Business and Finance*, *66*, 102050. DOI: 10.1016/j.ribaf.2023.102050

Loukoianova, E., Natalucci, F., Wang, D., & Kanada, S. (2024). *Fintech Applications for Boosting Climate Finance*. International Monetary Fund. DOI: 10.5089/9798400293740.066

Lu, J., Gulzar, F., & Lai, Y. (2024). Exploring the intersection between sustainable finance and achieving carbon neutrality in the transportation sector. *Heliyon*, *10*(15), e34467. DOI: 10.1016/j.heliyon.2024.e34467 PMID: 39145029

Marcatajo, G. (2023). Green claims, green washing and consumer protection in the European Union. *Journal of Financial Crime*, *30*(1), 143–153. DOI: 10.1108/JFC-11-2021-0240

Meyer, P. B., & Schwarze, R. (2019). Financing climate-resilient infrastructure: Determining risk, reward, and return on investment. *Frontiers of Engineering Management*, *6*(1), 117–127. DOI: 10.1007/s42524-019-0009-4

Mohd, S., & Kaushal, V. K. (2018). Green finance: A step towards sustainable development. *MUDRA: Journal of Finance and Accounting*, *5*(1), 59–74. DOI: 10.17492/mudra.v5i01.13036

Monasterolo, I., Mandel, A., Battiston, S., Mazzocchetti, A., Oppermann, K., Coony, J., Stretton, S., Stewart, F., & Dunz, N. (2024). The role of green financial sector initiatives in the low-carbon transition: A theory of change. *Global Environmental Change*, *89*, 102915. DOI: 10.1016/j.gloenvcha.2024.102915

Mupa, M. N., Pamful, E. E., Nnaji, J. C., & Adu-Boahen, J. O. A. N. N. A. (2024). Integrating ESG factors in investment decision-making for renewable energy projects. *Iconic Research and Engineering Journals*, *8*(2), 273–293.

Muth, D. (2023). Pathways to stringent carbon pricing: Configurations of political economy conditions and revenue recycling strategies. A comparison of thirty national level policies. *Ecological Economics*, *214*, 107995. DOI: 10.1016/j.ecolecon.2023.107995

Naeem, M. A., Karim, S., Rabbani, M. R., Bashar, A., & Kumar, S. (2023). Current state and future directions of green and sustainable finance: A bibliometric analysis. *Qualitative Research in Financial Markets*, *15*(4), 608–629. DOI: 10.1108/QRFM-10-2021-0174

Nicol, M., Shishlov, I., & Cochran, I. (2018). Green Bonds: Improving their contribution to the low-carbon and climate resilient transition. *I4CE–Institute for Climate Economics, Paris, 217.*

Nwariaku, H., Fadojutimi, B., Lawson, L. G. L., Agbelusi, J., Adigun, O. A., Udom, J. A., & Olajide, T. D. (2024). Blockchain technology as an enabler of transparency and efficiency in sustainable supply chains. *International Journal of Science and Research Archive*, *12*(2), 1779–1789. DOI: 10.30574/ijsra.2024.12.2.1454

Olovsson, C. (2018). Is climate change relevant for central banks? *Sveriges Riksbank Economic Commentaries*, *13*(13), 1–8.

Pan, J., Cross, J. L., Zou, X., & Zhang, B. (2024). To tax or to trade? A global review of carbon emissions reduction strategies. *Energy Strategy Reviews*, *55*, 101508. DOI: 10.1016/j.esr.2024.101508

Pana, T. N. (2021). Context & Challenges of the 2021 EU Green Bond Standard. *Available at SSRN* 3920262. DOI: 10.2139/ssrn.3920262

Pfaff, N., Utermarck, S., Altun, O., & Egorov, S. (2023). *Market integrity and greenwashing risks in sustainable finance*. International Capital Market Association.

Prasad, M. A., Loukoianova, M. E., Feng, A. X., & Oman, W. (2022). *Mobilizing private climate financing in emerging market and developing economies*. International Monetary Fund.

Qing, L., Abbas, J., Najam, H., Ma, X., & Dagestani, A. A. (2024). Investment in renewable energy and green financing and their role in achieving carbon-neutrality and economic sustainability: Insights from Asian region. *Renewable Energy*, *221*, 119830. DOI: 10.1016/j.renene.2023.119830

Raihan, A., Muhtasim, D. A., Farhana, S., Hasan, M. A. U., Pavel, M. I., Faruk, O., Rahman, M., & Mahmood, A. (2023). An econometric analysis of Greenhouse gas emissions from different agricultural factors in Bangladesh. *Energy Nexus*, *9*, 100179. DOI: 10.1016/j.nexus.2023.100179

Re, L. L., Lee, C., Cassisa, C., Zhang, W., & Moarif, S. (2020). *Implementing Effective Emissions Trading Systems: Lessons from International Experiences*. IEA Publications.

Reyer, C., Guericke, M., & Ibisch, P. L. (2009). Climate change mitigation via afforestation, reforestation and deforestation avoidance: And what about adaptation to environmental change? *New Forests*, *38*(1), 15–34. DOI: 10.1007/s11056-008-9129-0

Romano, A. A., Scandurra, G., Carfora, A., & Ronghi, M. (2017). *Climate finance as an instrument to promote green growth in developing countries*. Springer.

Sarkar, A. N., & Dash, S. (2011). Emissions trading and carbon credit accounting for sustainable energy development with focus on India. *Asia Pacific Business Review*, *7*(1), 50–80. DOI: 10.1177/097324701100700104

Shao, S., Hu, Z., Cao, J., Yang, L., & Guan, D. (2020). Environmental regulation and enterprise innovation: A review. *Business Strategy and the Environment*, *29*(3), 1465–1478. DOI: 10.1002/bse.2446

Shcherbakova, O. (2023). The role of green bonds in financing sustainable development. *Economy of Ukraine*, *66*(12), 745.

Shirai, S. (2023). *Green central banking and regulation to foster sustainable finance* (No. 1361). ADBI Working Paper.

Shrimali, G., & Jena, L. P. (2023). Enabling Finance for Net-Zero Energy Transitions. *Available at SSRN* 4659874.

Solwa, I. (2019). Renewable Energy IPPs in SSA–Effective use of Blended Finance.

Stavins, R. N. (2019). *Carbon taxes vs. cap and trade: Theory and practice*. Harvard Project on Climate Agreements.

Steuer, S., & Tröger, T. H. (2022). The role of disclosure in green finance. *Journal of Financial Regulation, 8*(1), 1–50. DOI: 10.1093/jfr/fjac001

Surayya, T. (2012). Innovative Financial Instruments and mechanisms for financing forest restoration and mitigating climate change: Select cases from India. *European Journal of Sustainable Development, 1*(2), 361–361. DOI: 10.14207/ejsd.2012.v1n2p361

Vishnoi, J. N. (2022). *The Role of the financial institutions in transition to low carbon economy* (Doctoral dissertation, Wien).

Vorster, S., Winkler, H., & Jooste, M. (2011). Mitigating climate change through carbon pricing: An emerging policy debate in South Africa. *Climate and Development, 3*(3), 242–258. DOI: 10.1080/17565529.2011.598367

Wang, G., & Zhang, Q. (2023). Transparency, Governance, and Innovation: Unveiling the Impact of Environmental Information Disclosure on Corporate Value and Green Innovation Capabilities. *International Journal of Operations and Quantitative Management, 29*(4), 54–70.

Wiser, R., Lantz, E., Mai, T., Zayas, J., DeMeo, E., Eugeni, E., Lin-Powers, J., & Tusing, R. (2015). Wind vision: A new era for wind power in the United States. *The Electricity Journal, 28*(9), 120–132. DOI: 10.1016/j.tej.2015.09.016

Yadav, A., Gyamfi, B. A., Asongu, S. A., & Behera, D. K. (2024). The role of green finance and governance effectiveness in the impact of renewable energy investment on CO2 emissions in BRICS economies. *Journal of Environmental Management, 358*, 120906. DOI: 10.1016/j.jenvman.2024.120906 PMID: 38636419

Yadav, A., Sharma, H., & Venkataramani, V. (2023). Mobilising Finance for Climate Action and Disaster-Resilient Infrastructure in Developing Countries.

Yang, F., Liu, W., Zhang, Y., Yang, G., & wala, T. (2024). Money grows on green energy: Financing a sustainable power future. *Heliyon, 10*(7), e28353. DOI: 10.1016/j.heliyon.2024.e28353 PMID: 38590910

Yilan, G., Cordella, M., & Morone, P. (2022). Evaluating and managing the sustainability of investments in green and sustainable chemistry: An overview of sustainable finance approaches and tools. *Current Opinion in Green and Sustainable Chemistry, 36*, 100635. DOI: 10.1016/j.cogsc.2022.100635

Ziolo, M., Bak, I., & Cheba, K. (2021). The role of sustainable finance in achieving sustainable development goals: Does it work? *Technological and Economic Development of Economy*, *27*(1), 45–70. DOI: 10.3846/tede.2020.13863

Zirek, D., & Unsal, O. (2023). Green bonds: Do investors benefit from third-party certification? *Global Finance Journal*, *58*, 100872. DOI: 10.1016/j.gfj.2023.100872

KEY TERMS AND DEFINITIONS

Climate Change: Climate change refers to long-term alterations in global or regional climate patterns, primarily caused by human activities such as burning fossil fuels, deforestation, and industrial processes. These activities increase the concentration of greenhouse gases like carbon dioxide and methane in the atmosphere, trapping heat and raising Earth's average temperature. This warming leads to melting glaciers, rising sea levels, more extreme weather events, shifts in ecosystems, and biodiversity loss. While natural factors can influence climate, the rapid pace of recent changes is largely human-induced. Addressing climate change requires global cooperation, policy reforms, sustainable practices, and technological innovation to mitigate its harmful effects.

Global Warming: Global warming refers to the long-term increase in Earth's average surface temperature due to the accumulation of greenhouse gases in the atmosphere, primarily from human activities like burning fossil fuels, deforestation, and industrial processes. These gases, such as carbon dioxide, methane, and nitrous oxide, trap heat from the sun, creating a "greenhouse effect" that disrupts the planet's climate system. As a result, global warming contributes to rising sea levels, extreme weather events, loss of biodiversity, and other environmental and socio-economic impacts. It is a central issue in climate change discussions and requires urgent global cooperation for mitigation and adaptation strategies.

Green House Gases: Greenhouse gases (GHGs) are atmospheric gases that trap heat from the sun, preventing it from escaping back into space, thereby warming the Earth's surface. This natural process is called the greenhouse effect, which is essential for maintaining temperatures suitable for life. However, excessive emissions from human activities—such as burning fossil fuels, deforestation, and industrial processes—intensify this effect, leading to global warming and climate change. Major greenhouse gases include carbon dioxide (CO_2), methane (CH_4), nitrous oxide (N_2O), and fluorinated gases. These gases vary in their heat-trapping abilities and atmospheric lifespans, contributing differently to environmental impacts and long-term climate disruption.

Green Finance: Green finance refers to financial activities that support environmentally sustainable development by promoting investments in projects and initiatives that benefit the environment. It includes funding for renewable energy, energy efficiency, pollution prevention, sustainable agriculture, and conservation of biodiversity. Green finance aims to integrate environmental, social, and governance (ESG) factors into investment decisions to drive long-term economic growth while reducing carbon emissions and environmental risks. Instruments of green finance include green bonds, green loans, and climate funds. By aligning financial flows with sustainable development goals, green finance plays a vital role in addressing climate change and promoting a transition to a low-carbon economy.

Renewable Energy: Renewable energy refers to energy derived from natural resources that are continuously replenished and virtually inexhaustible. These sources include sunlight, wind, rain, tides, waves, geothermal heat, and biomass. Unlike fossil fuels, renewable energy generates little or no greenhouse gas emissions, making it a sustainable and environmentally friendly alternative. It plays a crucial role in reducing carbon footprints, combating climate change, and enhancing energy security. Renewable energy technologies convert these natural sources into usable power for electricity, heating, and transportation. As global demand for clean energy rises, renewables offer a promising pathway toward a low-carbon and sustainable energy future.

Sustainable Development: Sustainable development is a holistic approach that meets the needs of the present without compromising the ability of future generations to meet their own needs. It integrates economic growth, social inclusion, and environmental protection, aiming to balance human well-being with the planet's health. This concept encourages responsible resource use, equitable access to opportunities, and long-term planning to ensure prosperity for all. Sustainable development promotes innovation, environmental stewardship, and social equity, addressing challenges such as climate change, poverty, and biodiversity loss. It is guided by principles of intergenerational equity, sustainability, and resilience, forming the foundation of global frameworks like the United Nations' SDGs.

Sustainable Finance: Sustainable finance refers to the process of taking environmental, social, and governance (ESG) considerations into account when making investment decisions in the financial sector. It aims to support economic growth while reducing pressures on the environment, addressing social inequalities, and promoting good governance. Sustainable finance integrates long-term value creation and risk management, encouraging investments in green technologies, renewable energy, sustainable agriculture, and inclusive projects. It aligns financial systems with the goals of sustainable development and climate action, including the Paris Agreement and the United Nations Sustainable Development Goals (SDGs), to foster a more resilient, equitable, and low-carbon global economy.

Chapter 8
Empowering Local Action: Strategies for Climate-Responsive Governance

Aryan Jhamb
https://orcid.org/0009-0008-9931-5219
Christ University, India

ABSTRACT

Governance explores the pivotal role of local governance in addressing the urgent challenges posed by climate change. As the effects of climate disruption become increasingly localized, cities, towns, and communities are on the front lines, facing the immediate impacts while also offering innovative solutions. This chapter examines the strategies that enable local governments to integrate climate resilience into their policies, governance structures, and day-to-day decision-making processes. It discusses the importance of building local capacity, fostering collaboration among stakeholders, and utilizing data-driven approaches for adaptive management. Through case studies and real-world examples, the chapter highlights how local leadership can empower citizens, engage in sustainable urban planning, and implement climate-mitigation practices. By enhancing the capacity for climate-responsive governance, this work underscores the need for a bottom-up approach in achieving global climate goals and creating resilient communities.

1. INTRODUCTION

Climate change presents an unprecedented global challenge with significant impacts on ecosystems, economies, and human health. The Intergovernmental Panel on Climate Change (IPCC) emphasizes the urgent need for immediate action, underscoring the dire consequences of inaction—rising sea levels, extreme weather

DOI: 10.4018/979-8-3373-1310-8.ch008

events, and biodiversity loss (IPCC, 2021). Despite international agreements like the Paris Accord, which aim to limit global warming and reduce greenhouse gas emissions, the implementation of climate policies often relies heavily on local governance. This reliance is particularly crucial as local governments are positioned to enact policies that reflect the specific needs and conditions of their communities. Local governments are uniquely equipped to tackle climate change through tailored solutions due to their proximity to the populations they serve. They control critical areas such as land use planning, transportation systems, and energy policies, making them pivotal in emission reduction and promoting sustainable practices (GSDRC, 2015). For instance, local authorities can implement zoning regulations that encourage green building practices or develop public transportation systems that reduce reliance on fossil fuels. However, despite their potential for impact, local governments face numerous challenges that hinder their effectiveness in addressing climate change.

Challenges Faced by Local Governments

One of the most significant barriers is limited financial resources. Many local governments operate on tight budgets that restrict their ability to invest in comprehensive climate initiatives (OECD). The financial constraints can severely limit the scope and scale of climate action plans, leading to inadequate responses to the pressing challenges posed by climate change. Furthermore, insufficient technical expertise can impede the development and implementation of effective climate strategies. Many local officials may lack the necessary training or resources to understand complex climate science or to design effective mitigation and adaptation plans. This gap in knowledge can result in poorly informed decisions that fail to address the underlying issues effectively. Political resistance also poses a challenge; some local leaders may prioritize short-term economic gains over long-term sustainability due to pressure from constituents or business interests (Bulkeley et al., 2019). This prioritization can lead to decisions that undermine environmental sustainability efforts, such as approving developments that increase emissions or degrade natural habitats.

Strategies for Empowering Local Governments

Addressing these barriers necessitates strategies that empower local governments and foster community engagement. Effective climate action requires a coordinated approach that aligns local initiatives with national and international goals. For example, central governments can support local efforts by providing funding mechanisms and technical assistance tailored to the unique challenges faced by municipalities (UNFCCC). Additionally, fostering collaboration between local governments and community organizations can enhance resource mobilization and ensure that climate actions are reflective of community needs. One effective strategy is the development of comprehensive local climate action plans that incorporate both mitigation and adaptation elements. Successful plans should be based on a thorough assessment of local vulnerabilities and should prioritize actions that enhance resilience while reducing emissions (GSDRC, 2015). For instance, a city might identify vulnerable neighbourhoods at risk of flooding and implement green infrastructure projects such as rain gardens or permeable pavements to manage stormwater effectively. Furthermore, incorporating indigenous knowledge alongside scientific approaches can create more robust adaptation strategies that foster community ownership and commitment (UNFCCC). Indigenous communities often possess valuable insights into sustainable land management practices honed over generations. By integrating these perspectives into formal planning processes, local governments can enhance their resilience strategies while respecting cultural heritage.

The Role of Community Engagement

Community engagement is vital for ensuring that climate action plans resonate with residents' needs and priorities. Engaging citizens through public consultations, workshops, and participatory planning processes fosters a sense of ownership over climate initiatives. When individuals feel involved in decision-making processes, they are more likely to support and participate in implementing those initiatives (IISD, 2020). Moreover, investing in social capital—defined as the networks and relationships within communities—can strengthen local governments' ability to respond effectively to climate change. Building trust among residents encourages collaboration on sustainability projects such as community gardens or neighbourhood clean-up events. These grassroots efforts not only contribute to environmental goals but also enhance social cohesion within communities.

Capacity Building for Local Governments

Building capacity within local governments is essential for enhancing their effectiveness in addressing climate change. Capacity building involves developing skills, knowledge, and resources necessary for implementing successful climate policies. Education and training programs tailored specifically for local officials can enhance understanding of climate challenges and solutions (IISD, 2020). Establishing dedicated units within local governments focused on climate action can bolster organizational capacity. These units can coordinate efforts across departments while ensuring that all relevant perspectives are considered in decision-making processes (OECD). Additionally, fostering inter-departmental collaboration enhances information sharing and promotes a culture of teamwork among staff members.

Collaboration Across Levels

Collaboration among various stakeholders is essential for effective climate governance. Local governments must work together with national authorities, civil society organizations, businesses, and community members to develop comprehensive strategies that address climate challenges. Creating multi-stakeholder platforms facilitates dialogue among diverse groups involved in climate action (UN DESA, 2024). These platforms serve as forums for sharing knowledge and resources while fostering mutual understanding among participants. Joint projects co-created through collaboration leverage diverse expertise from various sectors while ensuring stakeholder buy-in. Collaborative projects not only enhance the effectiveness of climate action but also foster a sense of ownership among participants (OECD). For example, partnerships between local governments and NGOs can lead to successful community-based adaptation initiatives that empower residents to take action against climate impacts.

Monitoring Progress

Monitoring progress is essential for ensuring accountability in climate governance. Establishing robust monitoring systems allows local governments to track their performance against established goals while providing data-driven insights into the effectiveness of interventions. Establishing baselines and indicators is fundamental for measuring progress in climate action initiatives (IISD, 2020). Baselines provide reference points against which changes can be assessed over time while indicators serve as measurable signs of progress toward specific goals. Engaging communities in evaluations through participatory approaches ensures inclusive insights that reflect the perspectives of those directly affected by climate initiatives. Participatory monitoring fosters ownership among stakeholders while enhancing the relevance of findings (GSDRC, 2015).

2. LITERATURE REVIEW

Climate governance encompasses the mechanisms and processes that guide societal responses to climate change. It aims to prevent, mitigate, and adapt to climate risks through transparent and inclusive decision-making. The significance of climate governance lies in its ability to create frameworks that align local initiatives with global climate goals, ensuring effective, resource-efficient, and accountable governance (UNDP, 2022). This alignment is crucial for mobilizing resources and fostering collaboration across different levels of government and society. The concept of climate governance has evolved to recognize the complexities of climate change as a multi-dimensional issue that requires coordinated action across various sectors and levels. Effective governance structures facilitate the integration of scientific knowledge into policy-making processes, enabling societies to respond more effectively to the challenges posed by climate change (GSDRC, 2015). By establishing clear roles and responsibilities among stakeholders, climate governance can enhance accountability and transparency in decision-making.

Multi-Level Governance Framework

Addressing climate change necessitates collaboration among local, national, and international stakeholders. Local governments are often at the forefront of implementing solutions tailored to their communities, while national authorities provide overarching frameworks that guide these efforts. International organizations play a critical role in fostering global coordination and setting ambitious targets for emissions reduction (UNICEF, 2020). The multi-level governance framework enhances problem-solving capacity and adaptability to climate challenges by promoting synergies among different actors. For instance, local governments can implement innovative practices based on their unique contexts while drawing on national policies for support and guidance (UNDP, 2022). This collaborative approach ensures that responses to climate change are comprehensive and inclusive, allowing for the integration of diverse perspectives.

Key Components of Governance

1. Policy Frameworks: Evidence-based strategies are essential for guiding climate responses. These frameworks should be adaptable and responsive to the evolving nature of climate science and its implications for society (IISD, 2020). Effective policy frameworks also consider socio-economic factors that influence vulnerability to climate impacts.
2. Institutional Arrangements: Dedicated bodies are necessary to coordinate stakeholder efforts effectively. These institutions should facilitate collaboration among various actors, including government agencies, non-governmental organizations (NGOs), and community groups (GSDRC, 2015). Establishing clear lines of communication can enhance information sharing and foster a culture of cooperation.
3. Stakeholder Engagement: Inclusive participation is critical for ensuring diverse perspectives are considered in decision-making processes. Engaging stakeholders from various sectors—such as business, civil society, and academia—can lead to more robust solutions that reflect the needs and priorities of different

communities (UNDP, 2022). Moreover, involving marginalized groups can help address social inequalities exacerbated by climate change.

4. Monitoring Mechanisms: Accountability frameworks are necessary to evaluate progress toward climate goals. Monitoring mechanisms should be designed to assess both the effectiveness of implemented policies and the overall impact on communities (IISD, 2020). Regular reporting can enhance transparency and build public trust in governance processes.

Challenges in Climate Governance

Despite the importance of effective climate governance, several challenges persist:

1. Fragmentation: Disjointed policies across governance levels hinder coordination efforts. The lack of coherence between local, national, and international policies can lead to inefficiencies and missed opportunities for collaboration (UNDP, 2022). Addressing fragmentation requires a concerted effort to align policies across different levels of government.
2. Resource Constraints: Financial limitations often restrict local governments' ability to implement comprehensive climate initiatives. Many municipalities struggle with budgetary constraints that prevent them from investing in necessary infrastructure or programs (OECD). Securing funding from national or international sources can help alleviate these constraints.
3. Political Will: Leadership commitment is often inconsistent across different levels of government. Political resistance can impede the implementation of ambitious climate policies or divert attention toward short-term economic gains rather than long-term sustainability (Bulkeley et al., 2019). Building political consensus around the urgency of climate action is essential for overcoming these barriers.

Policy Frameworks

Effective policy frameworks must be dynamic; they should evolve based on new scientific findings and socio-economic developments. Policymakers need access to up-to-date data on emissions trends, technological advancements in renewable energy, and adaptation strategies that have proven successful in other contexts (IISD, 2020). Furthermore, these frameworks must incorporate mechanisms for stakeholder feedback to ensure they remain relevant and effective over time. For instance, cities like San Francisco have implemented comprehensive sustainability plans that integrate community input through public workshops and surveys (City of San Francisco, 2021). This participatory approach not only enhances the legitimacy of policy decisions but also ensures that they reflect the priorities of residents.

Institutional Arrangements

Institutional arrangements must be designed not only for efficiency but also for inclusivity. This means creating spaces where diverse voices can contribute to policy development—especially those from marginalized communities who are often disproportionately affected by climate impacts (GSDRC, 2015). Collaborative platforms that bring together government representatives with NGOs, community leaders, and private sector actors can facilitate this inclusivity. For example, initiatives like the Global Covenant of Mayors for Climate & Energy provide a framework for local governments to collaborate on sustainability goals while sharing best practices across jurisdictions (Global Covenant of Mayors for Climate & Energy, n.d.). Such networks enhance institutional capacity by fostering collaboration among cities facing similar challenges.

Stakeholder Engagement

Engagement strategies should go beyond traditional public consultations; they must actively involve stakeholders in co-creating solutions. For example, participatory budgeting processes allow community members to decide how funds are allocated for sustainability projects within their neighbourhoods (UNDP, 2022). Such approaches empower residents while ensuring that initiatives are grounded in local realities. Moreover, engaging youth in climate discussions is critical as they represent future generations who will bear the brunt of current decisions. Programs aimed at educating young people about climate issues can foster a sense of responsibility and agency among future leaders (UNICEF, 2020).

3. ESTABLISHING GOVERNANCE STRUCTURES FOR CLIMATE ACTION

Cross-Functional Teams

Bringing diverse expertise together fosters innovation and enhances the effectiveness of climate governance. Cross-functional teams, which include urban planners, environmental scientists, community leaders, and other stakeholders, are essential for developing holistic solutions to climate challenges. These teams leverage the unique perspectives and skills of their members to create comprehensive strategies that address the multifaceted nature of climate change (UN DESA, 2024).

For instance, urban planners can contribute their knowledge of land use and infrastructure development, while environmental scientists can provide insights into ecological impacts and sustainability practices. Community leaders ensure that local knowledge and priorities are integrated into decision-making processes. This collaborative approach not only fosters innovation but also enhances the legitimacy and acceptance of climate policies among community members (OECD, 2024).

Moreover, cross-functional teams can facilitate the sharing of best practices and lessons learned from different sectors and regions. By collaborating across disciplines, these teams can identify synergies and potential trade-offs in climate action initiatives. For example, integrating green infrastructure into urban planning can mitigate flooding while enhancing biodiversity and improving public health (GSDRC, 2015). The involvement of multiple disciplines allows for a systems-thinking approach that addresses root causes rather than just symptoms, making climate governance more effective in the long run.

Additionally, cross-functional teams can improve policy adaptability. Climate change is dynamic, requiring constant reassessment and flexibility in governance approaches. A multidisciplinary team ensures that evolving scientific insights, technological advancements, and community needs are incorporated into climate action strategies. By integrating diverse perspectives, decision-making processes become more inclusive, increasing stakeholder buy-in and long-term commitment to sustainability initiatives.

Role Clarity

Defining roles within governance structures is crucial for enhancing accountability and ensuring effective implementation of climate policies. Clear role delineation helps to avoid overlaps and gaps in responsibilities among various stakeholders involved in climate action. Appointing dedicated climate leaders or coordinators within local governments can streamline implementation efforts and facilitate better communication among team members (UNDP, 2022).

Role clarity also extends to the relationships between different levels of government. Local authorities need to understand their responsibilities in relation to national policies and frameworks. This alignment ensures that local strategies are consistent with national goals while allowing for the flexibility necessary to address specific local conditions (OECD, 2024). For instance, in many countries, national governments provide guidelines for climate action that local governments must adapt to their contexts. By clearly defining these roles, local governments can better mobilize resources and engage stakeholders effectively.

Additionally, establishing performance metrics related to specific roles can enhance accountability within governance structures. By setting clear expectations and benchmarks for performance, local governments can monitor progress toward climate goals and make necessary adjustments to strategies as needed (IISD, 2020). This performance-based approach encourages continuous improvement and fosters a culture of accountability among stakeholders.

A well-defined governance structure also facilitates better collaboration between governmental agencies, private sector actors, and non-governmental organizations (NGOs). When roles are ambiguous, conflicts and inefficiencies may arise, delaying progress on climate initiatives. By ensuring that each entity understands its mandate, coordination improves, leading to more effective policy implementation and resource allocation.

Stakeholder Engagement

Transparent processes are fundamental for building trust among stakeholders involved in climate governance. Engaging the public through consultations and workshops ensures that policies reflect community priorities and values. Stakeholder engagement is not merely a procedural requirement; it is essential for fostering a sense of ownership over climate initiatives (UNDP, 2022).

Effective stakeholder engagement involves various methods, including public forums, surveys, focus groups, and participatory planning sessions. These platforms allow community members to voice their concerns, share their knowledge, and contribute ideas for climate action (OECD, 2024). For instance, cities like Portland have successfully implemented participatory budgeting processes that empower residents to decide how public funds are allocated for sustainability projects.

Moreover, engaging marginalized communities is critical for ensuring equity in climate governance. Vulnerable populations often bear the brunt of climate impacts yet have limited influence in decision-making processes. By actively involving these communities in discussions about climate policies, local governments can address social inequalities exacerbated by climate change (GSDRC, 2015).

Transparency in decision-making processes also enhances trust between local governments and citizens. Regular updates on progress toward climate goals, as well as clear communication about challenges faced along the way, foster a culture of openness that encourages ongoing engagement (IISD, 2020). This approach not only improves public support for climate initiatives but also ensures that policies are more effectively implemented by incorporating diverse perspectives and local knowledge.

Multi-Level Governance Integration

Aligning local strategies with national policies is essential for ensuring consistency in climate governance. Multi-level governance integration involves collaboration between local authorities and higher levels of government to create coherent frameworks for action (UN DESA, 2024). This alignment amplifies the impact of local actions by ensuring they contribute meaningfully to national and international climate goals.

National governments play a critical role in providing guidance and resources to support local initiatives. For example, national adaptation plans (NAPs) often outline specific actions that local governments should undertake to address climate vulnerabilities (OECD). By aligning their strategies with these broader frameworks, local authorities can secure funding opportunities and technical assistance from national programs.

Furthermore, multi-level governance integration fosters horizontal coordination among local governments within regions or networks. Collaborative efforts among municipalities can lead to shared resources and collective action on common challenges such as transportation emissions or waste management (GSDRC, 2015). For instance, regional partnerships can facilitate joint initiatives that leverage economies of scale while addressing specific regional needs.

In addition to vertical integration with national policies, horizontal collaboration among local governments enhances resilience against climate impacts. Intermunicipal cooperation allows neighbouring jurisdictions to share best practices and coordinate responses to shared risks such as flooding or heatwaves (OECD). This collaborative approach not only strengthens individual municipalities but also builds regional resilience.

Challenges in Establishing Governance Structures

Despite the importance of establishing effective governance structures for climate action, several challenges persist:

- Fragmentation: Disjointed policies across different levels of government can hinder coordination efforts. The lack of coherence between local initiatives and national frameworks may lead to inefficiencies or conflicting actions (UN DESA, 2024). Improved communication channels and policy alignment mechanisms are necessary to address this challenge.
- Resource Constraints: Many local governments face financial limitations that restrict their ability to implement comprehensive climate initiatives. Insufficient funding can impede the establishment of cross-functional teams

or stakeholder engagement processes (OECD). Securing international climate finance, public-private partnerships, and innovative financing mechanisms can help bridge this gap.

- Political Will: Inconsistent political commitment at various levels of government poses a significant challenge to effective governance structures. Political resistance may arise from competing interests or short-term economic considerations that overshadow long-term sustainability goals (Bulkeley et al., 2019). Political leadership and public advocacy are critical for maintaining momentum in climate governance.
- Capacity Building: Local governments may lack the technical expertise necessary for developing robust governance structures. Capacity-building initiatives are essential for equipping municipal staff with the skills needed for effective climate action planning and implementation (IISD, 2020). Knowledge-sharing platforms, training programs, and academic partnerships can enhance governance capacity.

4. ENGAGING COMMUNITIES IN CLIMATE ACTION

Awareness and Education

Effective community engagement in climate action begins with raising awareness and enhancing education about climate change. Tailored educational initiatives are essential for improving climate literacy, addressing specific demographic concerns, and promoting informed participation among community members. Many individuals still lack a comprehensive understanding of climate change and its implications for their daily lives, which can hinder their ability to engage meaningfully in climate action (UNDP, 2022).

Tailored Educational Initiatives

Educational programs should be designed to meet the unique needs of different demographic groups within the community. For instance, workshops can be organized for schools, local businesses, and community organizations to provide relevant information about climate impacts and sustainable practices. These initiatives can cover a range of topics, including renewable energy, waste reduction, sustainable agriculture, and biodiversity conservation (GSDRC, 2015). Interactive learning methods, such as hands-on activities and simulations, can enhance engagement and retention of information. For example, community members can participate in tree-planting events that not only educate them about the importance of trees in carbon sequestration but also allow them to take direct action (Eking International, 2024). Furthermore, utilizing local case studies can make the information more relatable and actionable.

Promoting Informed Participation

Informed participation is critical for effective climate governance. When community members understand the science behind climate change and its local impacts, they are more likely to engage in discussions and advocate for sustainable practices (IISD, 2020). Educational initiatives that incorporate local data on climate vulnerabilities can empower residents to identify specific issues affecting their communities and propose actionable solutions. Moreover, public forums and town hall meetings serve as platforms for disseminating information while encouraging dialogue among residents. These gatherings facilitate knowledge sharing and provide opportunities for individuals to voice their concerns and suggestions regarding climate policies (UN DESA, 2024). By fostering an environment of open communication, communities can build a collective understanding of climate challenges and collaboratively develop strategies to address them.

Grassroots Activism

Grassroots activism plays a vital role in engaging communities in climate action by empowering residents to take ownership of local initiatives. Encouraging local projects such as tree-planting campaigns or energy conservation programs fosters a sense of responsibility among community members while contributing to broader climate goals.

Local Initiatives

Grassroots movements often emerge from the recognition that local actions can have a significant impact on global climate change. For example, community-led tree-planting initiatives not only enhance urban green spaces but also contribute to carbon sequestration efforts (Eking International, 2024). Similarly, energy conservation campaigns can encourage residents to adopt sustainable practices such as reducing energy consumption or transitioning to renewable energy sources. Local initiatives also provide opportunities for residents to collaborate with one another and build strong social networks around shared environmental goals. This sense of community fosters collective action and encourages individuals to hold each other accountable for their commitments to sustainability (GSDRC, 2015).

Empowerment Through Participation

Grassroots activism empowers individuals by giving them a platform to express their concerns and take action on issues that matter most to them. When residents feel that their voices are heard and valued in the decision-making process, they are more likely to engage actively in climate initiatives (UNDP, 2022). Furthermore, grassroots movements often serve as catalysts for broader systemic change by advocating for policy reforms at higher levels of government. Successful local initiatives can inspire similar actions in neighbouring communities or even influence national policies aimed at addressing climate change (IISD, 2020).

Inclusivity

Inclusivity is a fundamental principle of effective community engagement in climate action. Involving marginalized voices ensures equitable policy-making that addresses the disproportionate impacts of climate change on vulnerable groups. Climate change does not affect all communities equally; marginalized populations often bear the brunt of its consequences due to existing social inequalities (Bulkeley et al., 2019).

Engaging Marginalized Communities

To ensure that all voices are heard in the decision-making process, it is essential to actively involve marginalized groups—such as low-income communities, indigenous peoples, and people with disabilities—in climate action discussions. This involvement can take various forms, including targeted outreach efforts that specifically engage these populations through workshops or focus groups designed to understand their unique challenges and perspectives (OECD). For instance, community engagement strategies should consider language barriers or cultural differences that may hinder participation from certain groups. Providing materials in multiple languages or incorporating culturally relevant examples can enhance accessibility and promote inclusivity (IISD, 2020).

Equitable Policy-Making

Inclusive decision-making processes lead to more equitable policies that address the needs of all community members. By incorporating diverse perspectives into climate action planning, local governments can develop strategies that prioritize social equity alongside environmental sustainability (GSDRC, 2015). Moreover, engaging marginalized communities in co-creating solutions fosters a sense of ownership over climate initiatives. When individuals feel empowered to contribute their insights and expertise, they are more likely to support the implementation of policies that reflect their values and priorities (UNDP, 2022).

Technology Utilization

In today's digital age, technology plays a crucial role in facilitating community engagement in climate action. Interactive tools and social media platforms can amplify engagement efforts by providing innovative ways for residents to connect with one another and share ideas.

Interactive Tools

Online platforms allow for real-time feedback on proposed policies or initiatives. Community members can participate in surveys or polls that gauge public opinion on specific climate actions. This data-driven approach not only informs decision-makers about community preferences but also encourages broader participation by making it easy for individuals to express their views (Commonplace, 2024). Additionally, technology can facilitate collaborative projects through platforms that enable residents to share resources or coordinate efforts on sustainability initiatives. For instance, neighbourhood apps can connect residents interested in starting community gardens or organizing clean-up events (Eking International, 2024).

Social Media Engagement

Social media serves as a powerful tool for raising awareness about climate issues while mobilizing communities for collective action. Campaigns on platforms like Instagram or Twitter can reach a wide audience quickly while fostering discussions around local climate challenges (IISD, 2020). Moreover, social media allows individuals to share their experiences with sustainability practices or showcase successful local initiatives. Highlighting these stories can inspire others within the community to take similar actions while reinforcing a culture of environmental stewardship.

Best Practices for Community Engagement in Climate Action

Effectively engaging communities in climate action requires strategic approaches that foster participation, inclusivity, and long-term commitment. The following best practices help ensure that climate initiatives are impactful and sustainable.

1. Tailor Educational Programs

Developing educational programs that meet the specific needs of different demographic groups enhances climate literacy and encourages meaningful participation.

- Demographic-Specific Content: Schools, businesses, and community centers should have customized learning materials addressing their unique needs. For example, schools may focus on teaching students about recycling and conservation, while businesses may receive training on reducing carbon emissions and sustainable supply chain practices.
- Localized Information: Communities respond better to climate education that directly connects with their experiences. A coastal town, for example, might benefit from information on rising sea levels and flood preparedness, whereas agricultural communities may require education on drought-resistant farming practices.
- Interactive Learning Approaches: Hands-on activities, such as tree-planting initiatives, clean-up drives, or climate simulations, enhance engagement and retention of information.

2. Utilize Diverse Engagement Methods

Different community members have varying levels of accessibility and preferences for engagement. Using multiple outreach methods ensures broader participation.

- Workshops and Town Halls: These in-person forums allow for direct discussions and real-time Q&A sessions. They help in explaining policies, gathering feedback, and fostering community involvement.
- Online Platforms: Virtual town halls, webinars, and educational videos expand access, especially for those unable to attend in-person meetings.
- Social media and Mobile Apps: Digital campaigns using platforms like Facebook, Instagram, and WhatsApp help spread awareness and mobilize community members. Mobile apps can provide updates on local sustainability initiatives and offer interactive ways to track individual contributions.

3. Foster Partnerships

Collaboration with local organizations strengthens engagement and expands the reach of climate initiatives.

- Schools and Universities: Educational institutions can integrate climate change topics into curricula and encourage students to participate in sustainability projects.

- Non-Governmental Organizations (NGOs): Partnering with environmental NGOs helps in organizing community projects, mobilizing volunteers, and accessing technical expertise.
- Local Businesses: Businesses can contribute by funding climate projects, promoting eco-friendly products, and reducing their environmental footprint.

4. Create Feedback Mechanisms

Ensuring that community members have a voice in climate governance builds trust and enhances participation.

- Surveys and Polls: Conducting surveys allows residents to share their priorities and concerns regarding climate policies and initiatives.
- Public Comment Periods: Governments and organizations should provide opportunities for feedback on proposed climate strategies.
- Community-Led Decision Making: Establishing local advisory groups helps ensure that community perspectives influence climate action plans.

5. Highlight Success Stories

Showcasing successful initiatives motivates communities to participate and fosters a sense of shared achievement.

- Media Coverage: Sharing stories through newspapers, social media, and local TV channels highlights positive environmental efforts.
- Recognition Programs: Awards for individuals and organizations that contribute significantly to sustainability efforts encourage continued participation.
- Community Events: Hosting events where local climate champions share their experiences inspires others to take action.

6. Ensure Accessibility

Removing barriers to participation ensures that climate initiatives are inclusive and equitable.

- Multilingual Resources: Providing educational materials and engagement platforms in multiple languages increases accessibility for diverse populations.
- Adapted Communication Strategies: Using sign language interpretation, braille materials, or simple visuals helps include individuals with disabilities.

- Flexible Participation Options: Offering both online and offline engagement opportunities ensures wider community involvement.

7. Encourage Collective Action

Fostering a culture of collaboration strengthens climate initiatives and builds community resilience.

- Neighbourhood Sustainability Projects: Community gardens, waste reduction drives, and renewable energy cooperatives bring residents together for shared environmental goals.
- Volunteer Programs: Engaging citizens in conservation efforts, such as river clean-ups or urban greening, fosters a sense of responsibility and long-term commitment.
- Policy Advocacy: Encouraging community members to engage in climate policy discussions at local, regional, and national levels strengthens collective efforts toward sustainability.

5. BUILDING CAPACITY FOR CLIMATE ACTION

Education and Training

Education and training are fundamental components of building capacity for effective climate governance. Programs designed for officials, leaders, and businesses enhance understanding of climate challenges and solutions, equipping stakeholders with the necessary knowledge to implement effective climate strategies.

Importance of Education in Climate Governance

The complexity of climate change necessitates a well-informed populace capable of understanding the scientific, social, and economic dimensions of this global challenge. Education plays a critical role in fostering climate literacy, which is essential for informed decision-making at all levels of governance (UNDP, 2022). For instance, training programs can cover topics such as greenhouse gas emissions, climate adaptation strategies, and sustainable practices in various sectors like agriculture, energy, and transportation. Targeted Training Programs Tailored educational initiatives can be designed to meet the specific needs of different stakeholders. For government officials, training can focus on policy formulation, implementation strategies, and the use of climate data for decision-making. Workshops can also be organized to enhance skills in stakeholder engagement and public communication regarding climate issues (GSDRC, 2015). For business leaders, training programs can emphasize the importance of corporate social responsibility (CSR) in relation to climate action. This includes educating businesses on sustainable practices that not only reduce their carbon footprint but also enhance their competitiveness in a rapidly changing market (OECD, 2024). Furthermore, educational initiatives can target local communities to raise awareness about climate change impacts and encourage grassroots participation in sustainability efforts. Community workshops can provide practical guidance on energy conservation techniques or waste management practices that contribute to local resilience against climate impacts (IISD, 2020).

Continuous Learning and Adaptation

Building capacity also requires a commitment to continuous learning and adaptation. Climate science is constantly evolving; therefore, ongoing education is necessary to keep stakeholders informed about the latest developments and best practices (UNFCCC). Establishing networks for knowledge sharing among professionals can facilitate this continuous learning process. For instance, online platforms can serve as repositories for educational resources, case studies, and best practices from various regions. These platforms can enable stakeholders to access relevant information and learn from each other's experiences in implementing climate action strategies (G20 Framework on Capacity Building).

Institutional Strengthening

Establishing dedicated units within organizations and fostering inter-departmental coordination are critical steps in bolstering organizational capacity for effective climate governance. Institutional strengthening involves creating frameworks that facilitate collaboration among various departments and agencies responsible for climate action.

Establishing Dedicated Units

Dedicated units or teams focused on climate change within governmental organizations can enhance accountability and streamline decision-making processes. These units can serve as focal points for coordinating climate-related initiatives across different sectors (UNFCCC). For example, a national climate change secretariat can oversee the implementation of national policies while ensuring alignment with international commitments such as the Paris Agreement. By concentrating expertise within dedicated units, organizations can develop specialized knowledge that informs policy development and implementation. These units can also facilitate training programs for staff members across different departments to ensure a comprehensive understanding of climate-related issues (IISD, 2020).

Fostering Inter-Departmental Coordination

Effective climate governance requires collaboration among various departments within an organization. Inter-departmental coordination enhances information sharing and ensures that policies are integrated across sectors. For instance, collaboration between departments responsible for environmental protection, urban planning, and public health can lead to more holistic approaches to addressing climate challenges (OECD). Establishing formal mechanisms for inter-departmental coordination—such as regular meetings or joint task forces—can promote collaboration and ensure that all relevant perspectives are considered in decision-making processes. This approach not only enhances organizational capacity but also fosters a culture of teamwork among staff members (GSDRC, 2015).

Knowledge Sharing

Knowledge sharing is vital for building capacity among stakeholders involved in climate governance. Collaborative platforms enable stakeholders to exchange best practices and co-develop innovative strategies that address shared challenges.

Collaborative Platforms

Creating platforms that facilitate knowledge sharing among government officials, businesses, civil society organizations, and academic institutions can enhance collective understanding of climate issues. These platforms can take various forms—such as workshops, conferences, online forums, or collaborative research projects—allowing stakeholders to share experiences and lessons learned from their respective contexts (UNDP, 2022). For example, regional networks can be established to connect municipalities facing similar climate challenges. These networks provide opportunities for local governments to collaborate on projects such as renewable energy initiatives or disaster preparedness plans (G20 Framework on Capacity Building). By sharing resources and expertise, participating municipalities can enhance their capacity to respond effectively to climate change.

Co-Development of Strategies

Knowledge sharing also enables stakeholders to co-develop innovative strategies tailored to local contexts. Collaborative approaches foster creativity by bringing together diverse perspectives that inform problem-solving processes (OECD). For instance, involving community members in the design of local adaptation plans ensures that strategies reflect the unique needs and priorities of those directly affected by climate impacts. Moreover, partnerships between academic institutions and local governments can facilitate research efforts aimed at understanding local vulnerabilities and identifying effective adaptation measures. Such collaborations not only enhance knowledge but also contribute to evidence-based decision-making processes (IISD, 2020).

Challenges in Building Capacity for Climate Action

Building capacity for effective climate governance is essential for fostering long-term resilience and sustainability. However, several challenges hinder the development and implementation of climate initiatives. Addressing these challenges requires strategic interventions, strong leadership, and collaborative efforts at local, national, and international levels. Below are some of the key challenges and potential solutions.

1. Resource Constraints

One of the most significant barriers to capacity building in climate action is the lack of financial and material resources. Many organizations, especially in developing countries, struggle with limited budgets, preventing them from establishing comprehensive climate education programs, investing in necessary infrastructure, or recruiting skilled personnel.

- Financial Limitations: Governments, NGOs, and community organizations often face budgetary constraints that limit their ability to implement climate-related policies and programs effectively. For instance, funding shortages may prevent the deployment of renewable energy projects or the development of climate adaptation strategies.
- Dependence on External Funding: Many climate initiatives rely on international donors, grants, or government subsidies. However, fluctuating funding availability can create inconsistencies in project implementation, making long-term sustainability difficult.

- Solution: Diversifying funding sources is crucial. Governments and organizations can seek partnerships with private-sector entities, apply for international climate funds (e.g., the Green Climate Fund), and encourage community-led fundraising efforts to secure the necessary financial support for capacity-building initiatives.

2. Lack of Technical Expertise

A shortage of technical knowledge and specialized skills presents another major challenge in building climate capacity. Many organizations lack adequately trained personnel to design, implement, and evaluate climate policies effectively.

- Skill Gaps: Technical knowledge in areas such as climate modelling, data collection, impact assessment, and policy development is often insufficient. This can hinder the ability of governments and institutions to develop science-based climate strategies.
- Limited Access to Training: Many local agencies and community-based organizations lack access to structured training programs that equip staff with the necessary expertise in climate governance, risk management, and sustainable development.
- Solution: Capacity-building efforts should prioritize continuous education and professional development. Governments and international organizations can facilitate training workshops, create knowledge-sharing networks, and provide online learning platforms to enhance the technical skills of climate professionals. Additionally, partnerships with universities and research institutions can help bridge expertise gaps.

3. Fragmented Governance Structures

The complexity of climate governance often results in fragmented decision-making processes, where multiple institutions and agencies operate in isolation rather than coordinating their efforts.

- Lack of Coordination: Climate action often involves multiple stakeholders, including local governments, national agencies, private sector actors, and civil society organizations. Poor communication and misalignment of goals can lead to inefficiencies and duplication of efforts.
- Policy Inconsistencies: Without a cohesive governance framework, climate policies may lack coherence, making it difficult to implement large-scale

sustainability projects effectively. Conflicting regulations between different government levels (local, regional, and national) can further hinder progress.

- Solution: Establishing clear communication channels and collaborative platforms is essential. Governments should implement integrated climate governance mechanisms, such as interagency working groups or regional climate councils, to improve coordination and information sharing among stakeholders. Digital tools, such as centralized data-sharing platforms, can also enhance collaboration.

4. Political Will

The effectiveness of climate capacity-building initiatives is heavily influenced by political commitment. However, inconsistent or weak political support can significantly hinder progress.

- Short-Term Political Agendas: Politicians often prioritize short-term economic gains over long-term climate action, particularly in regions where industries contributing to greenhouse gas emissions hold significant economic influence.
- Policy Reversals: Changes in government leadership may lead to shifts in climate priorities, resulting in policy reversals or delays in implementing existing climate strategies. This uncertainty can discourage investment in long-term capacity-building initiatives.
- Public Awareness and Engagement: In some cases, limited public awareness about climate change reduces political pressure for action, allowing policymakers to deprioritize sustainability efforts.
- Solution: Building a strong climate advocacy movement and increasing public engagement can help sustain political momentum. Civil society organizations, media outlets, and grassroots movements can play a crucial role in holding leaders accountable and ensuring that climate governance remains a priority across different administrations. Additionally, integrating climate considerations into national development strategies can help ensure long-term political commitment.

6. FOSTERING COLLABORATION IN CLIMATE ACTION

Collaboration is essential in addressing the multidimensional challenges of climate change. The interconnectedness of environmental, social, and economic systems means no single entity or sector can tackle these challenges in isolation. Governments, non-governmental organizations (NGOs), businesses, academics, and citizens must work together to develop and implement effective solutions. Collaboration fosters inclusivity, mutual learning, and the pooling of resources and expertise, enabling more robust and sustainable climate action.

This document expands on the roles and significance of Multi-Stakeholder Platforms (MSPs), Joint Projects, and Public-Private Partnerships (PPPs) as mechanisms for fostering collaboration in climate action.

Multi-Stakeholder Platforms (MSPs)

Multi-Stakeholder Platforms (MSPs) are organized, interactive forums that bring together a diverse range of stakeholders to engage in dialogue, decision-making, and collaborative implementation. Their primary aim is to address common challenges or achieve shared objectives, particularly in complex fields like climate action (CIFOR-ICRAF, 2024). Unlike traditional top-down governance, MSPs foster participatory agreements and consensus-building, enabling outcomes that are more equitable and effective.

MSPs are particularly valuable in the context of climate change because of the interconnectedness of its drivers and impacts. Environmental degradation, social inequalities, and economic vulnerabilities must be addressed together, and MSPs create structured environments for cross-sectoral collaboration. By facilitating the exchange of knowledge, resources, and best practices, they also build trust and mutual understanding (UNDP, 2022).

Characteristics of Effective MSPs

To achieve their potential, MSPs must be designed and managed effectively. The following characteristics are critical for success:

1. Clear Objectives

MSPs must have a well-defined mandate and clear objectives. These ensure that participants understand the platform's purpose and their roles within it. Clear objectives also provide a foundation for measuring progress and outcomes (CIFOR-ICRAF, 2024).

2. Equitable Representation

Inclusivity is key to the success of MSPs. Balanced representation ensures that diverse perspectives are incorporated, particularly those of marginalized or underrepresented groups. Such inclusivity enhances the legitimacy and fairness of decision-making processes (GSDRC, 2015).

3. Facilitated Dialogue

MSPs require skilled facilitators to manage discussions, navigate power dynamics, and ensure all voices are heard. Effective facilitation fosters open communication, minimizes conflicts, and enhances the quality of collaborative decision-making (IISD, 2020).

4. Adaptive Learning

MSPs must incorporate mechanisms for reflection and learning. Stakeholders should be encouraged to evaluate past experiences and adapt their strategies in response to new challenges or changing circumstances (UNFCCC, 2022).

5. Sustainability

The long-term viability of MSPs depends on sustained commitment from stakeholders and adequate funding. Building institutional capacity within participating organizations ensures that MSPs remain functional and effective over time (UNDP, 2022).

Examples of MSP Success in Climate Action

1. The Forest Stewardship Council (FSC):

FSC brings together environmental NGOs, businesses, and indigenous communities to develop sustainable forestry standards. This MSP has successfully promoted responsible forest management globally.

2. The NDC Partnership:

This global coalition supports countries in achieving their Nationally Determined Contributions (NDCs) under the Paris Agreement. It unites governments, international organizations, and civil society to provide technical and financial support.

Joint Projects

Overview and Importance

Joint projects are collaborative initiatives that bring together various stakeholders to address specific climate-related issues. These projects leverage the expertise, resources, and networks of multiple sectors, fostering innovation and ensuring broader buy-in for solutions.

Benefits of Joint Projects

1. Diverse Expertise By involving multiple stakeholders, joint projects can draw on a wide range of perspectives. For example, combining scientific research with indigenous knowledge often leads to innovative and context-sensitive solutions (OECD, 2022).
2. Increased Credibility Collaborative initiatives carry greater credibility than those led by a single entity. Multi-stakeholder involvement signals a unified and balanced approach, which can increase public trust and stakeholder confidence (GSDRC, 2015).
3. Shared Responsibility Joint projects promote accountability by distributing responsibilities among participants. Shared ownership of outcomes fosters greater commitment to project success (IISD, 2020).
4. Enhanced Resource Mobilization Collaborative projects often attract funding from diverse sources, including government grants, private investments, and international aid. Unified efforts are more appealing to funders seeking high-impact climate solutions (UNDP, 2022).

Examples of Successful Joint Projects

1. The Global Environment Facility (GEF):

GEF funds initiatives that address global environmental challenges. It has supported projects that combine government, NGO, and community efforts to promote sustainable practices.

2. Sustainable Energy for All (SE4ALL):

This initiative unites governments, businesses, and civil society to promote universal access to sustainable energy. It has catalyzed investments in renewable energy and energy efficiency projects worldwide.

3. Climate Resilience Projects in Small Island States:

Joint projects in countries like Fiji and the Maldives have brought together governments, NGOs, and the private sector to build resilient infrastructure, restore ecosystems, and prepare communities for climate impacts.

Public-Private Partnerships (PPPs)

The Role of PPPs in Climate Action

Public-Private Partnerships (PPPs) are strategic collaborations between government entities and private sector actors. They leverage the strengths of both sectors—government's regulatory authority and public service focus, paired with the private sector's efficiency, innovation, and investment capacity.

PPPs play a crucial role in advancing climate action by mobilizing resources, deploying innovative technologies, and building sustainable infrastructure.

Advantages of PPPs

1. Infrastructure Development Many governments lack the financial capacity to undertake large-scale climate infrastructure projects. PPPs enable the development of renewable energy facilities, resilient public transport systems, and waste management facilities by attracting private investment (OECD, 2022).
2. Technology Transfer Collaborations with private companies grant governments access to cutting-edge technologies that improve efficiency and reduce greenhouse gas emissions. For example, partnerships with tech firms can enhance solar energy systems or water resource management (UNFCCC, 2022).
3. Risk Sharing PPPs distribute financial and operational risks between the public and private sectors. This arrangement encourages both parties to pursue innovative solutions while mitigating potential losses (GSDRC, 2015).
4. Capacity Building Private sector partners often provide training and knowledge-sharing opportunities, equipping public sector personnel with the skills and tools needed for sustainable development (IISD, 2020).

Challenges Faced by PPPs

Despite their potential, PPPs face several challenges:

1. Alignment of Interests The public and private sectors often have different priorities. Governments focus on social and environmental outcomes, while private

companies seek profitability. Clear agreements that align these interests are essential (OECD, 2022).

2. Transparency Issues Transparency is critical for building trust in PPPs. Lack of clarity in decision-making processes can lead to skepticism from stakeholders and the public (UNDP, 2022).
3. Regulatory Frameworks Governments must establish supportive regulatory environments that facilitate PPPs while ensuring compliance with environmental and social standards. Weak regulations can undermine the effectiveness of partnerships (GSDRC, 2015).

Examples of PPP Success

1. Masdar City in the UAE:

This partnership between the government and private sector aims to build a zero-carbon city powered by renewable energy. The project demonstrates how PPPs can integrate innovation and sustainability.

2. RE100 Initiative:

This global collaboration involves corporations committing to 100% renewable energy usage. Governments and private companies work together to promote renewable energy adoption.

3. Clean Energy Investments in India:

Partnerships between the Indian government and private investors have accelerated the deployment of solar and wind energy projects, contributing significantly to the country's renewable energy targets.

Recommendations for Advancing Collaboration in Climate Action

To maximize the impact of collaborative approaches like Multi-Stakeholder Partnerships (MSPs), Joint Projects, and Public-Private Partnerships (PPPs), the following recommendations should be considered:

1. Promote Inclusivity: Ensure equitable representation of marginalized groups and vulnerable communities in all collaborative initiatives. Actively engage underrepresented voices through targeted outreach, capacity-building programs,

and participatory decision-making processes. Prioritizing inclusivity strengthens the legitimacy and effectiveness of climate solutions by incorporating diverse perspectives and local knowledge.

2. Establish Clear Objectives and Metrics: Define measurable goals and track progress to enhance accountability. Setting clear timelines, benchmarks, and key performance indicators (KPIs) helps stakeholders evaluate the success of climate initiatives. Regular progress assessments and publicly available reports ensure continuous monitoring and encourage improvements in implementation strategies.
3. Foster Transparency: Build trust by maintaining transparency in decision-making and resource allocation. Open communication channels, public disclosures, and stakeholder consultations help prevent conflicts of interest and ensure fairness. Transparency also enhances public confidence and increases stakeholder participation in climate action initiatives.
4. Strengthen Capacity-Building Efforts: Invest in training and knowledge-sharing to empower all stakeholders. Providing access to technical expertise, workshops, and educational programs helps communities and organizations implement effective climate strategies. Strengthening local capacities ensures long-term sustainability and promotes innovation in addressing climate challenges.
5. Encourage Adaptive Learning: Create feedback mechanisms to incorporate lessons learned and respond to emerging challenges. Establishing flexible, data-driven approaches allows organizations to refine strategies based on real-world experiences. Encouraging adaptive learning fosters resilience, ensuring climate action initiatives remain relevant and effective in a rapidly changing environment.

7. ENSURING ACCOUNTABILITY IN CLIMATE ACTION

Accountability is a fundamental principle in climate governance, forming the backbone of effective and transparent climate action. By systematically tracking progress, verifying results, and ensuring compliance with commitments, monitoring and evaluation (M&E) frameworks play a vital role in holding stakeholders accountable. The importance of accountability spans across multiple dimensions: fostering trust, improving governance, ensuring compliance, and driving continuous improvement. Below, we delve into the multifaceted significance of accountability in climate action and explore mechanisms to strengthen it.

Building Trust Among Stakeholders

Accountability fosters trust among the diverse stakeholders involved in climate action, including governments, non-governmental organizations (NGOs), private enterprises, and local communities. Transparent reporting and monitoring mechanisms enable stakeholders to see tangible progress toward climate goals, which can strengthen public confidence in climate initiatives. For example, clear and reliable reporting of carbon emissions reductions or progress on renewable energy targets reassures citizens and investors that their efforts and resources are yielding measurable results.

The Paris Agreement provides a prominent example of how accountability fosters trust on a global scale. By requiring countries to submit their Nationally Determined Contributions (NDCs) and report on their progress through a transparent framework, the agreement holds nations accountable for their commitments. This transparency not only builds trust between nations but also encourages greater ambition by allowing countries to learn from each other's successes and challenges.

Strengthening Governance Structures

Effective governance is impossible without accountability. In the realm of climate action, accountability ensures that resources are used efficiently and effectively, preventing mismanagement or misuse of funds intended for mitigation and adaptation efforts. Regular monitoring of projects and initiatives helps identify discrepancies, inefficiencies, or delays early, enabling corrective actions to be taken before they escalate into larger issues.

For example, climate finance mechanisms like the Green Climate Fund (GCF) emphasize robust accountability frameworks to ensure that funds are allocated to projects with the highest impact. Implementing stringent monitoring protocols and independent evaluations enhances the credibility of such funds, increasing their attractiveness to contributors and beneficiaries alike.

Encouraging Compliance with International and National Commitments

Accountability mechanisms are essential for ensuring compliance with international and national climate agreements. Without proper monitoring and reporting systems, commitments risk being reduced to mere rhetoric. For instance, the Enhanced Transparency Framework (ETF) under the Paris Agreement establishes a system for monitoring, reporting, and verifying countries' progress toward their climate commitments. This framework not only ensures compliance but also allows countries to identify gaps in their strategies and adjust their policies accordingly.

At the national level, governments can use M&E frameworks to ensure that subnational entities, such as states or municipalities, comply with climate policies. For example, India's Perform, Achieve, and Trade (PAT) scheme monitors the energy efficiency of industries and holds them accountable for meeting specified energy-saving targets. Non-compliance results in penalties, incentivizing adherence to commitments.

Documenting Successes and Challenges for Continuous Improvement

Accountability mechanisms facilitate a culture of continuous improvement by documenting both successes and challenges encountered during climate action. This documentation serves as a valuable resource for stakeholders to refine their approaches and learn from past experiences. For instance, if a renewable energy project fails to meet its targets due to unforeseen technical challenges, a thorough evaluation can identify the root causes and recommend solutions for future projects.

Moreover, accountability frameworks encourage stakeholders to share their lessons learned with others, fostering collective learning. Platforms like the UNFCCC's Climate Change Knowledge Portal enable stakeholders to exchange experiences and best practices, promoting innovation and collaboration in climate action.

Enhancing Public Participation and Social Equity

Accountability in climate action extends beyond institutional mechanisms; it also involves engaging the public in holding stakeholders responsible. Participatory M&E approaches empower communities to monitor and evaluate the implementation of climate initiatives, ensuring that they address local needs and priorities. This engagement not only enhances the effectiveness of interventions but also promotes social equity by giving marginalized groups a voice in decision-making processes.

For example, involving Indigenous communities in monitoring forest conservation projects ensures that these initiatives respect traditional knowledge and rights. Such participatory accountability mechanisms can prevent conflicts and promote more equitable outcomes.

Tools and Mechanisms for Accountability

To ensure accountability in climate action, a range of tools and mechanisms can be employed. These include:

1. Nationally Determined Contributions (NDCs): Countries submit their climate action plans under the Paris Agreement, outlining their targets and strategies. Regular reporting and review of these plans ensure accountability at the national and international levels.
2. Performance Metrics and Indicators: Establishing clear, measurable indicators helps track progress toward climate goals. For example, indicators such as reductions in greenhouse gas emissions, increases in renewable energy capacity, or improvements in climate resilience provide tangible measures of success.
3. Independent Audits and Evaluations: External evaluations by independent entities enhance the credibility of accountability mechanisms. These audits assess whether resources are being used effectively and whether objectives are being met.
4. Public Reporting and Communication: Transparent communication of progress and results ensures that stakeholders remain informed and engaged. Governments and organizations can publish annual reports, host public consultations, or use digital platforms to disseminate information.
5. Legal and Regulatory Frameworks: Binding legal agreements and regulations provide a robust foundation for accountability. For instance, carbon pricing mechanisms, such as cap-and-trade systems or carbon taxes, hold emitters accountable for their emissions.

Addressing Challenges to Accountability

While accountability is crucial for effective climate action, several challenges must be addressed to strengthen it:

1. Data Availability and Quality: Reliable data is essential for tracking progress and ensuring accountability. However, many countries and organizations face challenges in collecting and analysing high-quality data due to resource constraints or technical limitations. Investments in capacity-building and technology can help overcome this challenge.
2. Fragmentation of Efforts: The involvement of multiple stakeholders in climate action can lead to fragmented efforts and overlapping responsibilities. Clear roles and coordination mechanisms are needed to ensure accountability across different levels of governance.
3. Political and Institutional Barriers: Political will and institutional capacity are critical for enforcing accountability. Weak governance structures or lack of political commitment can undermine accountability mechanisms. Strengthening institutions and fostering political buy-in are essential for addressing this challenge.

4. Equity and Inclusivity: Ensuring accountability requires addressing power imbalances and promoting inclusivity. Marginalized groups often face barriers to participation in monitoring and evaluation processes. Inclusive approaches that engage diverse stakeholders are crucial for promoting social equity.

The Way Forward

Strengthening accountability in climate action requires a multi-pronged approach that combines robust M&E frameworks, stakeholder engagement, and transparent governance. By holding stakeholders accountable for their commitments, we can ensure that resources are used effectively, progress is tracked systematically, and climate goals are achieved equitably.

As the urgency to address climate change intensifies, the role of accountability in driving effective and sustainable action cannot be overstated. By embedding accountability into every aspect of climate governance, we can build a more resilient and equitable future for all.

8. CONCLUSION

Empowering local governance is central to addressing climate change effectively. As the impacts of climate change become increasingly severe and widespread, local governments are uniquely positioned to implement tailored solutions that reflect the specific needs and priorities of their communities. By combining clear governance structures, community engagement, capacity building, collaboration, and robust monitoring, local governments can lead the way in climate-responsive governance. This multifaceted approach not only mitigates environmental risks but also fosters resilience, equity, and sustainable development.

Local governments play a critical role in climate action for several reasons. First, they are closest to the communities they serve, allowing them to understand local vulnerabilities and strengths better than higher levels of government. This proximity enables local authorities to develop and implement policies that are contextually relevant and culturally appropriate (GSDRC, 2015). For instance, a city facing rising sea levels can tailor its adaptation strategies based on its unique geographical and socio-economic conditions. Moreover, local governments often have direct control over key areas such as land use planning, transportation systems, waste management, and energy policies—critical sectors that contribute significantly to greenhouse gas emissions (UNDP, 2022). By leveraging their authority in these areas, local governments can enact policies that promote sustainability and reduce emissions.

For example, implementing green building codes or enhancing public transportation systems can lead to significant reductions in urban carbon footprints.

Establishing clear governance structures is essential for effective climate action at the local level. This involves defining roles and responsibilities within local governments and ensuring that there are dedicated units or teams focused on climate initiatives. A well-structured governance framework facilitates coordination among various departments and stakeholders, enabling a more integrated approach to climate action (OECD, 2021). Effective multilevel governance is also vital for empowering local climate action. It requires vertical integration of national policies with local actions and horizontal coordination among different levels of government (UN DESA, 2024). By aligning local strategies with national objectives, local governments can ensure consistency in their efforts while maximizing the impact of their initiatives.

Engaging communities in the decision-making process is crucial for fostering ownership and commitment to climate action. Local governments can enhance community engagement through various methods such as public consultations, workshops, and participatory planning processes. When residents are actively involved in shaping policies that affect their lives, they are more likely to support and participate in implementation efforts (IISD, 2020). Investing in social capital—defined as the networks, relationships, and trust within communities—strengthens local governments' ability to respond effectively to climate change. By fostering strong connections among residents through community-led initiatives such as tree planting or energy conservation campaigns, local authorities empower citizens to take ownership of their environment while contributing to broader sustainability goals (GSDRC, 2015).

Building capacity within local governments is essential for enhancing their effectiveness in addressing climate change. Capacity building involves developing the skills, knowledge, and resources necessary for implementing successful climate policies. Education and training programs tailored to meet the specific needs of different stakeholders can enhance understanding of climate challenges and solutions (IISD, 2020). Establishing dedicated units within local governments focused on climate action can bolster organizational capacity. These units can coordinate efforts across departments while ensuring that all relevant perspectives are considered in decision-making processes (OECD). Additionally, fostering inter-departmental collaboration enhances information sharing and promotes a culture of teamwork among staff members.

Collaboration among various stakeholders is essential for effective climate governance. Local governments must work together with national authorities, civil society organizations, businesses, and community members to develop comprehensive strategies that address climate challenges. Creating multi-stakeholder platforms facilitates dialogue among diverse groups involved in climate action (UN DESA, 2024). These platforms serve as forums for sharing knowledge and resources while fostering mutual understanding among participants. Joint projects co-created through collaboration leverage diverse expertise from various sectors while ensuring stakeholder buy-in. Collaborative projects enhance the effectiveness of climate action while fostering a sense of ownership among participants (OECD). For example, partnerships between local governments and NGOs can lead to successful community-based adaptation initiatives that empower residents to take action against climate impacts.

Monitoring progress is essential for ensuring accountability in climate governance. Establishing robust monitoring systems allows local governments to track their performance against established goals while providing data-driven insights into the effectiveness of interventions. Establishing baselines and indicators is fundamental for measuring progress in climate action initiatives (IISD, 2020). Baselines provide reference points against which changes can be assessed over time while indicators serve as measurable signs of progress toward specific goals. Engaging communities in evaluations through participatory approaches ensures inclusive insights that reflect the perspectives of those directly affected by climate initiatives. Participatory monitoring fosters ownership among stakeholders while enhancing the relevance of findings (GSDRC, 2015).

While empowering local governance is crucial for effective climate action, several challenges must be addressed:

1. Resource Constraints: Many local governments face financial limitations that restrict their ability to implement comprehensive climate initiatives (OECD). Securing funding from national or international sources is critical for overcoming these constraints.
2. Political Will: Inconsistent political commitment at various levels of government may hinder the implementation of ambitious climate policies (Bulkeley et al., 2019). Building consensus around the urgency of addressing climate change is vital for securing support from political leaders.
3. Long Time Horizons: Climate change impacts often unfold over long time horizons, making it challenging to assess immediate outcomes from interventions (IIED). To address this challenge, monitoring frameworks should incorporate intermediate indicators that capture early signs of progress.

Investing in innovative and inclusive practices will pave the way for a sustainable future where local governments emerge as key drivers in the global fight against climate change. This investment includes not only financial resources but also time spent on building relationships within communities and fostering collaborative networks among stakeholders. Local governments must prioritize integrating climate considerations into all aspects of planning and development processes. This integration ensures that sustainability becomes a core value rather than an afterthought in decision-making frameworks. Furthermore, fostering knowledge exchange among municipalities can enhance collective learning while promoting best practices across different contexts (UNDP, 2022). By sharing experiences related to successful interventions or lessons learned from failures—local authorities can build resilience against future challenges posed by a changing climate.

REFERENCES

Bulkeley, H., (2019). *Local Governments and Climate Change: A Discussion Note*. GSDRC.

GSDRC. (2015). Local Governments and Climate Change. Retrieved from https://gsdrc.org/document-library/local-governments-and-climate-change/

ISD. (2020). Global Climate Change Governance: The Search for Effectiveness. Retrieved from https://www.iisd.org/articles/deep-dive/global-climate-change-governance-search-effectiveness-and-universality

IISD. (2020). Global Climate Change Governance: The Search for Effectiveness. Retrieved from https://www.iisd.org/articles/deep-dive/global-climate-change-governance-search-effectiveness-and-universality

Intergovernmental Panel on Climate Change (IPCC). (2021). [The Physical Science Basis.]. *Climatic Change*, 2021.

OECD. (2024). Climate Adaptation: Why Local Governments Cannot Do It Alone | OECD Policy Paper.

OECD. (n.d.). Climate Adaptation: Why Local Governments Cannot Do It Alone. Retrieved from https://www.oecd.org/en/publications/climate-adaptation-why-local-governments-cannot-do-it-alone_be90ac30-en.html

UN DESA. (2024). Multilevel Governance for Climate Change Mitigation: Effective Policy Frameworks at All Levels.

UNDP. (2022). What is Climate Governance? Retrieved from https://www.undp.org/sites/g/files/zskgke326/files/2022-09/Volume%20III%20What%20is%20climate%20governance.pdf

UNDP. (2022). What is Climate Governance? Retrieved from https://www.undp.org/sites/g/files/zskgke326/files/2022-09/Volume%20III%20What%20is%20climate%20governance.pdf

UNFCCC. (n.d.). Climate Change Adaptation Strategies for Local Impact. Retrieved from https://unfccc.int/resource/docs/2009/smsn/igo/054.pdf

UNICEF. (2020). What is Climate Governance? Retrieved from https://www.unicef.org/lac/en/reports/what-climate-governance

Chapter 9
Environmental Taxes and Carbon Footprint:
A Panel Causality Analysis of Selected EU Countries

Associate Professor Aytül Bişgin
https://orcid.org/0000-0003-2488-3541
Karamanoglu Mehmetbey University, Turkey

ABSTRACT

This study examines the causal relationship between environmental taxes and carbon footprint in the context of climate change mitigation, using data from ten major EU economies (Austria, Belgium, France, Germany, Ireland, Italy, Netherlands, Poland, Spain, Sweden) over 1994–2022. Employing panel causality analysis, the study first applies tests for cross-sectional dependence, slope heterogeneity, and panel unit roots. The Dumitrescu and Hurlin test results indicate a significant causality from environmental taxes to per capita carbon emissions in at least one country. The findings suggest that well-designed environmental taxes can play a critical role in reducing emissions. The study highlights the importance of revenue allocation and policy coherence and recommends future research on the effect of such taxes on ecological footprint components for more comprehensive sustainability assessment.

INTRODUCTION

Global warming is defined as the increase in the temperature of the earth's surface as a result of the greenhouse effect of gases released into the atmosphere by humans (Çokgezen, 2007:101). Today, countries face serious environmental challenges that threaten their sustainable development goals. The most prominent of these

DOI: 10.4018/979-8-3373-1310-8.ch009

is climate change. The problem of global warming and climate change concerns different societies and countries in spatial terms and in terms of time; it confronts present and future generations and has an intergenerational character (Doğan & Tüzer, 2011:158-159). Problems such as climate change, loss of biodiversity, and air and water pollution complicate the sustainable use of natural resources and put pressure on socioeconomic systems. The increase in these problems on a global scale has brought along the efforts of countries to strengthen their environmental management capacity. In recent years, with the rise in environmental awareness, governments have turned to sustainable policy instruments; practices such as adopting environmentally friendly technologies, developing green financing mechanisms, and tightening environmental regulations have come to the fore. As Karmaker et al. (2021) emphasize, the willingness and capacity to deal with environmental challenges has increased significantly, especially in developing countries. This shows that environmental problems are local and common issues that require global cooperation, and solutions are developed through multi-stakeholder approaches.

To mitigate the effects of greenhouse gas emissions, national economies have turned to practices such as environmental taxes, carbon capture and storage, carbon pricing, and carbon trading. Although these practices are seen as a potential option to prevent environmental degradation, the most widely used one is the environmental tax (Kılıç & Altıparmak, 2020). Environmental taxes, one of the most important economic instruments that can have a direct impact on preventing environmental pollution, aim to direct individuals and businesses towards more environmentally friendly choices by costing environmentally damaging activities. Environmental taxes ensure the integration of environmental costs into the economic system in line with the polluter pays principle (OECD, 2011; Tietenberg & Lewis, 2016). In this way, it is aimed to prevent environmental degradation and encourage environmentally friendly production and consumption behaviors. Many countries, especially industrialized states, widely apply environmental taxes to reduce carbon emissions, limit the consumption of natural resources, and support sustainable development policies. Carbon taxes, taxes on energy consumption, waste management taxes, and vehicle emission taxes are some of the prominent examples. By reflecting the actual cost of polluting activities, these taxes both correct the market mechanism and contribute to reducing environmental damage (Ekins, 2000). Moreover, environmental taxes are gaining importance as a regulatory element and a new public financing instrument. As the need to generate additional resources for public budgets increases in the fight against global emissions growth, environmental taxes have the potential to meet this need. Revenues can be used to finance environmental projects, renewable energy investments, or social support mechanisms, thus creating a holistic structure that serves both environmental and economic sustainability (OECD, 2011; IMF, 2019).

The agreements signed for environmental protection at the international level have been shaped to find solutions to ecological problems that have emerged at various periods throughout history and to support sustainable development goals. In this context, the first comprehensive global step on the environment was taken with the United Nations Stockholm Conference in 1972. This conference brought the understanding that environmental problems are international and require the common responsibility of all states to the global public opinion and paved the way for environmental awareness to be reflected in state policies. The Stockholm Declaration, published at the end of the conference, established the institutional basis for international cooperation in environmental protection and served as a guideline for subsequent initiatives in this field (Stockholm Conference, 1972). In the following period, remarkable steps were taken to address, particularly atmospheric environmental problems, the most important of which was the Montreal Protocol adopted in 1987. This protocol, which aims to gradually limit the production and consumption of harmful chemicals such as chlorofluorocarbons (CFCs) that cause ozone depletion, is recognized as one of the most successful global agreements for environmental sustainability. The Montreal Protocol has set a precedent for international environmental law thanks to its broad participation and implementation efficiency (Montreal Protocol, 1987). The 1992 United Nations Conference on Environment and Development, also known as the Rio Summit, emphasized the need to balance environmental protection and economic development. The United Nations Framework Convention on Climate Change (UNFCCC), signed as a result of this summit, formalized states' commitment to reduce greenhouse gas emissions and provided a basic framework for combating climate change (UNFCCC, 1992).

The Kyoto Protocol is the first binding international agreement to reduce greenhouse gas emissions, adopted in 1997 under the United Nations Framework Convention on Climate Change (UNFCCC) and entered into force in 2005. The agreement's main objective is to ensure that industrialized countries and transition economies reduce their greenhouse gas emissions by an average of 5.2% from 1990 levels between 2008 and 2012 (UNFCCC, 1998). The Protocol is based on the historical dimension of environmental responsibilities by targeting developed countries responsible for a significant portion of global carbon emissions. Accordingly, the ratification of the protocol by the countries responsible for at least 55% of the total global CO_2 emissions and their entry into the scope of the agreement were determined as a prerequisite for the protocol to enter into force (Bodansky, 2010). The Kyoto Protocol also includes market-based instruments such as carbon trading, clean development mechanism (CDM), and joint implementation (JI), known as flexibility mechanisms, thus enabling emission reductions to be achieved cost-effectively (Grubb, Vrolijk, & Brack, 1999). Although some major emitters (e.g., the United States) remained outside the protocol, the Kyoto Protocol was a landmark in the history of climate diplomacy, paving the way for subsequent international climate regimes such as the Paris Agreement (Rajamani, 2011).

The European Union (EU) has developed comprehensive policy mechanisms to achieve its sustainable development goals and play a leading role in tackling environmental threats. One of the most critical steps in this direction was the European Green Deal announced by the European Commission on December 11, 2019. With this policy package, the EU aims to become the world's first carbon-neutral continent by reducing net greenhouse gas emissions to zero by 2050 (European Commission, 2019). The Green Deal envisages structural transformations such as the transition to sustainable energy systems, circular economy-based production models, expansion of carbon pricing mechanisms, and promotion of nature-based solutions. One of the most recent steps supporting this process is the Carbon Neutrality Agreement for 2026. This agreement, adopted by EU member states in 2026, made it legally binding to achieve a 55% reduction in greenhouse gas emissions by 2030 and introduced the carbon border adjustment mechanism (CBAM) more comprehensively. The agreement also strengthened tax regulations to incentivize environmentally friendly production and mandated the pricing of carbon content in imported products to prevent carbon leakage. Thus, the EU has moved to a position of incentivizing carbon reduction not only in its domestic market but also in global supply chains. Indeed, as Karmaker et al. (2021) point out, the willingness and capacity of countries to deal with environmental challenges are increasing worldwide, and the EU stands out as a leading actor that embodies this capacity through legal, fiscal, and institutional instruments.

In this study, the impact of environmental taxes imposed on the 10 most prominent EU members in terms of GDP size (Austria, Belgium, France, Germany, Ireland, Italy, Netherlands, Poland, Spain, and Sweden) on carbon emissions in these countries is analyzed by panel causality analysis. The study starts with an introduction to the historical development of climate change and environmental policies. In the second part, the conceptual framework and indicators of carbon footprint are presented. The literature section presents national and international studies on this subject. The general situation of research in this field is given by comparing the findings and methodologies of these studies. The study, which continues with methods and findings, is concluded with a conclusion.

HISTORICAL DEVELOPMENT OF CARBON FOOTPRINT DATA OF SOME SELECTED EU MEMBER STATES

Carbon emission refers to the amount of carbon dioxide (CO_2) released into the atmosphere due to anthropogenic processes such as industrial activities, transportation, agriculture, and deforestation, particularly the combustion of fossil fuels (coal, oil, and natural gas). This gas is one of the main components contributing to global warming and climate change by creating a greenhouse effect in the atmosphere (IPCC, 2014). As carbon emissions are the primary determinant of long-term temperature increases, they are at the center of sustainable environmental policies. Increasing carbon concentration in the atmosphere affects temperature levels, leading to rising sea levels, increased frequency of extreme weather events, and ecosystem imbalances (NASA, 2020).

The negative environmental impacts of carbon emissions have forced countries to make their environmental policies more stringent and structural. Developed and developing countries have implemented national and international regulations to reduce greenhouse gas emissions. Global initiatives such as the Kyoto Protocol and the Paris Agreement encourage the reduction of carbon emissions through binding or voluntary targets. They are supported by carbon trading, carbon taxes, and renewable energy subsidies (UNFCCC, 2015). In addition, many countries are developing low-carbon strategies to reduce emissions intensity in the industry, energy, and transportation sectors and integrating these policies into legal frameworks (OECD, 2021).

Tackling carbon emissions shapes not only environmental sustainability but also economic transformation processes. Countries aim to reduce emissions and sustain economic development through green growth policies and clean energy investments. In this framework, market-based mechanisms such as carbon pricing encourage adopting more environmentally friendly production and consumption patterns by costing polluting activities (Stern, 2007). Controlling carbon emissions is central to preventing environmental degradation and achieving sustainable development goals.

A review of resource consumption patterns on a global scale is now inevitable, as the current level of consumption has reached an unsustainable point (WWF, 2012: 14). The Global Footprint Network (GFN) reports the resource consumption and biological capacity of more than 150 countries annually through National Footprint Accounts (NFA). In addition to the above-mentioned data, the GFN also provides carbon emissions per capita in countries. These calculations clearly reveal how much pressure countries exert on nature and the effects of this pressure on the environment.

The sub-components of the ecological footprint and the per capita carbon footprint data for the 10 EU member countries examined in the study (Austria, Belgium, France, Germany, Ireland, Italy, Netherlands, Poland, Spain, and Sweden) are presented and evaluated graphically below. However, since the data set for Ireland is not included in the Global Footprint Network database, the graph for this country could not be included in the study.

Figure 1. Carbon footprint per person in Austria (gha per person) (1961-2019)

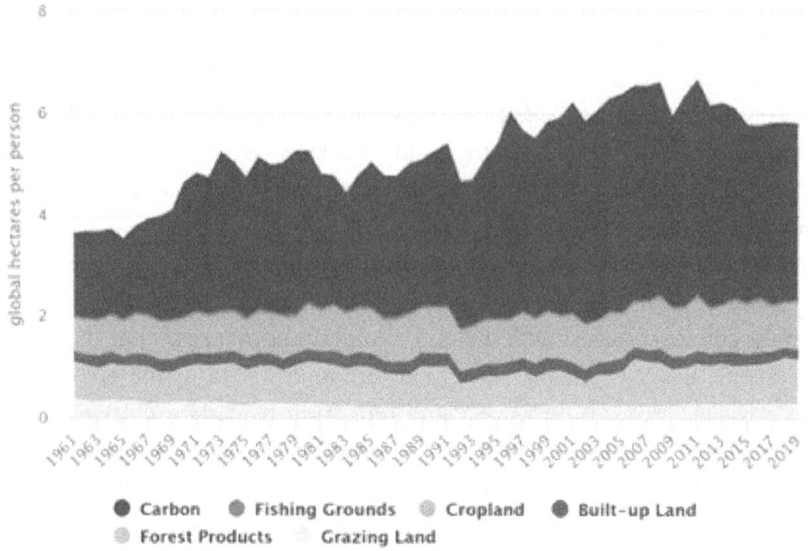

Source: *Global Footprint Network (GFN), 2025.*

The figure shows the subcomponents of Austria's ecological footprint from 1961 to 2019. The carbon (energy) footprint, shaded in blue, shows the most pronounced increase, from around 1.5 gha per capita in the 1960s to 3.5-4.0 gha in the early 2000s, accounting for more than half of the total footprint. On the other hand, the footprint of agricultural land (cropland) in orange fluctuated over the period, from 0.8 gha to 1.1 gha, while the footprint of forest products in light green increased from 0.7 gha to 0.9 gha. The red stripe (settlements) and the yellow stripe (pasture) have small, relatively constant shares of the total, while fishing grounds remain a component in the bottom layer, with little change in thickness. This distribution shows that the primary determinant of Austria's ecological pressure is energy-related carbon emissions, while pressure from land use (agriculture, forest products) is increasing more moderately. While the non-carbon categories have shown relative stability over time thanks to technological advances and efficiency gains in production processes, reducing the reliance on fossil fuels in energy systems, primarily through investments in renewable energy and energy efficiency measures, has been a decisive factor in reducing the country's overall ecological footprint.

Figure 2. Carbon footprint per person in Belgium (gha per person) (1961-2019)

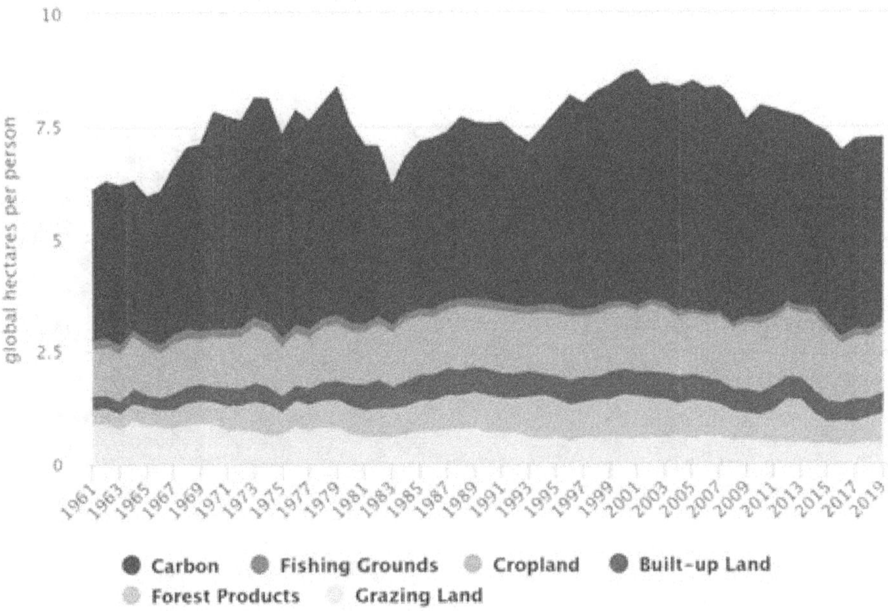

Source: *Global Footprint Network (GFN), 2025.*

The most dominant element of the graph is the carbon footprint, shown in blue, which increased from around 3.5 gha per capita in the 1960s to between 4.5-5.5 gha in the 1970s-1980s and to a peak of around 5.5 gha in the early 2000s. There was then a brief decline due to the 2008 financial crisis, after which it fluctuated between 4.5-5.0 gha in the second half of the 2010s and remained relatively stable until the end. The cropland footprint, shown in orange, started at around 1.0 gha in 1961 and reached 1.4 gha by the 1990s before declining slightly to around 1.2 gha in 2019. The share of forest products, colored in light green, increased from 0.5 gha to 0.9 gha and has recently remained between 0.6-0.8 gha. The shares of built-up land footprint, coded in red, and grazing land, shown in yellow, fluctuated between 0.3-0.6 gha and 0.4-0.7 gha, respectively, as small but consistent sub-components of the total. Fishing grounds are observed in the bottom layer of the graph as a minimal item spread over the whole period.

Figure 3. Carbon footprint per person in France (gha per person) (1961-2019)

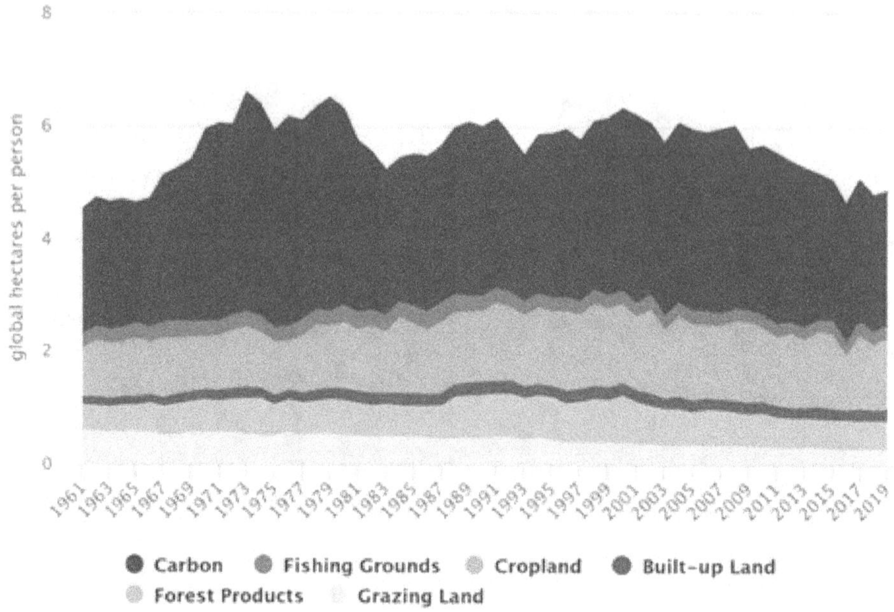

Source: *Global Footprint Network (GFN), 2025.*

The dark blue carbon footprint, which makes up the most significant slice of the graph, increased from approximately 2.5 gha per person in 1961 to 3.5–4.0 gha between 1970–1980 and to 4.0–4.2 gha in the early 2000s; it peaked in 2005–2008 (4.2 gha) and then gradually declined, dropping back to around 2.7 gha in 2019. The cropland footprint, shown in orange, increased from 0.8 gha in the 1960s to 1.1 gha in the 1990s and fluctuated in the 0.9–1.0 gha band after the 2000s. The share of forest products coded in light green increased from 0.6 gha to 0.8 gha and stabilized at 0.5–0.7 gha after 2010, while the red-striped built-up land remained stable in the range of 0.2–0.4 gha. The share of grazing land in yellow was 0.5 gha at the beginning of the period and decreased to 0.3 gha in 2019, while the contribution of fishing grounds remained negligible throughout the period.

Figure 4. Carbon footprint per person in Germany (gha per person) (1961-2019)

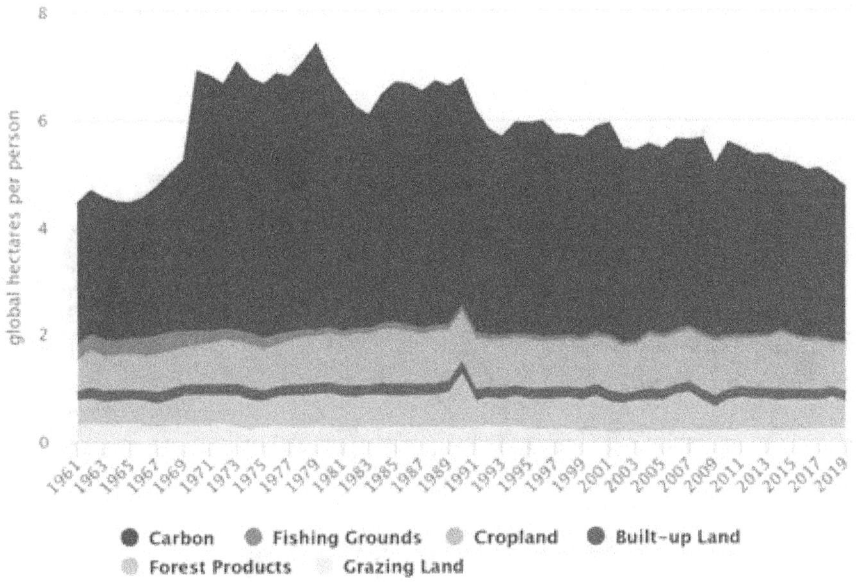

Source: *Global Footprint Network (GFN), 2025.*

The graph shows the subcomponents of Germany's ecological footprint from 1961 to 2019. The most dominant item, the carbon footprint (dark blue), was around 2.5 gha per capita in the early 1960s but rose to 7.0 gha towards the end of the 1970s due to the industrialization and energy consumption of the German economy; it saw a brief peak in 1990 when East and West Germany were reunited, and then gradually decreased, falling to 4.7 gha in 2019. The share of cropland, in orange, was around 0.6 gha in 1961 and nearly doubled to 1.1 gha in 1990. It made a significant jump in 1991 due to the reunification effect and then stabilized in the 0.8–0.9 gha band with a stable course. The forest products footprint, shown in light green, increased from 0.4 gha to 0.6 gha, while the built-up land and grazing land shares fluctuated slightly between 0.3–0.5 gha and 0.2–0.3 gha, respectively, indicating protected land use practices. The contribution of fishing grounds remained negligible throughout the period.

Figure 5. Carbon footprint per person in Italy (gha per person) (1961-2019)

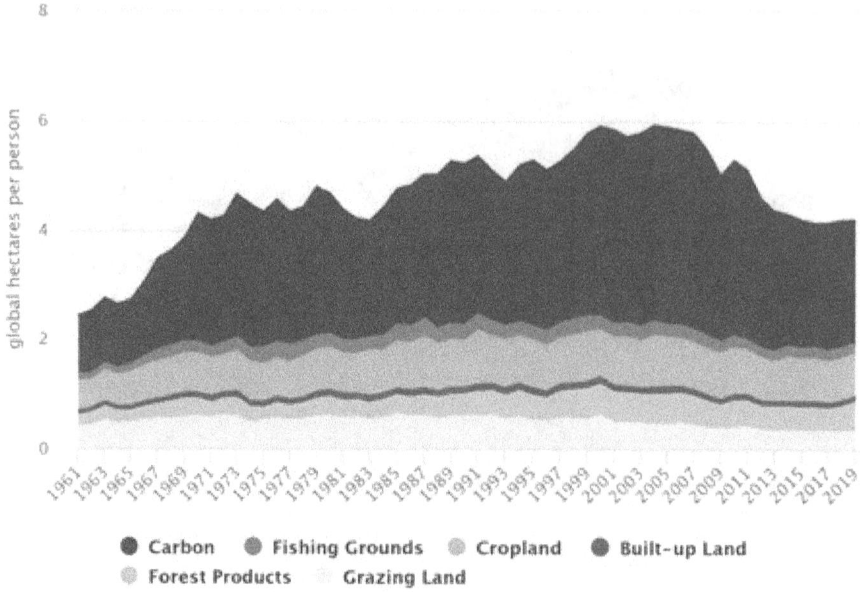

Source: *Global Footprint Network (GFN), 2025.*

The most dominant component of the graph is the dark blue carbon footprint, which starts at around 1.2 gha per capita in 1961 and climbs to 2.5-3.0 gha in the 1970s with industrialization and consumption growth. It peaked at 3.8 gha in the 1990s and early 2000s before declining to around 2.4 gha in 2019, driven by energy efficiency and renewable energy investments after 2005. The cropland footprint, shown in orange, starts at 0.6 gha in the 1960s, gradually increases to around 1.0 gha by the 1990s, and then stabilizes at around 0.8-0.9 gha. The share of forest products in light green rises from 0.3 gha to 0.6 gha before falling back to 0.3-0.4 gha in the 2010s, while the shares of built-up areas in red and grazing land in yellow show small fluctuations between 0.2-0.3 gha and 0.4-0.5 gha. The contribution of fishing grounds remains negligible over the whole period.

Figure 6. Carbon footprint per person in Netherlands (gha per person) (1961-2019)

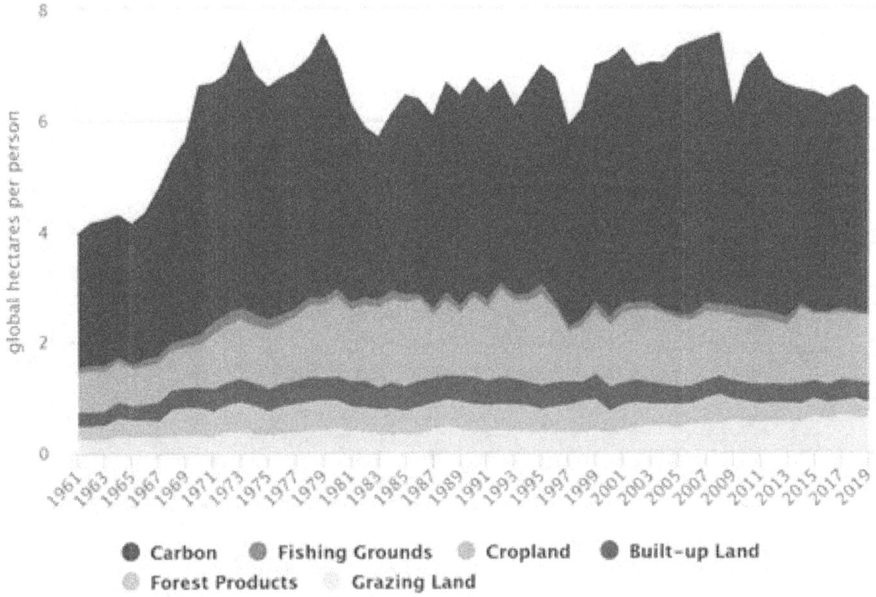

Source: *Global Footprint Network (GFN), 2025.*

Looking at the Netherlands' ecological footprint components for 1961-2019, the dominant category has always been the carbon footprint. The carbon share, shown in the dark blue region, increased from around 4.0 gha per capita in 1961 to 5.0-5.5 gha in the 1970s, experienced a brief decline in the mid-1980s, and then increased again throughout the 1990s and 2000s, peaking around 2008 at around 6.0-6.2 gha. After 2010, with the impact of energy efficiency and renewable energy practices, the carbon footprint has been on a gradual downward trend, reaching around 5.3 gha in 2019. The agricultural land (cropland) footprint, coded in orange, was around 0.5 gha in 1961, increased to 1.3 gha in the 1980s due to industrialization and export-oriented agricultural production, declined to around 1.1 gha in the 1990s and remained stable in the 1.0-1.2 gha band in the following period. The share of forest products in light green increased from 0.3 gha to 0.5 gha and stabilized around 0.4 gha after 2010, while the shares of built-up land in red and grazing land in yellow remained relatively stable with minor fluctuations between 0.4-0.6 gha and 0.3-0.5 gha. The contribution of fishing grounds remained negligible over the entire period, with a thin band just below the graph.

Figure 7. Carbon footprint per person in Poland (gha per person) (1961-2019)

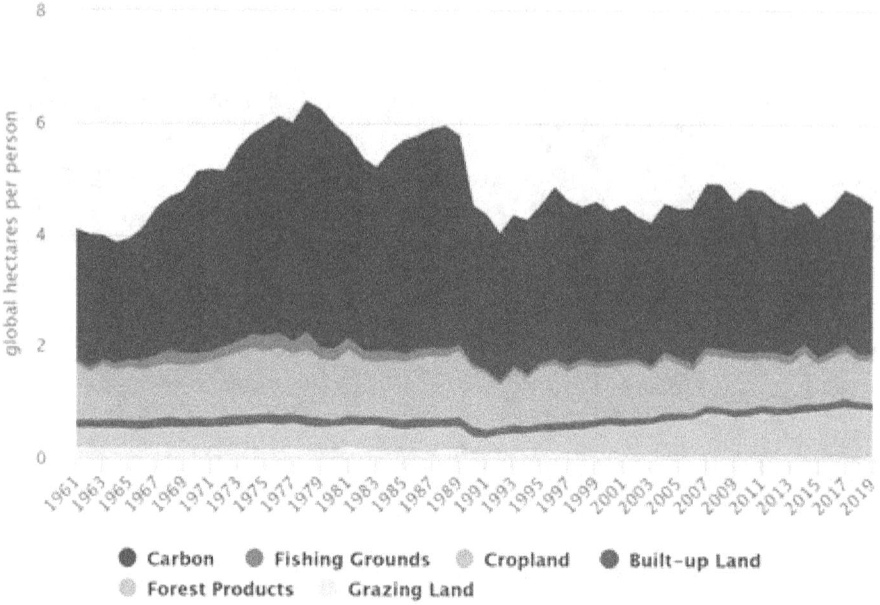

Source: *Global Footprint Network (GFN), 2025.*

The carbon footprint, shown in blue in the graph, increased from about 2.4 gha per capita in 1961 to 4.0-4.2 gha in the late 1970s due to industrialization and coal consumption, dropped sharply to 3.0 gha in the 1990s after the economic transformation, then fluctuated in the range of 3.5-3.8 gha throughout the 2000s before declining to about 2.7 gha in 2019. The orange cropland footprint started at 1.0 gha in the 1960s, increased to 1.4 gha in the 1970s-80s declined slightly in the 1990s and has recently stabilized at around 1.2 gha. The share of forest products in light green increased from 0.3 gha to 0.7 gha and remained in the 0.6-0.8 gha band, especially since the 2000s, driven by sustainable forestry practices. The shares of built-up land in red and grazing land in yellow remained relatively stable with minor fluctuations between 0.2-0.4 gha and 0.1-0.2 gha over the whole period, while the contribution of fishing grounds was negligible in the lower layer of the graph.

Figure 8. Carbon footprint per person in Spain (gha per person) (1961-2019)

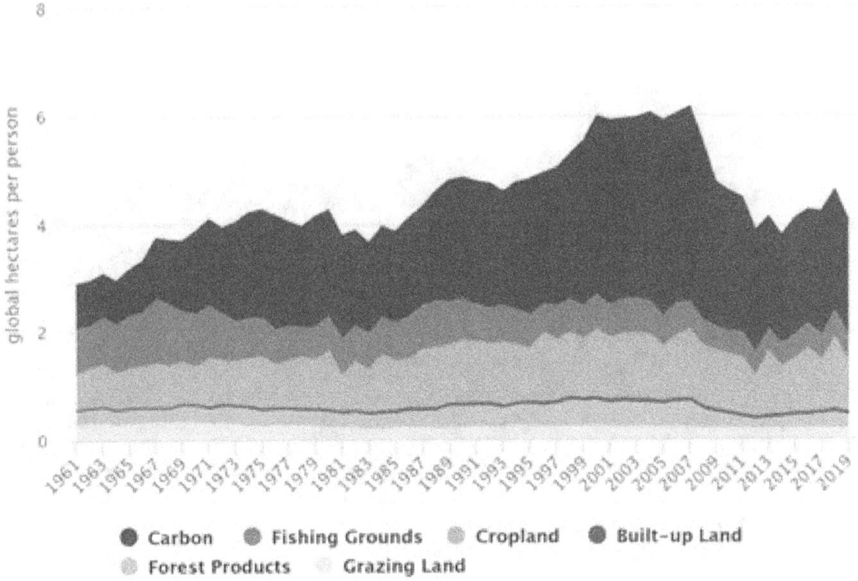

Source: *Global Footprint Network (GFN), 2025.*

The dark blue carbon footprint, which constitutes the largest share of the graph, starts at around 0.8 gha per capita in 1961 and reaches 2.0 gha in the late 1970s with the increase in industrialization, tourism, and energy consumption. It peaks at 3.5-3.8 gha in 2005-2008, with accelerating prosperity and infrastructure growth from the 1990s onwards, and then declines again to 2.2 gha in 2019 due to energy efficiency measures and the impact of the crisis period. The cropland footprint, shown in orange, increased steadily from 0.6 gha in 1961 to 1.2 gha in the 1990s before stabilizing at 1.0-1.1 gha after the 2000s. The share of fishing grounds in light blue declined from 0.7 gha in the early 1960s to 0.4 gha in the 1980s before increasing slightly to around 0.5 gha due to tourism and coastal resource management practices. The light green share of forest products increased moderately from 0.3 gha to 0.5 gha, while the shares of built-up land in red and grazing land in yellow remained in the range of 0.2-0.4 gha over the entire period; these components represent small but consistent sub-contributions to the total footprint.

Figure 9. Carbon footprint per person in Sweden (gha per person) (1961-2019)

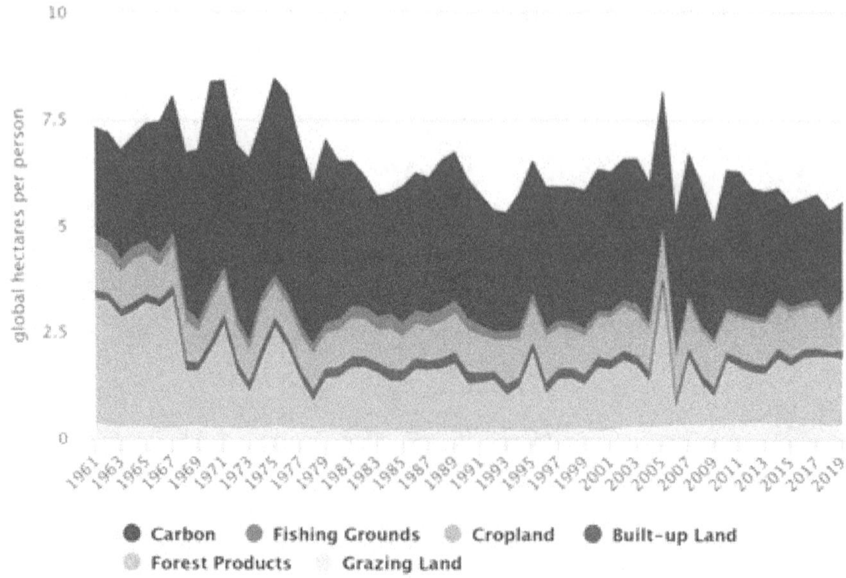

Source: *Global Footprint Network (GFN), 2025.*

The carbon footprint, shown in dark blue in the graph, starts at around 3.5 gha per capita in 1961, climbs to more than 4.0 gha in the early 1970s, reaches 4.5 gha in 1975-80, and then gradually declines to 3.5-4.0 gha. The share of cropland, in orange, ranges from around 0.8 gha in the 1960s to 1.0 gha in the 1980s and stabilizes in the 0.8-0.9 gha range after 1990. The forest products item in light green stood out as the most significant non-carbon component with 1.6 gha at the beginning of the period, declined rapidly to 1.0-1.2 gha between 1970-80, and then fluctuated between 0.8-1.0 gha, maintaining its position as the second largest sub-item. Built-up land in red contributed 0.2-0.4 gha, grazing land in yellow contributed 0.1-0.2 gha, and fishing grounds in light blue contributed a negligible contribution of around 0.1 gha over the whole period. The sudden peaks on the graph (e.g., 1973, 1977, and 2007) may be due to measurement or calculation changes in the data source.

LITERATURE

In recent years, theoretical and empirical studies on the impact of environmental taxes on carbon emissions have been increasing. Environmental taxes stand out as an essential policy tool for internalizing environmental externalities and reducing carbon emissions (Pigou, 1920; Ulucak et al., 2020). In this context, studies have shown that the findings on the impact of environmental taxes on reducing carbon emissions vary depending on countries, implementation methods, and sectoral scope. For example, although environmental taxes theoretically have the potential to reduce emissions, there are cases where their effectiveness in practice is low (Uğur, 2014; Ayvaz, 2025). In addition, evaluating tax revenues as budget revenues rather than environmental objectives or exemptions granted to energy-intensive sectors may prevent environmental taxes from providing the expected environmental benefits (Uğur, 2014; Ayvaz, 2025; Zhang & Baranzini, 2004). Some of the researchers who aim to evaluate the impact of environmental taxes on carbon emissions vary in scope and methodology. Ayvaz (2025) analyzed the relationship between environmental taxes, air transportation, tourism, fossil fuel consumption, and CO_2 emissions with the Panel Vector Autoregression (PVAR) model using data from 25 European countries for the period 1995-2021 and stated that the current taxes are insufficient in reducing emissions. In contrast, Atılgan (2025) revealed that environmental taxes significantly reduced CO_2 emissions in his panel cointegration analysis on G7 countries from 1994-2014. Uğur (2014) discussed the difficulties of global implementation of carbon-energy taxes and the fact that sectoral tax exemptions limit the effectiveness of these taxes in combating climate change within a conceptual framework.

Researchers agree that the effectiveness of environmental taxes depends not only on tax rates but also on tax design, revenue use, and scope of application (Atılgan, 2025; Ayvaz, 2025; Uğur, 2014). When evaluated in terms of findings, Atılgan (2025) found that environmental taxes have a statistically significant and negative effect on reducing carbon emissions in G7 countries. Similarly, many studies, such as Ulucak et al. (2020), Doğan et al. (2022), and Ghazouani et al. (2021), have concluded that environmental taxes reduce CO_2 emissions. However, Ayvaz (2025) found that existing environmental taxes are not effective enough in reducing emissions in the case of European countries. In his theoretical evaluations, Uğur (2014) stated that carbon-energy taxes are ineffective due to their use for fiscal purposes, global application differences, and sectoral exemptions.

In line with these findings, it is emphasized that in order for environmental taxes to be successful, not only the price mechanism but also the tax revenues should be directed to green technology investments, implementation unity should be ensured, and the tax structure should be implemented without any sectoral differences (Uğur, 2014; Ayvaz, 2025; Zhang & Baranzini, 2004).

DATA SET, METHODOLOGY AND EMPIRICAL FINDINGS 3.1. DATA SET

The dataset of the study, which aims to examine the impact of environmental taxes applied in the 10 most prominent EU members in terms of GDP (Austria, Belgium, France, Germany, Ireland, Italy, Netherlands, Poland, Spain, Sweden) on carbon emissions in the specified countries with panel causality analysis, covers the years 1994-2022 and consists of annual data. The data was obtained from the official website of the Global Footprint Network (2025) and the OECD database. In the analysis of the study, carbon emissions (CO2) and environmental tax (EnvTax) per capita of 10 EU countries were used. The analyses were carried out using the Stata program. Explanations regarding the variables used in the study are shown in Table 1.

Table 1. Variables and their descriptions

Variables	Symbols
Carbon Footprint per Person (gha)	CO_2
Environmental Tax	EnvTax

Methodology

The study aims to examine the effect of environmental taxes applied in the 10 most prominent EU members in terms of GDP size (Austria, Belgium, France, Germany, Ireland, Italy, Netherlands, Poland, Spain, and Sweden) on carbon emissions in the specified countries with panel causality analysis. Before performing the causality test, a correlation between units was examined with the Breusch Pagan LM test. Then, the homogeneity test results were obtained using the homogeneity tests developed by Swamy (1970). The variables' panel unit root test results were obtained using the panel unit root test by Im, Peseran, and Shin (CIPS) with an extended cross-section. Finally, the panel causality test results were provided with the test developed by Dumitrescu and Hurlin (2012).

$$CO_{2it} = \beta_0 + \beta_1 EnvTax_{it} + \varepsilon_{it}$$

In this model, i represents each panel unit (10 countries), and t represents the data period (1994-2022). CO2 is the dependent variable of this model and is used as an indicator of environmental performance. EnvTax is the explanatory variable.

Empirical Findings

Descriptive statistics for the variables are given in Table 2 below.

Table 2. Descriptive statistics of variables

Variable	Obs	Mean	Std. Dev.	Min	Max
CO_2	290	3.326	0.759	1.73	5.24
EnvTax	290	2.428	0.561	1.13	5.13

As seen in the table above, the number of observations for the variables is 290, and the average, minimum, and maximum values of the CO2 variable are 3.326, 1.73, and 5.24, respectively. The average, minimum, and maximum values of the EnvTax variable are 2.428, 1.13, and 5.13, respectively. Table 3 shows the inter-unit correlation results.

In this section, cross-sectional dependency tests (inter-unit correlation tests), which are pre-tests used in selecting first-generation and second-generation tests, should be performed. The presence of cross-sectional dependency indicates that second-generation tests should be continued. In this respect, the application of the test is critical. The presence of inter-sectional dependency is investigated with the "LM test of Breusch and Pagan 1980, CDLM and CD test Pesaran 2004 and LMadj" tests developed by Pesaran et al., 2008 (Örnek & Türkmen, 2019). The LM test is applied when the time dimension (T) is larger than the section dimension (N) (Pesaran, 2004: 4). CDLM shows distortions when N>T. Since the problem of increasing deviations comes to the fore as N increases, Pesaran (2004) tried to overcome this situation by developing the CD test used in the stage where N>T. Later, Pesaran et al. (2008) developed the deviation-corrected LMadj test.

Table 3. Cross section dependency test results

Test	Statistic	P-value
LM	283	0.000
LM adj	74.85	0.000
LM CD	14.25	0.000

The first test in the table above is the Breusch Pagan LM test. The basic hypothesis is that the residuals are not correlated between units. According to the results of this test, the basic hypothesis was rejected, and it was concluded that there was a correlation between units. Another pre-test is the homogeneity test. Homogeneity tests were developed by Swamy (1970). The null hypothesis of Swamy homogeneity tests, which test the homogeneity of slope coefficients, is that the slope coefficients are homogeneous. The results of the test are given in Table 4.

Table 4. Homogeneity test results

Slope Homogeneity Test	Statistic	P-Value
	3870.83	0.000

According to the homogeneity test results above, the basic hypothesis was rejected, and it was decided that the parameters were heterogeneous. After determining the pre-test findings, the second-generation tests for the unit root test were tested with the panel unit root test by Im, Peseran, and Shin (CIPS), whose horizontal section was expanded. The results of the test are given in Table 5 below.

Table 5. Cross-sectional Im, Peseran ve Shin (CIPS) Panel Unit Root Test Result

t-bar	cv10	cv5	cv1	Z[t-bar]	P-value
-1.989 -1.667	-2.210 -2.210	-2.330 -2.330	-2.570 -2.570	-0.745 0.302	0.228 0.619

Table 5 shows the CIPS test unit root test results of the carbon emission (CO2) per capita and the environmental tax (EnvTax) variable in the first row. Since the t-bar (CIPS) statistics are not greater than the given critical values in absolute value, neither series is stationary.

Dumitrescu and Hurlin (2012) extended the Granger causality test for heterogeneous panels. The basic hypothesis states that there is no causality from x to y for the entire panel, i.e., there is no homogeneous panel causality. Under the alternative hypothesis, the model is heterogeneous. In other words, some units may not have a causal relationship under the alternative hypothesis. The test results are given in Table 6 below.

Table 6. Dumitrescu and Hurlin Causality Test Results

Test	Statistic	p-value
w-bar	5.679	-
z-bar	10.464	0.000
z-bar tilde	8.837	0.000

According to the test results of Dumitrescu and Hurlin (2012), the basic hypothesis stating that there is no causal relationship between the environmental tax (EnvTax) variable and the per capita carbon emission (CO2) variable is rejected. A causal relationship is determined from the environmental tax (EnvTax) variable to the per capita carbon emission (CO2) variable in at least one unit. In other words, the changes or regulations made in the environmental tax (EnvTax) variable affect the per capita carbon emission (CO2). This result shows that the changes or regulations in environmental tax practices significantly affect carbon emissions. In other words, increasing the environmental tax or implementing new regulations can reduce per capita carbon emissions. The effectiveness of such policies can play an important role in achieving environmental sustainability goals and can shape the strategies of governments to reduce carbon emissions. In particular, with the direct effect of the environmental tax, companies, and individuals can be directed to more sustainable production and consumption behaviors, which can contribute to environmentally friendly changes in the long term. These findings are similar to those of some key studies in the literature (Pigou, 1920; Ulucak, Apergis, & Topcu, 2020; Shahzad, 2020). In these studies, it is emphasized that environmental taxes stand out as an essential policy instrument for internalizing environmental externalities and reducing carbon emissions. Additionally, it is noted that the findings on the impact of environmental taxes on lowering carbon emissions vary across countries, implementation types, and sectoral scopes. However, it is worth noting that there are also studies in the literature that yield different findings from those of the current study. For example, Ayvaz (2025) finds that existing environmental taxes are insufficient in reducing emissions in European countries. This result is likely based on the period and sample analyzed, which suggests that environmental taxes are not effective enough in reducing emissions.

CONCLUSION

Environmental pollution has become a significant problem both locally and globally today, threatening natural ecosystems, human health, and quality of life. Countries use different economic tools to combat environmental pollution, and environmental taxes stand out as one of the most effective solutions. Environmental taxes are becoming an important resource for limiting activities promoting pollution and financing environmental improvement. Globally, especially in developed countries, the use of environmental taxes is increasing in order to achieve carbon emission reduction targets. However, the fact that carbon emissions per capita continue to increase worldwide makes the effectiveness of these taxes even more critical. Countries see environmental taxes not only as an environmental policy tool but also as new financial resources that will sustainably support economic growth.

However, the increase in carbon emissions is particularly concentrated in industrialized countries, and developing countries are also greatly affected by this problem. These countries have difficulty preventing environmental destruction in the process of economic development. However, as the effects of global warming are increasingly felt, they are forced to take more precautions against this problem. Environmental problems such as climate change, rising sea levels, loss of agricultural productivity, increase in natural disasters, and loss of biodiversity are becoming more serious globally. For environmental taxes to be effective in limiting increasing carbon emissions, they should be used not only as a financial obligation but also as an incentive mechanism. In this context, the correct design of environmental taxes can be both an effective policy tool for protecting the environment and a strong source of financing that will help achieve sustainable development goals.

The data set of the study, which aims to examine the impact of environmental taxes applied in the 10 most prominent EU members in terms of GDP (Austria, Belgium, France, Germany, Ireland, Italy, Netherlands, Poland, Spain, Sweden) on carbon emissions in the specified countries with panel causality analysis, covers the period between 1994-2022 and consists of annual data. The data was obtained from the official website of the Global Footprint Network (2025) and the OECD database. In the analysis of the study, carbon emissions (CO2) and environmental tax (EnvTax) per capita of 10 EU countries were used. The analyses were carried out using the Stata program. Before the causality test, a correlation between the units was examined using the Breusch Pagan LM test. Then, the homogeneity test results were obtained using the homogeneity tests developed by Swamy (1970). The variables' panel unit root test results were obtained using the panel unit root test by Im, Peseran, and Shin (CIPS) with an extended cross-section. Finally, the panel causality test results were provided by the test developed by Dumitrescu and Hurlin (2012). As a result of the study, a causal relationship is determined from the environmental tax (EnvTax) variable to the carbon emission (CO2) per capita variable. In other words, it was concluded that the changes or arrangements made in the environmental tax (EnvTax) variable affect the carbon emission (CO2) per capita.

This finding is similar to various empirical studies that reveal the potential of environmental taxes to reduce emissions. In particular, studies such as Atılgan (2025), Ulucak et al. (2020), and Doğan et al. (2022) have also obtained statistically significant results regarding the carbon emission-reducing effect of environmental taxes. In this respect, the study aligns with the trends prominent in the existing empirical literature. However, some studies also find that environmental taxes are less effective than expected, mainly due to inadequacies in tax design or sectoral exemptions. For example, Ayvaz (2025) found that existing environmental taxes are not sufficiently effective in reducing emissions in European countries. This conclusion suggests that environmental taxes may not be adequately effective in reducing emissions, depending on the period and sample examined. Moreover, it shows that the environmental effectiveness of environmental taxes depends not only on whether they are implemented or not but also on the form of implementation, scope, and areas of use of revenues.

For environmental taxes to be more effective at the policy level, some basic strategic elements need to be considered. The most important are limiting sectoral exemptions, designing tax rates to reflect environmental costs, and directing the revenues obtained directly to green infrastructure projects and renewable energy investments. In addition, implementing complementary mechanisms such as the European Union's Carbon Border Arrangement (CBAM) with environmental taxes against the risk of carbon leakage will increase the effectiveness of these policies.

However, the study has some limitations. The examination is limited only to environmental tax and per capita carbon emission variables; subtypes of ecological taxes (energy, transportation, waste, etc.) are not separated. In addition, only carbon emissions are taken as the basis in the analysis; other components of the ecological footprint are excluded from the evaluation. In future studies, the impact of environmental taxes on the ecological footprint and its subcomponents can be examined by separating them at the country and sectoral levels. The ecological footprint is a holistic indicator that measures the total pressure of human activities on nature and plays a critical role in assessing environmental sustainability. In this context, analyses focusing on how environmental taxes affect other ecological components such as forests, agriculture, settlements, and fishing areas will enable a more holistic and healthy evaluation of the effectiveness of environmental policies.

REFERENCES

Atılgan, D. (2025). Çevre vergisi karbon emisyonlarını azaltmakta mı? Panel eşbütünleşme analizi. *Türk Tarim ve Doga Bilimleri Dergisi, 12*(1), 145–156. DOI: 10.30910/turkjans.1586719

Ayvaz, E. E. (2025). Çevre vergileri, hava taşımacılığı ve CO_2 emisyonları ilişkisi: Avrupa ülkeleri üzerine bir çalışma. *Erzurum Teknik Üniversitesi Sosyal Bilimler Enstitüsü Dergisi, 21*(21), 15–33. DOI: 10.29157/etusbed.1436615

Bodansky, D. (2010). *The art and craft of international environmental law*. Harvard University Press.

Çokgezen, J. (2007). Avrupa Birliği çevre politikası ve Türkiye. *Marmara Üniversitesi İ.İ.B.F. Dergisi, 23*(2), 91–115.

Doğan, B., Chu, L. K., Ghosh, S., Truong, H. H. D., & Balsalobre-Lorente, D. (2022). How environmental taxes and carbon emissions are related in the G7 economies? *Renewable Energy, 187*, 645–656. DOI: 10.1016/j.renene.2022.01.077

Doğan, S., & Tüzer, M. (2011). Küresel iklim değişikliği ile mücadele: Genel yaklaşımlar ve uluslararası çabalar. *Istanbul Journal of Sociological Studies*, (44), 157–194.

Ekins, P. (2000). *Economic growth and environmental sustainability: The prospects for green growth*. Routledge.

European Commission. (2019). *The European Green Deal*. https://commission.europa.eu/publications/communication-european-green-deal_en

Ghazouani, A., Jebli, M. B., & Shahzad, U. (2021). Impacts of environmental taxes and technologies on greenhouse gas emissions: Contextual evidence from leading emitter European countries. *Environmental Science and Pollution Research International, 28*(18), 22758–22767. DOI: 10.1007/s11356-020-11911-9 PMID: 33423194

Global Footprint Network. *Ecological footprint data explorer*. https://data.footprintnetwork.org/#/analyzeTrends?type=EFCpc&cn=210

Grubb, M., Vrolijk, C., & Brack, D. (1999). *The Kyoto Protocol: A guide and assessment*. Royal Institute of International Affairs/Earthscan.

IMF. (2019). *Fiscal policies for Paris Climate Strategies—From principle to practice*. International Monetary Fund. https://www.imf.org/en/Publications/Policy-Papers/Issues/2019/05/01/Fiscal-Policies-for-Paris-Climate-Strategies-From-Principle-to-Practice-46826

IPCC. (2014). [Synthesis report. Contribution of Working Groups I, II and III to the Fifth Assessment Report of the Intergovernmental Panel on Climate Change.]. *Climatic Change*, 2014.

Karmaker, S. C., Hosan, S., Chapman, A. J., & Saha, B. B. (2021). The role of environmental taxes on technological innovation. *Energy*, *232*, 121052. DOI: 10.1016/j.energy.2021.121052

Kyoto Protocol. (1997). *United Nations Framework Convention on Climate Change*.

Montreal Protocol. (1987). *United Nations Environment Programme*.

NASA. (2020). *The causes of climate change*. https://climate.nasa.gov/causes/

OECD. (2011). *Environmental taxation: A guide for policy makers*. OECD Publishing.

OECD. (2021). *Carbon pricing and climate policy: How to align policies and achieve net zero*. OECD Publishing.

OECD Data Explorer. *OECD environmental and carbon data*. https://data-explorer .oecd.org/

Örnek, İ., & Türkmen, S. (2019). Gelişmiş ve yükselen piyasa ekonomilerinde çevresel kuznets eğrisi hipotezinin analizi. *Journal of the Cukurova University Institute of Social Sciences*, *28*(3), 109–129.

Pesaran, M. H. (2004). General diagnostic tests for cross section dependence in panels. SSRN *Electronic Journal*. DOI: 10.2139/ssrn.572504

Pesaran, M. H., Ullah, A., & Yamagata, T. (2008). A bias-adjusted LM test of error cross-section independence. *The Econometrics Journal*, *11*(1), 105–127. DOI: 10.1111/j.1368-423X.2007.00227.x

Pesaran, M. H., & Yamagata, T. (2008). Testing slope homogeneity in large panels. *Journal of Econometrics*, *142*(1), 50–93. DOI: 10.1016/j.jeconom.2007.05.010

Pigou, A. C. (Ed), (1920). The economics of welfar. McMillan & Co, London.

Rajamani, L. (2011). The Cancun Climate Agreements: Reading the text, subtext and tea leaves. *The International and Comparative Law Quarterly*, *60*(2), 499–519. DOI: 10.1017/S0020589311000078

Rees, W. E. (2003). A blot on the land. *Nature*, *421*(6926), 898. DOI: 10.1038/421898a PMID: 12606978

Stern, N. (2007). *The economics of climate change: The Stern Review*. Cambridge University Press. DOI: 10.1017/CBO9780511817434

Stockholm Conference. (1972). *United Nations Conference on the Human Environment.*

Swamy, P. A. (1970). Efficient inference in a random coefficient regression model. *Econometrica, 38*(2), 311–323. DOI: 10.2307/1913012

Tietenberg, T., & Lewis, L. (2016). *Environmental and natural resource economics* (10th ed.). Pearson.

Uğur, S. (2014). Sera gazı emisyonlarının azaltımında karbon-enerji vergilerinin rolü. *FSM İlmî Araştırmalar İnsan ve Toplum Bilimleri Dergisi, 0*(3), 341–358. DOI: 10.16947/fsmiad.58262

Ulucak, R., Danish, , & Kassouri, Y. (2020). An assessment of the environmental sustainability corridor: Investigating the non-linear effects of environmental taxation on CO_2 emissions. *Sustainable Development (Bradford), 28*(4), 1010–1018. DOI: 10.1002/sd.2057

UNFCCC. (1998). *Kyoto Protocol to the United Nations Framework Convention on Climate Change.* https://unfccc.int/resource/docs/convkp/kpeng.pdf

UNFCCC. (2015). *The Paris Agreement.* https://unfccc.int/sites/default /files/english_paris_ agreement.pdf

United Nations Framework Convention on Climate Change. (1992). *UNFCCC.* https://unfccc.int/resource/docs/convkp/conveng.pdf

Zhang, Z., & Baranzini, A. (2004). What do we know about carbon taxes? An inquiry into their impacts on competitiveness and distribution of income. *Energy Policy, 32*(4), 507–518. DOI: 10.1016/S0301-4215(03)00152-6

Chapter 10
Income Inequality and Environmental Protection Expenditures Nexus Bootstrap Panel Rolling Window Causality Analysis

Nazlı Keyifli Sentürk
https://orcid.org/0009-0008-2649-1625
İzmir Kâtip Çelebi University, Turkey

Suat Hayri Sentürk
Gumushane University, Turkey & Ege University, Turkey

Birol Karakurt
Karadeniz Technical University, Turkey

ABSTRACT

The relationship between environmental protection expenditures and income inequality is undoubtedly a critical issue in terms of social justice and sustainable development. In societies with high income inequality, environmental protection expenditures are generally lower. This is because low-income individuals tend to have less motivation to develop awareness of environmental issues or to support environmental protection policies. On the other hand, in societies where income is more equitably distributed, environmental protection expenditures tend to increase. A key factor behind this is that a larger segment of society views environmental

DOI: 10.4018/979-8-3373-1310-8.ch010

sustainability as a public good and a necessary component of inclusive economic development, thereby generating collective demand in this direction. In this study, the relationship between income inequality and environmental protection expenditures in eight OECD countries was investigated using the Bootstrap Panel Rolling Window causality test.

INTRODUCTION

Industrialisation and globalisation have accelerated competitiveness. Countries are trying to realise their economic development goals in this environment. Economic growth targets must be sustainable in order to be desirable and at an ideal level. In fact, the wealth that emerges with economic growth should be distributed fairly among countries, regions and income groups, as well as environmental values (Değirmenci, 2020: 43). Undoubtedly, as a result of production or consumption activities, pollution, i.e. environmental pollution, may occur in many areas such as air, soil, water and so on. However, environmental pollution resulting from production and consumption activities may differ according to the level of countries' consideration of environmental values (Ertürk, 1998:106). In fact, if the increasing production and consumption activities increase environmental damage, the wealth created by economic development becomes detrimental to environmental values.

Although environmental problems are not a new phenomenon, there may be many reasons for these problems. Today, environmental problems arise as a result of production, consumption and distribution relations that shape the economic system. So much so that human beings have benefited from nature in order to meet their basic needs, but they have also caused serious damage to the environment with what they left to nature. In other words, both industrialised societies with high pollutant potential, which consume a lot of energy and resources, and the poor, who directly consume and exploit natural resources, can cause many environmental problems (Dağdemir, 2015:26). Governments, businesses and households are therefore responsible for combating pollution. In fact, for an effective fight against environmental pollution, governments need to involve all departments in the fight against environmental pollution with the help of various tools. These instruments include taxation, quotas, prohibitions, and environmentally friendly technologies (Değirmenci and Aydın, 2022:392). Undoubtedly, the efficiency of environmental expenditures, environmental taxes, bans and quotas are important in combating environmental pollution. In short, environmental protection is a very broad concept and covers not only the expenditures made by the state or the taxes collected, but also the rules that the private sector must comply with.

Measures for environmental protection first entered the agenda of developed countries, but these countries, due to various reasons such as low labor costs and weak control mechanisms, have shifted their production to developing countries where environmental regulations are not widely respected (Reardon and Vosti, 1995: 1497). Poor countries, on the other hand, have sacrificed their environmental policies and risked more pollution in order to ensure economic development by attracting foreign capital. However, when environmental pollution reached global dimensions, all countries around the world started to seek solutions for environmental pollution (Toprak, 2006:147).

The main purpose of environmental protection is to prevent, minimise or eliminate pollutants and pollution resulting from production and consumption activities or deterioration of the environment caused by other factors. Environmental protection expenditures are expenditures incurred for carrying out environmental protection activities. In other words, environmental protection expenditures are the sum of current and capital expenditures to combat pollution caused by the production and consumption of goods and services (Broniewicz, 2011:21).

As a matter of fact, even if the economic development of countries is the same, their environmental protection levels may differ. On the other hand, it is argued in the literature that an individual's education, age, race, gender, income and ideology are related to environmental policy (Blake et al., 1997; Newell and Green, 1997; Kahn and Matsusaka, 1997; Elliot et al., 1997; Magnani, 2000 and Bimonte, 2002). In fact, it is assumed that the income distribution in a country is a factor shaping the environmental policies of that country. When income is distributed fairly, individuals become richer and both environmental awareness and demands for a better quality environment may increase. This increase in demand may lead to pressures towards environmental regulations (Marsiliani and Renström, 2000, 1-2). Undoubtedly, income distribution can affect a country's expenditures for environmental protection. However, the level of a country's environmental protection expenditures (low-high) may not indicate the importance given to the environment by that country. Differences in environmental protection expenditures across countries may be due to the extent of environmental pollution or strategic reasons.

The economic phenomenon consists of individuals who make consumption decisions, diversify their incomes and care about the environment. As individuals become richer, their sensitivity to the environment increases and this situation triggers demands for environmental protection. While the tendency of individuals in more equitable societies to consume environmentally friendly goods is a valuable finding, it becomes more relevant in the context of this study when considered through its potential macroeconomic implications. Specifically, widespread pro-environmental preferences can shape public opinion, influence policy agendas, and ultimately result in increased public environmental protection expenditures. Thus, individual behavior may serve as a mechanism linking income distribution with environmental spending at the macro level (Cass et al., 2021: 4). As a matter of fact, such an environmental consumption tendency has a positive externality. In other words, it is observed that environmental public expenditures are negatively affected in societies or periods when income distribution is low, whereas environmental expenditures are positively affected when income distribution is fairly distributed (Vona and Patriarca, 2011: 2202). In fact, public policy consists of taxing the income of individuals and financing environmental public expenditures. In societies with a conscious and fair distribution of income, this situation occurs when individuals shape this process through the voting mechanism (Magnani, 2000: 432).

In societies where poverty is high, people are less interested in environmental degradation because they prioritise meeting their basic needs, that is, their survival, but as their level of welfare increases, their need to protect the environment becomes more pronounced (Morse, 2018:100). Undoubtedly, inequitable income distribution and environmental problems have long been recognised as obstacles to socioeconomic and environmental progress in many countries, both developed and developing. So much so that the Intergovernmental Panel on Climate Change also cites climate change and inequitable income distribution as the most urgent problems of the time (Ajide and Ibrahim, 2022: 1887).

Although environmental protection is on the agenda of most developed countries, it is observed that these countries have different levels of environmental protection even if their macroeconomic performances are similar (Marsiliani and Renström, 2000:1). For example, OECD countries such as France, Germany, Ireland, Italy, Luxembourg, Spain, Sweden and the United Kingdom have the highest environmental protection expenditure. The share of environmental protection expenditures in national income has shown a significant upward trend in the relevant countries in the 25-year period. While the share of environmental protection expenditures in national income was 0.65 on average in 1998 in the countries concerned, this ratio increased every year and reached 0.77 in 2022.

The causal relationship between income inequality and environmental protection expenditures is complex and context-dependent. In some countries, higher inequality suppresses environmental spending due to reduced public support and political resistance. In others, especially where environmental awareness and institutional capacity are strong, inequality may coexist with or even drive increased environmental investment. Moreover, environmental expenditures themselves can influence income distribution, either mitigating or reinforcing existing inequalities depending on the design and targeting of policies. When income inequality increases, individuals tend to focus more on immediate economic survival and daily life concerns rather than long-term collective issues such as environmental protection. As a result, environmental priorities may weaken across society. Based on this premise, the study employs the Bootstrap Panel Rolling Window Causality Analysis to examine the dynamic relationship over time between the aforementioned economic variables in the context of eight selected OECD countries. While various panel causality tests are commonly used in the literature for panel data analysis, this study specifically adopts the Bootstrap Panel Rolling Window Causality technique. Compared to other causality methods, this approach allows for the detection of hidden or time-varying causal relationships between variables that may not be captured when considering the entire time period as a whole.

LITERATURE REVIEW

The relationship between income inequality and public environmental protection expenditures has gained increasing attention in the literature, as scholars seek to understand how distributive justice may influence environmental policy outcomes. While some have examined the distributional impact of environmental policy (see Aubert and Chiroleu-Assouline, 2019), others have focused on the direction in which income distribution shapes environmental policy (see Boyce, 1994 and Magnani, 2000). The pioneering work on this issue was carried out by Boyce (1994), who put forward the public good choice theory by arguing that inequality of power and wealth generally increases environmental degradation.

Torras and Boyce (1998) empirically tested Boyce's (1994) study and analysed the impact of income redistribution on environmental quality by focusing on environmental policies in the period 1977-1991 by considering water quality indicators of 287 observation stations in 58 countries and observable air quality indicators of 18-52 cities in 19-42 countries and found that income redistribution affects the demand of society for environmental quality.

Ravallion et al. (2000) investigated the relationship between environmental quality and income inequality for 42 countries in the period 1975-1992. The study reveals that there is a negative correlation between emission intensity and income elasticity of many goods and that goods with higher income elasticity are generally those that produce less polluting emissions. It has been observed that as the income of the individual increases, the consumption of goods containing high levels of pollutants will shift towards the consumption of goods containing low levels of pollutants.

Bimonte (2002) measured the environmental protection status of a country's income distribution for 35 European countries by the ratio of protected areas to national land area and found that there is a negative relationship between income inequality and environmental protection.

Vona and Patriarca (2011) analyse the effect of the level of income distribution on the emergence of a new demand for green products, i.e. technological externalities, for European countries in the period 1999-2004. In the study, while income inequality in rich countries has a negative impact on the diffusion of technological innovations, the effective factor in poorer countries is per capita income.

Hao et al. (2016) empirically investigated the impact of income inequality on per capita carbon emissions in 23 Chinese provinces during the period 1995-2012. According to the Generalised Method of Moments (GMM) estimation results, per capita carbon emissions are higher in regions with widening income gaps.

Morse (2018) investigated the relationship between environmental quality, income and income inequality using 16 environmental performance indicators at national scale for 180 countries in the period 1995-2013. The study revealed that environmental health indicators are associated with income and income distribution, but indicators focusing on ecosystem vitality are not associated with income and income distribution.

Ridzuan (2019) analysed the relationship between environmental pollution and economic growth at the turning point of the environmental Kuznets curve in different periods of income inequality for 170 countries. In the study, it is determined that income inequality is an important factor in the transition to the turning point of the environmental Kuznets curve.

Magnani (2000) analysed the relationship between environmental protection expenditures, economic growth and income inequality in 19 OECD countries (Australia, Belgium, Canada, Denmark, Germany, Greece, Finland, France, Japan, Italy, Ireland, Netherlands, New Zealand, Norway, Portugal, Spain, Sweden, United Kingdom, USA) in the period 1980-1991 by using panel data method. The analysis revealed that there is a positive relationship between environmental protection expenditures and economic growth and a negative relationship between income inequality.

Değirmenci and Aydın (2020) analysed the relationship between environmental protection expenditures, income inequality and economic growth in 12 OECD countries (Austria, Belgium, Czech Republic, Spain, Finland, France, Germany, Greece, Ireland, Italy, Latvia, Lithuania) in the period 1995-2017. According to the results of Emirmahmutoğlu and Köse (2011) and Kónya (2006) panel causality tests, there is a unidirectional causality relationship between environmental protection expenditures and economic growth, and a bidirectional causality relationship between environmental protection expenditures and income inequality.

Wei and Zhao (2024) investigated the impact of environmental policy on income inequality in a study of 251 cities of China in the period 2010-2019. The study found that landscaping reduces per capita income and increases the urban-rural income gap. In addition, another finding of the study is that even if the government's environmental policies reduce pollution levels and provide environmental benefits, they will cause more income inequality as they will reduce the proportion of manufacturing industry and prevent entrepreneurs from investing.

In summary, while existing literature analyzing the relationship between income inequality and environmental protection expenditures generally employs panel econometric techniques such as OLS and GMM, as well as causality tests like those proposed by Emirmahmutoğlu and Köse (2011) and Kónya (2006), it is important to note that the causal relationship between these variables may change over time. In this regard, the Bootstrap Panel Rolling Window Causality test is applied in the present study. Furthermore, the selection of France, Germany, Ireland, Italy, Luxembourg, Spain, Sweden, and the United Kingdom is based on their data availability, policy diversity, and economic significance. These countries represent different welfare models and environmental policy approaches, allowing for a nuanced analysis of how income inequality interacts with public environmental protection expenditures across varying institutional contexts. The authors believe that this study is expected to make a significant contribution to the literature in terms of both the selected group of countries and the methodological approach employed.

DATA

In this study, 8 selected OECD countries (France, Germany, Ireland, Italy, Luxembourg, Spain, Sweden and United Kingdom) are subjected to econometric analysis and the period of analysis is based on 1998-2022. In the study, the time-varying effect of income inequality-environmental protection expenditures relationship was analysed with the Bootstrap Panel Rolling Window causality test. The period of analysis is chosen according to the data availability of the panel countries subject to the research. In other words, the data set cannot be further expanded due to data limitations. In the study, annual data set on income inequality and environmental protection expenditure variables is used. In the panel countries subject to the study, the Gini index as a measure of income inequality and the share of environmental protection expenditures in GDP are taken as basis. The data used in the study were obtained from the International Monetary Fund (IMF) and World Bank databases.

RESEARCH METHOD

In the study, cross-sectional dependence and heterogeneity tests—which are the prerequisites for the Bootstrap panel Granger causality test developed by Kónya (2006)—were first conducted. Then, the existence of a causal relationship between income inequality and environmental protection expenditures was examined using the Bootstrap panel Granger causality test. Finally, the time-varying causal relationship between the two variables was analyzed through the Bootstrap Panel Rolling Window Causality test.

In the sample of countries analyzed in this study, neglecting cross-sectional dependence can lead to biased analysis results. Therefore, before investigating causality relationships among the variables, it is necessary to test for cross-sectional dependence in the estimation model. To achieve this, the Lagrange Multiplier (LM) test, developed by Breusch and Pagan (1980), is used as a pioneering method to examine cross-sectional dependence. This test becomes more reliable when the time dimension (T) is smaller than the cross-sectional dimension (N). The formulation of the LM test statistic is presented in Equation (1):

$$LM = \sum_{i=1}^{N-1} \sum_{J=i+1}^{N} T_{ij}\, \hat{\rho}_{ij}^{2} \qquad (1)$$

The null hypothesis H_0 of the LM test states that "there is no correlation among residuals." The LM test statistic follows a X^2 distribution with n (n-1)/2 degrees of freedom. However, when the cross-sectional dimension (N) is larger than the time dimension (T), the LM test statistic may not be valid. In such cases, the scaled version of the test statistic proposed by Pesaran (2004) can be used. The test statistic is presented in Equation (2):

$$LM_s = \sqrt{\frac{1}{N(N-1)}} \sum_{i=1}^{N-1} \sum_{j=i+1}^{N} (T_{ij}\, \hat{\rho}_{ij}^{2} - 1) \qquad (2)$$

The scaled version of the Pesaran (2004) test statistic, denoted as LM_s, follows a N(0,1) distribution under the null hypothesis. However, when N is relatively larger than T, both LM and LM_s may exhibit significant size distortions. To address this, Pesaran (2004) developed an alternative test statistic, presented in Equation (3), where the null hypothesis H_0 states that "there is no correlation among residuals.":

$$CD = \sqrt{\frac{2}{N(N-1)} \sum_{i=1}^{N-1} \sum_{j=i+1}^{N} T_{ij}\, \hat{\rho}_{ij}^{2}} \qquad (3)$$

The CD test statistic developed by Pesaran (2004) has an asymptotic standard distribution. However, in cases where individual means differ from zero while the group mean is zero, the CD test statistic loses power and may produce biased results. To address this issue, Pesaran et al. (2008) proposed a bias-corrected LM test statistic. The bias-corrected LM test statistic (LM_{adj}) assumes the null hypothesis H_0, which states that there is no cross-sectional dependence. The formulation of the test statistic is presented in Equation (4):

$$LM_{adj} = \sqrt{\left(\frac{2T}{N(N-1)}\right)} \sum_{i=1}^{N-1} \sum_{j=i+1}^{N} \hat{\rho}_{ij} \frac{(T-k)\hat{\rho}_{ij} - \mu_{Tij}}{\sqrt{{}^{2}_{Tij}}} \tag{4}$$

LM_{adj} statistic follows an asymptotic standard normal distribution as $T \rightarrow \infty$ and $N \rightarrow \infty$. The null hypothesis H_0 for the test statistic is expressed as "there is no correlation among units."

Through the Bootstrap Panel Rolling Window causality method developed by Balcilar et al. (2010), the direction and magnitude of the bilateral relationship between economic growth, inflation, and external debt in the panel countries included in the study were first analyzed using the Bootstrap Panel Granger causality method developed by Kónya (2006). Subsequently, the causality relationship within subsamples was identified using the Bootstrap Panel Rolling Window approach. In other words, the primary objective of all these econometric tests is to reveal both the bilateral relationship between economic growth, inflation, and external debt for the entire panel and the time-based causality relationship over different periods.

The Bootstrap Panel Granger causality test developed by Kónya (2006) has several unique features: it does not require pre-tests such as unit root or cointegration, is based on seemingly unrelated regression (SUR) estimation that accounts for cross-sectional dependence, and does not require a common hypothesis for the entire sample since it uses Bootstrap critical values calculated for each unit individually based on Wald statistics. While applying this test, it is not necessary for the series in panel data analysis to be stationary or cointegrated. In order to apply the Kónya (2006) panel Bootstrap causality test, the model equation estimated using Zellner's (1962) SUR estimator is presented below in Figure 1 (Kónya, 2006, p. 981):

Figure 1. Model equation

$$Gini_{1,t} = \alpha_{1,1} + \sum_{i=1}^{IGini_1} \beta_{1,1,j}\, Gini_{1,t-j} + \sum_{i=1}^{IEnvironment_1} \gamma_{1,1,j} Environment_{1,t-j} + \varepsilon_{1,1,t}$$

$$Gini_{2,t} = \alpha_{1,2} + \sum_{i=1}^{IGini_1} \beta_{1,2,j}\, Gini_{2,t-j} + \sum_{i=1}^{IEnvironment_1} \gamma_{1,2,j} Environment_{2,t-j} + \varepsilon_{1,2,t} \quad (5)$$

$$Gini_{N,t} = \alpha_{1,N} + \sum_{i=1}^{IGini_1} \beta_{1,N,j}\, Gini_{N,t-j} + \sum_{i=1}^{IEnvironment_1} \gamma_{1,N,j} Environment_{N,t-j} + \varepsilon_{1,N,t}$$

and

$$Environment_{1,t} = \alpha_{2,1} + \sum_{i=1}^{IGini_1} \beta_{2,1,j}\, Gini_{1,t-j} + \sum_{i=1}^{IEnvironment_1} \gamma_{2,1,j} Environment_{1,t-j} + \varepsilon_{2,1,t}$$

$$Environment_{2,t} = \alpha_{2,2} + \sum_{i=1}^{IGini_1} \beta_{2,2,j}\, Gini_{2,t-j} + \sum_{i=1}^{IEnvironment_1} \gamma_{2,2,j} Environment_{2,t-j}\, \varepsilon_{2,2,t}$$
(6)

$$Environment_{N,t} = \alpha_{2,N} + \sum_{i=1}^{IGini_1} \beta_{2,N,j}\, Gini_{N,t-j} + \sum_{i=1}^{IEnvironment_1} \gamma_{2,N,j} Environment_{N,t-j} \varepsilon_{2,N,t}$$

ECONOMETRIC FINDINGS

The results of the cross-sectional dependence test are presented in Table 1. The test statistics for LM, LM_s, CD ve LM_{adj} which indicate cross-sectional dependence, demonstrate that the null hypothesis of "no cross-sectional dependence" is rejected at the 1% significance level for the panel countries analyzed in the study.

Table 1. Results of cross-section dependence tests

Test Statistic	Statistic Value	Probability (p-value)
LM	192.104***	0.000
LM_s	28.15***	0.000
CD	7.901***	0.000
LM_{adj}	3.588***	0.000

Note: (***) denotes statistical significance at the 1% level.

The results of the causal relationship between income inequality (gini) and environment protection expenditures (environment) are shown in Table 2.

Table 2. Bootstrap panel causality results

H_0= Gini does not cause environment.					
Countries	**Coefficient**	**Wald Sta.**	**Boots. Crit. Values**		
			%1	**%5**	**%10**
France	0.007	2.817	8.281	4.208	3.076
Germany	-0.009	1.799	14.939	8.646	6.711
Ireland	0.017	7.326***	4.795	3.088	1.973
Italy	-0.005	1.423	12.276	6.697	4.429
Luxembourg	0.026	24.199*	25.301	15.989	11.531
Spain	-0.022	13.797*	16.532	8.415	6.182
Sweden	0.007	2.987	29.630	18.423	14.393
United Kingdom	-0.012	0.586	12.763	7.455	5.507
H_0= Environment does not gini.					
Countries	**Coefficient**	**Wald Stat.**	**Boots. Crit. Values**		
			%1	**%5**	**%10**
France	-0.480	0.211	6.019	3.424	2.514
Germany	0.219	0.021	13.083	7.932	5.813
Ireland	1.368	2.426**	4.738	2.701	1.923
Italy	1.707	0.832	16.369	11.224	7.947
Luxembourg	-1.740	0.395	24.347	10.434	7.218
Spain	3.067	1.554	13.627	6.381	4.502
Sweden	1.003	0.099	25.322	16.381	12.773
United Kingdom	0.757	0.966	8.420	4.302	3.003

Note: (***), (**), (*) denote statistical significance at 1%, 5% and 10% levels, respectively.

The upper part of Table 2 shows the findings related to the test of the hypothesis "income inequality does not cause environmental protection expenditures" and the lower part shows the results related to "environmental protection expenditures do not cause income inequality". The null hypothesis 'income inequality does not cause environmental protection expenditures' is rejected at 1% significance level for Ireland and at 10% significance level for Luxembourg and Spain. Income inequality has a statistically significant and positive effect on environmental protection expenditures in Ireland and Luxembourg and a negative effect in Spain. There is no causal relationship from income inequality to environmental protection expenditures in France, Germany, Italy, Sweden and United Kingdom. According to the econometric findings, the null hypothesis that 'environmental protection expenditures do not cause income inequality' is rejected only for Ireland at 5% significance level. In this country, environmental protection expenditures positively affect income inequality. It is also observed that there is a bidirectional causal relationship between income inequality and environmental protection expenditures in Ireland. There is no statistically significant causal relationship between environmental protection expenditures and income inequality for the other seven countries (France, Germany, Italy, Luxembourg, Spain, Sweden and United Kingdom).

However, the interaction of the relationship between the variable may change over time. In this context, in order to test the causal relationship between income inequality and environmental protection expenditures, sample coefficients are rounded to test the causal relationship between these variables in sub-samples, and these unpredictable changes can be captured. In order to capture this unpredictable causal relationship, the Boostrap Rolling Window causality method was utilised.

The main hypothesis on the causality relationship states that there is no Granger causality between the variables. In the Boostrap Rolling Window causality test method, Boostrap p-values are collected from each sub-sample and these values are compared with the 10% critical value level. If the Boostrap p value is below the 10% critical value in any sub-period, the main hypothesis is rejected for the relevant sub-period. In the study, the sub-sample size is estimated by $[T(0.01+1.8/\sqrt{T})]$. In the function 'T' indicates the time dimension of the model. In a 28-year forecast model, the sub-sample size $[28(0.01+1.8/\sqrt{28})]=9.80$ is approximately 10 years. Since causal relationships may not remain constant over time-particularly during periods of crisis-working with time-varying sub-samples instead of a fixed model can help detect dynamic interactions. In this context, the sub-sample size determines how many periods of observation are included within the moving window over time in a panel dataset.For example, based on the dataset used in this study, which spans 28 years from 1998 to 2022, the sub-sample size for the Bootstrap Rolling Window causality test is determined using the estimation formula $[T(0.01+1.8/T)]$. Accordingly, the sub-sample size has been set at 10 years. In other words, the sub-sample size indicates the specific time periods in which the dynamic relationship between the two variables is observed.

Figure 2 and Figure 3 show the p-values of the Boostrap Rolling Window causality test for the detection of causality relations between income inequality and environmental protection expenditures in each of the panel countries of the study.

When the Boostrap Rolling Window causality results from income inequality to environmental protection expenditures presented in Figure 2 are evaluated, it is revealed that there is no causal link from income inequality to environmental protection expenditures for Sweden. In the panel countries of France, Germany, Ireland, Italy, Luxembourg, Luxembourg, Spain and the United Kingdom, the existence of latent hidden causality was detected in some periods. The main hypothesis 'income inequality is not the cause of environmental protection expenditures' was rejected in France in 2016 and 2022; Germany in 2004, 2005, 2006, 2007, 2008, 2016 and

2020; Ireland in 2022; Italy in 2016 and 2022; Luxembourg in 2004, 2005 and 2006; Spain in 2014, 2018 and 2020 and United Kingdom in 2014, 2020 and 2022. In these countries, there is a causal relationship from income inequality to environmental protection expenditures in these years.

These results can be attributed to the fact that, in France, the fuel tax hikes and the Yellow Vests protests in 2016 increased public opposition to environmental policies, while in 2022, the energy crisis in Europe and high inflation shifted the government's perspective on environmental spending within the public budget. In Germany, the deterioration in income distribution following the Agenda 2010 and Hartz IV reforms between 2004 and 2007 led to a prioritization of social expenditures over environmental spending. The 2008 global financial crisis caused cuts in environmental projects, while in 2016 and 2020, the aftermath of Brexit and the COVID-19 pandemic led to a renewed focus on social policy priorities. In Ireland, the rapid increase in tax revenues in 2022 enabled the expansion of environmental funds (e.g., for water infrastructure and energy efficiency). Additionally, the discourse around combating inequality was associated with support for green budgets. In Italy, in 2016 and 2022, environmental regulations-such as waste management and carbon permit schemes-were introduced under pressure from the EU. These developments may have triggered a noticeable negative or positive relationship between inequality and environmental expenditures. In Luxembourg, the austerity policies implemented in the aftermath of the Eurozone debt crisis in 2014 led to cuts in environmental spending. In 2018, political instability and budgetary constraints contributed to the postponement of environmental projects. In 2020, the COVID-19 crisis shifted priorities toward healthcare, thereby reducing the allocation for environmental spending. In the United Kingdom, the uncertainty surrounding the Brexit process in 2014 marked a period in which environmental expenditures were deferred. In 2020, the COVID-19 crisis drew full attention to the healthcare system, placing pressure on environmental budgets. In 2022, the energy crisis and inflation altered the direction of public spending, leading to a reprioritization in the environmental policy agenda.

When the Bootstrap Rolling Window causality findings from environmental protection expenditures to income inequality are evaluated in Figure 3, there is no hidden causality in the panel countries except Italy. Accordingly, a causal relationship was found from environmental protection expenditures to income inequality in Italy in 2010, 2011, 2012, 2013 and 2014.

Figure 2. Bootstrap rolling window causality test results from gini to environment (Bootstrap p-values)

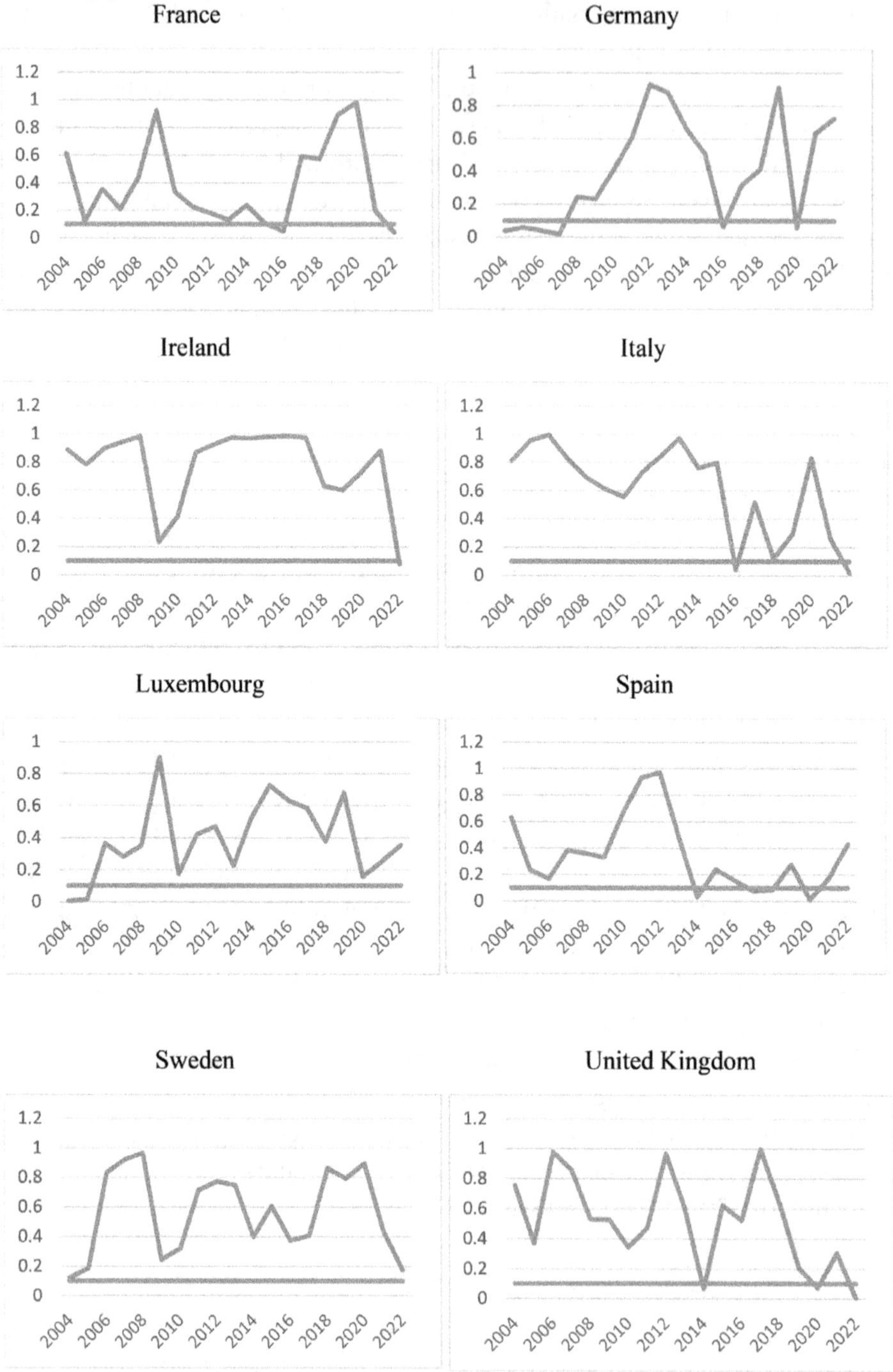

Figure 3. Bootstrap rolling window causality test results from environment to gini to (Bootstrap p-values)

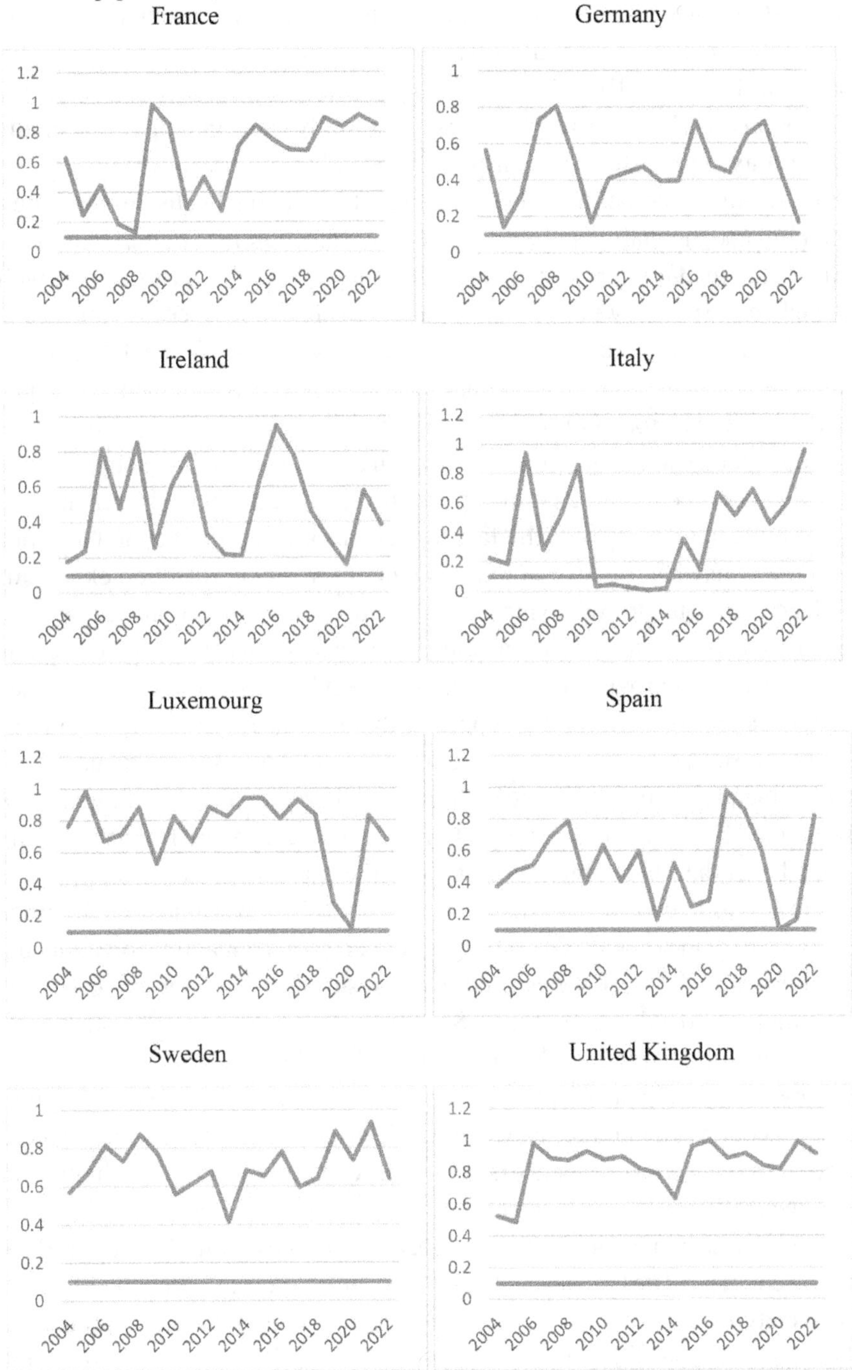

CONCLUSION

In this study, using the Bootstrap Panel Rolling Window causality test, the relationship between environmental protection expenditures and income inequality is analysed for the period 1998-2022 by using annual data on the basis of 8 selected OECD countries. The empirical findings of the study show that there is a unidirectional causality relationship between income inequality and environmental protection expenditures in Luxembourg and Spain, and a bidirectional causality relationship between income inequality and environmental protection expenditures in Ireland.

All of the panel countries subject to the study are OECD countries as well as developed countries. However, these countries exhibit differences in terms of their level of development, economic size, population size, environmental regulations and environmental policies. Therefore, the relationship between income inequality and environmental protection expenditures for the relevant panel countries may differ from country to country. For example, in Spain, income inequality was found to reduce environmental protection expenditures, while the opposite result was obtained in Luxembourg. In fact, the fact that environmental protection has a public characteristic may be effective in reducing environmental protection expenditures due to income inequality. When income inequality is high in a country, low-income groups may be preoccupied with meeting their basic needs and may not be able to prioritise issues that provide long-term and indirect benefits, such as environmental protection, thereby reducing public support in this area. The political power of wealthy groups may also be effective in reducing environmental protection expenditures due to income inequality. Indeed, since environmental protection expenditures are generally covered by tax increases, the rich may lobby against environmental policies that may harm their economic interests. Another factor is that income inequality may reduce public confidence, making it difficult for government environmental investments to gain support. The increase in unemployment and income inequality in the aftermath of the 2008 crisis led to the restriction of public budgets and the prioritisation of social expenditures. So much so that non-urgent issues such as environmental protection have been put on the back burner.

However, in very rich and prosperous countries such as Luxembourg, environmental protection expenditures can be prioritised even if income inequality increases. One of the reasons for this is that since per capita income is very high, even if income inequality increases, a large part of the society can comfortably meet their basic needs, so they may be willing to allocate resources for high-level needs such as the environment. The high level of education and strong environmental awareness of the society in this country cause individuals to see environmental protection as a prestige and responsibility. On the other hand, countries such as Luxembourg see environmental protection expenditures as an economic opportunity such as green energy and sustainable technology and consider these expenditures as investments in growth and new business areas. To summarise, in Spain, inequality is effective in reducing environmental expenditures because a large part of the society struggles with basic problems such as food, housing and health, while in Luxembourg, inequality may be effective in increasing environmental expenditures because the level of welfare is very high, the society is conscious and sees environmental expenditures as an economic opportunity.

Another empirical finding of the study is that there is a positive bidirectional relationship between income inequality and environmental protection expenditures in Ireland. The positive relationship observed in Ireland may stem from the high-income group's demand for environmental quality, EU regulations, or the government's strategy to respond socially to income inequality. In addition, the attempt to shift economic growth to environmentally friendly sectors in this country has increased expenditures for this sector.

Since policies aimed at reducing income inequality can also strengthen environmental policies, social and environmental policies should not be treated separately but rather addressed in an integrated manner. Environmental expenditures should be supported through fair taxation. Ensuring that the funds allocated for environmental spending are collected in a fair and transparent way-such as redistributing a portion of carbon tax revenues to low-income groups-can help mitigate inequality-related resistance to such policies. Moreover, approaches that define environmental protection not as a luxury but as a public right for all should be promoted.

When examined on a country-specific basis, it is observed that in Spain, as income inequality increases, environmental protection expenditures decrease. Therefore, public campaigns and environmental education programs aimed at increasing environmental awareness among low-income groups should be promoted. In Luxembourg, even though income inequality increases, environmental protection expenditures also increase. This positive trend should be maintained, and environmental spending should be viewed as an investment in innovative sectors such as green technologies and sustainable R&D. In the case of Ireland, there is evidence of a bidirectional relationship: income inequality affects environmental expenditures, and environmental expenditures also influence income inequality. The growth potential of environmental sectors should be utilized as a tool to combat inequality, and vocational training programs targeting low-income groups in green sectors should be encouraged.

In conclusion, addressing income inequality not only contributes to social justice but also enhances the public legitimacy and sustainability of environmental protection efforts. Therefore, integrated policy approaches that link social equity with environmental objectives are essential, particularly in countries where inequality may otherwise undermine environmental investments.

REFERENCES

Ajide, K. B., & Ibrahim, R. L. (2022). Environmental impacts of income inequality: Evidence from G7 economies. *Environmental Science and Pollution Research International*, *29*(2), 1887–1908. DOI: 10.1007/s11356-021-15720-6 PMID: 34363162

Balcilar, M., Ozdemir, Z. A., & Arslanturk, Y. (2010). Economic growth and energy consumption causal nexus viewed through a bootstrap rolling window. *Energy Economics*, *32*(6), 1398–1410. DOI: 10.1016/j.eneco.2010.05.015

Bimonte, S. (2002). Information access, income distribution, and the environmental kuznets curve. *Ecological Economics*, *41*(1), 145–156. DOI: 10.1016/S0921-8009(02)00022-8

Bimonte, S. (2002). Information access, income distribution, and the environmental kuznets curve. *Ecological Economics*, *41*(1), 145–156. DOI: 10.1016/S0921-8009(02)00022-8

Blake, D.E., Guppy, N., & Urmetzer, P. (1997). Canadian public opinion and environmental action: evidence from British Columbia. *Canadian Journal of Political Science/Revue canadienne de science politique*, 30(3), pp.451-472.

Boyce, J. K. (1994). Inequality as a cause of environmental degradation. *Ecological Economics*, *11*(3), 169–178. DOI: 10.1016/0921-8009(94)90198-8

Breusch, T., & Pagan, A. (1980). The lagrange multiplier test and its application to model specifications in econometrics. *The Review of Economic Studies*, *47*(1), 239–253. DOI: 10.2307/2297111

Broniewicz, E. (2021). Environmental protection expenditure in European Union. *Environmental Management in Practice*, CEE-M Working Paper.

Dağdemir, Ö. (2015). *Çevre sorunlarına ekonomik yaklaşımlar ve optimal politika arayışları (3. Baskı)*. Gazi Kitabevi.

Değirmenci, T. (2020). *Türkiye'de çevre koruma harcamalarının sürdürülebilir kalkınma boyutu. H. Yavuz (Edt.), Çevre ekonomisi ve mali ilişkiler üzerine seçme yazılar*. Sakarya Üniversitesi Yayınları.

Değirmenci, T., & Aydın, M. (2020). Çevre koruma harcamaları ile gelir dağılımı ve ekonomik büyüme arasındaki dinamik ilişkiler: Seçili OECD ülkeleri için panel nedensellik yaklaşımı. *Sosyoekonomi*, *28*(46), 391–406. DOI: 10.17233/sosyoekonomi.2020.04.18

Elliott, E., Seldon, B. J., & Regens, J. L. (1997). Political and economic determinants of individuals support for environmental spending. *Journal of Environmental Management, 51*(1), 15–27. DOI: 10.1006/jema.1996.0129

Ertürk, H. (1998). *Çevre bilimlerine giriş (3. Baskı).* Ceylan Matbaacılık.

Hao, Y., Chen, H., & Zhang, Q. (2016). Will income inequality affect environmental quality? Analysis basedon China's provincial panel data. *Ecological Indicators, 67*, 533–542. DOI: 10.1016/j.ecolind.2016.03.025

Kahn, M. E., & Matsusaka, J. G. (1997). Demand for environmental goods: Evidence from voting patterns on California initiatives. *The Journal of Law & Economics, 40*(1), 137–174. DOI: 10.1086/467369

Kónya, L. (2006). Exports and growth: Granger causality analysis on OECD countries with a panel data approach. *Economic Modelling, 23*(6), 978–992. DOI: 10.1016/j.econmod.2006.04.008

Magnani, E. (2000). The environmental kuznets curve, environmental protection policy and income distribution. *Ecological Economics, 32*(3), 431–443. DOI: 10.1016/S0921-8009(99)00115-9

Marsilliani, L., & Renström, T. I. (2000). *Inequality environmental Protection and growth.* Center for Economic Research, Durham University. DOI: 10.2139/ssrn.235083

Morse, S. (2018). Relating environmental performance of nation states to income and income inequality. *Sustainable Development (Bradford), 26*(1), 99–115. DOI: 10.1002/sd.1693

Newell, S. J., & Green, C. L. (1997). Racial differences in consumer environmental concern. *The Journal of Consumer Affairs, 31*(1), 53–69. DOI: 10.1111/j.1745-6606.1997.tb00826.x

Pesaran, M. H. (2004). *General diagnostic tests for cross section dependence in panels*, Cambridge Working Papers in Economics, 435, London: BP.

Ravallion, M., Heil, M., & Jalan, J. (2000). Carbon emissions and income inequality. *Oxford Economic Papers, 52*(4), 651–669. DOI: 10.1093/oep/52.4.651

Reardon, T., & Vosti, S. A. (1995). Links between rural poverty and the environment in developing countries: Asset categories and investment poverty. *World Development, 23*(9), 1495–1506. DOI: 10.1016/0305-750X(95)00061-G

Ridzuan, S. (2019). Inequality and the environmental Kuznets curve. *Journal of Cleaner Production, 228*, 1472–1481. DOI: 10.1016/j.jclepro.2019.04.284

Toprak, D. (2006). Sürdürülebilir kalkınma çevresinde çevre politikalari ve mali araçlar. *Süleyman Demirel Üniversitesi Sosyal Bilimler Enstitüsü Dergisi*, (2), 146–169.

Torras, M., & Boyce, J. (1998). Income, inequality, and pollution: A reassessment of the environmental kuznets curve. *Ecological Economics*, *25*(2), 147–160. DOI: 10.1016/S0921-8009(97)00177-8

Vona, F., & Patriarca, F. (2011). Income inequality and the development of environmental technologies. *Ecological Economics*, *70*(11), 2201–2213. DOI: 10.1016/j.ecolecon.2011.06.027

Xiaoyun, W., & Chuanmin, Z. (2024). Does Chinese environmental policy affect income inequality? Evidence from the central environmental protection inspection. *China & World Economy. Institute of World Economics and Politics*, *32*(3), 130–161.

ENDNOTES

1 Assoc. Prof. Dr., Izmir Katip Celebi University, Faculty of Economics and Administrative Sciences, Department of Finance, nazli.keyifli@ikc.edu.tr, ORCID: 0000- 0002-0589-8089.

2 Prof. Dr., Gumushane University, Faculty of Economics and Administrative Sciences, Department of Finance, suathayrisenturk@gmail. com, ORCID: 0000-0003-2930-8888.

3 Ege University, Ege Higher Vocational School, Accounting and Tax Application.

4 Prof. Dr., Karadeniz Technical University, Faculty of Economics and Administrative Sciences, Department of Finance, birolkarakurt@ktu.edu.tr, ORCID: 0000-0002-6325-2171.

Chapter 11
Innovative Tools for Local Climate Action Policy, Technology, and Collaboration

Ahsen Sacli
https://orcid.org/0000-0002-5665-2855
Ankara Hacı Bayram Veli Üniversitesi, Turkey

ABSTRACT

The increasing frequency of global environmental disasters is forcing the world to join the fight against climate change. The solution, whether on a global or local, lies in governance system that protects the world. However, with over half of the world's population living in cities, climate change poses increasingly challenging problems for urban. Due to globalization, local governments are becoming important. This study focuses on local practices and examining climate actions that integrate local governments with international climate policies governing the world. The study examines the foundations linking local government actions to global climate change in terms of political, technological, and cooperative dimensions. The political dimension defines climate change governance, the cooperative dimension is defined by international alliances, and the technological dimension is defined by international coalitions. Without collaboration among local governments, states, international agreements, alliances, and coalitions, sustainable solutions to climate change cannot be found.

DOI: 10.4018/979-8-3373-1310-8.ch011

INTRODUCTION

Climate change is one of the natural consequences of the increasing environmental problems that came to the world's agenda in the 1970s. Climate change, whose effects began to be seen as a result of environmental problems such as the increase in greenhouse gases and the decrease in sink areas, is a global problem and is not one of the problems that can be solved by a single country. The process of addressing this critical issue began in earnest with the signing of the United Nations Framework Convention on Climate Change in 1992, which entered into force in 1994. The Kyoto Protocol, signed on December 11, 1997, further established climate change as a legally binding issue on the international stage.

In general, mutual interdependence and solidarity are at the core of solving environmental problems. It is well known that countries must act together with other countries to solve these problems that they cannot solve on their own, and that they must behave in solidarity. A paradigm with lasting effects must be implemented to solve climate change. In this context, the paradigm of international governance on climate change has come to the fore. According to this paradigm, climate change governance is explained by a multi-level governance paradigm that encompasses not only states but also all non-state actors at the global level. Similarly, this paradigm is also valid at the national and local levels. It is stated that climate change governance, which is multi-level and multi-actor, will be successful if all actors fulfill their responsibilities and roles and ensure their continuous participation. In general, participation brings ownership of the issue. Civil society is particularly important in the fight against climate change because solutions are mainly based on persuasion due to their economic nature. Therefore, it is necessary and sufficient to focus on solving problems through multi-level governance. The prominent actors in climate change governance are states, cities and local governments, civil society organizations, and other elements. This section covers innovations that can be developed in terms of politics, technology, and cooperation in local climate actions.

The actors in local climate change governance include local authorities, municipalities, civil society, municipal companies, private companies, and other elements that vary according to the characteristics of the city. In particular, increasing urbanization worldwide, the growing proportion of the world's population living in cities, and the resulting increase in energy consumption, transportation-related environmental pollution, greenhouse gas emissions, and many other problems are causing significant problems. In this section of the book, local climate actions will be evaluated within the framework of innovative approaches based on politics, technology, and cooperation. The aim here is to provide a new perspective by addressing local climate actions and their components within the scope of innovative approaches.

1. URBAN CLIMATE CHANGE AND LOCAL CLIMATE ACTION

In the 21st century, the number and structure of cities continue to grow. According to 2024 data, 8.08 billion people, or 57.7% of the total population, live in cities (We Are Social, 2024). Therefore, the fact that more than half of the world's population lives in cities also means that greenhouse gas emissions will continue to increase. Cities, where economic and social life is concentrated, will also play a major role in reducing greenhouse gas emissions. For this reason, they must work towards transitioning to a zero-carbon economy. At this point, governments initially did not seem very willing to allow local governments to directly regulate or provide services to reduce greenhouse gas emissions or implement the necessary measures as part of climate change mitigation policies. However, as local governance styles have become dominant around the world, local climate governance has also begun to become effective.

Urban climate change is addressed through vertical and horizontal cooperation within the multi-level governance stages. Accordingly (Demirci, 2015, pp. 84-85):

1.*Horizontal Cooperation:* Horizontal cooperation in local climate change governance takes place in metropolitan cities through climate change governance and in international and national local networks. In particular, local boundaries can pose challenges for climate change adaptation preparations in metropolitan areas. Therefore, climate change adaptation strategies must be prepared and implemented on a broader or regional scale. Globally, there is already significant sharing of experience and practices in climate governance at the horizontal level among cities (Demirci, 2015, pp. 84-85).

2. *Vertical Collaborations*: In the context of vertical collaborations, which are as important as horizontal collaborations, it is necessary for states to establish a framework and conditions for local governments to implement climate policies (Zimmermann et al., 2010, p.84):

-Firstly, legislation that outlines the boundaries of the climate actions that local governments must implement must be in place to provide support,

-Appropriate regulations must be in place to ensure that local governments receive the financial support necessary to implement their regular climate action plans,

-Direct financial support must be provided to local governments for their local climate action plans,

-Local governments must be legally obligated to combat climate change (Zimmermann et al., 2010, p.84). Over time, many countries have introduced legal obligations at various levels according to their own systems.

The reason why multi-level governance practices are increasingly preferred in the fight against climate change is that it is believed that communication between all possible actors and institutions at the global, national, regional, and local levels will increase (Ozer, 2017, p.841). Multi-level governance is also supported by the Intergovernmental Panel on Climate Change (IPCC). It is emphasized that this will have a positive impact on adaptation to climate change and ensure effective urban management (IPCC, 2014, p. 538). The preference for multi-level climate governance in the fight against climate change by local governments provides advantages in every respect for the implementation of local climate action. First and foremost, a holistic approach is required to combat climate change. In addition, political will and voluntary action are essential. Success can be achieved through the efforts of all actors, including the state and local governments, with the support of legal regulations on mitigation, control, and adaptation.

2. LOCAL CLIMATE ACTION POLICY: CLIMATE CHANGE GOVERNANCE

Environmental problems are not issues that any state can keep within its own borders or solve on its own. These problems can only be solved through cooperation within the framework of mutual dependence between states at the international level. The dependence of states on each other is evident not only economically but also environmentally. This mutual dependence is particularly high in the case of global environmental problems such as climate change.

The effects of climate change, one of the most important issues for the sustainability of life on Earth in the 21st century, are rapidly spreading and becoming apparent. Of course, climate has never been stable in the millions of years of Earth's existence. However, with the Industrial Revolution, human pressure on nature has increased and accelerated climate change. With developments in science and technology, environmental pollution, depletion of natural resources, forests, and species, and the threat of natural life disappearing at an increasingly rapid rate have come to the fore. Due to these and many other destructive effects, all species on Earth, including humans, find themselves in a struggle to survive. Efforts to address global warming and climate change date back to the time when environmental issues first became a topic of international discussion. The climate change caused by the greenhouse effect was first recognized in the 1960s (Mengi, 1999, p. 44). Once recognized, efforts to address this issue began in the mid-1970s. Throughout this process, numerous action plans and strategies were developed, concrete steps were taken to address climate change, and progress was made, with the necessary mechanisms established to address this issue.

One of the important outcomes of the 1991 Rio Conference on Environment and Development was the Framework Convention on Climate Change, which entered into force in 1994. The Convention aims to eliminate, minimize, or maintain greenhouse gas emissions caused by human activities at levels that do not pose a threat to the climate. The Kyoto Protocol was adopted at the 3rd Conference of the Parties to the Convention in 1997. This marked the first real step toward preventing climate change under the Framework Convention on Climate Change. What is important here is that the Protocol established the mechanisms and responsibilities that must be implemented to reduce greenhouse gas emissions. The Kyoto Protocol aims to achieve a balance in greenhouse gas emissions in the atmosphere to prevent them from reaching harmful levels (Gormez, 2018, pp. 67-69). Within this framework, according to the flexibility mechanisms defined in the Protocol:

-**Emission Trading – ET:** Regulated in Article 17 of the Protocol.
-**Joint Implementation – JI:** Regulated in Article 6 of the Protocol.
-**Clean Development Mechanism – CDM:** Regulated in Article 12 of the Protocol (United Nations Framework Convention on Climate Change, 2025; Gormez, 2018, pp. 69-70).

The Conference of the Parties meets once a year, unless the participating countries decide otherwise in the previous year. Although the climate change process carried out through these conferences has not made much progress over time, the efforts of countries are increasing as the effects of climate change become increasingly severe. Although the re-elected President of the United States, Donald Trump, withdrew from the Climate Change Framework Agreement as he did during his previous presidency, city administrations in some states are voluntarily continuing their efforts to fulfill the requirements of this agreement. In this context, it is argued that much greater success can be achieved globally on climate change through local applications.

In the 21st century, human rights such as the right to the environment, the right to health, and the right to peace, which are considered solidarity rights within the scope of individual rights and freedoms, have come to the fore, and demands for decentralization and the right to participation have gained momentum. In a changing world influenced by globalization, the protection of national interests of states is also dependent on the organizational structure, authority, and responsibilities of local administrations, as well as the services they provide and their efforts to benefit society. The aim here is to ensure a sensitive balance between the principles of administrative integrity and locality. Balance can be achieved by ensuring the effective, economical, and efficient use of resources to create the most public value (Ozer, 2023, p.16).

Criticism of management processes at the national or international level is generally evaluated within the framework of the secrecy and authoritarianism brought about by the centralised structure in decision-making processes. Here, the concept of governance is used as an alternative. In addition to the traditional concept of management, there is talk of a more participatory structure with the introduction of control mechanisms. Climate change governance is a very important issue from a global, national, and local perspective. Climate change governance refers to a management process in which the public administration, the private sector, civil society, the state, and non-state organizations make decisions together. In summary, climate change governance demonstrates a multi-level form of governance. Multi-level governance involves horizontal and vertical cooperation. Cooperation between the state, the market (economic sphere), and civil society describes horizontal cooperation. Cooperation between top-down and bottom-up levels refers to vertical cooperation. Climate change governance requires a properly functioning system of control and institutional and administrative structures in order to solve problems that are as important as participation. What distinguishes climate change governance from other examples of governance is that it involves many different economic, environmental, and social variables. This situation makes compact cooperation between the state and all other actors a necessity (Ozer, 2017, pp. 833-851).

In general, urbanization rates have increased significantly worldwide since the mid-20th century, and this increase continues today. This increase has led to unhealthy and unplanned urbanization, in addition to exceeding nature's limits of self-renewal. Climate change is explained as significant changes that occur over time due to human activities, in addition to natural changes. These changes have emerged due to human activities in the 21st century and have become a threat to the lives of all ecosystems in the world. Extremely severe heat waves, violent storms, volcanic eruptions, increased impact of earthquakes, tsunamis, and many other natural events have caused significant problems in all infrastructure systems of cities, such as electricity, water, sewage, and transportation, and serious security problems have arisen. Therefore, the issue of climate change, which will greatly affect the future of local governments, is included in many global conferences, international agreements, and the United Nations (UN) Sustainable Development Goals (Ozer, 2023, pp. 17-18). Climate Action is one of the most important Sustainable Development Goals of the UN, with a target date of 2030 (United Nations, 2016). The comprehensive goal of the Paris Agreement, signed at the 21st Conference of the Parties to the United Nations Framework Convention on Climate Change in Paris in 2016, is to keep global average temperature increases below 2°C compared to pre-industrial levels. The comprehensive goal of the Paris Agreement, signed at the 21st Conference of the Parties, is to keep global average temperature increases below 2°C compared to pre-industrial levels and to limit global temperature increases to 1.5°C above pre-industrial levels. In addition, the Agreement aims to enhance the capacity to adapt to climate change, promote low greenhouse gas emissions development in response to climate change, and mobilize financial resources to address the existential threat posed by climate change (United Nations Framework Convention on Climate Change, 2025). Furthermore, the New Urban Agenda, adopted at the HABITAT III Summit held in Quito in 2016, has become one of the most important elements concerning the fight against climate change worldwide. The New Urban Agenda includes decisions such as minimizing disaster risks, building resilient and sustainable cities, reducing the negative effects of climate change, and creating climate-resilient cities (HABITAT III. 2016). C40 Cities, a global network of mayors from the world's leading cities committed to fighting the climate crisis, aims to set achievable and measurable targets to combat the effects of climate change, share information effectively, and collaborate within this framework. Initially, 18 cities came together under the leadership of the Mayor of London to form C20. Later, it became an organization with approximately 100 mayors as members under the name C40 (C40, 2025). It is stated that development should be aimed at the "biopolis" concept, which seeks to ensure that people live freely in a natural world, environmental sustainability is achieved, and zero pollution is targeted. The concept of biopolis refers to a shared life where humans and nature are in balance, environmental sustainability is achieved, and there is zero pollution (Ozer, 2023, p.18). The Climate Change Framework Convention Conference of the Parties (COP) announced the Climate Change Agreement to the world at COP 21, held in Paris in 2015. The Climate Change Plan was presented at COP 24, held in Katowice (Poland) in 2018. At COP 26, held in Glasgow, Scotland, in 2021, the Climate Change Plan was presented again, and a decision on global stock assessment was made. At COP 27, held in Sharm El-Sheikh, Egypt, in 2022, the transition to implementation of climate action began. At COP 28, held in Dubai (United Arab Emirates) in 2023, global inventory work was completed, and countries were invited to implement more ambitious climate action plans (United Nations Framework Convention on Climate Change, 2025). As a result, COP 28 has been the most critical conference since COP 21. Additionally, COP 28 emphasized the critical impact of local governments.

3. LOCAL CLIMATE ACTION COLLABORATION: INTERNATIONAL COOPERATIONS

Since the 1992 United Nations (UN) Conference on Environment and Development, local governments have been recognized as key stakeholders in global sustainability agenda efforts by nine stakeholder groups, including local governments. Local Governments and Municipal Authorities (LGMA), has represented local and regional government networks under the UN Framework Convention on Climate Change (UNFCCC). Since the first Conference of the Parties (COP) in 1995, LGMA has represented cities and regions and has been working to promote multi-level action on climate, nature, and desertification. Additionally, in 1995, parallel to COP 1, the Local Governments for Sustainability (ICLEI) held a Mayors' Summit on climate change, establishing a new tradition. When states gather to address climate change, representatives of local governments also convene (Zimmermann et al., 2010, p.83). After the Kyoto Protocol came into force in 2005, it became clear that local governments were rightly considered important policy actors in climate change policies implemented by states in climate change research in cities. The Local Government Climate Roadmap was launched at the 13th Conference of the Parties (COP 13 Local Governments Climate Roadmap, 2007) in 2007. Accordingly (United Nations Framework Convention on Climate Change, 2025):

1. Local climate strategies must be part of national climate strategies.
2. Local governments are the closest administrative units to the public, so climate action strategies are much more effective here.
3. Local governments in developing countries must demonstrate their commitment to reducing greenhouse gas emissions in their cities by 60% from 1990 levels, and by 80% in developed countries (United Nations Framework Convention on Climate Change, 2025).

There are also many international local government networks working on climate change. Some of these are (Demirci, 2015, p. 86):

-Local Governments for Sustainability (ICLEI)
-United Cities and Local Governments (UCLG)
-Climate Leadership Group (C40)
-World Mayors Council on Climate Change (WMCCC)
-Climate Alliance

With the Paris Agreement signed in 2015, local governments have been recognized as having a fundamental role in developing, implementing, transforming, and managing and directing climate action. For this reason, the Paris Agreement is considered a turning point for local governments. Since the Paris Agreement, LGMA has advocated that local and regional governments are key actors in shaping and implementing global climate strategies and that this role must be further strengthened. They argue that if successful at COP 30 in Belém (Brazil) in 2025, this will become a standard practice. Local governments use COPs to assess progress. The CHAMP process communicates the information obtained here to the states. The states then update this information and revise their NDCs before submitting them to the UN Climate Change (Local Governments and Municipal Authorities, 2025). While this process is ongoing, it is expected that every local government will have its own "Climate Action Plan," such as Local Agenda 21.

It is known that local governments have undergone a transformation to local governance systems since the 1990s. This transformation was initiated and sustained by New Public Management reforms. It is argued that local climate change mitigation efforts in cities can be classified as follows in terms of governance methods (Demirci, 2015, pp. 82-83):

1. Local governments managing their own actions as both consumers and role models
2. Local governments managing by facilitating non-state actors
3. Local governments managing by providing services as suppliers
4. Local governments managing by regulating as regulators

In this way, local governments can guide climate change mitigation efforts at the local level while fulfilling their various roles or responsibilities (Demirci, 2015, pp. 82-83).

In order to reduce greenhouse gas emissions, regulate infrastructure, provide services, and establish governance, the following factors must be fulfilled by local governments (Demirci, 2015, p.84; Bulkeley et al., 2011, p.127):

1. Policy makers who support local climate actions
2. Creating additional financial resources
3. The existing responsibilities and authorities of local governments
4. The framework established by the central and regional governments regarding the scope of responsibilities and authorities of local governments
5. Support provided by international local government networks
6. Making the climate change issue solvable at the local level

Therefore, strengthening local governance and increasing its capacity within the framework of reduction and adaptation policies at the local level is indispensable for cities (Demirci, 2015, p.84; Bulkeley et al., 2011, p.127).

In order to achieve the long-term goals set out in the Paris Agreement signed in 2015, it is necessary to rapidly adopt clean technologies that are currently in use, promote their widespread use, accelerate innovation, implement digital transformation, and facilitate access to newly developed technologies.

4. LOCAL CLIMATE ACTION COLLABORATION: INTERNATIONAL CLIMATE INITIATIVES AND COALITIONS

It is known that approximately 70% of greenhouse gas emissions caused by economic activities such as transportation around the world originate from cities (Intergovernmental Panel on Climate Change, 2022). Eight percent of global greenhouse gas emissions come from the transportation sector (Nature Climate Change, 2016). At the same time, since cities are home to more than half of the world's population, a significant portion of food consumption also takes place in cities. Food waste also produces a significant amount of methane gas, thus contributing greatly to greenhouse gas emissions. Cities are expected to reduce their global greenhouse gas emissions by half by 2030. To achieve this, cities are expected to develop and implement their own climate action plans. The Paris Agreement and all other international climate-related documents aim for all cities around the world to prepare climate action plans to "prepare for all climate-related impacts and develop solutions to ensure sustainability." The first step to achieve these goals is to collaborate with governments, and the second is to secure access to financial resources. Innovative technologies are needed to reduce the harmful effects of climate change and adapt to its requirements. It is known that local governments do not have the financial resources required to implement climate action plans, but they are one of the most fundamental elements in the fight against climate change (Kahraman & Acar, 2024, p. 20). This situation also concerns the private sector and the business world (Ozer, 2017, p. 838). States grant powers to local governments within the framework of certain laws. Therefore, each state has granted certain powers to city administrations regarding climate action plans within the framework of its own administrative system. Although there are significant differences in these powers depending on the city, city administrations can create significant impacts by forming unions or cooperating with each other. In addition, local governments have the capacity to reduce greenhouse gas emissions, develop urban and environmental plans, and reduce industrial waste and environmental pollution using their own resources. In this way, they can mitigate the negative effects of climate change. Moreover, they are less affected by the bureaucratic and political situation of the state and can therefore act more quickly than those who govern the state. Furthermore, local governments can adapt to change more easily and quickly. For this reason, they can be highly effective at both national and international scales. The COP 28 Local Climate Action Summit is particularly important as it is the first local government climate summit organized by the COP Presidency to highlight the central and important role of local governments in strengthening national efforts to reduce greenhouse gas emissions, address climate-related issues, and achieve greater and faster success in climate action. The Summit aims to bring together local and national government leaders globally to accelerate progress in addressing challenges by restructuring climate finance, increasing global climate action, accelerating the transition to renewable energy, and enhancing the climate resilience and adaptation of local governments (COP 28, 2023). Within this framework, the "High-Level Multi-Level Partnership Coalition for Climate Action (CHAMP)" was launched at the Summit to strengthen intergovernmental cooperation and solidarity. The Coalition will focus on National Determined Contributions (NDCs), National Adaptation Plans (NAPs), National Biodiversity Strategies and Action Plans (NBSAPs), and Long-Term Low Emission Development Strategies (LT-LEDS), as well as the implementation and monitoring of climate action plans through coalitions such as NDCs. The aim is to work together in solidarity and cooperation to ensure that climate action plans are implemented at the highest level through coalitions such as NDCs, to limit temperature increases to 1.5°C above pre-industrial levels, and to increase resilience. In addition, it has been announced that approximately $500 million in financing will be provided for city-focused climate investments in collaboration with CHAMP.

The following actions to be carried out by nation states to achieve the objectives set within CHAMP will be monitored (Local Governments Climate Summit, 2023):

1. Cooperating and collaborating with local governments on efforts to reduce greenhouse gas emissions, limit global temperature increase to 1.5 °C, adapt to climate change, and increase resilience, in line with the commitments made by states and the strategies developed, and to ensure the implementation and monitoring of these efforts.
2. Ensuring that local governments participate at the highest possible level in the entire process, including the review, redesign, further development, consolidation, and proper implementation of the strategies committed to and developed by states, and carrying out climate action opportunities to be developed at the national level for reduction and adaptation in cooperation and collaboration with local governments,
3. To contribute to the further development of NDCs prior to COP 30 in 2025 by establishing formal and informal processes and, to the extent possible, integrating basic information, objectives, and actions taken at the local and regional levels to reduce greenhouse gas emissions and adapt to climate change,
4. Ensure the participation of local governments in climate action plans, including those directly or indirectly related to NDCs, covering mitigation and adaptation, and secure the necessary support from the public and private financial sectors where required (including support for local, national, regional, and global policy reforms),
5. Implementation of the 2030 Agenda for Sustainable Development (United Nations, 2016). Voluntary National Review process (https://sustainabledevelopment.un

.org/vnrs/) is important to assess whether progress at the national and local levels is being made in a systematic and inclusive manner within the framework of implementing CHAMP requirements. These reviews should support multi-level climate action within the national policy framework and accelerate action where appropriate,

6. Bring together local leaders who have adopted CHAMP prior to both COP 29 and COP 30 in a global High-Level Political Dialogue on Multi-Level Climate Action, Share experiences gained during the implementation of CHAMP requirements and prepare a summary of the information compiled as a contribution to the Ministerial Meeting on Urbanization and Climate Change (Local Governments Climate Summit, 2023).

This initiative, approved by 63 countries, proposes a new process to assist local and regional leaders and the development of National Determined Contributions (NDCs) for CHAMP requirements, emphasizing that the solution to climate-related issues can be achieved through governance by directly referring to the multi-level governance paradigm in climate action plans.

5. LOCAL CLIMATE ACTION TECHNOLOGY: CLIMATE TECHNOLOGIES

The history of technology is the history of humanity. Technology, which originated from human needs, has continued to develop at an accelerating pace, especially since the 19th century. Thus, when humanity, which set out with the desire to dominate and control nature, faced environmental problems, it became clear that nature had won the battle against humanity. Production technologies, especially in the form required by capitalism, cause significant damage to the environment. The greatest damage, of course, comes from fossil energy sources. Due to all these and other damages, global warming caused by the accumulation of greenhouse gases in the atmosphere and climate change, which is a natural consequence of this, have become one of the most challenging problems humanity has faced and struggled to solve to date. In this context, innovative technologies are needed to reduce the harmful effects of climate change and to adapt to its requirements. This situation is also of interest to the private sector and the business world (Ozer, 2017, p. 838).

Just as environmental issues cannot be considered separately from technology, climate change cannot be evaluated separately from technology. In this context, the UN Framework Convention on Climate Change (UNFCCC) has facilitated the development and transfer of technology to developing countries. The development and dissemination of technology in this way has facilitated both the reduction of greenhouse gas emissions and adaptation to climate change. In order to achieve the long-term goals set out in the Paris Agreement signed in 2015, it is necessary to rapidly adopt and disseminate existing clean technologies, accelerate innovation, implement digital transformation, and disseminate newly developed technologies and facilitate access to them.

Technologies that are specifically mentioned and used in the fight against climate change are referred to as "climate technologies." Technologies for reducing greenhouse gas emissions and renewable energy technologies such as solar energy, wind energy, and hydroelectric energy are classified as climate technologies. Among these, early warning systems and drought-resistant technologies are also considered climate technologies. In addition to these, increasing energy efficiency and providing training on equipment use are also included in "soft climate technologies." The development, use, transfer, and internalization of new technologies are crucial for supporting national climate change actions. Technology has been given great importance since the very beginning of the UNFCCC process. After the original Convention was established in 1992, articles related to technology were added to the text. Technology transfer is also an important part of this process. Access to climate technologies requires large budgets, especially for developing countries. The work carried out under the UNFCCC facilitates this process for countries. In this context, work is being carried out on Technology Needs Assessments (TNA) for the Parties. According to this, developing countries, through the TNA process:

- Identify technological tools for climate change and national development,
- Build national capacity to ensure national sustainable development,
- Develop technology action plans to ensure the internalization and applicability of technology,

This process is supported financially by both the Global Environment Facility (GEF) and the Green Climate Fund (GCF) (United Nations Framework Convention on Climate Change, 2025).

At COP16 in 2010, the Technology Mechanism was established by the UNFCCC process to facilitate the process of developing, using, adopting, and transferring technology, or adopting and using transferred technology in the same manner, for developing countries. This mechanism consists of two bodies (United Nations Framework Convention on Climate Change, 2025):

1. Technology Executive Committee (TEC)

Implements the policies of the Technology Mechanism. It consists of 22 technology experts representing developed and developing country Parties. It meets several times a year to address technology policy issues. It works to solve technology-related problems for countries. It also develops supportive policies for the development and transfer of climate technologies (United Nations Framework Convention on Climate Change, 2025).

2. Climate Technology Centre and Network (CTCN)

It is the implementing body of the Technology Mechanism. It strives to meet technology-related needs, from development to accelerated transfer, by providing three core services.

-Provides technical assistance by assessing the requests of developing countries.
-Facilitates access to information on climate technologies.
-Enables all stakeholders in climate technology to communicate and develop cooperation through the established network.

CTCN is supported by the United Nations Environment Programme (UNEP) and has a communication network with all international, national, regional, local, and sectoral organizations. CTCN also has a responsibility to the Conference of the Parties through its advisory board and is governed by the decisions taken there (United Nations Framework Convention on Climate Change, 2025).

The Global Stocktake (GST), considered one of the key outcomes of COP 28 held in 2024, clearly emphasizes that accelerated technology transfer is the most fundamental tool for implementing climate action (Local Governments Climate Summit, 2023).

Decisions related to technology are made at all Conferences of the Parties. Various committees are established, reports from existing committees are reviewed, and issues are analyzed and evaluated. For example, COP 28 featured many technology-related sessions, including the COP Connect Series, part of the Technology and Entrepreneurship program; the Climate Technology Innovation Coalition; the Technology and Innovation Hub; the Climate Innovation Forum; and the NEX COP 28 Climate Technology Startup Accelerator (Local Governments Climate Summit, 2023; Kahraman & Acar, 2024). Similarly, at COP 29, the working reports of the Subsidiary Body for Scientific and Technological Advice were evaluated.

Many studies, analyses, and evaluations on technology and innovation are conducted at the Conferences of the Parties. The need to pay particular attention to this issue and its follow-up by the UNFCCC process is a significant step toward solving the climate change problem.

CONCLUSION

Climate change governance is an approach that addresses both supranational and subnational units. Although studies on supranational units have increased since the 1990s due to the impact of globalization, interest in subnational units has increased after the 2015 Paris Agreement. The increase in the number of cities, the diversification of climate change-related problems faced by cities (floods for some, high temperatures for others), and the greater vulnerability to climate change in underdeveloped/developing country cities with high population densities have triggered this process. Based on the understanding that the climate change problem cannot be addressed solely through policies and practices put forward by global and central governments, the direction of climate change governance has been determined from the bottom up, starting with local governments, followed by state, transnational, and global levels. The political dimension of local climate actions is shaped by climate change governance, the collaborative dimension by international climate change alliances, and the technological dimension by policies and practices of international coalitions.

The technology and innovations developed by local governments, which have all the information about the characteristics of the area, will contribute positively to the rapid adoption of these innovations by reaching the people who will benefit from basic services, as well as establishing trust between the administrators and the administered, thereby contributing to the consistency of policies and practices. For this reason, the state should support local governments in terms of finance and personnel. Local governments without financial support are seen as having their hands tied. The state should not withdraw from its responsibilities in climate change management, but rather support local governments, especially by guiding them in technology and innovation. Without the state's guidance, it does not seem possible for local governments to achieve success. Rather than providing services itself, the state should be evaluated in terms of efficiency and effectiveness by supporting local governments, which are the units closest to the public, financially and by guiding them in the provision of services. In short, the concept of "a state that steers rather than rows," which has become the slogan of the new public administration that paves the way for governance, should be parallel to climate change governance, which is inherent in governance and seen as one of its sub-branches.

The adaptation of internationally defined climate action plans to the national level and the implementation of these plans at the local level make local climate change planning extremely important. In this context, facilitating the participation of actors in climate change governance, supporting social participation, ensuring the participation of the private sector, and ensuring sustainability are of great importance in combating climate change. It is also important that climate-resilient local climate actions are innovative, adaptable, and sustainable. In particular, the use of renewable energy systems must be increased, and both the public and private sectors must comply with climate action plans in terms of greenhouse gas emissions. In this context, various model proposals and applications exist for local climate actions.

It is believed that climate change can be successfully combated at the local level through cooperation between all these factors within the framework of local climate change governance. The adequacy of implementing climate action plans at the local level within a multi-level governance framework is also possible through innovative approaches. Success achieved at the local level through cooperation will lead to success at the national level and cooperation between states will bring success at the global level. Technologically, the preference for environmentally friendly technologies, the use and widespread adoption of clean production systems, the preference for artificial intelligence-supported production, and an innovative approach to all these technologies are important stakeholders in the fight against climate change. In addition, many issues such as energy, transportation, and waste management must be made compatible with climate actions.

REFERENCES

C40 Cities. (2005). C40 Cities. https://www.c40.org/

Bulkeley, H., Schroeder, H., Janda, K., Zhao, J., Armstrong, A., Chu, S. Y., & Ghosh, S. (2011). The role of institutions, governance and urban planning for mitigation and adaptation. In Hoornweg, D., Freire, M., Lee, M. J., Bhada-Tata, P., & Yuen, B. (Eds.), *Cities and climate change: Responding to an urgent agenda* (pp. 125–159). The World Bank. DOI: 10.1596/9780821384930_CH05

COP 13 Local Governments Climate Roadmap. (2007). COP 13 Local Governments Climate Roadmap.

Demirci, M. (2015). Urban climate change governance. *Erciyes University Faculty of Economics and Administrative Sciences Journal*, *46*, 75–100.

Gormez, K. (2018). *Environmental issues*. Nobel Publications.

HABITAT III. (2016). New urban agenda. https://habitat3.org/documents-and-archive/new-urban-agenda/

Intergovernmental Panel on Climate Change. (2022). Climate change 2022: Mitigating climate change. https://www.ipcc.ch/report/ar6/wg3/

Kahraman, A. C., & Acar. (2024). Notes from COP 28. Marmara Municipalities Union.

Local Governments and Municipal Authorities. (2025). About the LGMA. https://www.cities-and-regions.org/about-the-lgma/

Local Governments Climate Summit. (2023). Local Governments Climate Summit. https://www.cop28.com/en/lcas

Mengi, A. (1999). *Global environmental issues and policies*. Mülkiyeliler Association Foundation Publications.

Nature Climate Change. (2016). Nature Climate Change. https://www.nature.com/articles/nclimate3169

Ozer, M. A. (Ed.). (2023). *Local governments and the future tomorrows*. Ekin Kitabevi Publications.

Ozer, Y. E. (2017). Analysis of actors in climate change governance and the situation of Türkiye. *International Journal of Management Economics and Business*, *13*(4), 833–851.

United Nations. (1997). Kyoto Protocol.

United Nations. (2016). Paris Agreement.

United Nations. (2016). Sustainable development goal 13: Climate action. https://sdgs.un.org/goals/goal13

United Nations Framework Convention on Climate Change. (2025). Climate change conferences of the parties (COP). https://unfccc.int/process/bodies/supreme-bodies/conference-of-the-parties-cop

We Are Social. (2024). Digital 2024: 5 billion social media users. https://wearesocial.com/uk/blog/2024/01/digital-2024-5-billion-social-media-users/

Zimmermann, M., Van Begin, G., & Cristobal, I. V. (2010). The international local government climate roadmap. In Van Staden, M., & Musco, F. (Eds.), *Local governments and climate change* (pp. 79–89). Springer.

KEYWORDS AND DEFINITIONS

Local Government: Constitutional institutions established by a state to meet the needs of smaller-scale communities within its own administrative structure, generally under the authority of the central government. Its general decision-making bodies are public legal entities elected by the people.

Climate Change: The climate system encompasses the entire surface of the Earth, including land, oceans, and the atmosphere, as well as all living things, and refers to everything that is in a continuous and interdependent relationship. Changes in average climate values over many years due to human activities and many other reasons are referred to as climate change.

Governance: It can be defined as the joint use of economic, political, and administrative resources by stakeholders in public or private institutions and organizations. Good governance means being able to make decisions jointly with stakeholders and implement those decisions correctly.

Technology: Briefly defined as applied science, technology has advanced alongside human history, including the development of today's digital technologies. Technology, which initially progressed according to human needs, entered a period in which science supported technology with the Second Scientific Revolution in the 19th century, after which science and technology together achieved extraordinary progress. In the 21st century, with digital technologies and artificial intelligence, combating climate change will be much easier and faster.

Innovation: Innovation means renewal or change. It is the process of making new creative ideas or inventions usable in areas such as economics, technology, services, and industry.

Interdependence: Due to globalization, the free movement of capital, the liberalization of international trade, and the universal nature of science and technology, countries have naturally become interdependent.

Solidarity: Interdependence makes it necessary for states to cooperate in solidarity.

Chapter 12
Leveraging Data for Climate Action:
Smart Governance in the Digital Age

Satya Subrahmanyam
https://orcid.org/0000-0003-0441-2742
Holy Spirit University of Kaslik, Lebanon

ABSTRACT

Climate change poses one of the most urgent challenges of our time, demanding innovative solutions. This chapter explores how data-driven smart governance can be a transformative tool in addressing climate change. By integrating technology, data, and participatory decision-making, smart governance allows for real-time insights and predictive modeling that enhance climate policies. Key technologies such as the Internet of Things (IoT), Artificial Intelligence (AI), and blockchain are essential in monitoring environmental data and optimizing resource use. The chapter discusses the challenges of data quality, privacy, and accessibility, and proposes a policy framework for effective climate governance. It also emphasizes the role of stakeholder engagement, international cooperation, and scalable solutions in shaping the future of climate action.

1. INTRODUCTION

The climate crisis represents one of the most significant existential challenges of the 21st century, characterized by rising global temperatures, intensifying natural disasters, biodiversity loss, and socio-economic disruptions. Scientific consensus confirms that anthropogenic activities—such as deforestation, fossil fuel combus-

DOI: 10.4018/979-8-3373-1310-8.ch012

tion, and industrial emissions—are driving climate change, posing grave threats to ecosystems and human societies alike (IPCC, 2023). Urgent action is required at all levels—local, national, and global—to mitigate greenhouse gas (GHG) emissions, adapt to climate impacts, and transition to sustainable development pathways. While traditional governance mechanisms have made incremental progress, the scale and complexity of the crisis demand innovative approaches.

Amid this crisis, the advent of digital technologies and the unprecedented availability of data have transformed the possibilities for climate action. Advanced tools such as artificial intelligence (AI), the Internet of Things (IoT), big data analytics, and blockchain offer transformative potential to enhance decision-making, improve resource efficiency, and enable real-time monitoring of climate-related indicators (Steele et al., 2022). Satellite imagery combined with machine learning can track deforestation patterns, while IoT sensors can monitor energy consumption and carbon footprints at micro and macro levels. Such technological advances highlight the pivotal role of data-driven solutions in combating climate change.

In this context, the concept of "smart governance" emerges as a critical framework for leveraging data and digital technologies to address climate challenges. Smart governance refers to the integration of technology, data, and participatory mechanisms to enhance decision-making, transparency, and accountability in public and private sector initiatives (Meijer & Bolívar, 2016). It moves beyond conventional governance models by harnessing data intelligence to create adaptive, efficient, and inclusive systems capable of addressing complex challenges like climate change. This chapter explores how smart governance can serve as a catalyst for transformative climate action by integrating digital technologies and data-driven insights into policy frameworks and implementation strategies.

The chapter has four key objectives. First, it aims to highlight the urgency of the climate crisis and the inadequacy of traditional governance mechanisms in responding to its scale. Second, it seeks to elucidate the role of data and digital technologies in addressing climate challenges, providing examples of their applications across mitigation, adaptation, and resilience-building efforts. Third, the chapter defines the principles and components of smart governance, emphasizing its relevance to climate action. Finally, it outlines the chapter's scope, which includes analyzing best practices, identifying challenges, and proposing pathways to optimize smart governance for climate resilience.

1.1 Overview of the Climate Crisis and the Urgency for Action

According to the Intergovernmental Panel on Climate Change (IPCC), the world's average temperature might increase by 1.5°C compared to pre-industrial levels in the 2030s if current emissions trends go unchecked (IPCC, 2023). Despite its seeming modest, this warming threshold has far-reaching consequences for ecosystems, economies, and human health. Hurricanes, wildfires, and floods are on the rise, and each year they displace millions of people and cost billions of dollars (UNDRR, 2022). The impact is felt most by vulnerable people, especially in developing countries. This worsens socioeconomic inequalities and hinders global attempts to fulfill the Sustainable Development Goals (SDGs) set forth by the United Nations.

Despite growing international recognition of these challenges, existing governance mechanisms often fall short in addressing the complexities of climate change. Fragmented policies, insufficient data integration, and limited stakeholder engagement impede coordinated action. The lack of real-time insights into climate trends hampers governments' ability to respond effectively to emerging threats. This underscores the need for a paradigm shift towards governance models that prioritize agility, innovation, and inclusivity.

1.2 The Role of Data and Digital Technologies in Addressing Climate Challenges

Data and digital technologies have emerged as indispensable tools in the fight against climate change. Their capacity to collect, analyze, and disseminate vast amounts of information enables stakeholders to better understand, predict, and respond to climate-related risks. Key applications of data and technology in climate action include:

Monitoring and Modeling Climate Systems: Satellite-based remote sensing, coupled with big data analytics, provides high-resolution insights into deforestation, glacial retreat, and ocean temperature variations. Climate models powered by AI offer predictive capabilities, enabling policymakers to anticipate extreme weather events and plan adaptive strategies (WMO, 2023).

Decarbonization and Energy Efficiency: IoT-enabled smart grids optimize energy consumption, reducing carbon footprints. Blockchain technology facilitates transparent carbon credit trading, ensuring accountability in emissions reduction efforts (Steele et al., 2022).

Sustainable Urban Planning: Data-driven urban management systems leverage AI and IoT to optimize waste management, reduce traffic congestion, and enhance energy efficiency in buildings. These systems contribute to the development of low-carbon cities, aligning with global sustainability targets.

Stakeholder Engagement and Decision-Making: Digital platforms enhance participatory governance by enabling citizens, businesses, and non-governmental organizations to contribute to climate policy dialogues. Crowdsourced data initiatives empower local communities to monitor and report environmental changes, fostering grassroots accountability (Meijer & Bolívar, 2016).

1.3 Defining "Smart Governance" in the Context of Climate Action

Smart governance represents an evolution in governance practices, characterized by the integration of digital technologies and data intelligence into decision-making processes. In the context of climate action, it involves:

Data-Driven Insights: Leveraging big data and AI to generate actionable insights for climate mitigation and adaptation strategies.
Collaborative Decision-Making: Facilitating multi-stakeholder collaboration through digital platforms, ensuring that diverse perspectives inform policy decisions.
Transparency and Accountability: Utilizing blockchain and other technologies to enhance the transparency of climate finance, emissions reporting, and policy implementation.
Adaptive Systems: Building governance structures capable of responding dynamically to emerging climate risks and scientific advancements.

Smart governance aligns with principles of sustainability, inclusivity, and resilience, making it particularly suited to addressing the multi-faceted challenges of climate change. It bridges the gap between data generation and actionable policy, fostering a proactive approach to climate governance.

1.4 Objectives and Scope of the Chapter

This chapter seeks to advance understanding of how smart governance can be leveraged for climate action in the digital age. The objectives are to:

1. **Highlight the Urgency of the Climate Crisis**: Provide a detailed analysis of the current state of the climate crisis, emphasizing the limitations of traditional governance models in addressing its scale and complexity.
2. **Demonstrate the Role of Data and Technology**: Illustrate how digital tools and data-driven approaches can enhance climate action, offering real-world examples of successful applications.
3. **Define Smart Governance**: Present a comprehensive definition of smart governance and its relevance to climate action, detailing its principles, components, and potential impacts.

4. **Propose Pathways for Implementation**: Analyze best practices, challenges, and opportunities for integrating smart governance into climate policies, providing actionable recommendations for policymakers, businesses, and civil society.

By looking into these goals, the chapter hopes to add to what is already known about data-driven climate governance and provides some ideas on how to use digital innovation for fair and sustainable results.

2. THE ROLE OF DATA IN CLIMATE ACTION

Data is pivotal in formulating effective strategies to combat climate change, enabling stakeholders to make informed decisions for both mitigation and adaptation. The advancement of data collection and analysis technologies has enhanced our capacity to monitor environmental changes in real-time, identify emerging threats, and develop evidence-based adaptive strategies. This section examines various types of climate-related data, their sources, and the critical importance of real-time data in climate action decision-making.

2.1 Types of Climate-Related Data

Addressing climate change necessitates a comprehensive understanding of environmental, social, and economic factors. These dimensions are captured through various data types, each serving specific roles in climate mitigation and adaptation strategies.

Weather Patterns and Climate Forecasts - Accurate knowledge of weather patterns and climate forecasts are essential for predicting and mitigating climate change impacts. Meteorological data—including temperature, precipitation, wind patterns, and ocean currents—forms the foundation for climate modeling and projections. Global climate models (GCMs) and regional climate models (RCMs) utilize historical weather data to simulate future climate scenarios under different greenhouse gas emission pathways. Such forecasts are crucial for developing policies related to disaster preparedness, agricultural planning, and water resource management (United Nations, n.d.).

Emissions Tracking and Environmental Monitoring - Monitoring greenhouse gas (GHG) emissions and environmental changes is vital for climate mitigation efforts. Data on emissions of gases like carbon dioxide (CO_2), methane (CH_4), and nitrous oxide (N_2O) provide insights into primary pollution sources and their geographic distribution. Environmental monitoring also includes observing deforestation, land degradation, and air and water quality. Technologies such as remote sensing and

IoT-enabled environmental sensors have revolutionized data collection, offering high spatial and temporal resolutions. These datasets inform strategies for reducing emissions and enforcing compliance with international agreements like the Paris Accord (U.S. Environmental Protection Agency, 2014).

Socioeconomic Data for Climate Adaptation - Climate change impacts disproportionately affect marginalized and vulnerable groups. Socioeconomic data—including population demographics, income levels, health statistics, and infrastructure vulnerability—plays a crucial role in understanding the human dimensions of climate change. This data aids policymakers in designing targeted interventions to enhance resilience among at-risk communities. Socioeconomic datasets can guide decisions on resource allocation for flood protection in low-income areas or the development of heat adaptation measures in urban centers disproportionately affected by rising temperatures (World Resources Institute, 2023).

2.2 Sources of Data

Climate-related data is sourced from a diverse array of platforms, each offering unique strengths in scope, resolution, and accessibility. These sources encompass advanced technological systems, collaborative initiatives, and institutional databases.

Satellite Imagery, IoT Sensors, and Drones - Technological advancements have expanded our capacity to monitor Earth's climate and ecosystems from multiple perspectives.

Satellite Imagery: Satellites operated by agencies like NASA and the European Space Agency (ESA) provide critical data on global temperature trends, sea-level rise, and land-use changes. NASA's Earth Observing System (EOS) captures data on atmospheric conditions, vegetation cover, and ocean health, contributing to comprehensive climate assessments (NASA, 2023).

IoT Sensors: The Internet of Things (IoT) enables localized data collection through interconnected sensors that monitor air quality, water levels, and energy consumption. Smart meters in urban infrastructure track energy usage, helping optimize resource efficiency.

Drones: Unmanned aerial vehicles (UAVs) or drones complement satellite observations by offering high-resolution, localized data on deforestation, glacier retreat, and coastal erosion. Drones are particularly valuable in regions inaccessible to traditional monitoring methods.

Open Data Platforms and Citizen Science Initiatives - The democratization of data has emerged as a powerful trend in climate action, enabling widespread participation in environmental monitoring.

Open Data Platforms: Organizations like the World Resources Institute (WRI) provide open-access databases on climate indicators, emissions inventories, and policy progress. These platforms promote transparency and allow researchers and policymakers to make evidence-based decisions (World Resources Institute, 2023).

Citizen Science Initiatives: Citizen science projects encourage public participation in data collection and environmental stewardship. The Globe Observer app developed by NASA enables citizens to contribute observations on cloud cover, vegetation, and mosquito habitats, enriching scientific datasets with ground-level insights. Such initiatives also foster environmental awareness and community engagement.

Governmental and Intergovernmental Databases - Governments and intergovernmental organizations maintain comprehensive databases instrumental in climate governance.

National Databases: Countries maintain emissions inventories and environmental statistics to track progress toward national climate goals. The U.S. Environmental Protection Agency (EPA) provides data on GHG emissions by sector, enabling targeted policy interventions (U.S. Environmental Protection Agency, 2014).

Intergovernmental Databases: Organizations like the United Nations aggregate data from multiple countries to produce global assessments. These databases serve as the foundation for international climate negotiations and scientific collaborations (United Nations, n.d.).

2.3 Importance of Real-Time Data in Decision-Making

Real-time data has become a game-changer in climate action, offering the ability to respond dynamically to rapidly evolving environmental conditions. Unlike static datasets, real-time data enables continuous monitoring and adaptive management. The following examples illustrate the transformative impact of real-time data on decision-making:

Disaster Preparedness and Response - Real-time data enhances the ability to anticipate and respond to natural disasters, minimizing their human and economic toll. Early warning systems for hurricanes, floods, and wildfires rely on real-time meteorological data to predict the trajectory and intensity of such events. This information allows authorities to issue timely evacuation orders, deploy emergency resources, and protect critical infrastructure (Meijer & Bolívar, 2016).

Energy and Resource Optimization - Real-time monitoring of energy usage and natural resources contributes to more efficient management practices, reducing waste and emissions. Smart grid systems use IoT sensors to track electricity demand in real-time, enabling utilities to optimize energy distribution and integrate renewable sources like solar and wind power. Similarly, real-time water quality monitoring systems help identify contamination events, ensuring timely intervention to protect ecosystems and public health.

Policy Implementation and Compliance - Real-time data improves the enforcement of climate policies by providing immediate insights into compliance levels. Satellite-based monitoring systems can track deforestation activities in near-real-time, enabling governments to identify and penalize illegal logging operations. Blockchain technology further enhances compliance by creating transparent, immutable records of emissions reductions and carbon credit transactions.

Enhancing Public Engagement - Real-time data fosters greater public engagement in climate action by providing accessible and relatable information. Interactive dashboards and mobile applications enable individuals to track their carbon footprints, monitor air quality in their neighborhoods, and participate in local environmental initiatives. This increased awareness and participation strengthen community-level resilience to climate impacts.

The role of data in climate action is indispensable. From monitoring weather patterns to tracking emissions and assessing socioeconomic vulnerabilities, data provides the foundation for evidence-based decision-making. The diverse sources of climate-related data—ranging from satellites and IoT sensors to citizen science initiatives and governmental databases—highlight the collaborative nature of data collection and its potential to drive innovation in climate governance.

Real-time data has proven essential in enabling adaptive, responsive, and effective climate action. Whether preparing for natural disasters, optimizing energy systems, enforcing compliance with environmental policies, or engaging the public, the immediacy of real-time data enhances our ability to address climate challenges dynamically. As the climate crisis intensifies, integrating data-driven approaches.

3. PRINCIPLES OF SMART GOVERNANCE FOR CLIMATE ACTION

Smart governance is essential for addressing the multifaceted challenges posed by climate change. By leveraging data and digital technologies, smart governance enables transparent, inclusive, and adaptive decision-making processes that ensure effective climate action. This section explores four key principles of smart governance: transparency and accountability in data use, citizen participation and inclusivity, evidence-based policymaking with adaptive frameworks, and the integration of public and private sector efforts through data sharing.

3.1 Transparency and Accountability in Data Use

Transparency and accountability are fundamental pillars of smart governance for climate action. Transparent use of data ensures that stakeholders—including governments, organizations, and citizens—can understand how climate-related decisions are made, fostering trust and enabling collaborative efforts.

Data Accessibility and Openness - Transparent governance mandates that climate-related data, such as greenhouse gas emissions, deforestation rates, and air quality, be made accessible to the public. Open data platforms, like the Global Forest Watch or the Climate Data Explorer, serve as crucial tools in this context. By providing open access to data, these platforms empower researchers, policymakers, and activists to assess environmental trends and hold decision-makers accountable for their actions (World Resources Institute, 2023).

Building Trust Through Data Integrity - Governance systems must ensure the reliability of climate-related data. This involves implementing stringent data validation protocols and adopting technologies such as blockchain for immutable data records. Blockchain can be used to track carbon credit transactions, ensuring that reported emissions reductions are genuine and verifiable (Meijer & Bolívar, 2016). Trust in data integrity reinforces public confidence in climate policies and initiatives.

Accountability in Policy Implementation - Transparent use of data also aids in monitoring the implementation of climate policies. Governments can utilize real-time data to track progress against national or global targets, such as those set under the Paris Agreement. Regularly publishing these findings enhances accountability and provides an evidence-based foundation for refining policies (UNFCCC, 2023).

3.2 Citizen Participation and Inclusivity in Governance

Climate action must be inclusive to ensure that all voices—particularly those from vulnerable and marginalized communities—are heard in decision-making processes. Citizen participation in governance fosters a sense of ownership, enhances policy effectiveness, and strengthens societal resilience to climate impacts.

Inclusive Decision-Making Processes - Smart governance leverages digital tools to enable participatory decision-making. Online platforms such as e-consultation portals or social media can gather public feedback on climate policies. In Norway, the government's "Climate Dialogues" initiative engages citizens in discussions about climate strategies, ensuring their perspectives are considered in national policymaking (OECD, 2021). Such approaches ensure that governance frameworks reflect diverse societal needs and priorities.

Citizen Science Initiatives - Citizen science empowers individuals to contribute to data collection and environmental monitoring. Programs like NASA's Globe Observer enable participants to document local environmental changes, such as deforestation or extreme weather events. This crowdsourced data not only enriches scientific databases but also raises awareness and fosters community engagement (NASA, 2023).

Addressing Social Inequities - Inclusive governance also requires addressing inequities exacerbated by climate change. Marginalized communities often face disproportionate risks, such as heightened vulnerability to natural disasters or inadequate access to climate adaptation resources. Governance mechanisms must prioritize the needs of these groups by incorporating socioeconomic data into policy design and allocating resources equitably (World Bank, 2023).

3.3 Evidence-Based Policymaking and Adaptive Frameworks

The complexity and uncertainty of climate change demand evidence-based policymaking supported by adaptive governance frameworks. This principle ensures that decisions are informed by robust scientific data and can be adjusted in response to new evidence or changing conditions.

Using Data for Evidence-Based Decisions - Smart governance leverages data analytics and modeling to predict climate impacts, evaluate policy outcomes, and identify optimal interventions. For example, climate simulations using global and regional climate models provide valuable insights into future temperature increases, sea-level rise, and extreme weather patterns. Policymakers can use this evidence to prioritize investments in renewable energy, disaster preparedness, or infrastructure resilience (United Nations, n.d.).

Adaptive Governance Models - Given the dynamic nature of climate change, governance systems must be flexible and responsive. Adaptive governance involves continuous monitoring, iterative policy adjustments, and stakeholder collaboration. Adaptive water resource management frameworks, such as those employed in the Colorado River Basin, use real-time data to manage water allocations based on current availability and demand (World Resources Institute, 2023).

Learning from Policy Experiments - Evidence-based governance also incorporates lessons from pilot projects and experiments. Cities implementing climate-resilient infrastructure projects can analyze outcomes to refine strategies before scaling them. Sharing these insights through networks like the C40 Cities Climate Leadership Group promotes knowledge exchange and accelerates global progress (C40 Cities, 2023).

3.4 Integration of Public and Private Sector Efforts Through Data Sharing

Effective climate governance requires seamless collaboration between public and private sectors. By sharing data and expertise, stakeholders can pool resources, enhance decision-making, and drive innovative solutions to climate challenges.

Collaborative Data Sharing Platforms - Digital platforms play a critical role in enabling data sharing between governments, businesses, and civil society organizations. For instance, the Open Energy Data Initiative facilitates the exchange of energy-related data to support renewable energy projects and efficiency improvements. Such platforms foster synergies across sectors and ensure that climate action is based on comprehensive, high-quality data (U.S. Department of Energy, 2023).

Private Sector Contributions - Private companies are increasingly adopting sustainable practices and sharing their climate-related data as part of corporate social responsibility (CSR) initiatives. Technology firms like Microsoft and Google provide open-access environmental datasets, including satellite imagery and AI-driven climate models. These contributions complement governmental efforts and accelerate innovation in areas such as emissions reduction and renewable energy deployment (Microsoft, 2023).

Public-Private Partnerships - Public-private partnerships (PPPs) are a cornerstone of integrated governance for climate action. PPPs combine public sector oversight with private sector efficiency and innovation, enabling large-scale projects such as renewable energy installations, smart city infrastructure, and climate-resilient urban planning. The Global Covenant of Mayors for Climate and Energy unites local governments and businesses to share data, set ambitious targets, and implement collaborative climate solutions (Global Covenant of Mayors, 2023).

The principles of smart governance—transparency and accountability, citizen participation, evidence-based policymaking, and public-private sector integration—are essential for addressing the complex challenges of climate change. Transparent use of data fosters trust and accountability, while inclusive governance ensures that all societal voices are represented. Evidence-based decisions grounded in scientific data enhance policy effectiveness, and adaptive frameworks allow governance systems to respond dynamically to emerging challenges. Finally, collaborative efforts between public and private sectors maximize resource efficiency and innovation.

As digital technologies continue to advance, their integration into governance systems offers unprecedented opportunities for tackling the climate crisis. By adhering to these principles, smart governance can drive transformative change, enabling societies to transition toward sustainable and resilient futures.

4. APPLICATIONS OF DATA IN SMART CLIMATE GOVERNANCE

Smart climate governance relies on the systematic application of data to guide decision-making, optimize resource management, and drive effective climate action. By leveraging digital technologies and analytical tools, governments, organizations, and communities can monitor environmental changes, design evidence-based policies, and build resilience to climate impacts. This section explores the critical applications of data in smart climate governance, categorized under four major areas: monitoring and reporting, urban planning and resilience, resource management, and policy design and implementation.

4.1 Monitoring and Reporting

Monitoring and reporting are foundational elements of smart climate governance. Accurate and timely data on environmental indicators provide critical insights for tracking progress, ensuring transparency, and holding stakeholders accountable.

Tracking Greenhouse Gas Emissions and Air Quality - To effectively implement mitigation methods, it is crucial to accurately monitor greenhouse gas (GHG) emissions, since they are a key driver of climate change. Emissions monitoring relies heavily on cutting-edge innovation such sensors connected to the Internet of Things (IoT) and satellite-based remote sensing. Scientists and politicians are able to track changes in emissions because to NASA's Orbiting Carbon Observatory-2 (OCO-2) that gathers accurate data of atmospheric carbon dioxide levels (NASA, 2023). Similarly, cities may address localized pollution concerns with the use of real-time data provided by air quality monitoring networks, which include pollutants like particulate matter and nitrogen oxides (NOx) (World Resources Institute, 2023).

Evaluating Progress Toward Climate Targets - Evaluating progress toward international climate objectives, including the targets set by the Paris Agreement, requires data-driven reporting. To determine whether national and sectoral statistics are in line with the worldwide goal of keeping warming below 1.5°C, tools such as the Climate Action Tracker are used (UNFCCC, 2023). In addition to promoting responsibility among states, these assessments educate global negotiators by revealing implementation shortcomings.

4.2 Urban Planning and Resilience

Urban areas, home to more than half of the global population, are both major contributors to and vulnerable victims of climate change. Data-driven approaches to urban planning and resilience-building enable cities to reduce emissions, adapt to climate impacts, and protect their populations.

Using Data for Smart Cities and Green Infrastructure Planning - Smart city initiatives integrate data analytics, IoT devices, and artificial intelligence (AI) to optimize urban systems and reduce environmental footprints. Cities like Singapore and Copenhagen use real-time data to improve public transportation efficiency, reduce energy consumption in buildings, and enhance waste management systems (Meijer & Bolívar, 2016). Green infrastructure, including urban forests, wetlands, and green roofs, helps reduce the impact of urban heat islands, boosts biodiversity, and cleans the air we breathe.

Disaster Preparedness and Early Warning Systems - Data is crucial for building urban resilience to climate-related disasters, such as floods, hurricanes, and wildfires. Early warning systems leverage meteorological data, satellite imagery, and predictive modeling to provide timely alerts to at-risk populations. India's Cyclone Warning Division uses advanced forecasting models to predict cyclone paths and intensities, enabling efficient evacuation and disaster response (World Bank, 2023). Such systems save lives, minimize economic losses, and strengthen communities' adaptive capacities.

4.3 Resource Management

Effective resource management is central to climate action, as it directly influences emissions reductions, conservation efforts, and the transition to sustainable systems. Data-driven tools enable the optimization of water, energy, and waste management, as well as the allocation of scarce resources.

Data-Driven Approaches to Water, Energy, and Waste Management - Data is becoming more and more important in climate governance to guarantee the sustainable use of natural resources. To decrease water waste in places such as Amsterdam and Las Vegas, smart water management systems use Internet of Things (IoT) sensors to monitor water levels, identify leaks, and improve distribution networks (U.S. Department of Energy, 2023). Data analytics are also used by smart grids to incorporate renewable energy sources, balance the supply and demand for power, and decrease transmission losses. According to the Global Covenant of Mayors (2023), data applications in trash management include models that forecast patterns in garbage creation and recycling bins that are Internet of Things (IoT) enabled, which enhance collection routes and recycling efficiency.

Leveraging AI and ML for Efficient Resource Allocation - In recent years, AI and ML have grown in prominence as potent instruments for optimizing resource use. Artificial intelligence systems are able to identify supply chain inefficiencies, maximize the production of renewable energy sources, and forecast the demand for resources. Data centers' energy usage may be cut by as much as 40% thanks to AI from Google's DeepMind, showing how AI can promote efficiency in several sectors (Microsoft, 2023). Not only do these advances save expenses, but they also help conserve resources and decrease pollution.

4.4 Policy Design and Implementation

The design and implementation of effective climate policies rely on robust data analysis, scenario modeling, and adaptive governance mechanisms. Data-driven approaches enable policymakers to assess potential impacts, refine strategies, and respond to emerging challenges in real time.

Scenario Modeling for Policy Impact Assessments - Scenario modeling uses data and computational simulations to evaluate the potential outcomes of various policy options. These models help policymakers understand trade-offs, identify unintended consequences, and prioritize interventions. Integrated assessment models (IAMs) like the Global Change Assessment Model (GCAM) analyze the interactions between energy systems, land use, and climate policies, providing insights into the feasibility of achieving emissions reduction targets (United Nations, n.d.). Scenario modeling also supports long-term planning for sectors such as energy, agriculture, and transportation.

Real-Time Feedback Loops for Adaptive Governance - Adaptive governance frameworks incorporate real-time data to monitor policy implementation and adjust strategies as needed. Feedback loops enable continuous learning and improvement, ensuring that governance systems remain responsive to evolving climate challenges. Carbon pricing mechanisms, such as the European Union Emissions Trading System (EU ETS), use real-time market data to set dynamic carbon prices, incentivizing emissions reductions while minimizing economic disruptions (OECD, 2021).

The applications of data in smart climate governance span a wide array of domains, from emissions tracking and urban resilience to resource optimization and evidence-based policymaking. By integrating data-driven approaches into governance frameworks, governments and organizations can enhance transparency, efficiency, and accountability, paving the way for effective climate action.

Monitoring and reporting systems provide critical insights into environmental trends and progress toward climate targets. Urban planning and resilience efforts benefit from data applications that optimize infrastructure, improve disaster preparedness, and protect vulnerable populations. Resource management innovations leverage data analytics, IoT devices, and AI to promote sustainability and efficiency. Finally, data-driven policy design and implementation enable governments to evaluate policy impacts, respond adaptively to emerging challenges, and align strategies with long-term climate goals.

As digital technologies continue to evolve, their role in smart climate governance will become increasingly indispensable. By harnessing the power of data, societies can address the complexities of climate change with precision, agility, and inclusivity, fostering a sustainable and resilient future.

5. CHALLENGES IN LEVERAGING DATA FOR CLIMATE ACTION

As the world intensifies its efforts to combat climate change, the role of data in informing, monitoring, and evaluating climate actions has become indispensable. However, the integration of data into climate governance is not without its challenges. These issues span technical, ethical, economic, and infrastructural dimensions, often complicating the efforts to deploy data-driven solutions effectively and equitably. This section delves into key challenges in leveraging data for climate action, including data gaps and accessibility, privacy and security concerns, ethical and equity issues, and technical and financial barriers.

5.1 Data Gaps and Accessibility

Data is the backbone of informed climate governance. However, significant disparities in data availability and accessibility hinder the effective implementation of data-driven climate solutions.

Disparities in Data Availability Between Developed and Developing Nations - One of the most significant challenges lies in the uneven distribution of climate-related data between developed and developing nations. Wealthier countries often have access to advanced satellite systems, sophisticated data analytics tools, and well-established databases. In contrast, many developing countries lack the resources and infrastructure to collect, process, and analyze climate data effectively (UNEP, 2022). While developed nations can use high-resolution satellite imagery to monitor deforestation, many developing nations rely on outdated or incomplete land-use records, leaving critical blind spots in global climate action.

Challenges of Fragmented Data Systems and Lack of Standardization - The absence of standardized protocols for climate data collection and sharing exacerbates the problem of fragmented data systems. Different organizations, governments, and researchers often use incompatible formats, methodologies, and metrics, making it challenging to integrate datasets and draw meaningful insights. Variations in emissions accounting methods across countries can create inconsistencies in global reporting systems like the United Nations Framework Convention on Climate Change (UNFCCC) (World Resources Institute, 2023). This fragmentation undermines the reliability of global climate models and complicates international cooperation.

5.2 Privacy and Security Concerns

The extensive use of data in climate governance raises critical questions about privacy and security. As more data is collected, stored, and analyzed, ensuring its protection becomes increasingly vital.

Balancing Transparency with Data Protection - Transparency is a cornerstone of smart climate governance, as it fosters accountability and public trust. However, the demand for openness must be balanced with the need to protect sensitive data. For example, detailed data on energy usage at the household level can help optimize energy efficiency programs but may also expose individuals to privacy risks if improperly anonymized (OECD, 2021). Striking this balance requires robust data governance frameworks that prioritize both transparency and data protection.

Risks of Cyberattacks on Critical Climate Data Systems - The growing reliance on digital technologies for climate action makes data systems increasingly vulnerable to cyberattacks. Critical infrastructure, such as weather forecasting systems and emissions monitoring networks, could be targeted by malicious actors, potentially disrupting climate initiatives. In 2021, for example, a cyberattack on the European Union's Emissions Trading System temporarily halted the issuance of carbon permits, highlighting the vulnerabilities in climate data infrastructure (European Commission, 2021). Such incidents underscore the importance of investing in cybersecurity measures to safeguard climate data.

5.3 Ethical and Equity Issues

The integration of data into climate governance must address ethical considerations and ensure that data-driven approaches are equitable and inclusive.

Ensuring Inclusivity in Data-Driven Governance - Data-driven climate solutions often risk excluding marginalized communities, particularly in regions where data collection is limited. Indigenous communities, whose traditional knowledge is invaluable for biodiversity conservation, are frequently overlooked in mainstream climate data systems (IPCC, 2022). This lack of inclusivity not only undermines the effectiveness of climate action but also perpetuates existing inequities. Addressing this challenge requires incorporating diverse perspectives and knowledge systems into data governance frameworks.

Avoiding Biases and Unintended Consequences of AI/ML Systems - Artificial intelligence (AI) and machine learning (ML) systems are increasingly used in climate governance to analyze data and predict outcomes. However, these systems are not immune to biases embedded in the data they process. AI models trained on data from urban areas may perform poorly when applied to rural or remote regions, leading to skewed policy recommendations (Rolnick et al., 2022). Additionally, the use of AI in resource allocation could inadvertently exacerbate existing inequalities if not carefully designed and monitored.

5.4 Technical and Financial Barriers

Implementing data-driven climate solutions requires significant technical expertise and financial investment, posing challenges for resource-constrained regions.

Costs of Implementing and Maintaining Advanced Data Infrastructure - Building and maintaining the infrastructure required for climate data collection and analysis can be prohibitively expensive, particularly for low-income countries. High-resolution satellite imagery, IoT devices, and advanced analytics platforms often come with steep costs, creating financial barriers to their adoption. While the

European Space Agency's Copernicus program provides valuable climate data, the costs associated with processing and utilizing this data can be a challenge for less affluent nations (ESA, 2023).

Lack of Technical Expertise in Data Analytics - Even when financial resources are available, a lack of technical expertise can hinder the effective use of climate data. Data analytics, AI, and ML require specialized skills that may not be readily available in many regions. This skills gap limits the ability of governments and organizations to interpret complex datasets, model climate scenarios, and develop evidence-based policies. Addressing this barrier necessitates investments in education, training, and capacity-building initiatives.

While data holds immense potential to drive climate action, realizing its full benefits requires addressing several critical challenges. Data gaps and accessibility issues, particularly in developing nations, hinder the equitable distribution of resources and insights. Privacy and security concerns pose significant risks to the integrity and reliability of climate data systems. Ethical and equity considerations must be at the forefront of data-driven governance to ensure inclusivity and fairness. Finally, technical and financial barriers limit the adoption and effective use of advanced data infrastructure, particularly in resource-constrained regions.

Overcoming these challenges will require a collaborative and multidisciplinary approach, involving governments, private sector actors, academia, and civil society. Investments in capacity-building, standardization, and cybersecurity are essential to enhance the accessibility, reliability, and security of climate data. Additionally, inclusive data governance frameworks must prioritize marginalized communities and address potential biases in AI and ML systems. By addressing these challenges, the global community can unlock the full potential of data to drive effective and equitable climate action, paving the way for a sustainable future.

6. CASE STUDIES OF SMART GOVERNANCE FOR CLIMATE ACTION

The integration of smart governance with data-driven strategies has emerged as a vital tool in addressing climate challenges. By leveraging advanced technologies, governments, organizations, and communities worldwide have implemented innovative initiatives to enhance resilience, reduce emissions, and promote sustainable development. This section explores successful examples of data-driven climate action, including smart city initiatives, predictive analytics for disaster resilience, and global data collaborations. It also delves into insights from failures or missed opportunities to highlight lessons for future initiatives.

6.1 Successful Examples of Data-Driven Climate Action

Smart City Initiatives for Sustainable Urbanization: Copenhagen - Copenhagen, Denmark, is widely regarded as a leader in sustainable urban development and smart city initiatives. The city's commitment to becoming carbon-neutral by 2025 has driven innovative data-driven strategies that integrate urban planning, mobility, and energy management (C40 Cities, 2022).

Among Copenhagen's most prominent endeavors is the Copenhagen Connecting program, which seeks to enhance city operations via the deployment of IoT sensors and the use of big data analytics. One example is the collection of real-time data on traffic patterns and energy use by sensors put on public transit and lighting. Reducing congestion, improving public transportation efficiency, and minimizing energy waste are all goals of this data analysis. Copenhagen has successfully reduced carbon emissions and ensured energy efficiency via the implementation of district heating systems driven by renewable energy sources (European Commission, 2023).

The success of Copenhagen's smart governance model lies in its integration of citizen engagement, public-private partnerships, and robust data-sharing frameworks. By involving residents in decision-making and leveraging technology, Copenhagen demonstrates how smart governance can lead to sustainable urbanization.

Predictive Analytics for Disaster Resilience: Cyclone Fani in India - Predictive analytics has proven critical in enhancing disaster preparedness and resilience. The successful management of Cyclone Fani, which struck India in May 2019, serves as a compelling example of data-driven disaster resilience.

The Indian Meteorological Department (IMD) utilized advanced weather modeling and predictive analytics to track the cyclone's path and intensity. Real-time satellite data and meteorological models provided accurate forecasts, enabling authorities to issue timely warnings and evacuation orders. Over 1.2 million people were evacuated from vulnerable coastal regions, significantly reducing casualties and damage (Gupta et al., 2020).

To better allocate resources, the Odisha government used GIS to pinpoint locations with a high risk of flooding. Communities were alerted via the use of mobile technology and social media platforms, which allowed for the dissemination of safety instructions and warnings. The data-driven governance's revolutionary potential in reducing the effects of climate-related catastrophes was on full display during Cyclone Fani's disaster management.

Data Collaborations in Global Climate Monitoring: Copernicus Programme - The Copernicus Programme, led by the European Union, is one of the most comprehensive initiatives for global climate monitoring. It integrates satellite imagery, ground-based observations, and advanced data analytics to provide actionable insights for climate action.

Copernicus plays a crucial role in tracking greenhouse gas emissions, monitoring deforestation, and assessing the impacts of climate change. For instance, its Sentinel satellites provide high-resolution data on land use, ocean temperatures, and atmospheric conditions. This data is freely accessible through open platforms, enabling researchers, policymakers, and organizations worldwide to develop evidence-based climate strategies (ESA, 2023).

One of the key strengths of the Copernicus Programme is its emphasis on international collaboration. By fostering partnerships with governments, academic institutions, and private sector actors, Copernicus has established a global network for climate monitoring. This collaborative approach ensures the availability of high-quality data and promotes shared accountability in addressing climate challenges.

6.2 Insights from Failures or Missed Opportunities

While data-driven climate governance has achieved remarkable successes, it is equally important to learn from instances where initiatives fell short of their objectives. These failures often highlight critical gaps in planning, execution, or inclusivity.

Missed Opportunities in Smart City Development - Smart city projects in some regions have faced criticism for prioritizing technology over inclusivity and equity. For example, in certain developing countries, smart city initiatives have focused heavily on infrastructure development without adequately addressing the needs of marginalized communities. A lack of stakeholder engagement and public consultation has led to the exclusion of vulnerable populations, undermining the broader goals of sustainability and resilience (UN-Habitat, 2022).

Fragmented data systems and insufficient interoperability have hindered the effectiveness of smart city projects. In some cases, incompatible data formats and siloed operations among government agencies have resulted in inefficiencies and duplication of efforts. These challenges underscore the importance of establishing standardized data protocols and inclusive governance frameworks.

Ethical Issues in Predictive Analytics - Predictive analytics, while transformative, can also raise ethical concerns when not implemented responsibly. Certain disaster prediction systems have been criticized for generating false alarms or overestimating risks, leading to unnecessary evacuations and economic losses. Moreover, the reliance on AI and machine learning algorithms can perpetuate biases if the underlying data is incomplete or skewed.

The case of Hurricane Harvey in 2017 illustrates some of these challenges. While predictive models successfully identified high-risk flood zones, limited data on vulnerable populations, such as undocumented migrants, resulted in gaps in evacuation planning. Addressing such disparities requires a more comprehensive approach to data collection and analysis, ensuring that predictive analytics accounts for diverse social and economic contexts (Rolnick et al., 2022).

Challenges in Global Data Collaborations - Global data collaborations, such as those led by intergovernmental organizations, often face geopolitical and logistical challenges. Disagreements over data ownership, access, and usage rights can impede the effectiveness of collaborative initiatives. While the Copernicus Programme has achieved significant successes, its data-sharing agreements with non-EU countries have occasionally faced delays due to bureaucratic hurdles and conflicting priorities.

The lack of adequate funding for data infrastructure in low-income countries limits their ability to fully participate in global initiatives. This disparity highlights the need for greater investment in capacity-building and equitable resource allocation to ensure that all nations can benefit from data-driven climate governance (UNEP, 2022).

Case studies of smart governance for climate action illustrate the transformative potential of data-driven initiatives in addressing climate challenges. Success stories such as Copenhagen's sustainable urbanization, Cyclone Fani's disaster resilience, and the Copernicus Programme's global climate monitoring demonstrate how innovative technologies, and collaborative approaches can drive meaningful progress. These examples highlight the importance of integrating data analytics, citizen engagement, and cross-sector partnerships into climate governance.

At the same time, insights from failures and missed opportunities underscore the critical need to address inclusivity, interoperability, and ethical concerns. Learning from these challenges will be essential for enhancing the effectiveness and equity of future initiatives. By adopting a balanced approach that combines technological innovation with robust governance frameworks, the global community can harness the full potential of data to achieve a sustainable and resilient future.

7. FRAMEWORK FOR EFFECTIVE DATA-DRIVEN CLIMATE GOVERNANCE

The escalating challenges of climate change call for innovative governance frameworks that are grounded in data-driven decision-making. Leveraging climate data effectively enables policymakers and stakeholders to gain actionable insights, foster collaboration, and enhance accountability. This section explores a framework for data-driven climate governance by outlining steps to establish a robust climate data ecosystem, build capacity in data analytics and artificial intelligence (AI), and create clear metrics for monitoring and evaluating climate action.

7.1 Building a Robust Climate Data Ecosystem

A robust climate data ecosystem is fundamental to data-driven governance, as it ensures the availability, accessibility, and reliability of data for informed decision-making. Strengthening frameworks for data collection and sharing is a crucial first step. Standardized data collection protocols are necessary to ensure the consistency and comparability of climate data across regions and sectors (IPCC, 2021). Modern technology allows for more accurate data collecting and the ability to track environmental indicators like deforestation and greenhouse gas emissions in real-time via the use of tools like internet of things (IoT) sensors and satellite imaging (World Bank, 2022).

Centralizing data repositories also plays a significant role in this ecosystem. National or international platforms can integrate diverse datasets, ensuring interoperability and easy access for policymakers, researchers, and other stakeholders. The transboundary nature of climate change necessitates international cooperation in data sharing. Multilateral agreements like the Paris Agreement emphasize the importance of transparent and collaborative data practices to support global climate action (UNFCCC, 2015).

Open data policies and cross-sector collaborations further strengthen the ecosystem. Governments should prioritize open data initiatives that make non-sensitive climate data publicly accessible. Such transparency not only fosters trust but also drives innovation by allowing stakeholders to develop data-driven solutions. Public-private partnerships (PPPs) are equally vital, as they bring together resources, expertise, and technology from diverse sectors to address climate challenges collaboratively. Platforms that facilitate knowledge exchange, such as the Global Partnership for Sustainable Development Data, are particularly effective in promoting cross-sector dialogue and the co-creation of solutions.

7.2 Developing Capacity in Data Analytics and Artificial Intelligence

Data analytics and AI hold immense potential for enhancing climate governance. However, the realization of this potential depends on the development of institutional and individual capacities. Investing in education and training is essential to build these competencies. Policymakers and government officials should be equipped with the skills to interpret and act on complex datasets through targeted training programs. Concurrently, academic institutions should integrate climate data science into their curricula to cultivate a pipeline of skilled professionals capable of addressing the unique challenges of climate analytics.

The application of AI in climate governance offers transformative possibilities. AI-powered predictive analytics can identify patterns and forecast future climate scenarios, providing policymakers with actionable insights for proactive decision-making (Rolnick et al., 2022). Machine learning models can predict the impact of extreme weather events, enabling targeted disaster preparedness. Similarly, optimization models powered by AI can enhance energy consumption, urban planning, and transportation systems, reducing carbon footprints. Real-time monitoring systems, driven by AI, provide immediate feedback on environmental changes, such as deforestation or pollution levels, allowing for swift intervention.

In addition to capacity-building, governments must enhance institutional infrastructure to integrate data analytics and AI into policymaking effectively. Establishing dedicated units or departments for climate data governance ensures a focused approach to interdisciplinary research and collaboration. These units can also foster partnerships with external stakeholders, further enhancing the impact of data-driven initiatives.

7.3 Establishing Metrics for Monitoring and Evaluating Climate Action

The effectiveness of climate governance hinges on the establishment of clear metrics for monitoring and evaluating progress. These metrics ensure accountability, identify gaps, and guide the refinement of strategies. Key performance indicators (KPIs) are essential for measuring the success of climate actions. Examples of relevant KPIs include emissions reduction targets, renewable energy adoption rates, and biodiversity indicators, such as changes in species populations or ecosystem health. Metrics that assess resilience, such as disaster preparedness scores and adaptation investments, also provide valuable insights into a region's capacity to cope with climate impacts.

Transparency and accountability are critical components of effective monitoring and evaluation (M&E) frameworks. Governments and organizations should publish regular progress reports that detail achievements, challenges, and priorities for future action. These reports should be data-driven and accessible to the public, fostering trust and enabling stakeholders to hold decision-makers accountable. Engaging communities and stakeholders in the evaluation process further enhances transparency by creating a feedback loop that informs decision-making.

Digital tools play a pivotal role in the M&E process. AI-driven platforms streamline the collection, analysis, and reporting of climate data, enabling more efficient and accurate evaluations. Blockchain technology offers secure and tamper-proof records of carbon credits and other climate-related transactions, ensuring the integrity of reported outcomes (OECD, 2021). These tools not only enhance the credibility of M&E processes but also improve their scalability and efficiency.

Data-driven climate governance offers a transformative approach to addressing the multifaceted challenges of climate change. By building a robust climate data ecosystem, governments and stakeholders can ensure the availability and accessibility of high-quality data. Developing capacity in data analytics and AI enables the effective application of these technologies to inform decision-making, while clear metrics for monitoring and evaluating climate action ensure accountability and continuous improvement. Together, these components create a holistic framework for climate governance that fosters transparency, collaboration, and innovation. As climate challenges intensify, embracing data-driven solutions will be crucial for achieving sustainable development goals and securing a resilient future for all.

8. THE FUTURE OF SMART GOVERNANCE IN THE DIGITAL AGE

As humanity faces the critical challenge of climate change, the role of technology in governance has never been more significant. The digital age presents unprecedented opportunities for creating smarter governance systems that leverage emerging technologies to address complex environmental issues. These innovations promise greater transparency, efficiency, and inclusivity in climate governance, enabling global collaboration and equitable solutions. This section explores the potential of emerging technologies such as blockchain and quantum computing, examines how interconnected data systems can foster international cooperation, and envisions a future where data-driven governance empowers sustainable climate solutions.

8.1 Emerging Technologies Shaping Climate Governance

The integration of advanced technologies into climate governance systems is transforming how nations and organizations respond to the global climate crisis. Two groundbreaking technologies, blockchain and quantum computing, have emerged as critical enablers in this transformation.

Blockchain for Transparent Carbon Markets - By making carbon markets more open and accountable, blockchain technology is changing the game when it comes to climate regulation. Some have voiced concerns about the lack of transparency and the potential for fraud in carbon trading systems, which enable companies and governments to exchange carbon credits as part of initiatives to reduce emissions

(OECD, 2021). To overcome these obstacles and provide a trustworthy record of transactions, blockchain uses a distributed and immutable ledger system.

By leveraging blockchain, carbon credits can be tracked throughout their lifecycle, ensuring their authenticity and preventing double counting. Platforms such as CarbonChain and IBM's blockchain-enabled environmental systems are already demonstrating the potential for this technology to improve market efficiency and stakeholder trust (World Economic Forum, 2023). Blockchain can facilitate cross-border carbon trading by standardizing verification processes, thereby promoting global cooperation and alignment with international climate agreements like the Paris Agreement (UNFCCC, 2015).

Quantum Computing for Complex Climate Modeling - Quantum computing, still in its nascent stage, holds transformative potential for climate governance by addressing the computational limitations of traditional systems. Climate modeling requires processing vast amounts of data to simulate complex environmental systems, which is often beyond the capacity of classical computing. Quantum computers, with their ability to perform multiple calculations simultaneously, can significantly accelerate this process (Aspuru-Guzik et al., 2022).

Quantum algorithms can optimize energy systems, model climate impacts with greater accuracy, and predict the effectiveness of mitigation strategies. By simulating intricate interactions between atmospheric, oceanic, and terrestrial systems, quantum computing enables policymakers to make more informed decisions. As quantum technology matures, it will likely become an indispensable tool for climate scientists and decision-makers seeking data-driven solutions to global challenges.

8.2 Potential for Global Collaboration Through Interconnected Data Systems

In the digital age, interconnected data systems are crucial for fostering global collaboration in climate governance. Climate change transcends national borders, making it imperative for countries and organizations to work together. Interconnected data systems enable the sharing of real-time information, enhance decision-making, and support coordinated action on a global scale.

The capacity to provide a uniform platform for monitoring and reporting climate-related data is a major advantage of linked data systems. International cooperation may be facilitated by initiatives like the Global Climate Observing System (GCOS) and the Earth System Grid Federation (ESGF) (IPCC, 2021). In order to provide a complete picture of the climate on Earth, these systems combine data from a variety of sources, including sensors on the ground and satellites.

Interconnected data systems empower under-resourced countries by providing access to critical information and tools for climate action. By reducing technological disparities, these systems promote equity and inclusivity in climate governance. Cloud-based platforms like Google Earth Engine enable developing nations to analyze satellite imagery and assess deforestation or disaster risks without requiring high-end computational infrastructure (World Bank, 2022).

To fully realize the potential of interconnected data systems, governments and organizations must prioritize data standardization and interoperability. International frameworks that establish common data formats, protocols, and governance structures are essential for ensuring that diverse datasets can be seamlessly integrated and utilized.

8.3 Envisioning a Data-Driven and Equitable Future

As digital technologies continue to evolve, the vision of a future where data empowers sustainable and equitable climate solutions becomes increasingly attainable. Such a future rests on three fundamental pillars: inclusivity, innovation, and accountability.

Inclusivity in Climate Governance - All stakeholders, especially disadvantaged populations and poor countries, should be able to participate in decision-making processes for a data-driven climate governance strategy to be inclusive. According to Fischer et al. (2022), more effective and culturally sensitive solutions may be achieved via participatory government models that combine local knowledge with modern data analytics. In order to supplement scientific data, indigenous groups often have priceless ideas about sustainable resource management.

Digital technologies should be designed to minimize barriers to access. Bridging the digital divide through capacity-building initiatives, such as providing training in data literacy and climate science, is essential for empowering communities to participate actively in governance.

Innovation as a Catalyst for Sustainability - Better systems of government will continue to emerge as a result of innovation. The advent of new technology like blockchain, AI, and the Internet of Things (IoT) may revolutionize current forms of government by making them more adaptable and responsive. In order for lawmakers to take a proactive approach to environmental challenges, sensors that are enabled by the Internet of Things may provide real-time data on air quality, water levels, and biodiversity (Georgiadis et al., 2023).

With the help of linked data systems and AI, we can detect trends in the weather and forecast them with unprecedented accuracy. Algorithms trained using historical data can improve resource allocation, predict the occurrence of severe weather, and assess the efficacy of policy measures.

Accountability Through Transparent Governance - Transparency is a cornerstone of effective governance. By leveraging digital technologies, governments and organizations can enhance accountability and build trust among stakeholders. Publicly accessible dashboards and platforms that track progress on climate targets, such as carbon emissions reductions or renewable energy adoption, enable citizens to hold decision-makers accountable (OECD, 2021).

Blockchain technology, as discussed earlier, further strengthens transparency by ensuring the integrity of data and transactions. For example, the use of blockchain in supply chain monitoring can verify that products labeled as "sustainable" meet environmental standards, reducing greenwashing and fostering consumer confidence.

The future of smart governance in the digital age holds immense promise for addressing the pressing challenges of climate change. Emerging technologies such as blockchain and quantum computing are reshaping climate governance by enhancing transparency, efficiency, and decision-making capabilities. Interconnected data systems enable global collaboration and equitable access to information, empowering all nations to participate in climate action.

By embracing inclusivity, innovation, and accountability, data-driven governance can create a sustainable and equitable future where the power of technology is harnessed to protect the planet and improve the lives of all its inhabitants. As nations and organizations navigate the complexities of the digital age, a shared commitment to leveraging these transformative tools will be crucial for achieving meaningful and lasting climate solutions.

9. CONCLUSION

The escalating challenges of climate change underscore the necessity of leveraging data as a cornerstone for effective climate action. In the digital age, data has become a critical resource for understanding environmental systems, predicting climate trends, and formulating informed policies. A robust data ecosystem, encompassing advanced technologies and interconnected systems, provides unparalleled opportunities for enhancing transparency, collaboration, and accountability in climate governance. By harnessing data effectively, policymakers can drive proactive measures that mitigate climate risks while promoting sustainable development.

Smart governance, powered by emerging technologies such as blockchain, artificial intelligence, and quantum computing, holds transformative potential for achieving climate goals. These innovations enhance the precision of climate modeling, streamline carbon market operations, and enable real-time monitoring of environmental changes. Furthermore, interconnected data systems facilitate global collaboration, breaking down geographical and technological barriers and empowering all nations—regardless of their resources—to contribute meaningfully to climate action.

However, realizing this vision requires a concerted effort from policymakers, technologists, and global stakeholders. Policymakers must prioritize the development of inclusive frameworks that ensure equitable access to data and technological resources. Technologists should focus on creating user-friendly, scalable solutions that address the diverse needs of communities worldwide. Meanwhile, global stakeholders must commit to fostering partnerships and sharing knowledge to maximize the impact of data-driven climate governance.

As the world stands at a crossroads, the integration of data into climate strategies offers a pathway to sustainable and equitable solutions. It is imperative for all actors to embrace this opportunity and work collaboratively to harness the power of data, not only to confront the climate crisis but to build a resilient and thriving future for generations to come.

REFERENCES

C40 Cities. (2022). Copenhagen: A global leader in climate action. Retrieved from https://www.c40.org

C40 Cities. (2023). About C40 cities. Retrieved from https://www.c40.org

Aspuru-Guzik, A., Cao, Y., & Romero, J. (2022). Quantum computing for the environmental sciences. *Nature Climate Change*, *12*(3), 204–211.

ESA (European Space Agency). (2023). Copernicus: Europe's eyes on Earth. Retrieved from https://www.esa.int

European Commission. (2021). Cybersecurity in the EU ETS: Lessons from the 2021 incident. Retrieved from https://ec.europa.eu

European Commission. (2023). Copenhagen's smart city initiatives. Retrieved from https://ec.europa.eu

Fischer, M., Lindberg, T., & Lenton, T. M. (2022). Participatory governance for climate resilience: Empowering local communities with data and technology. *Climate Policy*, *22*(4), 445–460.

Georgiadis, P., Paraschivoiu, M., & Wulff, A. (2023). AI and climate resilience: Leveraging technology for a sustainable future. *Climate Policy Journal*, *18*(3), 123–140.

Global Covenant of Mayors for Climate and Energy. (2023). About us. Retrieved from https://www.globalcovenantofmayors.org

Gupta, H., Sharma, M., & Mishra, R. (2020). Disaster management and predictive analytics: Lessons from Cyclone Fani. *International Journal of Disaster Risk Reduction*, *44*, 101–105.

Intergovernmental Panel on Climate Change (IPCC). (2021). *Climate change 2021: The physical science basis*. Cambridge University Press.

Intergovernmental Panel on Climate Change (IPCC). (2023). [*Synthesis Report*. Geneva: IPCC.]. *Climatic Change*, 2023.

IPCC. (2022). [Impacts, adaptation, and vulnerability. Retrieved from https://www.ipcc.ch]. *Climatic Change*, *2022*.

Meijer, A., & Bolívar, M. P. R. (2016). Governing the smart city: A review of the literature on smart urban governance. *International Review of Administrative Sciences*, *82*(2), 392–408. DOI: 10.1177/0020852314564308

Microsoft. (2023). AI for Earth. Retrieved from https://www.microsoft.com

NASA. (2023). Globe observer app. Retrieved from https://www.nasa.gov

NASA. (2023). NASA's Earth Observing System (EOS). Retrieved from https://www.nasa.gov

NASA. (2023). Orbiting Carbon Observatory-2. Retrieved from https://www.nasa.gov

OECD. (2021). Carbon pricing in action. Retrieved from https://www.oecd.org

OECD. (2021). Engaging citizens in climate policy. Retrieved from https://www.oecd.org

OECD. (2021). Enhancing transparency in climate action: Balancing privacy and accountability. Retrieved from https://www.oecd.org

Organisation for Economic Co-operation and Development (OECD). (2021). *Blockchain technologies as enablers of sustainable development goals*. OECD Publishing.

Rolnick, D., Donti, P. L., & Kaack, L. H.. (2022). Tackling climate change with machine learning. *Nature Climate Change*, *12*(3), 203–210.

Steele, J., Stoeckl, N., & McDonald, G. (2022). Blockchain applications for carbon markets and climate change mitigation. *Nature Climate Change*, *12*(3), 207–215.

UN-Habitat. (2022). Smart cities and inclusivity: Challenges and opportunities. Retrieved from https://www.unhabitat.org

UNEP (United Nations Environment Programme). (2022). The emissions gap report 2022. Retrieved from https://www.unep.org

UNFCCC. (2023). Paris agreement. Retrieved from https://unfccc.int

United Nations. (n.d.). Climate action and support trends. Retrieved from https://www.un.org/en/climatechange

United Nations. (n.d.). Scenario modeling for climate action. Retrieved from https://www.un.org/en/climatechange

United Nations Framework Convention on Climate Change (UNFCCC). (2015). *The Paris Agreement*. Retrieved from https://unfccc.int

United Nations Office for Disaster Risk Reduction (UNDRR). (2022). *Global Assessment Report on Disaster Risk Reduction 2022*. UNDRR.

U.S. Department of Energy. (2023). Open energy data initiative. Retrieved from https://www.energy.gov

U.S. Department of Energy. (2023). Smart water management. Retrieved from https://www.energy.gov

U.S. Environmental Protection Agency. (2014). Inventory of U.S. greenhouse gas emissions and sinks. Retrieved from https://www.epa.gov/ghgemissions

World Bank. (2022). *Harnessing technology for climate-smart development*. World Bank Group.

World Bank. (2023). Climate change knowledge portal. Retrieved from https://climateknowledgeportal.worldbank.org

World Bank. (2023). Disaster resilience initiatives. Retrieved from https://www.worldbank.org

World Economic Forum. (2023). Blockchain for sustainable and transparent carbon markets. *Global Future Council Reports*.

World Meteorological Organization (WMO). (2023). *State of the Global Climate 2022*. WMO.

World Resources Institute. (2023). Climate data explorer. Retrieved from https://www.wri.org

World Resources Institute. (2023). Data challenges in climate governance. Retrieved from https://www.wri.org

Chapter 13
Managing Climate Change Shocks:
South African Wine Industry and Cape Town's Adaptation to the "Day Zero" Water Crisis

Lucky Madikiza
https://orcid.org/0000-0003-4807-5004
University of South Africa, South Africa

ABSTRACT

In 2017, Cape Town faced a severe drought. The incident would possibly culminate in that city's taps being turned off by March 2018. The likely day of the city's water facilities completely running dry and leaving more than three million residents without water was labelled 'Day Zero'. Cape Town is part of the Western Province of South Africa which contains large stretches of vineyards which were highly affected by the drought. The South African wine industry, being one of the top major producers of wine in the world, stood to be severely affected. This chapter documents how the wine industry was impacted by this climate crisis and the adaptive mechanisms it implemented. Further, the incident provides a case example of how climate shocks may compel further futuristic thinking on the need to contemplate adaptive mechanisms by cities regarding managing future disruptive climate risks.

INTRODUCTION

Cape Town is part of the Western Province of South Africa which contains large stretches of vineyards which were highly affected by the drought. The South African

DOI: 10.4018/979-8-3373-1310-8.ch013

wine industry, being one of the top major producers of wine in the world, stood to be severely affected. This chapter documents how the wine industry was impacted by this climate crisis and the adaptive mechanisms it implemented. The industry employed various strategies to curtail the adverse impact of the drought including participating in research and innovating on growing practices. The discussion will also distil some lessons that could be learnt from the ordeal. Further, the incident provides a case example of how climate shocks may impact on agricultural industries and compels further futuristic thinking on the need to contemplate sustainable innovations to manage future disruptive climate risks.

IMPACT OF THE CLIMATE CRISIS ON WATER SECURITY: GLOBAL OVERVIEW

With the emergence of climate change crises ranging from floods and droughts, every global region has been prone to climate-associated disasters. It has been noted that over the past two decades, more than 80 metropolitan cities across the world have faced severe water shortages owing to droughts (Savelli et al., 2023). The issue of climate change has a large bearing on water scarcity being faced by cities. The Philippines' capital city region of Metro Manila has been experiencing severe water scarcity due to the recurrent El Niño phenomena, which is brought on by the anomalous warming of surface waters in the eastern Pacific Ocean (Lee et al., 2020). The El Nino phenomenon has also been a contributor to periodic droughts that have affected the southern African region including in South Africa.

The ongoing phenomenon of increasing urbanization is also set to place continuing additional pressure for water demand in cities across the world especially in developing country contexts whose cities are considered to have the lowest water security (Van Ginkel et al., 2018). Some cities in the global South are beginning to explore alternative water systems such as groundwater most notably in cities like Hermosillo (Mexico) and Mendoza (Argentina) amidst increasing urbanisation (Díaz-Caravantes et al., 2020). However, some regions are experiencing groundwater depletion challenges that make it a challenge to explore groundwater as an alternative option. For instance, groundwater has long been used in cities in the global South Asian region and so over-extraction of groundwater has seen steadily falling water tables in the Punjab regions of India and Pakistan (Sathre et al., 2022). The water security challenge remains one of the most formidable for growing urban centres to grapple with. Even with interventions like developing more cost-effective water infrastructure and technologies as well as progressive water pricing, it has been said that these technical-managerial remedies are insufficient to handle water emergencies if the underlying reasons are not found (Savelli et al., 2023, p. 933).

According to one study, in developing cities, like Nairobi, Jakarta, and Lima, it seems that independent of the environmental context, all municipal water functions eventually become impaired when cities experience extreme socioeconomic strain such as low GDP, urban slums, and a polluting industry, which in turn result in some instances in deteriorating water quality (Van Ginkel et al., 2018).

At best, the global water security challenge needs to be tackled from several fronts. Aside from effective and efficient water infrastructure management, there is a need to be mindful of polluting industrial practices as well as unsustainable water consumption tendencies, among other social factors. In one study, it is suggested that the only way to preserve available water resources is by altering privileged lifestyles and redistributing income and water resources more equally as well as re-orienting water management policies towards new political economic paradigms that prevent elite overconsumption and inequalities (Savelli et al., 2023, p. 933-934). In another study done in Kalkota India, findings suggested that water insecurity of a city is not only due to the malfunction or inadequacy of city's water system but also stems from the intersecting disadvantages, inequalities, and exclusion found in a society (Mukherjee et al., 2022). Even with regard to the water crisis that was experienced in Cape Town, one perspective suggests that the co-existing tensions around budgeting, inequality and climate change will need to be managed in ways that support more just futures and a more profound shift in the fundamental logics that are shaping water governance (Millington & Scheba, 2021, p. 129).

Given the complexity of handling climate change occurrences, some frameworks have been advanced in literature on how organisations may approach climate shock events. The next section provides an overview discussion on some of these frameworks before we delve into the City of Cape Town's experiences.

MANAGING CLIMATE CHANGE CRISES: THEORETICAL AND CONCEPTUAL FRAMEWORKS

Regarding managing organisations in the context of the climate change challenge, several theoretical and conceptual frameworks have been deemed useful. These frameworks assist organizations in developing a comprehensive plan to address the climate change issue, improving their capacity to react to or adjust to climate change events. The literature on ideas and theories related to climate change is enormous, and this discussion would not succeed in covering the vast work that is still continuously emerging in this area of research. For this discussion, a limited selection of frameworks is invoked in relation to the topic at hand. We discuss the Adaptative Management; Resilience; and integrated perspectives.

Adaptive Management Framework

Over time, both humans and animals have had to adapt to the effects of climate change. Events related to climate change over the past eras would have caused migration of humans and animals. The notion of adaption is considered to have originated from evolutionary biology (Bock, 1980; Simonet & Duchemin, 2010). Different definitions of the term have been published over time by academic institutions and policy settings, leading to a variety of interpretations of the term. It is also important to bear in mind that the adaptation has not always been adequately defined and has sometimes been described taxonomically rather than by defining its salient characteristics (Orlove, 2022, p. 538). Sometimes it may be used ambiguously such as when adaptation may connote a process of adapting and the condition of being adapted (Smit et al., 1999, p. 203). Even if a standard definition would be agreed upon, another issue of consideration would what would constitute successful adaptation, something that has been considered worthy of being researched in the adaptation corpus of literature (Guillén Bolaños et al., 2022, p. 2).

An overview of adaptation parlance will yield contrasting definitions. Cast in more general terms, "adaptation as a process is apparently about changing something (itself, others, the environment) so that it would be more suitable or fit for some purpose than it would have otherwise been " (Lints, 2010, p. 115). For the purposes of this chapter, the notion is being tied closer to the problem of climate change, hence it is imperative that a definition mention some aspect of environmental change. As such, two definitions of adaptation would be more applicable such as the one defining it as "adjustments in ecological-social-economic systems in response to actual or expected climatic stimuli, their effects or impacts" (Smit et al., 1999, p. 200), and another which defines it as "change in a system in response to some force or perturbation, in our case related to climate" (Smithers & Smit, 1997, p. 135).

The idea of adaptation has been examined by several academic fields, including sociology, geography, anthropology, and psychology. However, these fields typically approach concepts from different angles, with experts in their fields emphasizing aspects of adaptation that are relevant to their areas of study (Simonet & Duchemin, 2010). For instance, according to one definition, which considers health and well-being aspects, "adaptation has been defined as the process or adjustments through which people reduce the adverse effects of climate on their health and well-being, and take advantage of the opportunities that their climatic environment provides" (Eriksen et al., 2011, p. 8). In addition to the definitions, the finer operational specifics of adaptation would also require clarification. When it comes to adaptation, for example, a question arises as to the subject of the adaptation—who or what is adapting—and the object of the adaptation—what they are adapting to (Francesco & Carlo, 2018, p. 13). As regards the emphasis on this chapter, if we must delve into the question of who is adapting, then it is more important to examine how organizations, and particularly agricultural companies, would adapt to climate change than it is to focus on individuals.

The organisation-wide considerations of adaptation have been framed around the concept of the Adaptation Management Framework in management literature. The Adaptation approach is concerned with addressing the vulnerabilities introduced to habitats and livelihoods caused by climate change and concentrates on anticipating, managing, and adapting to these in the short and long-term. Adaptive management entails the identification and reduction, where possible, of uncertainty as it may be wrought by environmental variation, among other factors (Allen et al., 2011, p. 1340). Originally, adaptive management was developed as a means of addressing practical issues related to environmental and natural resource management and has been applied to ecosystem management issues with a range of ecosystem types, socioeconomic characteristics and risk implications (Gregory et al., 2006, p. 2410).

The adaptive management paradigm guides organizational efforts toward adaptation and as it does so, places emphasis on flexibility, adaptability and learning in response to climate change challenges, even as it acknowledges that uncertainty and change are inherent. It is adaptive because it recognizes that human intervention will always cause managed resources to change, that surprises are inevitable, that new uncertainties will surface, that active learning is a necessary method of sifting through uncertainty, and that social objectives must be met by policies that must also be flexible and constantly modified in order to accommodate these surprises (Gunderson, 1999, p. 1).

Given that Adaptation Management uses the descriptor of "adaption," a concept fraught with various interpretations as previously discussed, it is wont to attract varying understandings and therefore different implementation approaches. The method's inherent flexibility may be the cause of the confusion surrounding the term "adaptive management," which can be interpreted in a variety of ways along a spectrum of simplicity, such as the example of the "learning by doing" approach to the more explicit, which might involve, for instance, a rigorous process that should include careful planning and experimental design along with a methodical evaluation process that connects monitoring to management (Allen et al., 2011, p. 1342). Another definition explicates the process aspect when adaptive management is defined as " the process of defining and bounding the management problem, identifying and representing what we know through models of dynamics that identify assumptions and predictions so experience can further learning, identifying possible sources of uncertainty and identifying alternate hypotheses, and finally designing policies to allow continued resource management or production while enhancing learning" (Allen et al., 2011, p. 1340).

If Adaptive Management is said to deal with uncertainty, then that concept also needs to be unpacked as variants of uncertainty may entail imperfect knowledge, unpredictability and incomplete knowledge (Hasselman, 2017, p. 12). The other concepts embedded within the interpretations, which can have implications on how the management approach is operationalized (Gregory et al., 2006). For instance, some scholars consider the problematic concept of 'learning by doing' as what without a clear definition of what constitutes learning and how much of it is required (Gregory et al., 2006, p. 2411).

In a case study of how adaptive management was applied in water management in the Lachlan catchment, Murray–Darling Basin, Australia, amidst the degradation of the basin, different levels of learning were discernible with some learning being ad hoc whilst higher-level learning was also discernible when more active adaptive management of environmental flows was applied (Schoeman et al., 2019). Hence, the underpinning concepts of adaptive management such as learning can be ambiguous or open to varying interpretation, potentially making the implementation of adaptive management approaches a complex, if not an undertaking that can produce mixed results of successes or failures. Hence one of the key tenets for applying adaptive management is stakeholder consultation, so that there is some unison in understanding key concepts that pertain to its adaptive practices (Gregory et al., 2006; Stringer et al., 2006). In fact, lack of stakeholder engagement, together with learning not being used to modify management approaches, action procrastination and avoiding hard truths have been considered as factors that contribute to failure of adaptive management approaches (Allen et al., 2011).

One noteworthy aspect for Adaptative Management is that it should not be considered as a 'one-size fits all' intervention to complex problems in environmental management that exhibit ecological uncertainty (Gregory et al., 2006, p. 2412). It therefore is an approach that must be adapted itself to certain contexts which organisations face and which may differ according to country environments.

Resilience Framework

In contemporary discussions about global environmental change, resilience has emerged as a permeating theme in the necessity to cope with climatic disruptions (Brown, 2014). Principally, the concept of resilience is concerned about the toughness of organisations or institutions, including public sector organisations, amidst adversities and crises including natural disasters, climatic disasters, and pandemics such as Covid-19 as we have witnessed in recent years.

Building organizational resilience to climate-related shocks has been seen as crucial, and the Resilience Framework emphasizes this by highlighting capacities that are transformative and absorptive. The concept of resilience entails organisational responses to disruptive occurrences and stresses the capacity to recover in conditions of uncertainty, emergency and discontinuity as well as even developing a new skill in disruptive conditions (Xiao & Cao, 2017, pp. 1–2). Resilience is a concept that has been widely used in a variety of fields, such as organisational psychology, supply chain management, and ecology and although the context of any given field using the concept of resilience may vary, it is typically associated with the capacity to recover from disruptive events and return to a stable state (Bhamra et al., 2011, p. 5376). Although there are many ways to define resilience, at least three broad streams of conceptualization have been identified: resilience as an organizational characteristic, resilience as a result of an organization's actions, and resilience as a gauge of the disturbances an organization can withstand (Ruiz-Martin et al., 2018, p. 15).

After overcoming adversity, resilient organizations cultivate qualities like adaptability, change readiness, communication, knowledge, and interdependency knowledge which help an organization recognize and assess new threats and their potential effects on it, as well as build and maintain strong relationships with important stakeholders in its industry, and the ability to respond, recover, and grow from setbacks (Parsons, 2010, p. 18).

Beyond conceptualizations and definitions, the devil lies in the details of how resilience is really implemented or operationalized. The challenge of operationalizing the development of resilient organizations might occasionally arise from the incapacity to convert the notion of resilience into concrete organizational frameworks (McManus et al., 2008). Nonetheless, numerous frameworks and models have been proposed for enhancing or creating organisational resilience (Ruiz-Martin et al., 2018, p. 17).

Integrated Frameworks

Climate change events are often complicated, necessitating the use of multiple frameworks to make sense of them. First, while weather forecasters may predict floods or famines, the magnitude of their effects is frequently underestimated. At one level, their occurrence may be considered as a disaster necessitating quick adjustment, and at another, as a risk, if there has been some degree of anticipating their occurrence. As a result, frameworks have been developed that include both risk management and adaptability. As an example of this tendency, the Disaster Risk Reduction Framework (DRF) and the Climate Change Adaptation (CCA) Framework have been acknowledged as needing to be integrated and yet the two approaches originated and are enacted through different communities of research and practice (Islam et al., 2020).

As a standalone concept, Disaster Risk Reduction entails practices of reducing disaster risks through systematic efforts to analyze and manage causes of disasters and involves the creation and application of policies, strategies, and practices that minimize vulnerabilities and disaster risks throughout a society in order to avoid or limit negative impacts of hazards within the wider context of sustainable development (Begum et al., 2014, p. 364). On the other hand, Climatic Change adaptation (CCA) will entail adjusting to existing or predicted climatic influences in order to avert harm or capitalize on possibilities and in order to guarantee equitable outcomes, the process should ideally involve all stakeholders in the adaptation process (Pham & Saner, 2021, p. 1). Prospects for integrating DRR and CCA frameworks is that both are aimed at reducing vulnerability and in enhancing resilience (Begum et al., 2014, p. 362). Again, both aim to manage hydrometeorological hazards by reducing exposure and will be mindful on how to mitigate the effects of climate-related disasters and associated risks while also promoting long-term disaster management approaches (Forino et al., 2015, p. 373). Since both allude to resilience, this would suggest that both CCA and DRR could be blended with resilience approaches. For instance, Nelson (2011) asserts that resilience and adaptation are related concepts as they both capture the ways in which humans can proactively respond to forces of change. In turn, resilience can also be related to sustainability as a concept that is often centred around human, social and environmental capital. Conceptual interactions and linkages have been denoted on sustainability and its link to resilience and adaptation. As explained in Nelson et al.(2020, p. 312) changes in sustainability capital can impact system-wide adaptive capacity, potentially affecting crucial subsystems and components and which in turn can have an impact on system-wide resilience over time, as well as the creation and selection of adaptation/transformation methods required to mitigate the detrimental effects of hazards.

However, in practice, merging the climate change frameworks tends to have some difficulties since frameworks have evolved independently over time and have not always been deliberately integrative in their conceptual designs in terms of their objectives and goals. Practically, in terms of the DRR and CCA approaches, there has been some challenges in integrating the two perspectives as often there is no consensus on how to integrate them and there is the additional challenge that both emanate from different communities of research and practice and use different approaches (Islam et al., 2020; Keweloh, 2015). Again, merging resilience and adaption in practice may also emerge with challenges such as when adaptation responses can help to build resilience, but while also having the potential to undermine resilience as Nelson,(2011, p. 113) proclaims. It has been similarly observed that sustainability and resilience, whilst they have shared similarities, there could also be tensions between the two standing in the way of integration such as their disparate focuses on different goals, different assumptions of normality and different approaches in pursuing objectives (Roostaie et al., 2019, p. 140). To be more specific, the objectives of sustainability tend to focus on efficiency and resource management, whereas resilience tries to respond to external factors and performance systems during a crisis and regarding assumptions of normalcy, sustainability is concerned with stability and balance, whereas resilience is concerned with constant change and even chaos (Ibid). While there may be merits in integrating the various climate change frameworks, it is at the operational level that the various assumptions and objectives of the frameworks may need to be reconciled or bridged. Our next discussion, moving beyond the conceptual primer on climate change frameworks, investigates the practicalities undertaken of adjusting to a climate change event. In the instance of this chapter, the focus is on the City of Cape Town as well as its wine industry.

UNFOLDING OF WATER CRISIS IN CAPE TOWN AND THE CITY'S RESPONSES

Cape Town, one of the top tourism destinations in the world and the second biggest city in South Africa. In addition to being a popular travel destination, Cape Town and its Cape Winelands are one of South Africa's top export destinations worldwide for wine, a major industry in the area. It is important to recognize that South Africa is a water-scarce country with only one-third of South Africa receiving enough rain to support crop production, and with only 12% of the nation having fertile soil suitable for rainfed crop production (Naude & Naude, 2019, p. 2). The City of Cape Town represents the more conducive areas for agricultural production and has over the years had capabilities for wine and grape farming, deciduous and citrus fruits, and grains. The City and its surrounds often receive an annual rainfall of 520 mm usually between the months from June to August which are deposited in the city's 14 dams that have a capacity of more than 800 00 mL (Visser, 2018, p. 1).

In the years 2015-2018, Cape Town faced a drought crisis resulting in acute water shortage and severe implications for its tourist and wine producing sectors. The three-year drought began in 2015 with a hot and dry summer and very little winter rains, and persisted into a fairly dry 2016 that gave little time for recovery, and ultimately culminated with a record-breaking low rainfall in 2017 thereby making the water crisis even more severe than in previous years (Enqvist & Ziervogel, 2019, p. 7). Concerns about the water security of Cape Town have been forecasted in various research studies including one from the Water Research Commission in 1990 which predicted a likelihood of the city's water drying up in 2007 whilst the 2015-2018 drought period was also being forecasted by meteorologists a decade earlier (Visser, 2018). The Western Cape Province area in which Cape Town is located has been receiving a series of droughts over the past three decades although the one lasting up to 2018 was noted to be the worst. (Naude & Naude, 2019, p. 6). Reports surfaced during the three-year-long drought that Cape Town may face a "Day Zero" in the run-up to 2018, when the city's water taps will run dry and there would be "zero" water.

Intervention Plans

The City of Cape Town, faced with an oncoming water supply calamity, had to consider methods, and means to avoid a situation of zero water availability. Towards the end of 2017, the city had already experienced sparse rainfall, and entering 2018, a Critical Water Shortages Disaster Plan was formulated. The plan mapped out activities in three phases. The first phase entailed preservation restrictions entailing cutting water use by 40% and the second phase entailed disaster restrictions entailing water rationing whilst the last phase of disaster implementation would entail offering groundwater and bottled water in the instance when accessible surface water runs out (Enqvist & Ziervogel, 2019, p. 8).

Regulations on Water Restrictions

The City of Cape Town's municipality and the provincial government had to galvanise important interventions to water continuity amidst the impeding 'Day Zero' scenario forecasted to take place at any point in the year 2018. The provincial government imposed limits in January 2016 with the goal of attaining a 20% decrease in use for homes; however, the agriculture sector was not subject to these restrictions until May 2016, when the same percentage of registration was also seen (Ziervogel, 2019, p. 4). Further interventions, starting in 2017, included increased tariffs and penalties for high volume users, reduced pressure in the reticulation system and ramped-up installations of water management devices in households consuming more than the allowed quota (Enqvist & Ziervogel, 2019, p. 8). In terms of water restrictions, various levels were imposed with the highest being Level 6 (87 litres per person per day) and Level 6B (allowance of 50 litres per person per day) as well as prohibition of water use practices such as gardening, washing cars with tap water and outdoor pools (Matikinca et al., 2020).

Alternative Water Sources

Given the approaching disaster that would occur when the taps ran empty, the City of Cape Town considered other water sources. A plan was developed to pump up to 500 ML of water per day from new non-surface water sources until December 2021, and 200 ML per day until the end of 2018 from a combination of groundwater and treated wastewater (Nhamo & Agyepong, 2019, p. 5). The city also implemented a number of small groundwater abstraction schemes and seawater desalination units as well as a temporary wastewater treatment for re-use scheme (Taing et al., 2019, p. 532). Small streams and aquifers also became sources of water. However, water abstraction mechanisms such as desalination, which entails the extraction of water from the sea, renowned to be expensive, posed costs for the city. When the city considered containerized desalination plants, it was discovered that they were not cost-effective and would not provide enough volumes of water to have a measurable impact, so they were reduced to three temporary plants from the original nine, yielding approximately 16 million litres daily (Parks et al., 2019). The situation also proved financially punitive for Cape Town residents who also had to incur costs associated with installing water-efficient devices, grey water solutions and rainwater harvesting tanks while no tax savings or rebates were offered by the city council (Visser, 2018, p. 4).

Public Communication Campaign

An extensive public communication campaign around the various interventions to mitigate the arrival of 'Day Zero' by the City of Cape Town entailed the use of various communication platforms including social media. The media became one of the most visible communication outlets to highlight the impending 'Day Zero.' The public discourses on 'Day Zero' in the media and according to one research study, had thematic framings largely revolving around from it being considered an impending apocalyptic event and as a hoax by the more sceptical Cape Town citizens (Voci et al., 2020). In another study looking at public communication, it was discerned that the possibility of a water shortage was first mentioned in a more gentle way, along with suggestions for conserving water but when the public response was minimal in effecting behaviour change, communication was then ramped up, leading to the apocalyptic framing of Day Zero, which appealed to fear and anxiety (Noll, 2020). In another further study, whilst there was a doomsday clock narrative that was discernible in news media, there was also solidary campaigning whereby a sense of pride was being celebrated in warding off the doomsday scenario (Warner & Meissner, 2021). In terms of its doomsday warning, it does seem the 'Day Zero' had some effect. In a survey intended to collect feedback on the public awareness campaign, most participants saw the warning that Cape Town would probably run out of water entirely as a frightening and annoying prospect with one participant expressed frustration that they thought the government was using the Day Zero communication campaign as a "scare tactic" to get people to pay attention to the water situation, while another participant thought the messaging encouraged people to work together to conserve water throughout the crisis (Matikinca et al., 2020, p. 492). This suggests the public communication campaign had the effect of making people concerned about the doomsday scenario of zero water whilst also inducing behaviour change to conserve water.

Future Planning

In 2019, the City of Cape Town drafted a Water Strategy that takes a long-term approach to managing the city's water security. The five-pronged strategy was focused on safe access to water and sanitation for all (2) responsible water usage through pricing, regulation, active citizenship, and network management (3) Sufficient, dependable water from a variety of sources: surface, ground, desalination, and reuse (Water Resilient by 2030); (4) Shared benefits and risk management from regional water resources; and (5) transformation to a water-sensitive metropolis by 2040 (City of Cape Town, 2019). The strategy is noted to have integrated risk more directly than before and in its formulation, various departments with expertise in policy, strategy, resilience and finance were drawn in for stakeholder inputs given that it also considered climate change considerations (Ziervogel & Morgan, 2022). The aim of the strategy was to provide solutions through a collaborative approach that recognises the diverse stakeholders to ensure a more comprehensive understanding of water management and hence the City's Water and Sanitation Department was not the only one tasked with its implementation but instead, other departments were also involved (Ziervogel, 2019).

Tourism Sector Impact

The tourism sector and the agriculture industry, the main economic strongholds of Cape Town within the Western Cape Province, were impacted by the water crisis. In terms of tourism, it was discernable that passenger movements to Cape Town as one of the favourite global tourism destinations did decline although the number of visitors were still increasing. Examination of Cape Town's global passenger movement statistics revealed that, while the city's passenger movements increased during drought years, they probably did not increase as much as they would have in the absence of the severe drought with growth rates in 2017 and 2018 being 1.3% and 5.7% lower, respectively, than global rates (Drummond, 2019, p. 28). In the guest house establishments, water management interventions to deal with the drought of 2015-2018 which included using boreholes; collecting water dripping from air conditioners; re-use of laundry water for cleaning; planting drought-resistant plants in gardens; requesting guests to use water sparingly by using showers and collecting shower water to fill in the toilet cistern (Visser & Cloete, 2021). The fine-dining restaurants, a key sub-sector in tourism, also implemented water saving measures such as using water saving equipment, using grey water systems and boreholes and water tanks (Magida & Makoza, 2023). In the next section, we go into greater detail on the agricultural sector's response, namely wine production as a component of grape farming.

SOUTH AFRICAN WINE INDUSTRY'S RESPONSES

Drought Adaptation Measures

The grape farming and wine production sector has been vulnerable to changing climate change even before the 2018 Day Zero water crisis. For instance, in 2003 and 2004, there also lighter rains leading to the Cape Town City Council implementing level 2 water restrictions entailing reducing water use for watering of gardens or washing of cars. (Carter & Bayer, 2006, p. 8).

Droughts have been occurring in the past three decades in the Western Cape province with the words occurring in the five years leading up to 2018 as the year of the Day Zero water crisis (Naude & Naude, 2019, p. 6). So even before the year of the Day Zero Water Crisis of 2018, the agricultural sector in the Western Cape has had to grapple with frequent droughts and consider ways of adapting to sparse water conditions. Studies on climate change in South Western Cape region, the province in which Cape Town is located, have been showing increasing temperatures with warm days becoming warmer and occurring more frequently in the past decade (Vink et al., 2012, p. 5). South Africa in general, continues to remain vulnerable to frequent climatic extremes with warming projected to reach around 3 to 4 °C along the coast, and 6 to 7 °C in the interior by the end of the century (Chauke & Pongrácz, 2024, p. 2).

The farmers, given the propensity of drought experienced in previous years, have been adept at improvising on ways to adjust their farming practices amidst water scarcity. The farmers have been familiar with water stress adaptation strategies used in agriculture, and which also pertain to wine grape farming, which may include mulching, cover cropping, reducing tillage, investing in water-storage facilities, and improving irrigation systems (Riedo, 2019). During the Day Zero event, the wine industry implemented a variety of water-saving measures, including lowering water usage in vineyards and wineries and installing efficient irrigation systems (Wine Industry Adapts to Drought, 2017). Wine farmers also investigated more drought-resistant vines, including new clones of Grenache and cultivars such as Assyrtiko, Verdelho, Cabernet Sauvignon, and Touriga Nacional; implementing dry farming practices; and planting cover crops to ensure water retention and absorption when rain returns (South African Wine Industry Overcoming and Adapting to Day Zero Declaration, 2019). Farmers also reached out to the Western Cape Department of Agriculture and organisations such as GreenCape to adopt more climate-resilient practices such as intercropping, mulching and the adoption of soil moisture monitoring technologies (South Africa's wine industry has less room to hide from climate change than other regions, 2021).

Climate Change Conscious and Sustainable Wine Production

Beyond drought events, the South African wine industry is cognisant of climate change and within the ambit of its consideration, has been contemplating climate change responsive strategies.

From other observations in studies, the South African industry had appeared to be in a fairly robust position for dealing with the changes that are projected for the mid-21st century and with careful management and early investment into water saving strategies as well as informed cultivar choices, it was expected that there should be sufficient capacity to avoid major impacts on productivity in the industry (Carter & Bayer, 2006, p. 28). Sustainable Wine South Africa (SWSA) is the alliance between the Wine and Spirit Board (WSB), the Integrated Production of Wine (IPW) scheme and Wines of South Africa (WoSA). Together these organisations are driving the South African wine industry's commitment to sustainable, eco-friendly production. In 2014, the Smart Agriculture for Climate Resilience (SmartAgri) project was initiated and was aimed at creating a climate change response framework for the agricultural sector in the Western Cape and comprised six priorities in its action plan with entail conservation agriculture; restoring degraded landscapes; improved catchment management for water security and job creation; energy efficiency; and climate-proofing the Western Cape's agri-processing sector and the provision of an integrated knowledge system for climate smart practices (Theron & Hunter, 2022, p. 11). Another initiative, the Confronting Climate Change (CCC) initiative, was conceptualized by the South African fruit and wine industry in 2008 to enable South African growers and service providers to measure their carbon footprint, identify 'carbon hotspots', develop creative CO2 emission reduction solutions, and manage buyer and policymaker perceptions to ensure the industry's long-term viability (Our Sustainable Wine Journey, 2022).

In a further effort to instil sustainable eco-friendly agricultural practices, an organisation, Sustainable Wine South Africa (SWSA), has also been created. Key areas of focus for the organisation include advancing organic and biodynamic farming; sustainability; ethical trading; biodiversity and climate change awareness in the wine production sector. In 2010, the organisation introduced the world's first certification seal whereby producers must demonstrate compliance with established ethical and sustainability practices such as ensuring that grapes and wine were grown and prepared in accordance with strict environmental rules, conservation of biodiversity and treatment of waste water (Theron & Hunter, 2022, p. 13). Fair Trade certification has also been implemented in the wine industry (Overton et al., 2019). Whatever progressive strides have been made on sustainability, there are some challenges that the South African wine industry needs to address, such as the fact that efforts to measure and contain carbon emissions are still tentative and may cause problems if the European Union decides to apply carbon adjustment measures at the border for imports, the historic problem of labour conditions being still problematic, as well as the limited availability of organic and biodynamic wines which still constitute a small component of the total in South Africa (Das Nair et al., 2023, p. 7). Nonetheless, the industry has been associated with a longer history of engagement in sustainability standards and certification in wine (Ibid). In a further effort to instil sustainable eco-friendly agricultural practices, an organisation, Sustainable Wine South Africa (SWSA), has also been created.

POLICY AND PLANNING LESSONS LEARNT FROM THE CITY OF CAPE TOWN'S DAY ZERO WATER CRISIS

The opening discussion in this chapter discussed climate change frameworks that been considered in literature to manage climatic shocks or climate change events as they happen or as anticipated to occur. The case of Cape Town Day Zero water crisis exhibits evidence of the various climate change frameworks being invoked.

The Critical Water Shortages Disaster Plan initiative in 2018 could be evidence of a Disaster Risk Reduction Approach considering in entailed preservation restrictions (limiting water consumption), tentative disaster restrictions (intensive water rationing should preservation restrictions not be enough) and full-scale disaster implementation (extreme disaster scenario when there is no surface water). The plan was essentially adopted a risk-based approach to progressively manage a scenario when water would totally run out as an eventual scenario and inevitably, disaster planning would involve assessing risk events induced by a climate change issue. Some authors have discerned evidence of the Climate Change Adaptation Framework also being invoked which has been traced in South Africa's legislations such as the 2015 Disaster Management Amendment Act and in the 2018 Climate Change Bill as well as in efforts to build infrastructures that can withstand extreme weather events such as droughts (Nhamo & Agyapong, 2019).

In 2019, following the 2018 Day Zero Water Crisis, a Water Strategy was developed, focusing on five key elements: (1) Safe access to water and sanitation for everyone (2) Wise water usage through pricing, regulation, active citizenship, and network management (3) Sufficient, dependable water from varied sources such as surface, ground, desalination, and reuse to achieve water resilience by 2030 (4) Shared benefits and risk management from regional water sources and (5) Become a water-sensitive city by 2040 (City of Cape Town, 2019). While the first aspect emphasizes inclusion in water availability, the second and third elements aim to increase resilience, particularly in the face of adverse climatic conditions like drought. The fourth element addresses risk management, while the fifth factor demonstrates a commitment to sustainability. It can be discerned therefore that the Disaster Risk Adaption, Resilience and Sustainability Frameworks are also included in the strategy. This also illustrates that the frameworks need not be mutually exclusive and can be integrated, depending on the water management planning for climate change risk that is being adopted. For instance, sourcing water from varied sources such as groundwater has been considered as both a resilience and sustainability endeavour in relation to the Western Cape Water Supply System (Luker & Harris, 2019, p. 9). Yet again, it is important that integrated frameworks do not over-emphasize certain aspects at the expense of others.

Overall, regarding planning of climate change events, an integrated approach is more pragmatic since a climate shock may have various aspects to it such as being a disaster and yet a risk event if such an event is likely to happen in future. A composite approach may also be prompted by the fact that various stakeholders may be consulted, and each may bring their own perspectives to how a climate change crisis may need to be approached. In the Cape Town water crisis, there was evidence of intra-governmental stakeholder collaboration between the City of Cape Town as the municipal entity as well as water and the policy and planning departments although public consultation could have been better handled (Cochrane & Ziervogel, 2020). On the other hand, the public communication process was quite visible in creating awareness of the impending Day Zero scenario, although some views are that it tended to be alarmist with some Capetonians considering its framing as a comic apocalypse that that could be avoided through community engagement, which can be aided by media-driven education (Voci et al., 2020, p. 9). Granted, although the Day Zero communication campaign was impactful in inducing behaviour change and reducing household water usage, some officials expressing that the campaign and the panic it caused, came at a high price (Ziervogel, 2019, p. 17).

For water infrastructure management policy and planning practitioners, there are pertinent lessons that could be lent from the Cape Town experience. The fact that the City of Cape Town had to look beyond conventional methods of water supply and look for alternative sources such as groundwater, stormwater and water reuse, is key in building resilience in the water sector (Rodina, 2019, p. 5). Diversification of water supply sources would need to be an ongoing effort for the city.

CONCLUSION

The discussion profiled and uncovered events leading to and during the event of Cape Town's Day Zero water crisis event. Such an event can hardly be considered an isolated experience of Cape Town. Increasingly, evidence is showing increases in climate change related events such as droughts globally and which often have adverse effects to agricultural production. Other countries may want to take heed of the Cape Town experience in the water crisis especially when faced with conditions of water scarcity wrought by climate change. It is hoped that this discussion in this chapter can provide some baseline of insights to draw from should countries, facing similar water crisis scenarios, contemplate as they navigate the various nuances of climate change events that are frequently recurring in this century.

Beyond the City of Cape Town, policy considerations on a national climate change policy must consider the issue of climate risk events such as harsh droughts and what actionable steps must be undertaken in such scenarios. There should be ongoing contemplations and reviews of the policy environment on climate change which are futuristic in outlook given the climate change is always going to be a factor of consideration in our societies.

REFERENCES

Allen, C. R., Fontaine, J. J., Pope, K. L., & Garmestani, A. S. (2011). Adaptive management for a turbulent future. *Journal of Environmental Management, 92*(5), 1339–1345. DOI: 10.1016/j.jenvman.2010.11.019 PMID: 21168260

Begum, R. A., Sarkar, M., Jaafar, A. H., & Pereira, J. J. (2014). Toward conceptual frameworks for linking disaster risk reduction and climate change adaptation. *International Journal of Disaster Risk Reduction, 10*, 362–373. DOI: 10.1016/j.ijdrr.2014.10.011

Bhamra, R., Dani, S., & Burnard, K. (2011). Resilience: The concept, a literature review, and future directions. *International Journal of Production Research, 49*(18), 5375–5393. DOI: 10.1080/00207543.2011.563826

Bock, W. J. (1980). The Definition and Recognition of Biological Adaptation. *American Zoologist, 20*(1), 217–227. DOI: 10.1093/icb/20.1.217

Brown, K. (2014). Global environmental change I: A social turn for resilience? *Progress in Human Geography, 38*(1), 107–117. DOI: 10.1177/0309132513498837

Carter, S., & Bayer, J. L. (2006). *IR-06-043 The Projected Influence of Climate Change on the South African Wine Industry* (Interim Report IR-06-043). International Institute for Applied Systems Analysis Schlossplatz 1, Laxenburg, Austria.

Chauke, H., & Pongrácz, R. (2024). The validation of climate in the wine-growing region of the Western Cape of South Africa. *GEM: International Journal on Geomathematics, 15*(1), 4. DOI: 10.1007/s13137-023-00244-7

Cochrane, L., & Ziervogel, G. (2020). The Cape Town Drought: Lessons from the City that Nearly Ran Out of Water – Gina Ziervogel. *NokokoPod, 3*, 1–16.

Das Nair, R., Chisoro, S., & Ponte, S. (2023). *Sustainability in the South African wine industry: Status, opportunities, and challenges*. CBS, Centre for Business and Development Studies.

Díaz-Caravantes, R. E., Zuniga-Teran, A., Martín, F., Bernabeu, M., Stoker, P., & Scott, C. (2020). Urban water security: A comparative study of cities in the arid Americas. *Environment and Urbanization, 32*(1), 275–294. DOI: 10.1177/0956247819900468

Drummond, K. (2019). *The relationship between droughts and the tourism industry: A case study on Cape Town, South Africa* [Honours Thesis. 3113, Western Michigan University]. https://scholarworks.wmich.edu/honors_theses/3113

Enqvist, J. P., & Ziervogel, G. (2019). Water governance and justice in Cape Town: An overview. *WIREs. Water*, *6*(4), e1354. DOI: 10.1002/wat2.1354

Eriksen, S., Aldunce, P., Bahinipati, C. S., Martins, R. D., Molefe, J. I., Nhemachena, C., O'Brien, K., Olorunfemi, F., Park, J., Sygna, L., & Ulsrud, K. (2011). When not every response to climate change is a good one: Identifying principles for sustainable adaptation. *Climate and Development*, *3*(1), 7–20. DOI: 10.3763/cdev.2010.0060

Forino, G., Von Meding, J., & Brewer, G. J. (2015). A Conceptual Governance Framework for Climate Change Adaptation and Disaster Risk Reduction Integration. *International Journal of Disaster Risk Science*, *6*(4), 372–384. DOI: 10.1007/s13753-015-0076-z

Francesco, B., & Carlo, C. (2018). *DEFINING ADAPTATION: An Analysis of Adaptation as a Response to Climate Change*. Copenhagen Consensus Center. http://www.jstor.com/stable/resrep16322.7

Gregory, R., Ohlson, D., & Arvai, J. (2006). DECONSTRUCTING ADAPTIVE MANAGEMENT: CRITERIA FOR APPLICATIONS TO ENVIRONMENTAL MANAGEMENT. *Ecological Applications*, *16*(6), 2411–2425. DOI: 10.1890/1051-0761(2006)016[2411:DAMCFA]2.0.CO;2 PMID: 17205914

Guillén Bolaños, T., Scheffran, J., & Máñez Costa, M. (2022). Climate Adaptation and Successful Adaptation Definitions: Latin American Perspectives Using the Delphi Method. *Sustainability (Basel)*, *14*(9), 5350. DOI: 10.3390/su14095350

Gunderson, L. (1999). Resilience, Flexibility and Adaptive Management—- Antidotes for Spurious Certitude? *Conservation Ecology*, *3*(1), art7. DOI: 10.5751/ES-00089-030107

Hasselman, L. (2017). Adaptive management; adaptive co-management; adaptive governance: What's the difference? *Australasian Journal of Environmental Management*, *24*(1), 31–46. DOI: 10.1080/14486563.2016.1251857

Islam, S., Chu, C., Smart, J. C. R., & Liew, L. (2020). Integrating disaster risk reduction and climate change adaptation: A systematic literature review. *Climate and Development*, *12*(3), 255–267. DOI: 10.1080/17565529.2019.1613217

Keweloh, R. (2015). Integrating Disaster Risk Reduction and Climate Change Adaptation in Theory and Practice. *IFHV Working Paper, 5*(3).

Lee, H., Son, J., Joo, D., Ha, J., Yun, S., Lim, C.-H., & Lee, W.-K. (2020). Sustainable Water Security Based on the SDG Framework: A Case Study of the 2019 Metro Manila Water Crisis. *Sustainability (Basel)*, *12*(17), 6860. DOI: 10.3390/su12176860

Lints, T. (2010). The Essentials of Defining Adaptation. *4th Annual IEEE Systems Conference*, 113–116.

Luker, E., & Harris, L. M. (2019). Developing new urban water supplies: Investigating motivations and barriers to groundwater use in Cape Town. *International Journal of Water Resources Development*, *35*(6), 917–937. DOI: 10.1080/07900627.2018.1509787

Magida, N., & Makoza, F. (2023). The Effect of Water Crisis on Fine Dining Restaurant Entrepreneurs in the Cape Town Metro. *Tourism and Leisure, 12*.

Matikinca, P., Ziervogel, G., & Enqvist, J. P. (2020). Drought response impacts on household water use practices in Cape Town, South Africa. *Water Policy*, *22*(3), 483–500. DOI: 10.2166/wp.2020.169

McManus, S., Seville, E., Vargo, J., & Brunsdon, D. (2008). Facilitated Process for Improving Organizational Resilience. *Natural Hazards Review*, *9*(2), 81–90. DOI: 10.1061/(ASCE)1527-6988(2008)9:2(81)

Millington, N., & Scheba, S. (2021). Day Zero and The Infrastructures of Climate Change: Water Governance, Inequality, and Infrastructural Politics in Cape Town's Water Crisis. *International Journal of Urban and Regional Research*, *45*(1), 116–132. DOI: 10.1111/1468-2427.12899

Mukherjee, S., Sundberg, T., Sikdar, P. K., & Schütt, B. (2022). An Integrated Quantitative Assessment of Urban Water Security of a Megacity in the Global South. *Frontiers in Water*, *4*, 834239. DOI: 10.3389/frwa.2022.834239

Naude, R. T., & Naude, M. J. (2019). Impact of climate change and extreme weather conditions on wine growing within the Stellenbosch region. *Journal of Contemporary Management*, *16*(2), 111–134. DOI: 10.35683/jcm19041.0029

Nelson, D. R. (2011). Adaptation and resilience: Responding to a changing climate. *Wiley Interdisciplinary Reviews: Climate Change*, *2*(1), 113–120. DOI: 10.1002/wcc.91

Nelson, K., Gillespie-Marthaler, L., Baroud, H., Abkowitz, M., & Kosson, D. (2020). An integrated and dynamic framework for assessing sustainable resilience in complex adaptive systems. *Sustainable and Resilient Infrastructure*, *5*(5), 311–329. DOI: 10.1080/23789689.2019.1578165

Nhamo, G., & Agyepong, A. O. (2019). Climate change adaptation and local government: Institutional complexities surrounding Cape Town's Day Zero. *Jàambá*, *11*(3). DOI: 10.4102/jamba.v11i3.717 PMID: 31308900

Noll, D. (2020). *Day Zero: Communicating the water crisis in Cape Town, 2015-2018—A qualitative study about the communication during the water crisis and how the public responded to the messages* [Bachelor's Thesis]. Karlstads Universiteit.

Overton, J., Murray, W. E., & Howson, K. (2019). Doing good by drinking wine? Ethical value networks and upscaling of wine production in Australia, New Zealand, and South Africa. *European Planning Studies*, *27*(12), 2431–2449. DOI: 10.1080/09654313.2019.1628181

Parks, R., Mclaren, M., Toumi, P. R., & Rivett, P. U. (2019). Experiences and lessons in managing water from Cape Town. *Grantham Institute Briefing Paper No.29.*

Parsons, D. (2010). Organisational Resilience. *Australian Journal of Emergency Management*, *25*(2), 18–20.

Pham, H., & Saner, M. (2021). A Systematic Literature Review of Inclusive Climate Change Adaption. *Sustainability (Basel)*, *13*(19), 10617. DOI: 10.3390/su131910617

Rodina, L. (2019). Water resilience lessons from Cape Town's water crisis. *WIREs. Water*, *6*(6), e1376. DOI: 10.1002/wat2.1376

Roostaie, S., Nawari, N., & Kibert, C. J. (2019). Sustainability and resilience: A review of definitions, relationships, and their integration into a combined building assessment framework. *Building and Environment*, *154*, 132–144. DOI: 10.1016/j.buildenv.2019.02.042

Ruiz-Martin, C., Lopez-Paredes, A., & Wainer, G. (2018). What we know and do not know about organizational resilience. *International Journal of Production Management and Engineering*, *6*(1), 11. DOI: 10.4995/ijpme.2018.7898

Sathre, R., Antharam, S. M., & Catena, M. (2022). Water Security in South Asian Cities: A Review of Challenges and Opportunities. *CivilEng*, *3*(4), 873–894. DOI: 10.3390/civileng3040050

Savelli, E., Mazzoleni, M., Di Baldassarre, G., Cloke, H., & Rusca, M. (2023). Urban water crises driven by elites' unsustainable consumption. *Nature Sustainability*, *6*(8), 929–940. DOI: 10.1038/s41893-023-01100-0

Schoeman, J., Allan, C., & Finlayson, C. M. (2019). Exploring the Multiple Meanings of Adaptive Management: A Case Study of the Lachlan Catchment in the Murray–Darling Basin. *Environmental Management*, *64*(4), 470–482. DOI: 10.1007/s00267-019-01203-5 PMID: 31511922

Simonet, G., & Duchemin, E. (2010). The concept of adaptation: Interdisciplinary scope and involvement in climate change. *Surveys and Perspectives Integrating Environment and Society, 3*(1), 1–9.

Smit, B., Burton, I., Klein, R. J. T., & Street, R. (1999). THE SCIENCE OF ADAPTATION: A FRAMEWORK FOR ASSESSMENT. *Mitigation and Adaptation Strategies for Global Change, 4*(3-4), 199–213. DOI: 10.1023/A:1009652531101

Stringer, L. C., Dougill, A. J., Fraser, E., Hubacek, K., Prell, C., & Reed, M. S. (2006). Unpacking “Participation” in the Adaptive Management of Social–ecological Systems: A Critical Review. *Ecology and Society, 11*(2), art39. DOI: 10.5751/ES-01896-110239

Taing, L., Chang, C. C., Pan, S., & Armitage, N. P. (2019). Towards a water secure future: Reflections on Cape Town's Day Zero crisis. *Urban Water Journal, 16*(7), 530–536. DOI: 10.1080/1573062X.2019.1669190

Theron, H., & Hunter, J. J. (2022). Mitigation and Adaptation Practices to the Impact of Climate Change on Wine Grape Production, with Special Reference to the South African Context. *South African Journal of Enology and Viticulture*. DOI: 10.21548/43-1-4735

Van Ginkel, K. C. H., Hoekstra, A. Y., Buurman, J., & Hogeboom, R. J. (2018). Urban Water Security Dashboard: Systems Approach to Characterizing the Water Security of Cities. *Journal of Water Resources Planning and Management, 144*(12), 04018075. DOI: 10.1061/(ASCE)WR.1943-5452.0000997

Vink, N., Deloire, A., Bonnardot, V., & Ewert, J. (2012). Climate change and the future of South Africa's wine industry. *International Journal of Climate Change Strategies and Management, 4*(4), 420–441. DOI: 10.1108/17568691211277746

Visser, G., & Cloete, J. (2021). Guest Houses, Drought and Water Management Strategies in the Cape Winelands, South Africa. *African Journal of Hospitality, Tourism and Leisure, 10*(10(5)), 1549–1561. DOI: 10.46222/ajhtl.19770720.178

Visser, W. P. (2018). A perfect storm: The ramifications of Cape Town's drought crisis. *The Journal for Transdisciplinary Research in Southern Africa, 14*(1). DOI: 10.4102/td.v14i1.567

Voci, D., Bruns, C. J., Lemke, S., & Weder, F. (2020). Framing the End: Analyzing Media and Meaning Making During Cape Town's Day Zero. *Frontiers in Communication, 5*, 576199. DOI: 10.3389/fcomm.2020.576199

Warner, J. F., & Meissner, R. (2021). Cape Town's "Day Zero" water crisis: A manufactured media event? *International Journal of Disaster Risk Reduction, 64*, 102481. DOI: 10.1016/j.ijdrr.2021.102481

Ziervogel, G. (2019). *UNPACKING THE CAPE TOWN DROUGHT:LESSONS LEARNED* [Report for Cities Support Programme undertaken by African Centre for Cities].

Ziervogel, G., & Morgan, G. (2022). The Adaptive Governance Capacities of the City of Cape Town Built in Response to Extreme Events. *Journal of Extreme Events, 09*(01), 2250005. DOI: 10.1142/S2345737622500051

Chapter 14
Rethinking Citizen Participation in the Implementation of Environmental Projects:
Toward a Unified Conceptual Understanding

Latifa Trabelsi
Higher Institute of Technological Studies of Bizerte, Tunisia

Manel Hakim Masmoudi
https://orcid.org/0000-0001-7902-7094
University of Sfax, Tunisia

ABSTRACT

This chapter proposes a clearer and more holistic framework for understanding citizen engagement not only in decision-making but also in the implementation of environmental initiatives. Through a structured conceptual review of key theories, typologies, and related concepts, we highlight the overlooked role of citizens during the implementation phase, where authority-led action often dominates. We examine how this gap undermines inclusive governance and explore how implementation fits within the broader environmental project lifecycle. The chapter also analyzes related constructs such as co-production, co-design, co-creation and involvement clarifying their complementary roles in participatory implementation. Drawing from previous research, we underscore how sustained engagement fosters community ownership and

DOI: 10.4018/979-8-3373-1310-8.ch014

long-term change. The integrative framework we propose refines the understanding of citizens' participation across project phases, advocating for a more inclusive and precise conceptualization to enhance future research and governance effectiveness.

INTRODUCTION

> *"What we have ignored is what citizens can do and the importance of real involvement of the people involved – versus just having somebody in Washington make a rule."*
>
> *— Elinor Ostrom*

In an era characterized by growing environmental challenges and the persistent need to adopt more sustainable development pathways, citizen engagement has become an essential pillar of environmental governance. The implementation of pro-environmental initiatives, in particular, increasingly depends on inclusive and participatory mechanisms that move beyond traditional top-down approaches. As highlighted by Gelderman et al. (2017), many municipalities and local governments are actively formulating comprehensive sustainability agendas that embed environmental principles across various domains of public service — from waste reduction, recycling, and reuse, to the protection of vital natural resources such as water, soil, biodiversity, and non-renewable assets. Within this evolving landscape, involving citizens not only enhances the perceived legitimacy of environmental actions but also strengthens their effectiveness by tapping into local knowledge, fostering collective responsibility, and reinforcing community ownership.

Scholars emphasize that initiatives grounded in shared governance and collective responsibility are more likely to foster meaningful behavioral change and generate lasting environmental benefits. This shift reflects a growing recognition that sustainable solutions must be co-designed and implemented with the active involvement of the communities they intend to serve. As Baxter (2022) observes, the emergence of collaborative environmental projects is driven by several converging factors — notably, citizens' increasing aspiration to contribute to solving societal challenges, the perceived inadequacies or limitations in governmental responses, and the appeal of local-level initiatives that offer closer engagement, stronger influence over outcomes, and greater transparency in the decision-making process. These dynamics highlight the importance of inclusive mechanisms that empower communities not only to participate, but also to shape and monitor the environmental actions that directly affect them.

Flexible and transparent decision-making is crucial for addressing the complex and evolving nature of environmental challenges, as it lays the foundation for more inclusive, adaptive, and legitimate forms of environmental governance (Miller &

Wyborn, 2020). In this context, citizen and stakeholder participation has moved from the margins to the core of sustainability-oriented policies and research. Action-oriented and locally grounded approaches, particularly in energy and sustainability transitions, demonstrate how participatory processes can empower communities to actively shape, implement, and monitor environmental initiatives (Bergmann et al., 2021). Beyond the local scale, participatory governance supports long-term transformation through collective visioning, experimentation, and social learning, which are essential for navigating complex sustainability transitions (Sengers et al., 2019; Huttunen et al., 2022).

This conceptual inquiry emerges from three key observations in the existing literature. First, despite widespread recognition of the value of participatory approaches in pro-environmental initiatives, the role of citizens remains significantly underexplored. Participation is often framed through the lens of stakeholders, experts, or policymakers, while the public's involvement is either vaguely defined or entirely overlooked (Sengers et al., 2019). Second, there is a persistent emphasis on citizen participation during the decision-making phase, whereas their role in the implementation of environmental initiatives receives considerably less attention. This gap is critical, as the success of environmental governance depends not only on inclusive decision-making but also on how effectively policies and programs are carried out on the ground. Finally, the conceptual landscape surrounding participation is marked by considerable ambiguity. A variety of related terms—such as engagement, involvement, co-production, co-design, and co-creation—are frequently used interchangeably with "participation," despite nuanced differences. This terminological overlap contributes to conceptual confusion and hinders theoretical and practical clarity in understanding citizen roles in environmental action.

This article seeks to explore the distinction between citizen participation in decision-making processes and their participation in the implementation of environmental projects, emphasizing the implications of each for effective environmental governance and management. By delving into these differences, our objective is to advance a more nuanced understanding of the ways in which citizen participation can improve the effectiveness of environmental initiatives and promote the adoption of sustainable practices within communities.

To address these gaps and provide greater conceptual clarity, this chapter is structured as follows. We begin by exploring the theoretical foundations of citizen participation in environmental projects, distinguishing it from political participation and reviewing various typologies, levels, and mechanisms of engagement. We then shift focus to argue for a clearer conceptualization of citizen participation specifically during the implementation phase—an often overlooked yet critical stage in environmental governance. This includes a discussion of where implementation fits within broader project lifecycles and an illustrative case of participatory forest management. Next, we examine a range of closely related concepts—such as co-creation, co-design, and co-production—that, while often used interchangeably with participation, offer valuable distinctions, particularly in the context of implementation. Finally, the chapter culminates in a conceptual framework that seeks to unify and refine the understanding of citizen participation in the implementation of environmental initiatives.

THEORETICAL FOUNDATIONS OF CITIZEN PARTICIPATION IN ENVIRONMENTAL PROJECT

The growing urgency of environmental challenges has brought citizen involvement to the forefront of sustainability and governance efforts. Participation is more than a normative principle—it serves as a concrete channel through which citizens interact with public institutions, engage with administrative processes, and contribute to shaping policy directions in ways that reflect their own interests and concerns. Rather than being passive observers, individuals can take on active roles in influencing decisions that impact their lives and surroundings (Oliveira & Ckagnazaroff, 2023). This section aims to clarify the conceptual boundaries of citizen participation in contrast to other forms of engagement, present its key typologies and levels, and explore the mechanisms through which it is operationalized—laying the groundwork for a deeper understanding of its unique role in environmental projects.

A Bird's-Eye Perspective on the Concept of Citizen Participation

The term *participation* finds its roots in the Latin word *participatio*, meaning "to take part" or "to share" (Oliveira & Ckagnazaroff, 2023). Historically, it has emerged as a key notion in the evolving relationship between citizens and public institutions, reflecting growing societal demands for transparency, inclusion, and accountability (Oliveira & Ckagnazaroff, 2023). Yet, this concept remains complex and multidimensional, and it can be approached from at least three distinct but complementary perspectives. From a moral and ethical standpoint, citizen participation is seen as a voluntary act rooted in compassion, shared responsibility, and openness to change—an expression of humanistic values and support for others (Birch, 2017). From a governance and institutional lens, it is viewed as a mechanism within open government systems, where public institutions actively collaborate with citizens to improve services and address collective challenges (Wijnhoven et al. 2015; Millard, 2018). In these systems, government institutions actively engage with and leverage societal resources, to drive positive changes and improvements within society. This approach emphasizes the importance of collaboration between citizens and governmental bodies in addressing social issues and enhancing public services.

Finally, from a critical and power-based perspective, participation is interpreted as a means of redistributing power, enabling historically excluded individuals and communities to influence decision-making processes that directly affect their lives (Arnstein, 1969).

These three perspectives together lay the groundwork for a deeper and more unified understanding of citizen participation, particularly within the context of implementing environmental projects.

From the Start: Distinguishing Citizen Participation and Political Participation

Scholarly debates on the concept of participation have largely centered around two principal lines: political participation and citizen (or social) participation. These two categories represent distinct but interconnected ways through which individuals and groups engage with societal and governance processes (Oliveira & Ckagnazaroff, 2023, Richardson, 2005).

Political participation is defined as the actions taken by private citizens to influence or support government and politics (Milbrath, 1981). As such, it is often associated with representative democracy, where indirect engagement occurs through elected representatives (Cornwall & Gaventa, 2017). This involves individuals or groups competing for political power and negotiating the selection of political goals and the strategies to achieve them. Therefore, political participation includes a range of activities related to political life, such as mobilizing support for or opposition to political authorities, engaging in movements to claim rights, lobbying, and political action by various actors (including elites, religious institutions, and the military). It also encompasses electoral activities, such as voting, running for office, participating in political parties, working on electoral campaigns, and holding political office.

The current demand for participation extends beyond traditional modes of political participation. The direct participation of civil society in decision-making and policy formulation has become a central demand in the pursuit of citizenship and the guarantee of universal rights for all (Dagnino, 2003).

However, implementing direct participation requires a fundamental transformation in power relations between the state and society.

Citizen or social participation, on the other hand, occurs through the redistribution of power. Unlike political participation, citizen participation challenges the conception of a strictly representative democracy that restricts participation to electoral activities. It is a concept related to citizenship and democratic governance (Oliveira & Ckagnazaroff, 2023, Cornwall & Gaventa, 2017).

Citizen participation involves including individuals in political decisions as well as in the formulation, implementation, and evaluation of public policies. The objective is not to supplant the state but to enable citizens to take part directly in governance, encouraging politicians to heed their constituents and bureaucrats to be more responsive to those they serve.

Several scholars contend that meaningful citizen participation not only enriches the decision-making process but also fosters social stability by cultivating a sense of community and promoting acceptance and respect for governance (Callahan, 2007).

Exploring the Typologies of Citizen Participation

Various typologies have been developed to categorize forms of citizen participation, each based on different criteria such as the degree of power or involvement, the direction of communication flows, the objectives, the degree of voluntariness, etc. This proliferation certainly reflects the complexity and multidimensional nature of citizen participation; however, the sheer number and diversity of these classification systems can sometimes create confusion. With overlapping categories, differing terminologies, and varying conceptual frameworks, it becomes challenging for researchers and practitioners alike to navigate and apply these models consistently.

The most commonly cited typology of citizen participation organizes it along a continuum reflecting varying degrees of citizen power and involvement. A classic example is Arnstein's (1969) "ladder of participation," which identifies multiple rungs representing levels from nonparticipation to full citizen control. This model, sometimes depicted as a "wheel of participation" (Davidson, 1998), has inspired numerous subsequent continuums, including those proposed by Pretty (1995), Tippett et al. (2007), Lawrence (2006), Kindon et al. (2007) (This typology will be thoroughly examined in the upcoming paragraph, levels of citizen participation).

Another important typology, proposed by Rowe and Frewer (2000), categorizes participation based on the direction of communication flows between citizens and authorities. This framework distinguishes between one-way communication, where information flows only from the authorities to the public, and two-way communication, which involves interactive exchanges allowing citizens to provide feedback, express opinions, and influence decision-making. This typology highlights the importance of communication dynamics in shaping the quality and effectiveness of citizen participation.

Additionally, some frameworks differentiate citizen participation based on their underlying theoretical foundations, primarily distinguishing between normative and pragmatic participation. Normative participation is rooted in democratic ideals and the intrinsic rights of citizens to be involved in governance processes. It emphasizes participation as a fundamental democratic value, promoting inclusiveness, empowerment, and social justice (Thomas, 1993). This perspective views participation not merely as a means to an end but as an essential component of citizenship and democratic legitimacy. On the other hand, pragmatic participation is more instrumental and outcome-driven, focusing on the practical benefits of involving citizens to improve policy effectiveness, enhance decision quality, or increase acceptance of public initiatives. From this viewpoint, participation is a tool to achieve specific goals such as conflict resolution, better resource management, or efficient service delivery (Innes & Booher, 2004). The distinction between these two approaches highlights the diverse motivations behind participatory processes and the varying expectations placed on citizen involvement. Understanding this theoretical divide is crucial for designing participation mechanisms that align with the desired democratic or pragmatic outcomes.

Other typologies focus on the underlying objectives of participation, categorizing different forms based on the specific purposes they aim to achieve. Common purposes include empowerment, consultation, and information gathering. Empowerment means giving citizens more control and influence. Consultation involves asking for opinions but not giving decision power. Information gathering focuses on collecting data from the public. Researchers like Tippett et al. (2007) have studied these differences.

Adding to these perspectives, participation can also be categorized based on the degree of voluntariness, spanning a spectrum from spontaneous and voluntary forms to those that are induced, provoked, imposed, or granted. Spontaneous participation involves self-organized groups without formal structures, while voluntary participation sees citizens creating groups with self-defined goals and methods, such as trade unions. In contrast, induced participation occurs when external actors persuade citizens to participate, and provoked participation involves external agents forming groups with objectives different from the participants' . Imposed participation obliges individuals to participate in activities, like religious rituals and granted participation involves organizations or public agents giving decision-making power to subordinates or citizens, such as in participatory budgeting structures (Oliveira & Ckagnazaroff, 2023).

Another important classification is proposed by Wijnhoven et al. (2015), who categorize citizen participation into three main types: "citizen sourcing," where citizens support daily public administrative tasks such as reporting road problems; "collaborative democracy," where citizens provide input on general policy decision-making; and "citizen ideation and innovation," where citizens actively contribute to identifying societal problems and developing solutions. Within this last category, scholars further distinguish between top-down (government-led) and bottom-up (citizen-led) participation approaches (Susha & Gronlund, 2014).

This range of possible classifications can be broadly extended and expanded. Indeed, the list of participation types can be enriched with distinctions such as direct or indirect, complete or partial, intensive or extensive, effective or ineffective, among others.

These varied categories reflect the multifaceted nature of citizen participation. Recognizing this diversity is essential for developing nuanced frameworks that accurately describe and assess participation across different contexts and objectives.

Levels of Citizen Participation: Arnstein's Ladder and Inspired Models

As mentioned previously, one of the most common typologies of participation is the one based on different degrees or levels of participation along a continuum. Numerous alternative terms have been suggested for the different rungs of the ladder.

Abviously, Arnstein's (1969) ladder is widely recognized as a pivotal work for classifying the various levels of citizen participation. It provides a robust conceptual foundation by depicting participation as a continuum, ranging from nonparticipation, through stages of tokenism, to full citizen control (Kiss et al. 2022).

Emerging from the urban planning discipline Arnstein's ladder has been repeatedly reworked, applied and adopted in various urban domains, including ecological and environmental planning), environmental projects, green infrastructure planning and recently in a review of Nature-Based-Solution (Tippett et al., 2007; Wilker et al., 2016; Puskás et al., 2021).

Arnstein's Ladder (1969): Levels of Participation

In her seminal article published over fifty years ago, Arnstein (1969) laid the foundation for a widely recognized and enduring classification of citizen participation. Her model, often depicted as an eight-rung ladder, illustrates a continuum ranging from total non-participation to full citizen empowerment and control over decision-making processes. Each rung represents a specific degree of citizen influence and engagement in planning and governance (figure 1).

Figure 1. Arnstein's ladder of participation

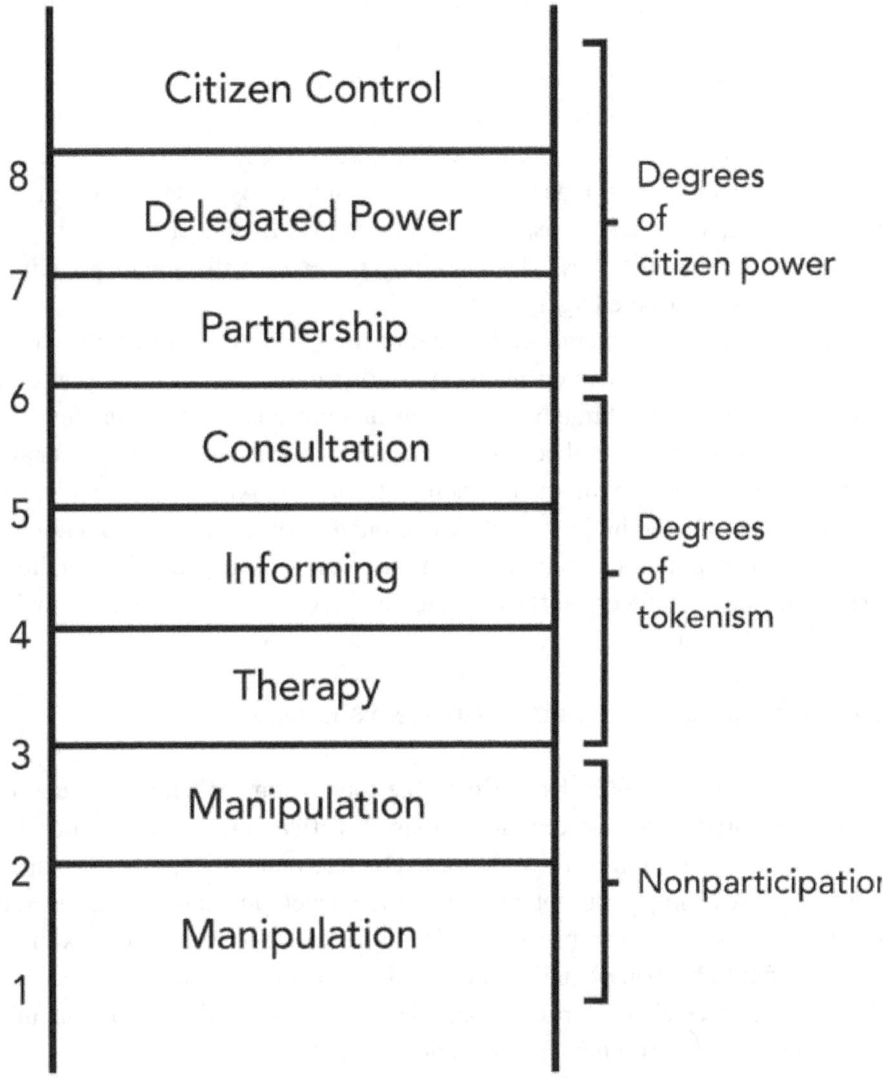

The two lowest rungs—manipulation and therapy—represent disguised forms of non-participation. At these levels, citizens are not genuinely involved in decisions; rather, they are used as instruments to legitimize decisions already made. Manipulation involves steering or influencing citizens without granting them real power, while therapy aims to "cure" or "correct" citizen behavior under the guise of participation. These stages are essentially top-down and paternalistic, with power firmly held by authorities.

The middle rungs—informing, consultation, and placation—correspond to what Arnstein calls "tokenism." Here, citizens may express their opinions and be heard, but their ability to affect decisions is limited or nonexistent. Informing is a one-way process where authorities communicate decisions or information to citizens without expecting feedback. Consultation invites citizens to share their views, for example through surveys or public meetings, but there is no guarantee their input will influence outcomes. Placation goes a step further by allowing citizens to advise or participate in certain aspects of planning, yet final decision-making power remains with the authorities. These levels can create an illusion of participation, where citizens' roles are largely symbolic.

True citizen power emerges at the top three rungs: partnership, delegated power, and citizen control. Partnership involves the redistribution of power between citizens and authorities, often through negotiation or joint committees, enabling shared decision-making. Delegated power grants citizens dominant authority over specific projects or issues, giving them concrete decision-making roles. Finally, citizen control represents the highest level, where citizens exercise full and autonomous control over planning and management processes. Reaching this level indicates a mature democratic process, reflecting a society's commitment to genuine inclusion and empowerment.

Other Typologies Inspired by Arnstein's Ladder

Arnstein's work (1969) has become the foundation for numerous subsequent studies on participation, leading to various classifications of participation levels. Arnstein's classification is still widely cited by many authors working on participation, often providing a conceptual basis for the development of new measurement tools and classification frameworks. Although Arnstein's original work emerged from the field of urban planning, it continues to inspire many subsequent studies that categorize levels of participation, whether applied within urban planning or adopted in other disciplines (Puskás et al. 2021).

Davidson's (1998) model, called the 'wheel of participation,' assesses community involvement in planning through four main levels: inform, consult, participate, and empower, each further divided into three sub-categories that reflect different degrees of engagement.

In the context of spatial and environmental planning, Tippett et al. (2007) propose that the rungs of the ladder could perhaps be usefully reconceptualised as falling into five major processes of participation in planning wich are inform (information provision and communication); design (active engagement in developing options and plans), consult (consultation to elicit response to options and plans); deliver (community devolved implementation and management of plans) and monitor (review of effectiveness, learning cycle to incorporate learning from implementation).

Lawrence (2006) also developed a synthesis of participatory approaches based on Arnstein (1969) and Pretty et al. (1995) works. This synthesis distinguishes between only four types of participation, discarding the more overtly exploitative types to leave a spectrum from consultative (all power held by the centre) to transformative (participants control the process). The two intermediate types of participation involve sharing of knowledge and labour in relationships that are either functional (decisions still remain with the centre) or collaborative (central and local actors share decisions).

Similarly, Kindon et al. (2007) proposed a 'participation continuum' ranging from co-option to collective action. This continuum illustrates varying levels of participant engagement in decision-making processes, from tokenistic involvement where participants have little real power (co-option), to genuine collaborative efforts where participants actively share power and work together to achieve common goals (collective action).

The concept of co-option refers to situations where participants' voices or activities are absorbed or controlled by more powerful actors, often limiting meaningful influence. On the other hand, collective action represents autonomous mobilization by participants who define priorities and act collectively to effect change.

Wilker et al. (2016) adopt a typology of participation in green infrastructure planning that emphasizes the need for inclusive and interactive engagement throughout the project lifecycle. Their typology is structured around different levels or ranks of participation, which range from more passive forms to active, collaborative involvement. Their framework aligns with common participation typologies that include information sharing, consultation, collaboration, co-decision and empowerment.

In their study of government portals, Sandoval-Almazán and Gil-García (2012) proposed a classification closely aligned with Arnstein's Ladder of Participation. They describe a participation continuum that ranges from basic "information display" to more advanced stages such as "opportunities for collaboration." This continuum reflects a hierarchy of citizen engagement, where the depth and nature of involvement intensify progressively. At the lower end, government institutions mainly provide information or services to the public. As participation levels increase, citizens are invited to interact with government entities, then to actively engage, and ultimately to collaborate with authorities in decision-making processes (De Jong et al. 2019).

Here too, this range of possible typologies, all inspired by Arnstein's Ladder of Participation, can be broadly extended and expanded. Indeed, applying this ladder in different contexts reveals the emergence of diverse approaches which, although varied in form and terminology, remain fundamentally rooted in Arnstein's original logic, a logic based on a progressive gradation of power granted to participants. Depending on cultural, political, or organizational specificities, analytical frameworks adapt and reinterpret this ladder to better reflect local realities while preserving the core idea of a participation continuum. This flexibility explains the enduring relevance and richness of Arnstein's model in the study of participatory processes worldwide.

Mechanisms of Participation

Citizen participation can take place through various mechanisms, which may be more or less formalized, and can serve deliberative, advisory, or purely informative functions. Drawing on the works of Webler and Renn (1995), Rowe and Frewer (2000), Dagnino and Teixeira (2014) and Oliveira and Ckagnazaroff (2023) provided a synthesis of the main mechanisms of citizen participation that are public hearings, advisory committee, management councils, deliberative consultation, the participatory budget and public forums.

Public hearings represent one of the most widely used mechanisms for citizen participation and are implemented by numerous governments around the world. These official meetings provide an opportunity for the public to receive information and share their views on the specific topic under discussion (Webler & Renn, 1995). However, some scholars criticize public hearings as a participatory tool, arguing that many are conducted primarily to fulfill legal obligations, often exclude key stakeholders, and can be dominated by particular interest groups rather than fostering genuine citizen involvement. Critics therefore describe these hearings as frequently symbolic and superficial exercises. Consequently, the success of public hearings largely depends on factors such as the institutional framework, procedural rules, stakeholder inclusion, available resources, and the balance of power among participants (Oliveira & Ckagnazaroff, Orr 2013, Webler & Renn, 1995).

Another participation mechanism is the advisory committee, a purpose-built group of stakeholders who convene regularly to discuss specific issues and share ideas. Orr (2013) highlights that these committees can offer valuable external expertise and support governments in carefully evaluating the impacts of public policies. Unlike management boards, advisory committees lack formal authority or decision-making power; their role is solely to provide advice when requested. The effectiveness of an advisory committee as a participatory tool largely depends on factors such as how often it is consulted, the diversity and expertise of its members, and their ability to contribute meaningfully.

Although advisory committees consist of carefully chosen stakeholders, Oliveira and Ckagnazaroff (2023) classify them as mechanisms of citizen participation. However, it is important to recognize the distinction between stakeholders and citizens, as emphasized by Webler et al. (1995), who define stakeholders as political actors with a vested interest in the issue at hand.

Another form of participation mechanism is the management councils. These councils are collegiate bodies that bring together representatives from the state and civil society to collaborate in the formulation, implementation, and evaluation of public policies. These management councils serve not only as forums for expressing opinions but also as spaces for concrete action, aiming to identify public demands and interests and translate them into public policy (Dagnino & Teixeira, 2014). Since these councils can be either advisory or deliberative, their effectiveness as mechanisms for citizen participation largely depends on the selection process of their members and the extent of their influence (Oliveira & Ckagnazaroff, 2023).

Deliberative consultation is a participatory mechanism designed to gauge public opinion by engaging citizens with relevant information on specific issues. It seeks to understand how informed discussion can influence attitudes and preferences (Oliveira & Ckagnazaroff, 2023). According to Orr (2013), this process typically begins by selecting a random sample of citizens from a defined geographic area, who are then asked to respond to an initial set of questions. From this larger group, a representative subset is invited to participate in a deliberative event. In this event, the participants receive information about the issues addressed in the questions applied to the larger group and are divided into small groups to discuss the theme. At the end of the event, the participants are asked to answer another set of questions. Finally, their answers are compared to the answers the larger group had provided at the beginning of the process, examining whether opinions have changed due to the discussions held at the event (Orr, 2013).

The participatory budget is another mechanism of citizen participation that originated in 1989 in Porto Alegre (RS), Brazil, and subsequently expanded to hundreds of municipalities across the country as well as to various regions worldwide, including Argentina, Uruguay, France, and Spain. While its implementation varies, participatory budgeting fundamentally constitutes a democratic process of deliberation and decision-making, where citizens have the opportunity to influence or determine the allocation of a portion of a municipality's or region's budget through periodic assemblies and direct negotiation phases with government authorities (Oliveira & Ckagnazaroff, 2023, Cabannes, 2004).

Although participatory budgets are commonly described as deliberative frameworks, this characterization does not always hold true in practice. It has been pointed out that in many cases, participatory budgeting functions primarily as an advisory mechanism. Consequently, its effectiveness as a tool for citizen participation largely depends on the political will of governments to act on citizens' demands and on the active involvement of civil society (Calisto Friant, 2019).

The last mechanism of citizen participation is public forums. These forums are open meetings held either in physical venues accommodating stakeholders or online via digital platforms. During these events, government representatives deliver formal presentations on topics of public interest, followed by a debate where participants engage through questions and comments. This mechanism facilitates the dissemination of information and the resolution of contentious issues through public discussion (Orr, 2013).

PARTICIPATION IN IMPLEMENTATION OF ENVIRONMENTAL PROJECT AS A DISTINCT CONCEPT

Environmental projects often require citizen participation beyond planning and decision-making, yet involvement during the implementation phase remains under-theorized and undervalued. This gap stems from two key issues. First, implementation is typically treated as a domain of project management, separating it from governance and reducing opportunities for participatory engagement. Second, the persistent conflation of environmental governance and project management blurs the boundaries between strategic and operational stages, making it difficult to define where and how participation should occur during implementation.

This section addresses these gaps by repositioning implementation as a dynamic and participatory phase within the broader continuum of environmental governance. Rather than viewing it as a purely technical step, we argue that implementation involves negotiation, interaction, and co-production among stakeholders. By revisiting foundational concepts in environmental governance and management, we situate implementation as a critical moment for reinforcing inclusive governance practices.

To illustrate this perspective, we examine the Mangrove Forest Management Project as a case study, showing how active citizen participation during implementation can reshape project outcomes. Ultimately, this approach highlights implementation as a crucial phase that not only links strategy to action, but also strengthens legitimacy, responsiveness, and sustainability in environmental governance.

Environmental Management/Environmental Governance

The literature remains divided regarding the conceptual boundaries between environmental governance and environmental management. A substantial part of the literature either conflates the two or treats them as interchangeable frameworks.

Some authors argue that traditional management approaches are not effective in dealing with complex environmental problems. Instead, they call for more inclusive and participatory approaches that involve local knowledge, dialogue, and continuous learning—approaches often referred to as governance (Armitage et al., 2011).

This confusion is also seen when governance and management are proposed together as solutions to environmental challenges, without explaining how they are connected or which one should come first (Richter et al., 2023)

The second group of scholars, however, argues that environmental governance and environmental management are not synonymous but rather distinct and complementary concepts (Richter et al., 2023). While environmental management generally refers to the operational dimension, planning, implementing, and controlling environmental activities, governance encompasses the broader system of rules, institutions, power dynamics, and stakeholder relationships that guide and shape environmental decision-making. It involves not only public authorities, but also private actors, civil society, and local communities, emphasizing participation, accountability, and adaptability. In line with this perspective, Lockwood (2010) considers management to be a "product of applied governance," suggesting a directional relationship in which governance sets the framework within which management operates (Richtel et al., 2023). Yet, this relationship is not strictly one-way: in practice, management often generates feedback through monitoring and evaluation, which can inform and reshape governance arrangements.

Despite these distinctions, confusion persists in the literature when governance and management are proposed simultaneously as solutions to environmental problems without clarifying their interconnection or sequencing. Although both are necessary, the absence of a clear framework complicates the task for policymakers, who need guidance on where and how to intervene (Lemos & Agrawal, 2006). Clarifying this distinction is particularly valuable for understanding the positioning of the implementation phase—not as a purely managerial or technical task, but as a moment situated at the interface between governance and management. Recognizing implementation in this light allows us to reframe it as a participatory and adaptive process shaped by broader governance structures and capable of influencing them in return.

Environmental Governance

Environmental governance encompasses the institutions, structures, and processes that define who makes decisions regarding the environment, how those decisions are made, for whom, and the resulting actions and their impacts (Lockwood et al., 2010; Bennett & Terre Satterfield, 2018). This multidimensional framework includes formal rules such as laws, policies, and tenure systems, alongside informal norms like cultural contexts, power relations, and social practices, all of which shape human interactions and influence environmental management outcomes (North, 1990; Biermann et al., 2017: Bennett & Terre Satterfield, 2018). Structures refer to formal entities such as governmental agencies, co-management committees, and civil society organizations, as well as informal networks of actors that embody governance capacities, facilitating coordination, participation, and effective decision-making (Bodin, 2017). The governance processes themselves—comprising policy formulation, negotiation of interests, conflict resolution, information dissemination, and enforcement—drive both decision-making and the practical realization of these decisions in the field (Cash et al., 2019). Among these processes, the implementation phase is particularly critical yet often overlooked; it transforms governance decisions into concrete actions through stakeholder engagement, adaptive management, and ongoing monitoring that allow for iterative adjustments based on observed outcomes (Olsson et al., 2004; Armitage et al., 2011). Understanding this phase is essential because it reveals how governance mechanisms translate into effective environmental management, influencing both the legitimacy and long-term sustainability of interventions (Lockwood, 2010). Thus, integrating an analysis of institutions, structures, and processes with a focused attention on implementation offers a comprehensive framework for improving governance and fostering resilient, sustainable environmental stewardship.

Environmental Management

The ultimate goal of applying project management principles lies in delivering outcomes that meet or exceed the expectations of all stakeholders—whether individuals, communities, or organizations. Successfully completing a project means fulfilling the entire scope of work with high quality, adhering to agreed timelines, and staying within budgetary constraints. Achieving these objectives not only generates satisfaction among clients and beneficiaries but also builds trust, enhances reputations, and fosters stronger relationships among all parties involved. When projects are executed effectively, the benefits ripple beyond immediate deliverables: they empower communities, drive organizational growth, and contribute to sustainable development. In essence, successful project implementation creates a win-win scenario where both the implementers and recipients realize tangible value and long-lasting positive impacts.

Environmental management encompasses this practical dimension by strategically coordinating resources, planning, and operational activities to achieve defined environmental objectives (Lockwood et al., 2010; Armitage et al., 2011). Recent studies underscore the importance of adaptive and integrative approaches that merge scientific knowledge with local insights and encourage active stakeholder participation to address the complexity and uncertainty inherent to environmental challenges (Armitage et al., 2009; Olsson et al., 2004). Effective environmental management should balance competing social, economic, and ecological demands, while promoting resilience and sustainability.

Within this context, project management practices serve as essential tools for the methodical planning, coordination, and monitoring of environmental initiatives, ensuring that projects meet their intended quality standards, timelines, and budgets. Due to the multifaceted nature of environmental projects, often involving technical challenges and diverse stakeholder interests, flexible, adaptive project management approaches that incorporate iterative learning cycles are increasingly recognized as vital for success (Armitage et al., 2009).

From a governance standpoint, environmental management is positioned as the implementation arm that brings governance frameworks to life. Governance defines the institutional structures, decision-making authority, stakeholder participation, and accountability mechanisms that set the parameters for action (Lockwood et al., 2010; Lemos & Agrawal, 2006). Environmental management operationalizes these rules through concrete activities such as planning, executing, and evaluating environmental programs (Ostrom, 1990; Duit et al., 2010). The implementation phase, therefore, acts as a critical nexus where policy intentions are translated into practice. Its effectiveness hinges on seamless coordination among multiple actors, continuous monitoring, and an adaptive capacity that allows for adjustments based on feedback and evolving circumstances (Armitage et al., 2009; Cash et al., 2019).

As a consequence, implementation lies at the heart of environmental management, acting as the crucial phase where governance frameworks are translated into concrete actions. Effective environmental outcomes depend on the seamless integration of governance structures with management processes, with implementation bridging policy and practice.

Figure 2. Translating policy into practice: The implementation phase in environmental project lifecycles

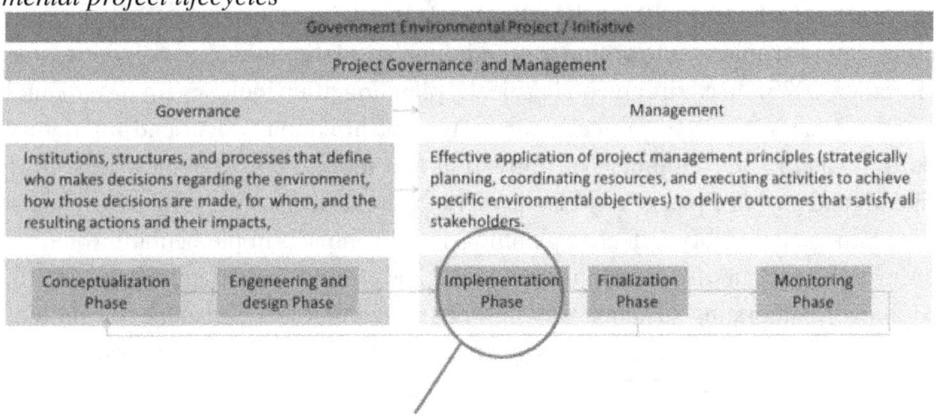

Where Does the Implementation Phase Fit In?

According to many scholars such as Sholarin and Awange (2016), project implementation is situated as the third critical phase within the standard project management lifecycle, preceded by initiation and planning, and followed by closure. It marks the transition from theoretical design to practical execution, where plans and strategies are translated into tangible outcomes.

Initiation (Conceptual Phase)

The initiation phase begins when a problem or opportunity is identified, prompting background analysis to define project objectives, preliminary scope, and constraints. For environmental and energy projects, this phase typically includes feasibility studies, site surveys, baseline data collection, and stakeholder mapping to validate site-specific conditions and assess initial viability. Early regulatory alignment and environmental sensitivity are essential at this stage to ensure long-term project success (Sholarin & Awange, 2016). Incorporating sustainability principles and stakeholder expectations from the outset helps prevent costly changes later and supports more resilient project frameworks (Kivilä et al. 2017). The integration of environmental and social impact considerations has thus become standard practice in modern project planning (Silvius & Schipper, 2015).

Planning Phase (Engineering and Design)

The planning phase translates the project concept into detailed tasks, deliverables, and execution strategies. This stage involves developing comprehensive engineering designs and using tools such as the Work Breakdown Structure (WBS), Gantt charts, and risk matrices to optimize scheduling, cost estimation, and risk mitigation (Kerzner, 2025). In environmental projects, planning often includes Environmental Impact Assessments (EIA), technical design specifications, legal and regulatory permits, and the development of mitigation and monitoring strategies. In sectors like oil and gas, this phase may also involve exploration activities and assessments of commercial viability. Effective planning directly impacts implementation quality, ensuring alignment with environmental, social, and technical constraints (Pinto, 2020). Moreover, integrating sustainability into planning processes enhances stakeholder acceptance and legitimacy, especially in projects involving sensitive ecosystems (Martens & Carvalho, 2017).

Execution Phase (Implementation)

The execution phase, or implementation, is where the actual transformation occurs—planning is translated into concrete actions. Resources—human, financial, and technical—are mobilized, and construction or remediation activities commence, it is the phase during which the actual transformation of project objectives into tangible outcomes occurs (Project Management Institute, 2021). In environmental projects, this may involve drilling, site remediation, infrastructure development, and implementation of environmental control measures. This stage demands rigorous performance monitoring to ensure alignment with time, cost, and quality objectives (Martinsuo, 2013). Adaptive management practices are crucial to respond to unexpected ecological or stakeholder challenges, particularly in dynamic or sensitive environments (Sholarin & Awange, 2016). Furthermore, strong leadership, real-time decision-making, and effective stakeholder coordination are critical drivers of successful implementation outcomes (Mir & Pinnington, 2014).

Controlling Phase (Monitoring and Evaluation)

Running in parallel with the execution phase, the Controlling Phase involves real-time tracking, performance measurement (cost, time, quality, environmental impact), and the identification of deviations from the original plan. This phase requires adaptive responses and corrective actions to keep project outcomes aligned with objectives (Williams & Brown, 2012). In environmental projects, monitoring is often legally mandated and includes continuous data collection, field inspections, and third-party audits. Obaid and Abu-Naser (2023) emphasize the growing use of data analytics and smart monitoring technologies to enhance control efficiency, especially under conditions of uncertainty.

Closure Phase (Finalisation)

The closure phase signifies the official end of an environmental project, involving the delivery of results, resource release, documentation of lessons learned, and verification of compliance with environmental regulations. This stage may include site remediation, decommissioning, and the start of long-term monitoring. It follows a structured process—Preliminary and Detailed Site Investigations, risk assessment, and remediation (Gregory et al. 2006). Stakeholder engagement and transparent communication are essential. Effective closure enhances sustainability, fosters institutional learning (Silvius & Schipper, 2015), and reduces future risks (Sholarin & Awange, 2016).

Example of Mangrove Forest Management Project

This project offers a compelling example of participatory environmental management, clearly illustrating the three core phases of planning, implementation, and monitoring. Implemented as part of CPF Company's environmental CSR strategy, the *Mangrove Forest Management Project* in Thailand mobilized local communities in every stage of mangrove conservation and restoration. To systematically assess this engagement, Sattayapanich et al. (2022) developed a structured measurement scale that captures the frequency and depth of community participation across these three dimensions.

During the planning phase, residents collaborated with the company and other stakeholders to co-develop the project's vision, rules, and resource allocation strategies. Their participation helped ensure that local needs and priorities were reflected in the interventions. This stage also fostered knowledge-sharing on mangrove ecosystems and restoration techniques. In this context, community involvement was measured using items such as: *"I have been invited by CPF Company to take part in decision-making processes for project development and planning"* and *"I have expressed my desires on the ecological management project to CPF Company (e.g., expected outcomes, types of activities, needed skills)"*.

In the implementation phase, community members were actively involved in practical tasks such as collecting seedlings, planting mangrove trees, managing waste, and preparing conservation areas. They also contributed through donations of resources and by raising public awareness. Participation in this phase was assessed using items like: *"I have participated in planting mangrove trees"* and *"I have donated assets, materials, or resources (e.g., food, planting tools, boats)"*.

The monitoring phase focused on ensuring long-term success by involving locals in the ongoing observation of forest conditions, identifying problems, and taking corrective actions. Community members reported environmental damage, replanted trees, and helped prevent human disturbance. Participation here was captured using items such as: *"I have replanted mangrove trees when I noticed damage or death"* and *"I have reported forest destruction or human invasion to responsible authorities"*.

This case highlights how bottom-up engagement, when supported by structured measurement and inclusive processes, can enhance the sustainability and effectiveness of environmental projects.

RELATED CONCEPTS AND THEIR RELEVANCE TO PARTICIPATION IN PROJECT IMPLEMENTATION

As previously noted, citizen participation continues to be mainly conceptualized as a deliberative or consultative process, primarily occurring during the pre-implementation stages of governance and closely linked to decision-making.

However, our literature review reveals that citizen participation during the implementation phase remains largely under-theorized and undervalued, despite its clear practical significance for the success and legitimacy of public policies.

Moreover, there is a notable interchangeability and sometimes ambiguous use of related concepts such as co-creation, co-design, co-production, engagement, and involvement. This conceptual diversity prompts us to delve deeper into these related fields to identify robust theoretical foundations. Our aim is to consolidate and clarify both the scope and the boundaries of the concept of participation in the implementation phase.

How "Co-Creation", "Co-Design", and "Co-Production" Enhance the Understanding of Participation in Project Implementation

Although the concepts of co-creation, co-design, and co-production originally emerged from distinct research fields—such as production, marketing, design and public administration—they are now frequently used interchangeably in the literature (Voorberg et al. 2015; Sanders & Stappers, 2008; Osborne et al. 2013).

This overlap often leads to conceptual ambiguity, as few studies have clearly distinguished between these three terms (Ramaswamy & Ozcan, 2018). Consequently, the boundaries between co-creation, co-design, and co-production remain blurred, with many authors adopting these concepts without fully addressing their specificities or original disciplinary contexts (Bovaird & Loeffler, 2012).

Furthermore, these concepts have consistently been linked to cooperation and collaboration, emphasizing the active participation of diverse stakeholders—including users, citizens, service providers, and policymakers—in joint processes aimed at generating value or improving outcomes (Prahalad & Ramaswamy, 2000; Sanders & Stappers, 2008). This participatory nature is central to understanding their role in promoting inclusive and democratic approaches across various fields. It is precisely this participatory aspect that we aim to explore here to shed light on the concept of "participation in implementation." In the following sections, we examine the origins of the three concepts, emphasizing their interchangeable use. We then present the conceptualization proposed by Vargas et al. (2022), which provides clear distinctions between co-creation, co-design, and co-production to focus on the elements that align with the concept of participation during the implementation phase.

Co-creation, Co-Design and Co-Production: From Diverse Roots to Conceptual Interchangeability

The concept of co-creation originated within the fields of production and marketing economics, where consumer participation was initially integrated into the supply chain to reduce costs (De Koning et al. 2016). The idea that involving customers could also enhance customer satisfaction was introduced later and customer participation has also been highlighted as a means of differentiation. The term "value co-creation" emerged to describe a proactive initiative by customers dissatisfied with existing options, prompting them to take action (Prahalad & Ramaswamy, 2000).

The idea that customers seek a dialogue with firms led to the proposal of guidelines for "co-creating the voice of the customer." Meanwhile, Western economies are shifting towards a service-dominant logic, where consumers no longer simply purchase goods or services, but rather products that deliver a service, with value derived from the overall customer experience. In this context, consumers buy an experience, with the product or service serving as an artifact. Consequently, it is argued that within a service-dominant logic—as opposed to a goods-dominant logic—the customer is always a co-creator (Jaworski & Kohli, 2014).

Parallel to the transformations in production and marketing economics, similar shifts occurred within the field of design. In this field, co-creation has its origins in Human-Centered Design (HCD) and participatory design—also known as co-operative design and later termed co-design—which seeks to actively involve all stakeholders throughout the design process. These movements emerged in the 1970s in Scandinavia, where joint decision-making and collaborative work practices began to gain prominence (De Koning et al. 2016).

In participatory design, participants are valued as essential contributors, offering their expertise and knowledge as critical resources. This is why co-creation is often closely linked to participatory design. Participatory design is defined as an approach that emphasizes people's active involvement as co-designers in the design process. Co-design is defined as the facilitation of exchange between people who experience products, interfaces, systems and spaces and people who design for experiencing (Mattelmäki & Visser 2011)

The term "co-production" finds its scholarly origins in the public sector, in the work of Ostrom and other economists who studied collaboration between public departments and citizens (Brandsen & Honingh 2018).

Ostrom (1996) defined co-production as "the process through which inputs used to produce a good or service are contributed by individuals who are not in the same organization". Co-production implies that citizens can play an active role in producing public goods and services of consequence to them. Ostrom focuses on the fact that, in the ways in which they interact with the public in service delivery, public agencies can either elicit synergistic, cooperative behaviour from clients that improves the overall quality of service delivery, or may fail to do so.

This concept of co-production highlights the active role of citizens, who are no longer passive recipients of public services but rather engaged participants involved in the design, management, and delivery of these services.

Ostrom's pioneering research challenged the traditional top-down model of public service delivery by emphasizing that effective and efficient public services often depend on cooperation between service providers and users. This collaboration mobilizes complementary resources, knowledge, and skills, leading to improved service outcomes.

The voluntary nature of citizen involvement in co-production has been emphasized as a factor that can enhance the quality and responsiveness of public services. Furthermore, the concept has been expanded to include the institutionalization of co-production, where citizen participation becomes a regular, organized practice embedded within public policy and governance frameworks. This shift reflects a broader transformation in public administration towards "new public governance," which moves away from hierarchical, bureaucratic models in favor of more participatory and collaborative approaches. Co-production thus represents a significant evolution in public sector theory and practice, recognizing that public value is co-created through partnerships between governments and citizens. This approach not only improves service delivery but also fosters greater trust, accountability, and empowerment among stakeholders (Joshi & Moore, 2004).

In the last decade, all these terms–co-production, co-design, co-creation– have appeared widely in scientific literature, in professional magazines, websites of product development companies, design research and market research agencies and also in reports of public organisations. In these writings people show examples of how their version of these concepts has been applied.

As mentioned previously, these concepts are now frequently used interchangeably in the literature. This interchangeability has been notably highlighted by Voorberg et al. (2015), who, in their systematic review, observe that the boundaries between co-creation and co-production are often blurred, with these terms being used to describe similar collaborative processes in the design and delivery of public services. Sanders and Stappers (2008), experts in design, emphasize that co-design is sometimes regarded as synonymous with co-creation, underscoring that both terms refer to forms of collaboration where stakeholders actively participate in defining and realizing solutions. Similarly, Osborne et al. (2013) discuss in the public administration domain the tendency to use co-production and co-creation interchangeably, noting that this reflects a broader shift toward more participatory and collaborative approaches in public governance.

A recent systematic review by Lee et al. (2024) also reveals that the concepts 'co-production', 'co-creation', and 'co-design' are defined similarly and used interchangeably in public space literature. Authors who use these concepts simultaneously in their articles often do not distinguish between them. Also, while these concepts are becoming popular in the field of public space, there is no article that discusses the differences among them. Lee et al. (2024) note that, for future research, it would be relevant to delineate co-production, co-creation and/or co-design more clearly from other related concepts such as participation.

Co-Creation, Co-Design and Co-Production Boundaries and Link to Participation in Implementation

The distinction proposed by Vargas et al. (2022) between co-creation, co-design, and co-production is emphasized here, as it stands out for clearly delineating the boundaries between these three concepts. This framework, applied to public health, helps in identifying specific phases where connections to participatory practices during implementation can be established.

Vargas et al. (2022) view co-creation as a concept aligned with participatory research, engaging stakeholders in co-design or co-production approaches, and enabling them to construct a shared agenda that facilitates collective action and the creation of useful solutions. Building on the work of De Koning et al. (2016), who analyzed and clustered various models of co-creation into a six-step continuous research cycle, Vargas et al. (2022) used this framework to prototype the use of co-creation as an overarching concept that encompasses both co-design and co-production. They represent co-creation as a linear process, while emphasizing that it entails continuous improvement of outputs or outcomes through incremental changes and transformative innovation (figure 2 below).

Co-creation is a collaborative approach for creative problem solving. It emphasizes diverse stakeholders at all parts of an initiative process, beginning with determining and defining the problem through to the final stages of a project. It engages stakeholders before the problem is defined and considers that stakeholders are not necessarily on opposite sides but can collaborate to find new shared values and opportunities for mutual benefit in defining the problem and subsequent solutions. "Since co-design serves as a tool for co-creation, and co-creation involves different stakeholders in the co-production of a service, codesign and co-production can be positioned under the umbrella of co-creation" (Vargas et al. 2022).

Co-creation promotes the creation of value that can take multiple forms: for the individual it can be psychological (e.g. feelings of appreciation) or economic (e.g.

greater earnings). For organisations, value is most often economic (e.g. lower costs, improved productivity) or a social good.

Co-design describes active collaboration between stakeholders in designing solutions to a prespecified problem. It encourages citizen participation to formulate or improve specific concerns. Co-design often takes the form of group problem-solving, typically occurring after the problem has been clearly defined; consequently, it is not always formally documented.

Vargas et al. (2022) outline four key stages related to co-design.

The first stage in the co-design process is "Identify". It involves identifying the relevant structures and stakeholders connected to the issue at hand. The primary focus is to pinpoint opportunities for value creation and potential solutions to the problems. It is essential to recognize all stakeholders who should be included in the process.

The second stage is "Analyse". This step involves examining the stakeholder network to identify both shared and conflicting values. The goal is to systematically clarify decision-making processes and options, the nature of relationships, and the obligations tied to these interactions. It also includes understanding how stakeholders engage with one another and gathering relevant experiences and ideas that can inform potential solutions.

"Define" is the third step in the co-design process where participants prioritise problems, next steps and actions built on the rights, obligations, and ideas identified in previous steps.

The last step in co-design is "design" of initiatives. This includes defining clear goals, outlining the actions required to reach those goals, establishing evaluation methods, and allocating the necessary resources and assets. To get the right initiative, stakeholders must collaborate closely and be willing to adjust their perspectives and values based on insights gained through their interactions.

Co-production engages stakeholders in the implementation of a previously agreed solution (strategy) to a previously agreed problem and focuses on how to allocate resources and assets within these constraints, to achieve better outcomes. Co-production occurs after the initiative was designed and takes place at the point of implementing the initiative.

A widely used definition of co-production is provided by Brandsen and Honingh (2016) who view co-production as a relationship between the employees of an organization and citizens which requires direct and active inputs from these citizens to the work of the organization. The professional is a paid employee of the organization, whereas the citizen receives compensation below market value or no compensation at all. While Brandsen and Honingh (2018) consider co-production as involving citizens in just one stage, the co-implementation phase, according to Vargas et al. 2022, two stages are related to co-creation. The first one is "Realise" where stakeholders, during the implementation, test the designed strategies and gather information. This realisation stage can remain continuous or occur in stages where testing ideas are reevaluated. It is essential in this step to build structures that enable continual dialogue between stakeholders to implement ideas and generate further ideas for improvement and future implementation.

The final stage in the co-production (and co-creation) process is the "Evaluate" phase. During this step, the proposed outcomes are reviewed alongside the methods used in earlier stages. It involves reflecting on the insights gained from diverse stakeholders, considering environmental changes, and identifying strategies to ensure sustainability, such as resource management, forming new partnerships, and building capacity.

Among all the stages mentioned in the co-production process, it is the "Realise" phase within the co-creation framework that closely aligns with the concept of participation during the implementation phase.

This stage involves the active engagement of stakeholders in putting plans or designs into action, ensuring that collaborative efforts move beyond planning and design to tangible outcomes. Participation at this point is crucial because it transforms shared ideas and co-designed solutions into real-world applications, fostering ownership, accountability, and responsiveness among all involved parties. By emphasizing stakeholder involvement during implementation, the "Realise" phase embodies the practical realization of co-created value, bridging the gap between conceptual collaboration and effective service delivery or policy enactment.

Figure 3. Co-production as the space where participation in implementation materializes within the co-creation process

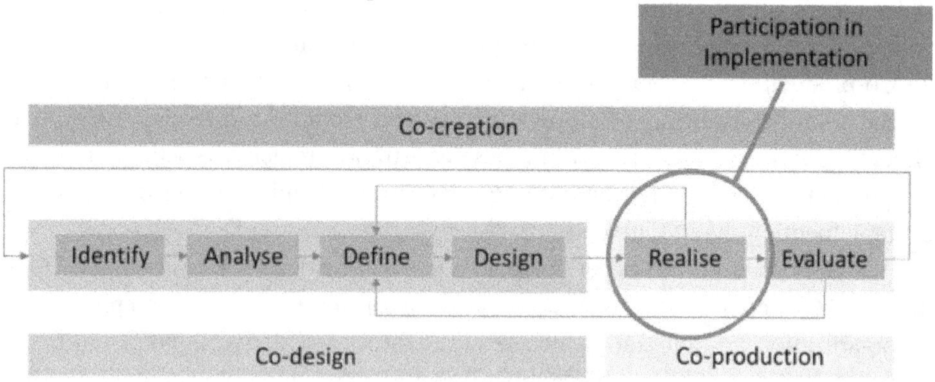

Source: Adapted from Vargas et al. 2022

Insight to Achieve From the Concept of Customer Involvement

Why focus here on the concept of Customer Involvement? Simply because customer involvement is often discussed alongside related concepts such as participation, engagement, value co-creation, customer co-creation, customer integration and open innovation. These terms are frequently used interchangeably or even as synonyms in both academic and practical contexts (Katona, 2022). This conceptual overlap suggests that a deeper exploration of involvement could shed light on the concept of participation, particularly in the context of implementation.

The concept of customer involvement has been explored across various scientific fields, including knowledge management, marketing, innovation and new product development, entrepreneurship, and information systems. In particular, within the innovation and new product development domain, customer involvement has assumed a more central and strategic role, with companies increasingly viewing customers not merely as sources of information but as active partners and co-creators in the development of new products and services.

In this field, prior research highlights the benefits and risks of customer involvement. On the one hand, involving customers can improve outcomes such as innovation speed and customer satisfaction. On the other hand, involving customers can cause challenges such as lower product innovativeness, information overload from customer opinions and process delays (Sadanha et al. 2017).

Across all the fields mentioned, various aspects of customer involvement are examined; however, there is a lack of a clear and comprehensive typology that defines its domain, scope, and boundaries. The coexistence of multiple, often fragmented conceptualizations of customer involvement hinders the effective consolidation and advancement of knowledge within the domain (Katona, 2022).

Nevertheless, our focus is not to unify the proposed conceptualizations of involvement or to discuss its impact on performance; rather, our focus is on how the concept of customer involvement can shed light on participation in the implementation process.

Customer Involvement in the Advertising Field

The field of marketing and advertising has offered a rich foundation of conceptualizations of customer involvement. The most classic conceptualizations of customer involvement in this field consider involvement as a subjective psychological state that reflects the importance and personal relevance of a product or service to a customer and explains consumer behavior. It affects the level of brand loyalty, brand discrimination, product selection, and purchase decisions (Cheung & To, 2011).

Andrews et al. (1990) proposed a comprehensive framework for understanding customer involvement in advertising by categorizing it into four major types: attention/processing strategies, personal/situational involvement, audience/process involvement, and enduring/product involvement. This classification helps clarify the diverse ways involvement can manifest in consumer behavior, capturing a spectrum that ranges from fleeting, momentary attention to advertisements or products, to deeper, long-term personal relevance and commitment. For instance, attention/processing strategies focus on how consumers allocate cognitive resources during brief encounters with marketing stimuli, reflecting immediate engagement levels. In contrast, enduring/product involvement reflects a sustained psychological connection with a product category or brand, influencing consistent preferences and loyalty over time.

Customer Involvement in Innovation and New Product Development

In innovation and New Product Development fields, customer involvement is defined as the extent to which a firm interacts with representatives of one or more customers in developing products (Carbonell et al., 2009), it refers to the interaction between customers and the design process" (Kaulio 1998).

A systematic approach to understanding customer involvement in product development may also need to consider both longitudinal and lateral dimensions of customer involvement, as proposed by Kaulio (1998).

The longitudinal dimension describes the points of interaction at different phases of the product design process and the lateral dimension taps into the depth of customer engagement in the product design process (Kim et al. 2018).

Kaulio (1998) described this dimension with three categories: "design for" where a product is designed on behalf of the customers based on knowledge learned about the customers; "design with" in which a product reflects the customers' preferences, needs, and requirements with diverse solutions; and "design by" where customers are proactively involved and take the lead to develop their product.

In the new product development literature, two types of customer involvement are often distinguished: Customer involvement as an information source (CIS), also named information intensive customer involvement (ICI), and customer involvement as co-developers (CIC), also named product-focused customer involvement (PCI) (Mahmood et al., 2024, Nambisan, 2002; Cui & Wu, 2017).

CIS (or ICI) is a type of involvement focused on gathering information from customers via customer opinions and feedback and structured inquiry mechanisms such as focus groups. Customer involvement of such type takes customers as the most important resource towards information, feedback and opinions (Nambisan, 2002).

In CIS, customers play the role of information providers, sharing knowledge on what they need or want for a new product. The internal NPD team takes the responsibility of applying information gathered from customers to the NPD process and design products that meet customers' needs (Cui &Wu, 2017, Mahmood et al., 2024).

In CIC (or PCI), customers assume an active, participatory role as co-developers of new products. This involvement often takes the form of key customers driving product development or customizing products to meet specific needs (Nambisan, 2002). Customers collaborate closely with internal NPD experts throughout the development process, transforming NPD into a partnership where both parties jointly create new products.

In CIC, customers maintain close and frequent collaboration with internal NPD experts over an extended timeframe. They participate in joint problem-solving activities with the NPD team and make direct contributions to product design. Additionally, customers are actively involved in decision-making processes alongside NPD staff, including choices related to product feature design, specification of interface requirements, and setting priorities and metrics for the development process (Cui &Wu, 2017, Mahmood et al., 2024).

CIS and CIC encounter distinct challenges within the new product development process. The primary challenge for CIS lies in obtaining sufficient customer information and effectively utilizing it. Since NPD employees control what customer data is collected and how it is interpreted, their dominant role can limit the incorporation of creative insights from customers. CIC addresses this limitation by giving customers a more active role in development; however, it introduces new difficulties. The volume of detailed customer input complicates information processing, while the increased need for coordinating customer involvement adds complexity to NPD management, potentially causing miscommunication and conflicts that may hinder the effective use of customer contributions. These challenges in both CIS and CIC can affect how successfully customer inputs are transformed into innovative new products (Cui &Wu, 2017).

Can Customer Involvement Really Shed Light on Participation in Implementation?

The conceptualizations of customer involvement within marketing and advertising do not offer the specific insights necessary to fully grasp the concept of participation. In contrast, the field of new product development provides a more relevant and comprehensive framework for understanding this concept, particularly through the lens of CIC, which emphasizes active customer participation. However, it is important to note that the type of participation discussed in this context primarily relates to involvement in the design of new products, closely aligning with the concept of co-design introduced in the previous paragraph. This highlights that while CIC offers valuable perspectives on participation, its focus remains on the early stages of product development rather than on the broader implementation process (Figure below).

Figure 4. Customer involvement as co-developer (CIC): Engaged at the start, absent in implementation

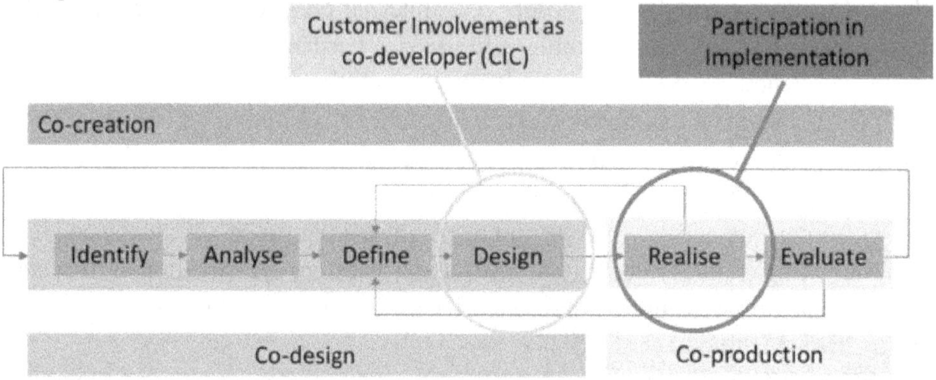

Source: Adapted from Vargas et al. 2022

CONCLUSION: A CALL FOR A MORE HOLISTIC UNDERSTANDING OF CITIZEN PARTICIPATION

The concept of citizen participation is still predominantly framed as an act occurring in the pre-implementation phases of project governance and management. This framing is well documented in the literature, which often situate citizen participation within the early stages of policy formulation, planning, and decision-making. Our literature review reveals that the prevailing approach views citizen participation through both a governance and institutional lens, as well as from a critical, power-based perspective. Within this framework, citizen participation is considered a vital mechanism of open government systems, where public institutions actively collaborate with citizens to improve public services and address collective challenges. The abundant literature relating to typologies, levels, and mechanisms of citizen participation discussed earlier largely falls within this framework, focusing primarily on participation in decision-making processes (Arnstein, 1969, Oliveira & Ckagnazaroff, 2023; Wijnhoven, et al. 2015; Millard, 2018).

This predominant focus on the pre-implementation phases tends to overlook the critical role that citizen participation can and should play throughout the entire project lifecycle, including implementation, monitoring, and evaluation. Expanding participation beyond decision-making stages is essential to ensure accountability, responsiveness, and the effective realization of public policies and projects. Thus, we call for a more holistic understanding of citizen participation that integrates it fully into project governance and management processes (OCDE, 2023, Nabatchi et al. 2017)

Our literature review, particularly concerning the interplay between governance and management, provides enhanced clarity on the conceptualization of citizen participation within project contexts. Drawing from project management literature—which we prefer to frame as *project governance and management* to emphasize the dual dimensions of oversight and operational execution—this process can be understood as comprising two distinct but interconnected subprocesses.

The first subprocess involves the early phases of conceptualization, engineering, and design. During these stages, citizen involvement is primarily characterized as participation in decision-making, aligning with Arnstein's (1969) notion of citizen participation. This phase is critical for incorporating diverse stakeholder inputs, shaping project objectives, and ensuring legitimacy and responsiveness in policy formulation. The engagement of citizens at this stage is often seen as a means to redistribute power and foster inclusivity in governance structures.

The second subprocess, more closely associated with project management than governance, encompasses the phases of implementation, finalization, and monitoring. While traditionally underemphasized in participation literature, these phases are equally vital for sustaining citizen engagement and ensuring accountability throughout the project lifecycle. Participation during implementation allows citizens to contribute to adaptive management, problem-solving, and quality assurance, thereby enhancing project outcomes and legitimacy. Finalization and monitoring phases provide opportunities for citizens to evaluate project impacts and hold authorities accountable, reinforcing transparency and trust in public institutions.

Recognizing citizen participation as a continuum across both governance and management subprocesses challenges the conventional focus on pre-implementation decision-making alone and highlights the importance of sustained, multi-phase engagement for effective and democratic project delivery.

Referring to the extensive literature on the concepts of co-creation, co-design, and co-production, participation in the implementation phase can be equated with co-production, and more specifically with the so-called "realization" stage (Ostrom, 1996; Vargas, 2022). Although these notions are sometimes differentiated by terminological nuances, they largely converge in their approach to the active involvement of citizens and users throughout the lifecycle of public policies or projects.

Thus, participation in implementation, far from being a marginal or neglected concept, is fully addressed in the literature under the term co-production. This "realization" stage implies close collaboration between public institutions and citizens to adapt, execute, and adjust actions according to real needs, thereby enhancing the effectiveness, legitimacy, and quality of public interventions.

It may therefore be more a matter of jargon differences than a major conceptual divergence: participation in the implementation phase fits fully within the framework of co-production, highlighting the importance of continuous and active citizen involvement beyond the decision-making phases alone.

Figure 5. Citizen participation in focus: A holistic model for conceptual clarification and implementation positioning

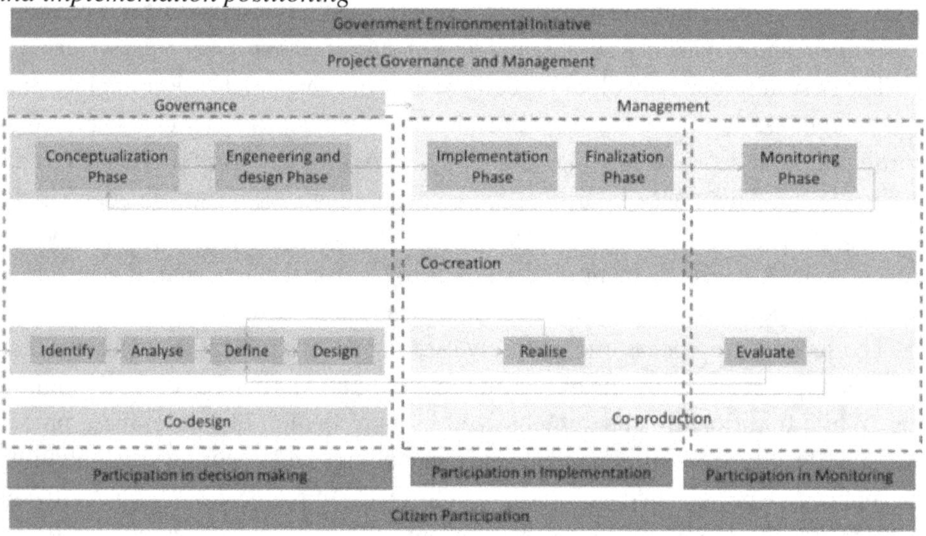

REFERENCES

Andrews, J. C., Durvasula, S., & Akhter, S. H. (1990). A framework for conceptualizing and measuring the involvement construct in advertising research. *Journal of Advertising*, *19*(4), 27–40. DOI: 10.1080/00913367.1990.10673198

Armitage, D., Berkes, F., Dale, A., Kocho-Schellenberg, E., & Patton, E. (2011). Co-management and the co-production of knowledge: Learning to adapt in Canada's Arctic. *Global Environmental Change*, *21*(3), 995–1004. DOI: 10.1016/j.gloenvcha.2011.04.006

Armitage, D., Plummer, R., Berkes, F., Arthur, R. I., Charles, A. T., Davidson-Hunt, I. J., Diduck, A. P., Doubleday, N. C., Johnson, D. S., Marschke, M., McConney, P., Pinkerton, E. W., & Wollenberg, E. K. (2009). Adaptive co-management for social-ecological complexity. *Frontiers in Ecology and the Environment*, *7*(2), 95–102. DOI: 10.1890/070089

Arnstein, S. R. (1969). A ladder of citizen participation. *Journal of the American Institute of Planners*, *35*(4), 216–224. DOI: 10.1080/01944366908977225

Baxter, S., Barnes, A., Lee, C., Mead, R., & Clowes, M. (2023). Increasing public participation and influence in local decision-making to address social determinants of health: A systematic review examining initiatives and theories. *Local Government Studies*, *49*(5), 861–887. DOI: 10.1080/03003930.2022.2081551

Bennett, N. J., & Satterfield, T. (2018). Environmental governance: A practical framework to guide design, evaluation, and analysis. *Conservation Letters*, *11*(6), e12600. DOI: 10.1111/conl.12600

Bergmann, M., Schäpke, N., Marg, O., Stelzer, F., Lang, D. J., Bossert, M., Gantert, M., Häußler, E., Marquardt, E., Piontek, F. M., Potthast, T., Rhodius, R., Rudolph, M., Ruddat, M., Seebacher, A., & Sußmann, N. (2021). Transdisciplinary sustainability research in real-world labs: Success factors and methods for change. *Sustainability Science*, *16*(2), 541–564. DOI: 10.1007/s11625-020-00886-8

Biermann, F., Kanie, N., & Kim, R. E. (2017). Global governance by goal-setting: The novel approach of the UN Sustainable Development Goals. *Current Opinion in Environmental Sustainability*, *26-27*, 26–31. DOI: 10.1016/j.cosust.2017.01.010

Birch, S. (2017). The ethics of political participation: Engagement and democracy in the 21st century. *Res Publica (Liverpool, England)*, *23*(1), 1–12. DOI: 10.1007/s11158-017-9389-7

Bodin, Ö. (2017). Collaborative environmental governance: Achieving collective action in social-ecological systems. *Science*, *357*(6352), eaan1114. DOI: 10.1126/science.aan1114 PMID: 28818915

Bovaird, T., & Loeffler, E. (2012). From engagement to co-production: The contribution of users and communities to outcomes and public value. *Voluntas*, *23*(4), 1119–1138. DOI: 10.1007/s11266-012-9309-6

Brandsen, T., & Honingh, M. (2016). Distinguishing different types of co-production: A conceptual analysis based on the classical definitions. *Public Administration Review*, *76*(3), 427–435. DOI: 10.1111/puar.12465

Brandsen, T., & Honingh, M. (2018). Definitions of co-production and co-creation. In *Co-production and co-creation* (pp. 9–17). Routledge. DOI: 10.4324/9781315204956-2

Cabannes, Y. (2004). Participatory budgeting: A significant contribution to participatory democracy. *Environment and Urbanization*, *16*(1), 27–46. DOI: 10.1177/095624780401600104

Calisto Friant, M. (2019). Deliberating for sustainability: Lessons from the Porto Alegre experiment with participatory budgeting. *International Journal of Urban Sustainable Development*, *11*(1), 81–99. DOI: 10.1080/19463138.2019.1570219

Callahan, K. (2007). Citizen participation: Models and methods. *International Journal of Public Administration*, *30*(11), 1179–1196. DOI: 10.1080/01900690701225366

Carbonell, P., Rodríguez Escudero, A. I., & Pujari, D. (2009). Customer Involvement in New Service Development: An Examination of Antecedents and Outcomes. *Journal of Product Innovation Management*, *26*(5), 536–550. DOI: 10.1111/j.1540-5885.2009.00679.x

Cash, D. W., (2019). Knowledge systems for sustainable development. *Proceedings of the National Academy of Sciences of the United States of America*, *116*(14), 6570–6578. DOI: 10.1073/pnas.1817180116 PMID: 12777623

Cheung, M. F., & To, W. M. (2011). Customer involvement and perceptions: The moderating role of customer co-production. *Journal of Retailing and Consumer Services*, *18*(4), 271–277. DOI: 10.1016/j.jretconser.2010.12.011

Cornwall, A., & Gaventa, J. (2017). Participation in governance. In *International development governance* (pp. 405–413). Routledge. DOI: 10.4324/9781315092577-22

Cui, A. S., & Wu, F. (2017). The impact of customer involvement on new product development: Contingent and substitutive effects. *journal of product innovation management, 34*(1), 60-80. DOI: 10.1111/jpim.12333

Dagnino, E. (2003). Citizenship in Latin America: An introduction. *Latin American Perspectives, 30*(2), 211–225. DOI: 10.1177/0094582X02250624

Dagnino, E., & Teixeira, A. C. C. (2014). The participation of civil society in Lula's government. *Journal of Politics in Latin America, 6*(3), 39–66. DOI: 10.1177/1866802X1400600303

Davidson, S. (1998). Spinning the wheel of empowerment. *Planning (Chicago, Ill.), 1262*(3), 14–15.

De Jong, M. D., Neulen, S., & Jansma, S. R. (2019). Citizens' intentions to participate in governmental co-creation initiatives: Comparing three co-creation configurations. *Government Information Quarterly, 36*(3), 490–500. DOI: 10.1016/j.giq.2019.04.003

De Koning, J. I. J. C., Crul, M. R., & Wever, R. (2016, May). Models of co-creation. In Service design geographies. Proceedings of the servdes. 2016 conference (Vol. 125, pp. 266-278). Linköping, Sweden: Linköping University Electronic Press. DOI: 10.1111/0272-4332.00064

Gelderman, C. J., Semeijn, J., & Vluggen, R. (2017). Development of sustainability in public sector procurement. *Public Money & Management, 37*(6), 435–442. DOI: 10.1080/09540962.2017.1344027

Gregory, R., Ohlson, D., & Arvai, J. (2006). Deconstructing adaptive management: Criteria for applications to environmental management. *Ecological Applications, 16*(6), 2411–2425. DOI: 10.1890/1051-0761(2006)016[2411:DAMCFA]2.0.CO;2 PMID: 17205914

Huttunen, S., Ojanen, M., Ott, A., & Saarikoski, H. (2022). What about citizens? A literature review of citizen engagement in sustainability transitions research. *Energy Research & Social Science, 91*, 102714. DOI: 10.1016/j.erss.2022.102714

Innes, J. E., & Booher, D. E. (2004). Reframing public participation: Strategies for the 21st century. *Planning Theory & Practice, 5*(4), 419–436. DOI: 10.1080/1464935042000293170

Jaworski, B. J., & Kohli, A. K. (2014). Co-creating the voice of the customer. In *The service-dominant logic of marketing* (pp. 109–117). Routledge.

Jeppesen, L. B. (2005). User toolkits for innovation: Consumers support each other. *Journal of Product Innovation Management, 22*(4), 347–362. DOI: 10.1111/j.0737-6782.2005.00131.x

Joshi, A., & Moore, M. (2004). Institutionalised co-production: Unorthodox public service delivery in challenging environments. *The Journal of Development Studies, 40*(4), 31–49. DOI: 10.1080/00220380410001673184

Katona, Á. (2022). Customer involvement: A systematic literature review of the definition, measurement, and innovation outcomes. *Vezetéstudomány/Budapest. Management Review, 53*(5), 59–69.

Kaulio, M. A. (1998). Customer, consumer and user involvement in product development: A framework and a review of selected methods. *Total Quality Management, 9*(1), 141–149. DOI: 10.1080/0954412989333

Kerzner, H. (2025). *Project management: a systems approach to planning, scheduling, and controlling*. John Wiley & Sons.

Kim, J., Park, J., & Glovinsky, P. L. (2018). Customer involvement, fashion consciousness, and loyalty for fast-fashion retailers. *Journal of Fashion Marketing and Management, 22*(3), 301–316. DOI: 10.1108/JFMM-03-2017-0027

Kindon, S., Pain, R., & Kesby, M. (2007). *Participatory action research approaches and methods. Connecting people, participation and place*. Routledge. DOI: 10.4324/9780203933671

Kiss, B., Sekulova, F., Hörschelmann, K., Salk, C. F., Takahashi, W., & Wamsler, C. (2022). Citizen participation in the governance of nature-based solutions. *Environmental Policy and Governance, 32*(3), 247–272. DOI: 10.1002/eet.1987

Kivilä, J., Martinsuo, M., & Vuorinen, L. (2017). Sustainable project management through project control in infrastructure projects. *International Journal of Project Management, 35*(6), 1167–1183. DOI: 10.1016/j.ijproman.2017.02.009

Lawrence, A. (2006). 'No personal motive?' Volunteers, biodiversity, and the false dichotomies of participation. *Ethics Place and Environment, 9*(3), 279–298. DOI: 10.1080/13668790600893319

Lee, D., Feiertag, P., & Unger, L. (2024). Co-production, co-creation or co-design of public space? A systematic review. *Cities (London, England), 154*, 105372. DOI: 10.1016/j.cities.2024.105372

Lemos, M. C., & Agrawal, A. (2006). Environmental governance. *Annual Review of Environment and Resources, 31*(1), 297–325. DOI: 10.1146/annurev.energy.31.042605.135621

Lockwood, M. (2010). Good governance for terrestrial protected areas: A framework, principles and performance outcomes. *Journal of Environmental Management, 91*(3), 754–766. DOI: 10.1016/j.jenvman.2009.10.005 PMID: 19896262

Mahmood, F., Marwat, I. U. K., & Shah, S. A. (2024). Customer Involvement and Sustainable Product Innovations: The Role of Design Excellence and Technological Capability. *The Lighthouse Journal of Social Sciences, 3*(1), 146–160.

Martens, M. L., & Carvalho, M. M. (2017). Key factors of sustainability in project management context: A survey exploring the project managers' perspective. *International Journal of Project Management, 35*(6), 1084–1102. DOI: 10.1016/j.ijproman.2016.04.004

Martinsuo, M. (2013). Project portfolio management in practice and in context. *International Journal of Project Management, 31*(6), 794–803. DOI: 10.1016/j.ijproman.2012.10.013

Mattelmäki, T., & Visser, F. S. (2011). Lost in co-x. *Proceedings of the IASDR2011.*

Milbrath, L. W. (1981). Political participation. In *The Handbook of Political Behavior* (Vol. 4, pp. 197–240). Springer US. DOI: 10.1007/978-1-4684-3878-9_4

Millard, J. (2018). Open governance systems: Doing more with more. *Government Information Quarterly, 35*(4), S77–S87. DOI: 10.1016/j.giq.2015.08.003

Miller, C. A., & Wyborn, C. (2020). Co-production in global sustainability: Histories and theories. *Environmental Science & Policy, 113*, 88–95. DOI: 10.1016/j.envsci.2018.01.016

Mir, F. A., & Pinnington, A. H. (2014). Exploring the value of project management: Linking project management performance and project success. *International Journal of Project Management, 32*(2), 202–217. DOI: 10.1016/j.ijproman.2013.05.012

Nabatchi, T., Sancino, A., & Sicilia, M. (2017). Varieties of Participation in Public Services: The Who, When, and What of Participation in Service Design and Delivery. *Public Administration Review, 77*(5), 766–776. DOI: 10.1111/puar.12765

Nambisan, S. (2002). Designing virtual customer environments for new product development: Toward a theory. *Academy of Management Review, 27*(3), 392–413. DOI: 10.2307/4134386

Obaid, T., & Abu-Naser, S. S. (2023). Big data analytics in project management: a key to success. International Journal of Academic Engine e ring Re search (IJAER), ISSN: 2643-9085, Vol. 7 Is s ue 7, July - 2023, Page s: 1-8

Oliveira, D. J. S., & Ckagnazaroff, I. B. (2023). Citizen participation as one of the principles of open government. *Cadernos Gestão Pública e Cidadania, 28*, e84867. DOI: 10.12660/cgpc.v28.84867

Olsson, P., Folke, C., & Berkes, F. (2004). Adaptive co-management for building resilience in social–ecological systems. *Environmental Management, 34*(1), 75–90. DOI: 10.1007/s00267-003-0101-7 PMID: 15383875

Organisation de coopération et de développement économiques (OCDE). (2023). *Exploring new frontiers of citizen participation in the policy cycle.* https://www.oecd.org/governance/exploring-new-frontiers-citizen-participation.pdf

Orr, S. K. (2013). *Environmental policymaking and stakeholder collaboration: Theory and practice.* Taylor & Francis. DOI: 10.1201/b16131

Osborne, S. P., Radnor, Z., & Nasi, G. (2013). A new theory for public service management? Toward a (public) service-dominant approach. *American Review of Public Administration, 43*(2), 135–158. DOI: 10.1177/0275074012466935

Ostrom, E. (1996). Crossing the great divide: Coproduction, synergy, and development. *World Development, 24*(6), 1073–1087. DOI: 10.1016/0305-750X(96)00023-X

Ostrom, E. (1996). Crossing the Great Divide: Coproduction, Synergy, and Development. *World Development, 24*(6), 1073–1087. DOI: 10.1016/0305-750X(96)00023-X

Pinto, J. K. (2020). *Project management: achieving competitive advantage.* Pearson.

Prahalad, C. K., & Ramaswamy, V. (2000). Co-Opting Customer Competence. *Harvard Business Review, 78*(1), 79–87.

Pretty, J. N. (1995). Participatory learning for sustainable agriculture. *World Development, 23*(8), 1247–1263. DOI: 10.1016/0305-750X(95)00046-F

Project Management Institute. (2021). *A guide to the project management body of knowledge (PMBOK® guide)* (7th ed.).

Puskás, N., Abunnasr, Y., & Naalbandian, S. (2021). Assessing deeper levels of participation in nature-based solutions in urban landscapes–A literature review of real-world cases. *Landscape and Urban Planning, 210*, 104065. DOI: 10.1016/j.landurbplan.2021.104065

Ramaswamy, V., & Ozcan, K. (2018). What is co-creation? An interactional creation framework and its implications for value creation. *Journal of Business Research, 84*, 196–205. DOI: 10.1016/j.jbusres.2017.11.027

Richardson, L. (2005). Social and political participation and inclusion. In *A more equal society?* (pp. 93–16). Policy Press., DOI: 10.2307/j.ctt9qgqd0.10

Richter, J., Basten, D., Michalik, B., Rosenkranz, C., & Smolnik, S. (2023). Opening the black box of knowledge management mechanisms: Exploring knowledge flows at a consultancy. *Kybernetes, 52*(13), 1–28. DOI: 10.1108/K-08-2022-1118

Rowe, G., & Frewer, L. J. (2000). Public participation methods: A framework for evaluation. *Science, Technology & Human Values, 25*(1), 3–29. DOI: 10.1177/016224390002500101

Sanders, E. B. N., & Stappers, P. J. (2008). Co-creation and the new landscapes of design. *CoDesign, 4*(1), 5–18. DOI: 10.1080/15710880701875068

Sandoval-Almazan, R., & Gil-Garcia, J. R. (2012). Are government internet portals evolving towards more interaction, participation, and collaboration? Revisiting the rhetoric of e-government among municipalities. *Government Information Quarterly, 29*, S72–S81. DOI: 10.1016/j.giq.2011.09.004

Sattayapanich, T., Janmaimool, P., & Chontanawat, J. (2022). Factors affecting community participation in environmental corporate social responsibility projects: Evidence from mangrove forest management project. *Journal of Open Innovation, 8*(4), 209. DOI: 10.3390/joitmc8040209

Sengers, F., Wieczorek, A. J., & Raven, R. (2019). Experimenting for sustainability transitions: A systematic literature review. *Technological Forecasting and Social Change, 145*, 153–164. DOI: 10.1016/j.techfore.2016.08.031

Sholarin, E. A., & Awange, J. L. (2016). *Environmental project management: principles, methodology, and processes*. Springer.

Silvius, A. G., & Schipper, R. (2015). A conceptual model for exploring the relationship between sustainability and project success. *Procedia Computer Science, 64*, 334–342. DOI: 10.1016/j.procs.2015.08.497

Susha, I., & Grönlund, Å. (2014). Context clues for the stall of the Citizens' Initiative: Lessons for opening up e-participation development practice. *Government Information Quarterly, 31*(3), 454–465. DOI: 10.1016/j.giq.2014.02.005

Thomas, J. C. (1993). Public involvement and governmental effectiveness: A decision-making model for public managers. *Administration & Society, 24*(4), 444–469. DOI: 10.1177/009539979302400402

Tippett, J., Handley, J. F., & Ravetz, J. (2007). Meeting the challenges of sustainable development—A conceptual appraisal of a new methodology for participatory ecological planning. *Progress in Planning, 67*(1), 9–98. DOI: 10.1016/j.progress.2006.12.004

Vargas, C., Whelan, J., Brimblecombe, J., & Allendera, S. (2022). Co-creation, co-design and co-production for public health: A perspective on definitions and distinctions. *Public Health Research & Practice, 32*(2). DOI: 10.17061/phrp3222211 PMID: 35702744

Voorberg, W. H., Bekkers, V. J., & Tummers, L. G. (2015). A systematic review of co-creation and co-production: Embarking on the social innovation journey. *Public Management Review, 17*(9), 1333–1357. DOI: 10.1080/14719037.2014.930505

Webler, T., Kastenholz, H., & Renn, O. (1995). Public participation in impact assessment: A social learning perspective. *Environmental Impact Assessment Review, 15*(5), 443–463. DOI: 10.1016/0195-9255(95)00043-E

Webler, T., & Renn, O. (1995). A brief primer on participation: philosophy and practice. In *Fairness and competence in citizen participation: Evaluating models for environmental discourse* (pp. 17–33). Springer Netherlands. DOI: 10.1007/978-94-011-0131-8_2

Wijnhoven, F., Ehrenhard, M., & Kuhn, J. (2015). Open government objectives and participation motivations. *Government Information Quarterly, 32*(1), 30–42. DOI: 10.1016/j.giq.2014.10.002

Wilker, J., Rusche, K., & Rymsa-Fitschen, C. (2016). Improving participation in green infrastructure planning. *Planning Practice and Research, 31*(3), 229–249. DOI: 10.1080/02697459.2016.1158065

Williams, B. K., & Brown, E. D. (2012). *Adaptive management: the US Department of the Interior applications guide*. US Department of the Interior, Adaptive Management Working Group.

Chapter 15 The Role of Smart City Applications in Creating Sustainable and Climate-Responsive Cities in the EU: The Case of Amsterdam

Gulnar Bayramova
https://orcid.org/0009-0007-1149-0868
The Graduate School, Manisa, Turkey

Ahmet Uçar
Faculty of Economics and Administrative Sciences, Manisa, Turkey

ABSTRACT

The formation of sustainable and resilient cities has become a critical focus in modern urban governance. Rapid urbanization, infrastructure deficiencies, unplanned expansion, and social inequalities challenge the pursuit of sustainability in urban contexts. In response, smart city applications have emerged as effective instruments for addressing these issues. This chapter focuses on the Amsterdam Smart City (ASC) Program as a key case study aligned with EU sustainability goals. Using a literature review and qualitative content analysis, the study explores the impact of ASC's initiatives on environmental and social resilience. It assesses how ASC's multi-stakeholder governance supports sustainability and evaluates specific smart solutions, such as lighting and energy-efficient buildings, in achieving carbon reduction and circular urbanism goals.

DOI: 10.4018/979-8-3373-1310-8.ch015

INTRODUCTION

Today, cities are becoming overcrowded and growing in an unhealthy and unplanned manner due to rapid urbanization and intensive migration movements. This rapid population growth has brought with it serious challenges for urban management, adversely affecting economic and social life and leading to a significant decline in quality of life. At the same time, issues such as increasing energy consumption, carbon emissions and overuse of natural resources have made cities both vulnerable and influential actors against climate change. Under these circumstances, urban administrations have had to develop solutions that are not only sustainable but also climate-responsive. At this point, within the framework of the opportunities offered by developments in information and communication technologies, smart city applications are considered as one of the prominent solution models. These applications enable the more efficient, sustainable, and climate-sensitive management of urban infrastructure while also helping to reduce environmental impacts and enhance the resilience capacity of cities.

Smart cities stand out today as a technology-driven urban management approach that offers digital and data-based solutions to the complex problems faced by city administrations. Within this vision, data is collected, analyzed, and integrated into management processes across numerous domains—from transportation systems and infrastructure services to street and avenue information, as well as addressing and zoning data to participatory applications. This makes it possible to generate faster, more effective, and sustainable solutions for the needs of individuals living in the city. Smart cities also rely on an integrated technology system that brings together public institutions, private sector actors (both small and large-scale enterprises), financial institutions, and citizens in a digital environment. In this context, smart cities facilitate the digital execution of transactions between citizens and the state, thereby saving time and resources (Uçar, Şemşit & Negiz, 2017: 1786).

This transformation is not only a technological advance, but also a structural transformation process that aims to make cities more resilient in terms of energy efficiency, environmental sustainability and climatic resilience. In this context, the concept of 'smart city' represents a model in which digital technologies are used in an integrated manner with urban management and innovative approaches to solutions are developed. According to Gül and Atak Çobanoğlu (2017: 1544), a smart city is an urban structure where information technologies are actively utilized in municipal services, technological participation is ensured, and which possesses an environmental and innovative framework, while also being a setting where both individuals and institutions are involved in active learning and adaptation processes.

The smart city concept is not only a technology-driven transformation, but also includes a multi-layered urban governance model that supports sustainable development. Within this framework, various policy initiatives have been implemented at the European Union (EU) level as well as worldwide. Particularly as information and communication technologies gained strategic importance in economic systems, the EU has shifted towards a new understanding of development, formulating new strategic frameworks to both sustain economic growth and enhance societal participation through digitalization. This transformation process aims not only at the development of digital infrastructures but also at increasing the environmental and social resilience of cities.

The work on smart cities in the EU gained momentum with the Seventh Framework Program covering the years 2007-2013 and became concrete with the Europe 2020 Strategy published in 2010. This strategy prioritized innovation, competitiveness, and employment, aiming for stronger and sustainable growth against the economic crisis. European Innovation Partnerships (EIPs) facilitated the promotion of innovative solutions in sectors including energy, transportation, and digital technologies. In this context, programs like the "Smart Cities and Communities Innovation Partnership" (SCC) seek to diminish energy consumption and greenhouse gas emissions while enhancing quality of life (European Commission, 2012: 2).

In this context, the European Union has placed environmental sustainability and building resilient cities at the center of its policy agenda. The Union's goal of becoming a climate-neutral continent by 2050 raises the question of how smart cities can contribute to this vision. Accordingly, technological innovations, the active participation of citizens in decision-making processes, and innovative urban planning approaches are among the fundamental tools for the EU to achieve its sustainability targets. The "Manifesto for a New Urbanity," prepared in collaboration between the European Union and the Council of Europe, stands as a significant document that concretizes these objectives. The Manifesto emphasizes the continuous improvement of urban life, alongside priorities such as strengthening communication infrastructure, creating smart urban areas, and ensuring environmental sustainability (Gül & Atak Çobanoğlu, 2017: 1551).

Different practices and strategies in the member cities reflect the EU's vision of sustainable and smart cities. The City of Amsterdam is an important example of this, with its innovative Smart City strategies and projects aiming at environmental sustainability. The city has adopted a governance model that not only achieves its environmental goals by using information and communication technologies but also prioritizes the active participation of citizens in the process. This approach has enabled the realization of innovative projects in cooperation with the public, private sector, and civil society and has made Amsterdam a pioneer in the field of smart cities in the international arena. Continuous monitoring, evaluation, and open communication mechanisms have increased the effectiveness of the strategy and enabled rapid adaptation to the changing needs of the city.

This research was conducted using a qualitative research design based on a literature review. In this context, it aimed to understand the impacts of smart city applications in Amsterdam on sustainability and urban resilience. The study was carried out in three main stages.

First, through a literature review on smart cities, key themes such as smart city concepts, citizen participation, technology integration, and sustainability were addressed, and research gaps in this field were identified.

In the second stage, the European Union's sustainability goals were examined, and how smart cities contribute to these goals was analyzed. Within this framework, the EU's sustainability vision, which integrates strategies such as carbon emission reduction and the circular economy, was discussed in detail.

In the third stage, the case of Amsterdam was analyzed; the city's smart city strategies, projects, and governance models related to sustainability and digitalization were examined. Furthermore, the role of digital technologies in Amsterdam, citizen participation, and social impacts were also evaluated.

Through the example of Amsterdam, this study offers generalizable lessons for other cities within the European Union and provides significant recommendations for transforming smart city practices into more resilient and sustainable urban structures.

1. EXPLORING SMART CITIES

1.1. The Concept of the Smart City

The concept of "smart city" has garnered significant interest from both academic and practical perspectives due to the increasing urbanization and the challenges it entails in recent years. While there are various definitions, smart cities generally aim to improve urban living and operational efficiency through the integration of technology, data and governance.

The concept of smart cities first emerged in the 1960s with the term "cybernetically planned cities." During this period, the first theoretical discussions about the integration of technology into urban planning began. In the 1980s, the concept of "networked cities" was introduced, aiming for cities to become more connected and collaborative through digital networks and communication systems (Yimsek & Yakar, 2023:51). In the 1990s, an approach known as the "smart growth movement" was adopted, which emphasized the importance of carrying out urban growth in a more strategic and sustainable manner (Çetin & Çiftçi, 2019:135).

It is quite difficult to clearly define the concept of "smart city" in the literature (Kaygısız and Aydın, 2017: 56). Definitions of this concept focus on various perspectives by addressing the issue from different angles. However, it should not be forgotten that smart city definitions are constantly evolving as technological needs in cities change and develop on a daily basis (Alkan, 2015: 72). According to Gül and Çobanoğlu (2017: 1544), a smart city is a model in which information technologies are used in urban services, technology provides participation, and perpetually learning individuals and institutions support an eco-friendly structure. This dynamic nature of the smart city concept requires continuous research and adaptation to effectively meet the needs of urban populations. Especially, the integration of new technologies will play a critical role in shaping the urban fabric of the future.

In this context, some definitions specifically emphasize the fundamental role of Information and Communication Technologies (ICTs) in the integration of city-wide services. For example, Washburn and Sindhu (2010: 2) define smart cities as follows:

> *"The use of ICT [makes] the critical infrastructure components and services of a city—which include city administration, education, healthcare, public safety, real estate, transportation, and utilities—more intelligent, interconnected, and efficient."*

On the other hand, there are approaches that offer a more holistic perspective. Caragliu, Del Bo, and Nijkamp (2011: 50) provide an example of this definition:

> *"a city may be called 'Smart' when investments in human and social capital and traditional and modern communication infrastructure fuel sustainable economic growth and a high quality of life, with a wise management of natural resources, through participatory governance."*

These definitions indicate that smart cities are shaped not only by technological infrastructures but also by multi-dimensional elements such as social capital, sustainable development, and participatory governance.

Within this context, attention has also been drawn to the ideological and structural dynamics underlying the emergence of the concept. Vanolo (2014: 888) noted that the concept of smart cities arose from the combination of "smart city" and "smart growth" discourses. In this context, it is seen that information and communication technologies support urban growth and reshape the structure of cities. Additionally, it is stated that multinational companies play an important role in the widespread adoption of this concept. On the other hand, Yılmaz (2015: 4-5) approached the concept of smart cities in a simpler way and linked it to the modernization process. According to him, smart cities are a modernization effort that enables cities to use their resources more efficiently and provide better services to urban residents.

Today, smart cities aim to achieve sustainable development goals and improve quality of life by integrating technology and innovative approaches. In this process, it is envisioned to go beyond traditional governance models and manage physical, social, and digital systems in an integrated manner. Smart cities aim to systematically predict current and future problems, integrate digital innovation with physical planning, and establish effective collaborations among different stakeholders (Akıllı Şehir Nedir).

Smart cities are next-generation cities based on the integration of technology. This integration is achieved through technologies such as the Internet of Things (IoT), Cloud Computing, and Geographic Information Systems (GIS). These technologies aim to enhance the quality of life, support environmental sustainability, and create innovative solutions. In this process, information and communication technologies support urban growth, reshaping the city structure (Akıllı Şehir Nedir, n.d.).

The European Parliament highlights the need for new and innovative methods to manage the complexities and challenges faced by cities due to increasing urbanization. These methods aim to address critical issues such as problems arising from overcrowding, energy consumption, resource management, and environmental protection. In this context, smart cities emerge not only as a strategy to manage the urban life of the future in a more innovative way but also as a key strategy for addressing issues such as poverty, inequality, unemployment, and energy management (European Parliament ,2014: 17).

At the core of smart cities lies the integration of information and communication technologies with human and social capital to achieve larger and more sustainable economic growth and enhance the quality of life. Smart cities are shaped around six main axes: Smart Economy, Smart Mobility, Smart Governance, Smart Environment, Smart Living, and Smart Citizens (European Parliament, 2014: 18). These axes are methods used to make the development of cities more sustainable and livable. It is expected that smart cities, by offering services across a wide range from digital infrastructure to social services, will ensure that cities operate more efficiently and sustainably.

1.2. Core Dimensions of Smart Cities

Smart cities aim to make urban life more efficient and sustainable by using digital technologies and infrastructures. However, they are not just limited to technological infrastructure. Elements such as citizen participation, sustainability, and governance also play a significant role in the success of these cities. The effective integration of these elements is the key to ensuring that smart cities are efficient, sustainable, and livable.

1.2.1. Citizen Participation in Smart Cities

Citizen participation in smart cities is one of the most important components of successful implementations. Studies indicate that citizen engagement is crucial for the success of smart city initiatives. For instance, Sontiwanich et al. (2022)

emphasize that the demographic profile and public knowledge significantly influence citizen engagement in smart city projects, highlighting the need for inclusive governance structures that empower citizens. Similarly, Xu and Zhu (2020) argue that citizen satisfaction is foundational to smart city development, suggesting that effective participation mechanisms can bridge the gap between citizen expectations and project outcomes.

However, the literature also reveals contradictions regarding the nature of citizen participation. While some studies advocate for a bottom-up approach that prioritizes citizen input, others highlight the prevalence of top-down initiatives that may marginalize citizen voices. For example, Johnson et al. (2020) categorize citizen engagement practices into traditional civic engagement and digital transactions, noting that the latter often lacks genuine citizen involvement. This dichotomy raises questions about the effectiveness of current engagement strategies and the extent to which they truly reflect citizen needs.

Moreover, Cardullo and Kitchin introduce the concept of a *"scaffold of smart citizen participation,"* which critiques the technocratic framing of citizen roles in smart cities (Cardullo & Kitchin, 2018:2). Their analysis of Dublin's smart city initiatives underscores the necessity of rethinking citizen participation frameworks to ensure they are genuinely inclusive and responsive to community needs.

1.2.2. Technological Frameworks in Smart Cities

The technological infrastructure of smart cities involves the integration of digital technologies to make the functioning of cities more efficient. Kitchin (2013) discusses the pervasive nature of digital technologies in urban environments, describing smart cities as composed of *"everyware"* (Greenfield, 2006:9) that facilitates real-time monitoring and management of urban processes. This technological infrastructure is often positioned as a solution to various urban challenges, from traffic management to energy efficiency.

However, the reliance on technology raises concerns about the digital divide and the potential exclusion of marginalized communities. Shin et al. (2021) highlight that advanced smart city innovations can exacerbate existing inequalities if not implemented with a focus on inclusivity. This concern is echoed by Ismagilova et al. (2020) who argue that data privacy and security issues must be addressed to foster trust and engagement among citizens. The challenge lies in balancing technological advancement with ethical considerations and equitable access.

Furthermore, the methodologies employed in studying smart city technologies vary widely. Case studies, such as Pla-Castells et al.'s (2015) examination of the traffic management system in Valencia, provide in-depth insights into specific implementations. In contrast, theoretical approaches, like those of Joss et al. (2017), focus on analyzing the broader implications of smart technologies on urban governance. This methodological diversity enriches the discourse but also highlights the need for standardized frameworks to evaluate smart city initiatives comprehensively.

1.2.3. Sustainability and Smart Cities

Sustainability is a recurring theme in smart city literature, with many studies linking smart city initiatives to broader sustainability goals. Luque-Vega et al. (2020) emphasize the importance of shifting smart city strategies towards more holistic approaches, as seen in their SCOPPE methodology, which integrates citizen happiness and participation into urban planning. This perspective aligns with the Sustainable Development Goals (SDGs), which emphasize the importance of inclusive and sustainable urbanization.

However, the literature reveals a tension between the concepts of smart cities and sustainability. Ahvenniemi et al. (2017) note that while there is a growing recognition of the need for sustainable practices, the linkage between smart city initiatives and sustainability outcomes remains underexplored. This gap suggests that future research should focus on developing frameworks that explicitly connect smart city strategies with sustainability metrics.

Additionally, the role of local communities in driving sustainable smart city initiatives is highlighted by Preston et al. (2020) who emphasize the untapped potential of citizen engagement in achieving energy efficiency and sustainability. This underscores the importance of fostering collaborative governance models that empower citizens to contribute to sustainability efforts actively.

1.2.4. Governance Models in Smart Cities

Governance models play a crucial role in shaping the implementation and success of smart city initiatives. The literature indicates a shift towards more collaborative governance structures that prioritize citizen engagement and co-creation. For instance, Park and Fujii (2023) discuss the importance of civic engagement in developing smart city solutions, emphasizing the need for participatory governance frameworks. This aligns with the findings of Bayerl (2023), who argues that public acceptance of integrated technological solutions is contingent upon transparent governance practices.

However, the effectiveness of these governance models is often contingent upon the political and institutional contexts in which they operate. As highlighted by Lim et al. (2018) while many smart cities claim to prioritize citizen-centric approaches, the reality often reflects a more transactional relationship between citizens and local authorities. This discrepancy raises questions about the authenticity of citizen engagement in smart city governance and the potential for technocratic governance to undermine democratic processes.

In summary, the future of smart cities is closely related to the social and environmental impacts of technology. Next-generation smart city projects should use technology not only to provide a more efficient way of life but also to ensure that societies develop in a more just, sustainable, and inclusive manner. In the future, to achieve a more equitable and ethical structure, new methods must be developed to strengthen citizen participation in these cities.

2. THE ROLE OF SMART CITY POLICIES FOR A SUSTAINABLE EUROPE

Smart cities represent a comprehensive approach that combines technological innovations, sustainable development principles, and a participatory governance model, aiming to make urban life more livable, efficient, and inclusive. The rapid growth of cities intensifies the pressure on infrastructure and necessitates the development of new solutions that prioritize energy and resource efficiency. According to the projection by the United Nations Conference on Trade and Development (UNCTAD, 2017), 70% of the world's population will reside in cities by 2050, which makes smart city policies not just a choice but a necessity.

In parallel with these global trends, the European Union (EU) has positioned smart city approaches as one of the main policy instruments in achieving sustainability and climate resilience goals. Especially within the scope of the Seventh Framework Program conducted between 2007 and 2013, significant investments were made in this field. Furthermore, smart city projects received concrete support through the European Innovation Partnerships initiated by the European Commission in 2010 (Uçar, Şemşit & Negiz, 2017: 1789).

The European Green Deal, announced in December 2019, institutionalized the EU's sustainability vision. With the aim of making Europe the first climate-neutral continent by 2050, the European Green Deal foresees major changes in areas such as reducing greenhouse gas emissions, speeding up the switch to renewable energy, and using resources in a circular way (Delivering the European Green Deal, 2021). The key to this transformation is restructuring cities according to sustainable urban planning principles.

The European Green Deal is aligned with the UN Sustainable Development Goals, which aim to address a range of environmental, social, and economic challenges around the world. The European Green Deal stresses the importance of climate action and includes other sustainability aspects such as biodiversity conservation, resource efficiency, and social inclusion. The EU's commitment to the Sustainable

Development Goals is evident in its policies to create a circular economy, which aims to minimize waste and reuse and recycle resources. This approach is particularly relevant for EU agricultural policies, which are being restructured to ensure sustainable food systems in line with the European Green Deal and the Sustainable Development Goals (United Nations Department of Economic and Social Affairs, n.d.).

On the other hand, green infrastructure and nature-based solutions are another key pillar of the EU's vision for sustainable cities. These approaches are vital for urban planning and design, facilitating the creation of healthy and resilient landscapes and communities. The EU Green Infrastructure Strategy and the EU Climate Change Adaptation Strategy emphasize the need to integrate ecological principles in urban development and aim to ensure that cities are resilient to environmental pressures such as extreme weather events and biodiversity loss. By integrating green spaces and promoting biodiversity in urban environments, the EU aims to both improve the quality of life of its citizens and address climate-related challenges (Tóth, 2022: 1-2).

The implementation of the European Green Deal is expected to bring major changes in the EU economy. It aims to create new employment opportunities in green sectors, make energy use more efficient, and encourage investments in renewable energy technologies (Jäger-Waldau, Kougias, Taylorand & Thiel, 2020: 5).

But transitioning to a sustainable economy is not without challenges. As the economies and environments of EU member states are different, they need different measures to ensure that all areas can effectively contribute to and benefit from the European Green Deal. Moreover, geopolitical tensions, such as the COVID-19 pandemic and the war in Ukraine, have made this process even more difficult and highlighted the need for a policy framework that is resilient and adaptable to external shocks (Szczubełek, 2022: 175).

The EU's commitment to sustainability also includes its economic policies, especially through the Circular Economy Action Plan. This plan aims to contribute to a more sustainable economic model by promoting waste minimization and resource reuse. The transition to a circular economy is important for urban resilience because it encourages the smart use of resources and makes urban activities less harmful to the environment. This commitment is further reinforced by the EU's emphasis on sustainable public procurement, which aims to create a market for sustainable products and services by aligning public spending with environmental objectives (European Commission, n.d.).

Moreover, smart city approaches are seen to play a strategic role in achieving the European Union's sustainability and urban resilience goals. European cities are in a pioneering position in applying developments in information and communication technologies (ICTs) in urban management, as they are the centers of early industrialization. Europe's high level of economic development has enabled faster adoption of these technological transformations at the local level, and cities such as Paris, Copenhagen, Barcelona, London, Berlin, and Amsterdam have become prominent examples of smart city applications. While each of these cities has the basic elements of smart city architecture in general, they have also developed differentiated strategic approaches in line with their unique social, cultural, and institutional structures (Gül and Çobanoğlu, 2017: 1552).

Smart city applications offer digital solutions for more efficient use of resources, reduction of greenhouse gas emissions, and more efficient operation of existing urban infrastructures. In this framework, developments in areas such as the integration of transport systems, improving waste and water management, and promoting participatory governance models are directly in line with the European Union's vision of sustainable urbanization. The "Smart Cities Marketplace" initiative, supported by the European Commission, strengthens cooperation between cities, the private sector, and other stakeholders, enabling the dissemination of smart solutions developed in areas such as energy efficiency, circular economy, and sustainable transport in different cities.

In this context, the Europe 2020 Strategy emphasized that smart and sustainable cities will make a significant contribution to the continent's long-term development goals. The development of next-generation digital infrastructures and the application of innovative technologies in different areas of urban life not only ensure the creation of environmentally friendly and livable cities but also support economic growth and encourage the active participation of citizens in decision-making processes. Such policies not only serve to solve problems at the local level but also contribute to the creation of a culture of "good practice" based on knowledge sharing and cooperation across Europe (European Parliament, 2014: 59-60).

The European Union's sustainability vision represents a comprehensive approach that aims to build resilient and liveable cities while tackling environmental crises. By blending digital innovations with green infrastructure and participatory governance, smart city policies play a critical role in the goal of making cities climate-neutral by 2050. This transformation process offers a model not only for Europe but also for sustainable urbanization efforts on a global scale.

3. AMSTERDAM: SMART CITY STRATEGY AND APPLICATIONS

Amsterdam, like other cities in Europe, has faced the main challenges of the 21st century, such as urbanization, energy efficiency, environmental sustainability, and digitalization. In this context, the smart city concept has been adopted to improve the quality of urban life and address these challenges. Amsterdam's smart city strategy has become a model for cities in the transformation process, making it a leading city in this field. This strategy reveals the necessary steps to create a successful smart city and how these steps are organized (Mora & Bolici, 2017: 252).

3.1. Development of Amsterdam's Smart City Strategy

3.1.1 Phase One: Initiating the Process

Amsterdam's smart city strategy was launched in 2007 as the "Amsterdam Smart City Program." This initiative was developed through collaboration between the Amsterdam Innovation Motor (AIM), the Liander energy network operator, and the Amsterdam Municipality. AIM, established in 2006, aims to support Amsterdam's development in the information society. Liander is part of Alliander, the largest energy company in the Netherlands, which is responsible for building, maintaining, and managing energy networks for the distribution of natural gas and electricity across the country. These three organizations formed various working groups and launched the smart city strategy (Mora and Bolici, 2017: 254-255).

This initiative is based on the potential of information and communication technologies (ICT) to improve the functioning of cities and is backed by a strong political commitment. The belief that technology can play a key role in solving environmental problems and creating a sustainable urban environment is the motivation that has shaped this strategy. The smart city strategy provided the opportunity to achieve Amsterdam's strategic goals faster, and these policies have remained stable despite changes in the municipal administration (Mora and Bolici, 2017: 255).

3.1.2 Phase Two: Planning

In the planning phase of Amsterdam's smart city strategy, the strategy was focused on addressing environmental issues with the goal of significantly reducing carbon emissions, in line with the city's priorities, and aligning with various strategies across Europe. During this process, a strategy was adopted that is in harmony with the New Amsterdam Climate Program, which aims to reduce CO2 emissions by 40% by 2025. The goals of the strategy include reducing energy waste and carbon emissions, promoting sustainable economic growth, and encouraging citizens to adopt more sustainable lifestyles. Four core principles were adopted in the development of the strategy: community effort, economic feasibility, technology push and demand pull, and knowledge dissemination. Furthermore, the Living Lab methodology was used to engage citizens in projects and encourage behavioral change. To ensure the correct implementation of the projects, the Amsterdam Smart City Foundation was established, and various working groups were formed. External consultants, such as Accenture, and independent research institutes, like TNO, collaborated to ensure the success of the strategy. A clear procedure for the selection, production, and monitoring of projects was developed (Mora & Bolici, 2017: 256-257).

3.1.3 Phase Three: Project Development

Amsterdam's smart city strategy is based on continuous ICT-based projects that bring new applications, services, devices, and technological infrastructures to the city. The Amsterdam Smart City Foundation has played a central role in coordinating and effectively implementing these projects. Projects are evaluated primarily based on criteria such as feasibility, cost, and CO2 reduction potential, and the most suitable partners are determined for the approved projects. The project development process progresses to an execution stage, where the roles of partners are clarified, and the foundation monitors and supports the entire project. Between 2009 and 2011, 16 projects were completed, and today, the number has exceeded 70. More than 160 organizations, including various network operators, public institutions, construction companies, universities, financial institutions, and telecommunications and ICT companies, are actively supporting these projects (Mora & Bolici, 2017: 258-259).

3.1.4 Phase Four: Monitoring and Evaluation

Progress monitoring and outcome evaluation in Amsterdam's smart city strategy are periodically carried out through collaboration between the Amsterdam Smart City Foundation and project partners. Value cases that assess energy savings and CO2 emission reduction potential for each project are included in a report published alongside the project's results. This report provides a comprehensive evaluation of the progress toward the overall goals of the strategy. Additionally, the strategy is managed with a dynamic approach, and over time, areas have been expanded through continuous review and improvement (Mora & Bolici, 2017: 259-260).

3.1.5 Phase Five: Communication

In Amsterdam, all knowledge and experiences gained are widely shared. Knowledge sharing is carried out continuously and horizontally by the Amsterdam Smart City Foundation starting from the beginning of the third stage. The aim here is not just promotion but also to encourage the formation of new collaborations. Conferences are one of the main communication tools for spreading information about the strategy and promoting the work done. The Amsterdam Smart City Foundation has participated in more than 50 national and international conferences. These events provide in-depth information on topics such as the goals, priorities, strategic principles, financial strategy, activity plans, and results achieved through projects (Mora & Bolici, 2017: 260).

This exploratory study shows that Amsterdam is a pioneer in the smart city movement, and its success stems from an approach based on strategic urban planning principles. While shaping the smart city strategy, the city followed a path based on strategic thinking, collaboration, and inclusive criteria, organizing its activities within a strategic framework. The lesson to be drawn from the Amsterdam example is that strategic urban planning is an effective tool for managing the process of cities becoming smart environments (Mora & Bolici, 2017: 261).

3.2. Amsterdam's Smart City Governance Model and Innovation Ecosystem

The success of Amsterdam's smart city strategy is based not only on technological infrastructure but also on the development of governance and innovation ecosystems. Governance in the Amsterdam Smart City (ASC) model varies across different levels and projects. At the program level, the ASC platform presents a horizontal management structure where private sector organizations play a leading role but collaborate with public and information organizations. This governance model, while connected to more formal rules and evaluation systems among the actors in complex projects, does not follow a centralized decision-making process. Instead, public actors in the projects are partners in collaboration, not decision-makers. These flexible structures allow for various citizen participation models, enabling citizens to play an active role as project initiators in socially innovative participation. Thus, the ASC governance model fosters strong collaboration and citizen participation both strategically and operationally (Capra, 2016: 33).

This governance model also forms the foundation of Amsterdam's innovation ecosystem. Amsterdam Smart City (ASC) functions as a "project factory" that implements projects aimed at increasing sustainability and quality of life. This platform brings together various stakeholders to initiate important projects. The Amsterdam Economic Board and cultural platforms like Pakhuis de Zwijger also contribute to the innovation ecosystem. The success of the projects relies on the establishment of clear goals, user participation, and viable business models. However, the scalability of solutions and the need for systematic learning across projects are other key aspects. This process strengthens Amsterdam's innovation capacity, making the city a more resilient and sustainable smart city (Van Winden et al., 2016: 16-18).

Amsterdam's innovation system is based on a platform with a restructured innovation framework, unlike traditional models, consisting of roles and relationships among actors. In this system, nine different categories of actors establish dynamic, one-way, or two-way relationships at various innovation stages, based on their interests and resources. The use of an open web-based platform enables transparent and fair information sharing among actors, allowing a broader range of participants to be involved in the system. However, although the innovation field is geographically limited to the Amsterdam urban area, more research is needed to understand how different types of digital technologies and their usage forms could influence the organization of this system. Facilitating dynamic actor interactions through an open web-based platform presents a new way to develop innovation systems in urban areas (Putra & Knaap, 2018: 247).

3.3. Main Projects and Innovations in Amsterdam

Amsterdam is a global leader in shaping sustainable urban solutions and the smart city concept by implementing innovative projects and technologies aimed at a sustainable future. This leadership stems from various innovation centers and projects across the city. For example, the Amsterdam Institute for Metropolitan Solutions (AMS) conducts research in areas such as urban water, energy, waste, food, and data management, developing smart city applications and establishing Amsterdam as a hub for data-driven urban solutions. These efforts also aim to cultivate future smart city professionals by offering master's programs (Akıllı Şehir, n.d.).

Additionally, the Amsterdam ArenA Innovation Center focuses on developing innovative solutions related to smart city and smart stadium concepts, improving customer experience, security, and sustainability. Similarly, the Smart Citizens Lab brings together citizens, scientists, and designers to contribute to data collection and awareness-raising activities about the city through open platforms. Furthermore, the Vehicle-to-Grid (V2G) technology in Amsterdam allows electric vehicles to supply power to the grid, optimizing the storage and use of renewable energy. The GridFriends project aims to promote renewable energy sharing in residential projects in Amsterdam, with the goal of making the region energy-neutral. The city also offers an extensive transportation network with 800 kilometers of bicycle lanes and over 200,000 parking spaces to encourage bicycle use (Akıllı Şehir, n.d.).

These projects and innovations are concrete examples of Amsterdam's smart city vision. For instance, the city's smart lighting poles collect data on traffic, air pollution, and parking spaces through Wi-Fi and sensors, adjusting brightness based on environmental conditions to save energy (Akıllı Şehir, n.d.). Such projects support the city's sustainability goals.

Another significant project is the IoT Living Lab. This initiative, which develops interactive solutions supported by the Internet of Things (IoT), aims to improve the quality of life in the city using open data and developer-friendly platforms. Since 2015, many sectoral solutions have been developed through workshops, hackathons, and pilot projects. The IoT Living Lab also helps other cities establish open IoT infrastructures and aims to create social impact through a new initiative called the City Innovation Exchange Lab (CITIXL) (Van der Veen, 2016).

On the other hand, the Circular Amsterdam Platform aims to implement circular economy principles in the city, promoting the reuse of natural resources and minimizing waste. The goal is to reduce the use of new raw materials by half by 2030 and create a fully circular economy by 2050. In this regard, Amsterdam has defined more than 70 actions for the 2023-2026 period, allocating over 14 million euros for these projects (City of Amsterdam).

Finally, the City-Zen project aims to realize a fossil-free city vision in Amsterdam and Grenoble. This EU-funded program seeks to overcome the technical, economic, and social barriers encountered during the integration of innovative energy solutions. Research on how clean energy sources such as solar, wind, biomass, and geothermal energy can be utilized in urban life is contributing to a new energy vision for the city (Somers, 2016).

CONCLUSION

As European Union (EU) cities contend with the compounding pressures of rapid urbanization and climate change, the limitations of traditional urban management have become starkly clear. This study has argued that smart city applications offer a necessary evolution, representing a new strategic paradigm for urban development. By systematically integrating technological innovation with sustainability and climate adaptation goals, this approach directly confronts the complex challenges of unplanned growth, resource scarcity, and social inequality, moving beyond the insufficient responses of conventional models.

In this context, the European Union, in line with its vision of becoming a climate-neutral continent by 2050, has adopted smart city applications as key tools to create sustainable and resilient cities through strategic initiatives such as the European Green Deal, Circular Economy Action Plan, and Smart Cities Marketplace. Smart city applications aim to optimize resource use, reduce carbon emissions, and increase energy efficiency through digital technologies, while strengthening urban resilience against the negative impacts of climate change through the integration of green infrastructure and nature-based solutions. This strategic approach encompasses not only environmental goals but also social inclusion and economic transformation, thereby presenting an integrated sustainability vision.

The EU's holistic approach emphasizes the importance of considering smart city applications as a tool for sustainability. However, to achieve these goals, not only technological investments but also social impacts need to be taken into account. The success of smart cities depends not only on the digitalization of infrastructures but also on policies that promote active citizen participation in decision-making processes, transparency, and social justice.

Amsterdam is a successful example of the EU's smart city vision. The Smart City Programme, launched in 2007, is designed to reduce carbon emissions, improve energy efficiency, and balance economic growth with environmental sustainability. It aims for a 40% CO_2 reduction by 2025, in line with the New Amsterdam Climate Program. Projects include circular economy, renewable energy, smart transportation, and data-based infrastructure management. The Amsterdam Smart City Foundation encourages citizen participation through the Living Lab, while the capacity to adapt to climate change is enhanced through technologies such as IoT and artificial intelligence.

Amsterdam's innovative governance model offers an effective coordination and participation mechanism between the private sector, public institutions, and citizens through its decentralized and horizontal collaborative structure. This model supports a vision of sustainability that strengthens social inclusion and complements technology and people-centered practices. The innovation ecosystem in the city stands out as a platform where different actors establish dynamic relationships and share information transparently, thus facilitating the scalability and dissemination of applications.

Concrete applications in Amsterdam provide vivid examples of how the smart city vision is aligned with environmental and social goals. Smart lighting systems, vehicle-to-grid energy technology, open data platforms, and sustainable transportation infrastructure are all examples of projects that contribute to reducing carbon emissions while improving energy efficiency. Circular economy initiatives focus on waste reduction and natural resource conservation, supporting the efficient use of resources. Furthermore, programs such as City-Zen, supported by the European Union, aim to build cities less dependent on fossil fuels by speeding up the integration of renewable energy sources into city life.

All these developments show how Amsterdam has integrated both mitigation and adaptation strategies with smart city practices in the face of climate change. In this context, increasing social participation and making governance structures flexible and inclusive are as critical as strengthening technological infrastructures in tackling the environmental challenges that cities face. By achieving this balance, the Amsterdam model sets an example for urban transformation in line with both sustainability and climate sensitivity.

For future work, a comprehensive analysis of the impacts of Amsterdam's smart city projects on long-term sustainability and climate resilience is needed. In particular, the opportunities and risks of digital technologies (IoT, artificial intelligence, and big data analytics) in the context of climate adaptation and social equity should be examined in detail. Furthermore, measuring the tangible contribution of circular economy and renewable energy practices on urban carbon footprint and resource efficiency is crucial for the EU's sustainability goals. Studies on the effects of large-scale policy frameworks, such as the European Green Deal, on the governance processes of smart city initiatives and their adaptability across cities will improve the understanding and development of multi-layered governance dynamics.

In conclusion, the Amsterdam example and the EU's all-encompassing strategies reveal that smart city applications play a necessary role in creating sustainable and climate-sensitive urban systems. These approaches enable cities not only to improve their technological infrastructure, but also to build balanced and inclusive models along the axes of social justice, economic vitality and environmental resilience. This comprehensive model developed in Europe is an important reference point to guide cities globally in adapting to the climate crisis and realizing sustainable development.

REFERENCES

Ahvenniemi, H., Huovila, A., Pinto-Seppä, I., & Airaksinen, M. (2017). What are the differences between sustainable and smart cities? *Cities (London, England), 60*, 234–245. DOI: 10.1016/j.cities.2016.09.009

Akıllı şehir nedir? (2024). Akıllı Şehirler Portalı. Erişim tarihi: 18 Kasım 2024, https://www.akillisehirler.gov.tr/akilli-sehir-nedir/

Akıllı şehir. (2024). Amsterdam Akıllı Şehir. Erişim tarihi: 1 Aralık 2024, https://www.akillisehir.com/idet/4/31/amsterdam

Alkan, T. (2015). Akıllı kentler ya da 21. yüzyıl şehirleri. *Bilişim Dergisi, 182*, 70–77.

Bayerl, P. S. (2023). Securing the smart city: Patterns of public acceptance for integrated technological solutions. *Proceedings of the International Smart City Conference (ISC)*. DOI: 10.1109/ISC257844.2023.10293633

Capra, C. F. (2016). The smart city and its citizens. *International Journal of E-Planning Research*, *5*(1), 20–38. DOI: 10.4018/IJEPR.2016010102

Caragliu, A., Del Bo, C., & Nijkamp, P. (2009). Smart cities in Europe. *3rd Central European Conference in Regional Science—CERS, 2009*, 45–59.

Cardullo, P., & Kitchin, R. (2019). Being a 'citizen' in the smart city: Up and down the scaffold of smart citizen participation in Dublin, Ireland. *GeoJournal*, *84*(1), 1–13. DOI: 10.1007/s10708-018-9845-8

Çetin, M., & Çiftçi, Ç. (2019). Literatüre göre dünya ve ülkemizden örneklerle akıllı kent kavramının irdelenmesi. *Ulusal Çevre Bilimleri Araştırma Dergisi*, *2*(3), 134–143.

City of Amsterdam. (n.d.). *Circular economy*. Erişim tarihi: 24 Kasım 2024, https://www.amsterdam.nl/en/policy/sustainability/circular-economy/

Dinar Wahidayat Putra, Z., & Van Der Knaap, W. G. M. (2018). Urban innovation system and the role of an open web-based platform: The case of Amsterdam smart city. *Journal of Regional and City Planning*, *29*(3), 234–249. DOI: 10.5614/jrcp.2018.29.3.4

European Commission. (2012). Smart cities and communities- European innovation partnership. Accessed date: 19 October 2024, https://eur-lex.europa.eu/legal-content/EN/TXT/?uri=CELEX:52012DC0470

European Commission. (2021). Delivering the European Green Deal. https://commission.europa.eu/strategy-and-policy/priorities-2019-2024/european-green-deal/delivering-european-green-deal_en

European Commission. (n.d.). Circular economy action plan Accessed date: 29 November 2024, https://environment.ec.europa.eu/strategy/circular-economy-action-plan_en

European Parliament. (2014). Mapping smart cities in the EU. Accessed date 15 November 2024, https://www.europarl.europa.eu/RegData/etudes/etudes/join/2014/507480/IPOL-ITRE_ET(2014)507480_EN.pdf

Greenfield, A. (2006). *Everyware: The dawning age of ubiquitous computing*. New Riders.

Gül, A., & Atak Çobanoğlu, Ş. (2017). Avrupa'da akıllı kent uygulamalarının değerlendirilmesi ve Çanakkale'nin akıllı kente dönüşümünün analizi. *Süleyman Demirel Üniversitesi İktisadi ve İdari Bilimler Fakültesi Dergisi, 22*(Kayfor 15 Özel Sayısı), 1543–1565.

Ismagilova, E., Slade, E. L., & Kitchin, R. (2020). Security, privacy, and risks within smart cities: Literature review and development of a smart city interaction framework. *Information Systems Frontiers*, *22*, 243–264. PMID: 32837262

Jäger-Waldau, A., Kougias, I., Taylor, N., & Thiel, C. (2020). How photovoltaics can contribute to GHG emission reductions of 55% in the EU by 2030. *Renewable & Sustainable Energy Reviews*, *126*, 109836. DOI: 10.1016/j.rser.2020.109836

Johnson, C. L., Bozeman, B., & He, Z. (2020). Canadian smart cities: Are we wiring new citizen-local government interactions? *The Canadian Geographer. Geographe Canadien*, *64*(4), 548–561. DOI: 10.1111/cag.12623

Joss, S., Cook, M., & Dayot, Y. (2017). Smart cities: Towards a new citizenship regime? A discourse analysis of the British smart city standard. *Journal of Urban Technology*, *24*(4), 29–49. DOI: 10.1080/10630732.2017.1336027

Kalkınma Bakanlığı, T.C. (2015). *Bilgi toplumu stratejisi ve eylem planı (2015-2018).*

Kaygısız, Ü., & Aydın, S. Z. (2017). Yönetişimde yeni bir ufuk olarak akıllı kentler. *Mehmet Akif Ersoy Üniversitesi Sosyal Bilimler Enstitüsü Dergisi*, *9*(18), 56–81.

Kitchin, R. (2014). The real-time city? Big data and smart urbanism. *GeoJournal*, *79*(1), 1–14. DOI: 10.1007/s10708-013-9516-8

Lim, Y., & Lee, H. (2018). Citizen participation in building citizen-centric smart cities. *Geografia: Malaysian Journal of Society and Space*, *14*(4), 112–127. DOI: 10.17576/geo-2018-1404-04

Luque-Vega, L. F., Michel-Villarreal, R., & Lopez-Neri, E. (2020). Smart cities oriented project planning and evaluation methodology driven by citizen perception—IoT smart mobility case. *Sustainability (Basel)*, *12*(15), 6214. DOI: 10.3390/su12177088

Manville, C., Cochrane, G., Cave, J., Millard, J., Pederson, J. K., Thaarup, R. K., Liebe, A., Wissner, M., Massink, R., & Kotterink, B. (2014). *Mapping smart cities in the EU*. [Rapor]. European Parliament, Policy Department A: Economic and Scientific Policy. https://www.europarl.europa.eu/RegData/etudes/etudes/join/2014/507480/IPOL-ITRE_ET(2014)507480_EN.pdf

Mora, L., & Bolici, R. (2017). How to become a smart city: Learning from Amsterdam. A. Bisello, D. Vettorato, R. Stephens, & P. Elisei (Ed.), *Smart and sustainable planning for cities and regions: Results of SSPCR 2015* (ss. 251–266). Springer.

Park, Y., & Fujii, H. (2023). Civic engagement in a citizen-led living lab for smart cities: Evidence from South Korea. *Urban Planning*, *8*(2), 269–282. DOI: 10.17645/up.v8i2.6361

Pla-Castells, M., Laredo, J. L. G., & Pujol, A. V. (2015). Use of ICT in smart cities: A practical case applied to traffic management in the city of Valencia. *Proceedings of the International Conference on Smart Cities and Smart Infrastructure*. DOI: 10.1109/SCSP.2015.7181559

Preston, J., Waugh, L., & Jones, T. (2020). Citizen engagement for co-creating low carbon smart cities: Practical lessons from Nottingham City Council in the UK. *Energies*, *13*(24), 6615. DOI: 10.3390/en13246615

Somers, E. (2016). *City-zen*. Amsterdam Smart City. Erişim tarihi: 2 Aralık 2024, https://amsterdamsmartcity.com/updates/project/city-zen

Sontiwanich, S., Tuan, A., & Chaisawat, M. (2022). An unsustainable smart city: Lessons from uneven citizen education and engagement in Thailand. *Sustainability (Basel)*, *14*(20), 13315. DOI: 10.3390/su142013315

Szczubełek, G. (2022). Perspectives for energy development in Poland and in the European Union in the context of the European Green Deal: Renewable energy and energy efficiency. *Olsztyn Economic Journal*, *17*(2), 175–190. DOI: 10.31648/oej.8934

Tóth, A. (2022). Planning and designing green infrastructure across landscapes and scales. *Acta Horticulturae et Regiotecturae*, *25*(1), 1–7. DOI: 10.2478/ahr-2022-0001

Uçar, A., Şemşit, S., & Negiz, N. (2017). Avrupa Birliği akıllı kent uygulamaları ve Türkiye'deki yansımaları. *Süleyman Demirel Üniversitesi İktisadi ve İdari Bilimler Fakültesi Dergisi, 22*(Kayfor 15 Özel Sayısı), 1785–1798.

United Nations Conference on Trade and Development [UNCTAD]. (2017). *Handbook of statistics*. https://unctad.org/system/files/official-document/tdstat42_en.pdf

United Nations Department of Economic and Social Affairs. (2024). *The 17 goals*. Sustainable Development. Erişim tarihi: 20 Kasım. https://sdgs.un.org/goals

Van der Veen, E. (2016, 15 Ocak). *IoT living lab*. Amsterdam Smart City. Erişim tarihi: 25 Kasım 2024, https://amsterdamsmartcity.com/updates/project/iot-living-lab

Van Winden, W., Oskam, I., van den Buuse, D., Schrama, W., & Van Dijck, E. J. (2016). *Organising smart city projects: Lessons from Amsterdam*. Hogeschool van Amsterdam.

Vanolo, A. (2014). Smartmentality: The smart city as disciplinary strategy. *Urban Studies (Edinburgh, Scotland), 51*(5), 883–898. DOI: 10.1177/0042098013494427

Washburn, D., Sindhu, U., Balaouras, S., Dines, R. A., Hayes, N., & Nelson, L. E. (2010). *Helping CIOs understand "smart city" initiatives*. Forrester Research. http://public.dhe.ibm.com/partnerworld/pub/smb/smarterplanet/forr_help_cios_und_smart_city_initiatives.pdf

Xu, W., & Zhu, Q. (2021). Evaluating the impact mechanism of citizen participation on citizen satisfaction in a smart city. *Environment and Planning. B, Urban Analytics and City Science, 48*(8), 2419–2436. DOI: 10.1177/2399808320980746

Yimsek, F. S., & Yakar, M. (2023). A general glimpse into the smart cities. *Türkiye Arazi Yönetimi Dergisi, 5*(1), 49–56. DOI: 10.51765/tayod.1270504

Chapter 16
Climate Change and Urban Poverty Solutions to Prevent Poverty

Emre Cengiz
https://orcid.org/0000-0002-3854-6215
Gumushane University, Turkey

ABSTRACT

Cities are at the centre of economic, social and political life. As a result, cities are constantly growing and their populations are rapidly increasing. As is well known, references are made to the industrial revolution when defining today's cities. With the industrial revolution, cities have become major centres for rural-to-urban migration. With this wave of migration, the construction of informal structures in cities and the struggle for survival of new urban poor in these settlements can be observed. This process is continually observable in developed, developing and underdeveloped countries at different times and to varying degrees. As a result of cities themselves becoming commodified in a world continually impacted by climate change, the urban poor's struggle to survive has been intensified by factors such as increasing rent prices, reduced access to basic services, diminishing access to food and clean water, and settlement instability.

INTRODUCTION

The concept of poverty is a concept that has existed throughout human history. From tribal life to urban antecedents, from ancient cities to today's post-modern cities, we can see poverty emerging in urban space in every period. The factors of poverty of each period may be different. Especially after the industrial revolution,

DOI: 10.4018/979-8-3373-1310-8.ch016

the population of cities started to increase dramatically. As a result of the increase in job, social, etc. opportunities offered by cities, cities have become centres of attraction. However, not all people migrating to cities could have these opportunities. However, cities cannot provide the desired opportunities to the people who migrate and poverty emerges. Since the industrial revolution, classes and their living spaces have continued to differentiate and segregate sharply within the urban space. Cities have become places where prosperity and misery are experienced by different classes at the same time.

We can see that the concept of 'sustainability' has become important in the studies of both the UN and non-governmental organisations after the environmental awareness started to be formed after the middle of the twentieth century, when the damages caused by humans to the environment became visible over time and started to be negatively affected by this situation. The concept of 'sustainability', which is important for the continuity of humanity and the universe, has been discussed through various meetings and conferences and has clearly stated the responsibilities of state administrations and individuals in terms of the rights of future generations and what needs to be done.

Since the late 20th century, rapidly changing climate and increasing natural disasters have started to cause a unique challenge for the world and humanity. The increase in greenhouse gases, the occurrence of heat waves, rising sea levels due to the melting of glaciers, irregularities in climate systems (Cengiz& Uğur, 2024:1) cause problems for human life.

After the effects of global warming and climate change, which is one of the important fields of study in recent periods, have started to be felt in urban space, we can see that the segments struggling to live in cities are more affected. In this study, solution proposals for preventing the increasing urban poverty after climate change will be discussed and the proposals will be analysed.

1. URBAN POVERTY

After the industrial revolution, people started to migrate to industrial cities. After the second half of the 20th century, we can see that the number of people living in cities started to increase rapidly. After 1980, globalisation and neo-liberal economic policies that started to take effect in the world accelerated the urbanisation process even more. As a result, when we look at the 21st century world, the number of people living in cities has become more than the number of people living in rural areas. While approximately 1 billion people lived in cities in 1960, we can see that more than 4.6 billion people live in cities in 2023. Figure 1 shows the increase in the urban population from 1960 to 2023. The urban population increased by 1.82%

from 2019 to 2020, by 1.57% from 2020 to 2021, and by 1.65% from 2022 to 2023, reaching a total urban population of 4,614,425,176 people.

Figure 1. Urban population

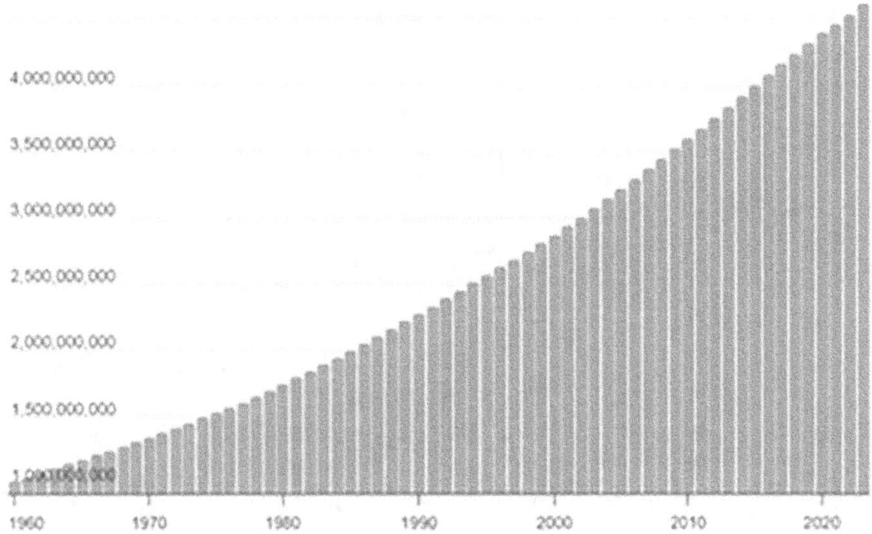

Source: *World Urban Population, https://www.macrotrends.net/global-metrics/countries/wld/world/urban-population#:~:text=World%20urban%20population%20for%202023,a%201.82%25%20increase%20from%202019.*

In light of this information, Figure 2 shows the continent-based urban population from 1950 to 2020. When the figure is analysed, it can be seen that the urban population rates of the African and Asian continents have increased approximately 3 times. Considering the 70-year period, this represents a large urban population increase. Considering the high number of underdeveloped and developing countries in both continents, it can be said that urban problems and urban poverty are experienced intensely. In addition, in Latin America and the Caribbean, the urban population has doubled in the last 70 years. Again, the intensity of urban poverty and urban problems can be mentioned in this continent. In Europe and North America, which are the pioneers of industrialisation and development, the urban population has increased by approximately 20% in 70 years. Although we cannot talk about the intensity of urban poverty and urban problems due to rapid urbanisation in the last 70 years in these two continents, which have already completed their industrialisation and urbanisation, we can say that urban poverty and urban problems are on the agenda with different dimensions.

Figure 2. Continent-based urban population (1950-2020)

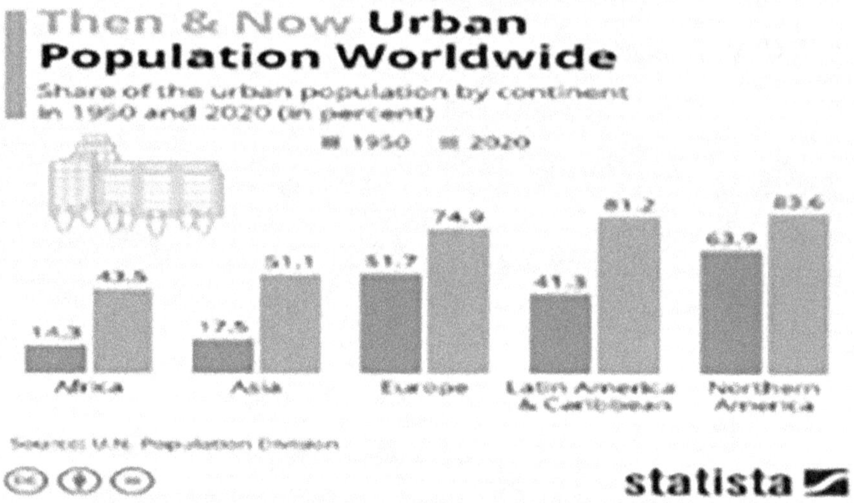

Source: *Katharina Buchholz (2020). Then & Now: Urban Population Worldwide, https://www.statista.com/chart/23349/share-of-urban-population-by-continent/*

Ever since the industrial revolution, urban poverty has been experienced in different dimensions on urban space. Workers migrating to industrialised cities have become/attempted to become the new inhabitants of the city and have struggled to hold on to the city in the spaces/housing assigned to them in the urban space. The misery and poverty experienced in cities where planned and unplanned industrialisation programmes have been implemented show different dimensions. In unplanned industrialised cities, workers, the new inhabitants of the cities, took care of themselves in terms of housing. In planned industrialised cities, on the other hand, they continued their struggle to hold on to the city in designated spaces and dwellings.

Although there are urban spatial differences in planned or unplanned industrialised cities, both have been negatively affected by global warming and climate change. Cities became uninhabitable as a result of environmental problems becoming more evident and especially the increase in air pollution (1940s and 1950s). After the 1970s, there have been discussions on the fact that there is a close relationship between the environment and the economy and that there should be an absolute harmony between environmental protection and development in order to eliminate the environmental hazards that arise as a result of this relationship. It is difficult to say that the necessary practices have been implemented until the 1990s (Uğur&Cengiz,2023:3). This situation has led to the destruction of the environment and cities, which are an important part of the environment.

Figure 3. New urban dwellers after the industrial revolution

Source: *Jacob Riis. Yard in Jersey Street (now gone) Where Italians Live in the Worst Slums, before 1898. Modern gelatin silver printing out paper. Museum of the City of New York. Gift of Roger William Riis https://brewminate.com/jacob-riis-and-how-the-other-half-lives-poverty-in-19th-century-america/#google_vignette*

The concept of poverty is generally defined in economic terms. It refers to access to the calories required for the individual to continue his/her daily life. In other words, the poverty line means that the individual has the financial power / income level to purchase food at the minimum calorie level determined in order to continue his/her life. It is worth noting that the poverty line is different for each country, but the poverty line for each country is expressed in US dollars (Jandhyala & Tilak, 2002: 191).

In the press release titled 'statement of Commitment for action to eradicate poverty adopted by administrative committee on coordination' published by the UN in 1998, poverty is defined as follows:: "fundamentally, poverty is a denial of choices and opportunities, it is a violation of human dignity. It means lack of basic capacity to participate effectively in society. It means not having enough to feed and clothe a family, not having a school or a clinic to go to, not having the land on which to grow one's food or a job to earn one's living, nor having access to credit. It means insecurity, powerlessness and exclusion of individuals, households and communities. It means susceptibility to violence and it often implies living on marginal and fragile environments, not having access to clean water and sanitation." (UN, 1998)

Bassett (2009) state that poverty is a universal and global problem. Stating that the numbers and percentages of poor citizens are different, he states that no country in the world has been able to eliminate poverty. Although everyone wants to eradicate poverty, there are differences of opinion about the method to eradicate poverty. He states that personal responsibility, state intervention, free market, religious and social institutions have a role in preventing poverty.

For many years, poverty was generally expressed as a 'natural' condition. Poverty, which emerged as a result of rapid urbanisation after the industrial revolution, first became a social and political problem in the cities of England, and with the spread of industry in Europe and the USA, it became a social and political problem in industrialised places/cities. The economic and social transformations experienced for about half a century have paved the way for the emergence of new dimensions of poverty and impoverishment in almost all cities of the world (Benassi & Morlicchio, 2019:1).

Urban poverty differs according to the development levels of countries. In underdeveloped countries, the poverty rate in both rural and urban areas is high. In undeveloped countries, the poverty rate in rural areas and rural areas is higher than the poverty rate in urban areas. In developing countries, improvements in urban areas, transport and infrastructure works are increasing. However, the poor in urban areas have higher living costs. The cost of living in rural areas is relatively lower. In general, 75 per cent of the world's poor continue to live in rural areas (Ravellion, 2007:16).

There is a positive relationship between the increase in the rate of urbanisation and the reduction of poverty. It should not be forgotten that it is an indisputable reality that new urban problems emerge in cities where both poverty and urbanisation rates are high (Ravellion, 2007:16). In urban areas with high rates of poverty and rapid urbanisation, urbanisation problems also arise. Poverty and wealth are sharply separated in urban space. It is possible to see the urban space differentiated according to income groups. The segregation in urban space can sometimes be seen with a road and sometimes with a wall. While one side can benefit from both urban services and the advantages offered by the city, the other side has problems in accessing urban services and cannot benefit from the advantages offered by the city. Figure 4 clearly demonstrates this situation.

Figure 4. Inequality from a bird's eye view by Johnny Miller

Source: *Unequal Scenes, https://www.millefoto.com/unequalscenes*

2. EFFECTS OF CLIMATE CHANGE ON CITY AND URBAN LIFE

Urban areas are becoming increasingly dangerous places for urban dwellers due to climate change. In order to understand the effects of climate change, which threatens urban areas and urban dwellers, it is necessary to indicate to what extent urban functions, which are at the centre of human life, are affected. When we look at the direct effects of climate change on the lives of urban dwellers; poverty, food and water scarcity, negative impacts on livelihoods and one of the most important ones is the forced displacement of urban dwellers. These direct impacts affect urban dwellers socially, economically and psychologically (Gasper et al, 2011:150).

Today, the environmental problems faced by developing countries are more serious than ever before. In fact, the concept of 'brown agenda' has started to be used to emphasise these environmental problems. The brown agenda refers to inadequate water resources and sanitation, solid waste disposal problems, management problems of both urban and industrial wastes and air pollution. These problems generally appear as problems related to poverty (UN-Habitat, 2010:165).

Since natural disasters and climate change cause increasing problems on cities, the following environmental targets have been developed for cities (UN-Habitat, 2010:165):

- Protection and improvement of environmental health
- Protection of water, soil and air quality from pollution and contamination
- Reducing the impact of urban areas on natural resources at regional and global scales
- Prevention and mitigation of the impacts of natural disasters and climate change on urban areas

Table 1 shows the economic and social impacts of climate change on cities.

Table 1. Social and economic impacts of climate change in cities

Social Impacts		Economic Impacts	
Health effects	Climate change may cause sudden and permanent negative effects on the physical and psychological health of urban dwellers. Severe weather events, extreme temperatures and infectious diseases can be given as examples.	**Industry**	Both infrastructure and superstructure are adversely affected due to natural events/ disasters caused by climate change. Industrial activities will be adversely affected and production will be damaged.
Food and water scarcity	Food resources may be at risk as a result of decreasing and increasing precipitation. Floods and water scarcity may have negative effects on agricultural areas.	**Retail and commercial services**	Climate change will negatively affect transport infrastructure, workforce and supply chain of products.
Livelihood impacts	Climate change will make it more difficult for urban dwellers to make a living. Especially the poor will have to compromise on both nutrition and social expenditures.	**Tourism**	Climate change will negatively affect tourism seasons and tourism destinations. Especially coastal areas will be more affected by this situation.
Displacement	In recent years, it is stated in researches that climate change has effects on migration. Especially the poor are more affected by this situation.	**Insurance**	The insurance sector may be vulnerable to increasing natural disasters due to climate change. In addition, increasing property values in places where natural disasters are likely to occur will also put the insurance sector in a difficult situation.
Distributional impacts within and among cities	It is known that both gender and age are important in terms of being affected by climate change. Women or men, young or old are affected by climate change in different dimensions.		

Source: Rebecca Gasper, Andrew Blohm, Matthias Ruth (2011). Social and economic impacts of climate change on the urban environment, Current Opinion in Environmental Sustainability, Volume 3, Issue 3, May 2011, Pages 150-157 (created by the author from this article)

3. CLIMATE CHANGE, URBAN POVERTY AND SOLUTIONS

The poor, and especially the poor living in high-risk areas, are more vulnerable to the impacts of climate change. In general, the poor are more dependent on climate-sensitive resources such as water and food (Wilbanks et al, 2007: 359). Therefore, the poor have difficulties to continue their lives with the effects of climate change.

In cities where climate change is experienced and its effects are beginning to be seen, it is important to produce policies to meet the needs of the poor. It is important to provide basic services in addition to action plans where climate change hazards are tried to be minimised (Hague et al, 2014:117-118).

The effects of climate change differ according to the location of the cities. In addition, the poor living in urban areas are also affected differently according to the cities they live in. Table 2 shows the effects of climate change according to the characteristics of the cities and the problems that the poor may face.

Table 2. Climate change, urban space and the urban poor

Urban space	Affected by climate change	Urban poor
Coastal Cities	Sea level rises in coastal cities. Coastal erosion and floods are observed.	Fishing and tourism will suffer. cities and agricultural areas may be flooded and people living there will lose their jobs and habitats and will find it difficult to continue their lives.
Dryland Cities	Water scarcity is observed in dryland cities. Sandstorms become more frequent. Poor infrastructure can worsen.	Access to water is difficult. Long water queues are formed by urban dwellers who want to obtain water from water sources. Access to water is even more difficult in poorer areas. Boreholes cannot be drilled or existing boreholes cannot be restarted when they fail.
Inland and High Altitude Cities	Rainfall patterns are changing.	Floods adversely affect all areas of life by damaging infrastructure, superstructure and cultivated agricultural areas.

Source: Climate Cahnge and The Urban Poor (2009). IEDD (The table was created by the author from the mentioned study.)

Policies should be developed to protect urban dwellers, especially the urban poor, living in cities affected by climate change. It is important who the actors and stakeholders will be while formulating policies. Central government, local government, local people, non-governmental organisations, private sector and academicians should act together in the process and decisions should be taken by emphasising the importance of participation. The possible effects of climate change on the city should be investigated and a preparatory study should be carried out on what the possibilities may be. Policies to be implemented for the preparatory work should be determined. While determining the policies, studies for the urban poor, who are more vulnerable to climate change, should be detailed. When the policies are implemented, a unit should be established to investigate the effectiveness and efficiency of the policies. In this way, the failing aspects of the policies can be identified and problems can be eliminated with rapid intervention.

Figure 5. Climate change, urban poverty and solution proposal scheme

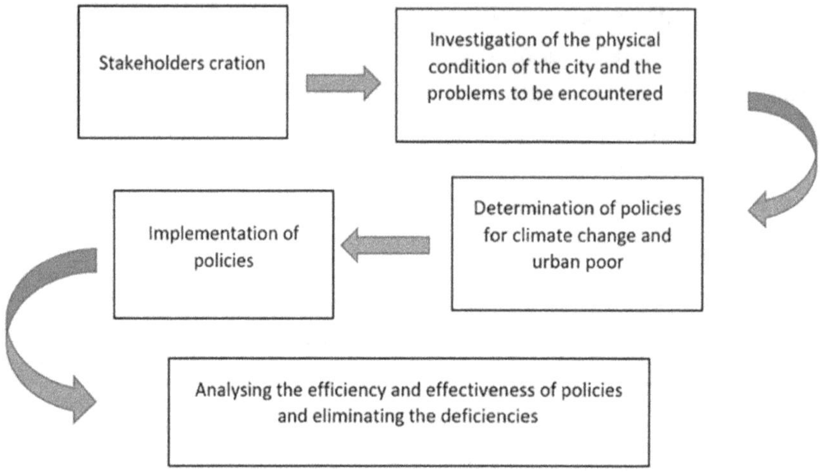

Source: *Created by the author*

CONCLUSION

Since climate change exists in our lives as a reality, people need to adapt to this situation. In other words, the joint struggle of central governments, local governments, city administrations, people living in urban and rural areas, non-governmental organisations and all experts working in this field is important in this respect.

Poverty is a concept that has existed from the first human communities living on earth to today's post-modern world. The reason why poverty is an important field of study in today's world is the rapid growth of cities after the industrial revolution, the increase in the rate of urbanisation and the consequent increase in both vital and environmental problems. Especially since 1971, when the world's self-renewal rate was exceeded, urban spaces, where a large part of the world population lives, have become more challenging spaces in terms of human life.

After 1980, the globalised world and the neo-liberal economic policies that started to be implemented have caused cities to become more prominent. Cities have ceased to be places where commodities are produced and have turned into commodities. In this situation, which David Harvey calls the second cycle of capital, capital has started to transform cities (For detailed information, see Harvey, 1985). Both the negative effects of climate change and capital's efforts to transform cities have left cities with insurmountable problems. This process is even more difficult for the urban poor who struggle to survive in cities. The urban poor, who struggle to survive in the city, face a life far away from the advantages offered by the city and basic urban services.

Undoubtedly, all people living in cities are affected by the impacts of climate change. However, the urban poor, as a disadvantaged group, are more vulnerable in terms of access to food and clean water. Therefore, while combating climate change, this situation of the urban poor should be taken into account and policies should be developed for the urban poor.

REFERENCES

Bassett, D. L. (2009). Global ruralism. Journal of Gender, Race & Justice. University of Alabama Public Law Research Paper No. 1346991.

Benassi, D., & Morlicchio, E. (2019). New urban poverty. In Orum, A. (Ed.), *The Wiley Blackwell encyclopedia of urban and regional studies*. John Wiley & Sons Ltd., DOI: 10.1002/9781118568446.eurs0466

Buchholz, K. (2020). Then & now: Urban population worldwide. https://www.statista.com/chart/23349/share-of-urban-population-by-continent/

Cengiz, E., & Uğur, Ö. (2024). İklim değişikliği ve doğal afetler. In Cengiz, E., & Uğur, Ö. (Eds.), *Doğal afetler ve iklim değişikliği: Politik süreçler, kurumlar ve uygulamalar* (pp. 1–12). Nobel Bilimsel Eserler.

Gasper, R., Blohm, A., & Ruth, M. (2011). Social and economic impacts of climate change on the urban environment. *Current Opinion in Environmental Sustainability, 3*(3), 150–157. DOI: 10.1016/j.cosust.2010.12.009

Haque, A. N., Dodman, D., & Hossain, M. M. (2014). Individual, communal and institutional responses to climate change by low-income households in Khulna, Bangladesh. *Environment and Urbanization, 26*(1), 112–129. DOI: 10.1177/0956247813518681

Harvey, D. (1985). *Urbanisation of capital.* Blackwell.

Jandhyala, B., & Tilak, G. (2002). Education and poverty. *Journal of Human Development, 3*(2).

Macrotrends. (n.d.). World urban population. https://www.macrotrends.net/global-metrics/countries/wld/world/urban-population

Ravallion, M. (2007). *Urban poverty. Finance & Development.* International Monetary Fund.

Uğur, Ö., & Cengiz, E. (2023). İklim değişikliği siyaseti. In Uğur, Ö., & Cengiz, E. (Eds.), *İklim değişikliği siyaseti: Riskler, fırsatlar ve yeni vizyonlar* (pp. 1–12). Nobel Bilimsel Eserler.

UN-Habitat. (2010). The state of Asian cities 2010/11.

Unequal Scenes. (n.d.). https://www.millefoto.com/unequalscenes

United Nations. (n.d.). Statement of commitment for action to eradicate poverty adopted by Administrative Committee on Coordination [Press release ECOSOC/5759].

Wilbanks, T. J., Romero Lankao, P., Bao, M., Berkhout, F., Cairncross, S., Ceron, J.-P., Kapshe, M., Muir-Wood, R., & Zapata-Marti, R. (2007). Industry, settlement and society. In Parry, M. L., Canziani, O. F., Palutikof, J. P., van der Linden, P. J., & Hanson, C. E. (Eds.), *Climate change 2007: Impacts, adaptation and vulnerability. Contribution of Working Group II to the Fourth Assessment Report of the Intergovernmental Panel on Climate Change*. Cambridge University Press.

Compilation of References

Abbas, J. (2020). Impact of total quality management on corporate green performance through the mediating role of corporate social responsibility. *Journal of Cleaner Production*, *242*, 118458. DOI: 10.1016/j.jclepro.2019.118458

Abbass, K., Qasim, M. Z., Song, H., Murshed, M., Mahmood, H., & Younis, I. (2022). A review of the global climate change impacts, adaptation, and sustainable mitigation measures. *Environmental Science and Pollution Research International*, *29*(28), 42539–42559. DOI: 10.1007/s11356-022-19718-6 PMID: 35378646

Adams, H., Adger, W. N., Ahmad, S., Ahmed, A., Begum, D., Matthews, Z., Rahman, M. M., Nilsen, K., Gurney, G. G., & Streatfield, P. K. (2020). Multi-dimensional well-being associated with economic dependence on ecosystem services in deltaic social-ecological systems of Bangladesh. *Regional Environmental Change*, *20*(42), 2–16. DOI: 10.1007/s10113-020-01620-x

Adger, W. N. (2003). *Social aspects of adaptive capacity*. Climate Change, Adaptive Capacity and Development, 29-49. DOI: 10.1142/9781860945816_0003

Adger, W. N., Dessai, S., Goulden, M., Hulme, M., Lorenzoni, I., Nelson, D. R., Naess, L. O., Wolf, J., & Wreford, A. (2009). Are there social limits to adaptation to climate change? *Climatic Change*, *93*(3-4), 335–354. DOI: 10.1007/s10584-008-9520-z

Adger, W. N., Hughes, T. P., Folke, C., Carpenter, S. R., & Rockström, J. (2009). Social-ecological resilience to coastal disasters. *Science*, *309*(5737), 1036–1039. DOI: 10.1126/science.1112122 PMID: 16099974

Aggarwal, K., Doshi, R., Mijwil, M. M., Hiran, K. K., Gök, M., & Bala, I. (2024). A Simple Way to Comprehend the Difference and the Significance of Artificial Intelligence in Agriculture. *Intelligent Techniques for Predictive Data Analytics*, 189-201.

Agrawal, A. (2008). *The role of local institutions in adaptation to climate change.* Social Dimensions of Climate Change: Equity and Vulnerability in a Warming World, 173-198. https://hdl.handle.net/10986/28274

Agrawal, R., Agrawal, S., Samadhiya, A., Kumar, A., Luthra, S., & Jain, V. (2024). Adoption of green finance and green innovation for achieving circularity: An exploratory review and future directions. *Geoscience Frontiers*, *15*(4), 101669. DOI: 10.1016/j.gsf.2023.101669

Ahsan, M. N., & Warner, J. (2014). The socioeconomic vulnerability index: A pragmatic approach for assessing climate change led risks-A case study in the south-western coastal Bangladesh. *International Journal of Disaster Risk Reduction*, *8*, 32–49. DOI: 10.1016/j.ijdrr.2013.12.009

Ahvenniemi, H., Huovila, A., Pinto-Seppä, I., & Airaksinen, M. (2017). What are the differences between sustainable and smart cities? *Cities (London, England)*, *60*, 234–245. DOI: 10.1016/j.cities.2016.09.009

Ajide, K. B., & Ibrahim, R. L. (2022). Environmental impacts of income inequality: Evidence from G7 economies. *Environmental Science and Pollution Research International*, *29*(2), 1887–1908. DOI: 10.1007/s11356-021-15720-6 PMID: 34363162

Akıllı şehir nedir? (2024). Akıllı Şehirler Portalı. Erişim tarihi: 18 Kasım 2024, https://www.akillisehirler.gov.tr/akilli-sehir-nedir/

Akıllı şehir. (2024). Amsterdam Akıllı Şehir. Erişim tarihi: 1 Aralık 2024, https://www.akillisehir.com/idet/4/31/amsterdam

Akın, G. (2006). Küresel Isınma, Nedenleri ve Sonuçları. *Ankara Üniversitesi Dil ve Tarih-Coğrafya Fakültesi Dergisi*, *46*(2), 29–43. DOI: 10.1501/Dtcfder_0000000673

Al Mahmud, R., Titumir, M., Afrin, T., Islam, M. S., Titumir, R. A. M., Afrin, T., Islam, M. S., Onneshan, U., & Dhaka, B. (2020). *Traditional Knowledge, Institutions and Human Sociality in Sustainable Use and Conservation of Biodiversity of the Sundarbans of Bangladesh.* Managing Socio-ecological Production Landscapes and Seascapes for Sustainable Communities in Asia. Springer (pp. 67–91).

Alam, M., & Bhardwaj, S. (2020). *Temporal variation in glacier's area and identification of glacial lakes in Sikkim.* DOI: 10.1007/978-981-15-2097-6_8

Alam, A. (2012). Some Indian bryophytes known for their biologically active compounds. *International Journal of Applied Biology and Pharmaceutical Technology*, *3*(2), 239–246.

Alamgir, M., & Cheng, M. C. (2023). Do green bonds play a role in achieving sustainability? *Sustainability (Basel), 15*(13), 10177. DOI: 10.3390/su151310177

Ali, A. M. (2006). Rice to shrimp: Land use land cover changes and soil degradation in Southwestern Bangladesh. *Land Use Policy, 23*(4), 421–435. DOI: 10.1016/j.landusepol.2005.02.001

Ali, B. (2020). Community participation and quality of life in naturebased tourism: Exploring the antecedents and moderators. *Journal of Hospitality & Tourism Research (Washington, D.C.), 20*(10), 1–32. DOI: 10.1177/1096348020980094

Alkan, T. (2015). Akıllı kentler ya da 21. yüzyıl şehirleri. *Bilişim Dergisi, 182*, 70–77.

Allen, C. R., Fontaine, J. J., Pope, K. L., & Garmestani, A. S. (2011). Adaptive management for a turbulent future. *Journal of Environmental Management, 92*(5), 1339–1345. DOI: 10.1016/j.jenvman.2010.11.019 PMID: 21168260

Allen, C., Metternicht, G., & Wiedmann, T. (2018). Initial progress in implementing the Sustainable Development Goals (SDGs): A review of evidence from countries. *Sustainability Science, 13*(5), 1453–1467. DOI: 10.1007/s11625-018-0572-3

Almansi, F. (2010). *The relationship between disaster risks and urban planning in Argentina,* Case study prepared for ISDR Global Assessment Report 2011.

Almansi, F., & Hardoy, J. (2013). *Climate change impact in coastal areas of the Rio de la Plata River: Actions for prevention and adaptation*. DOI: 10.4324/9781315658735

Almansi, F., Hardoy, J., & Pandiella, G. (2014). Impacto del cambio climático en ciudades costeras del estuario del Rio de la Plata. *Medio Ambiente y Urbanización, 79*, 217–237. DOI: 10.1080/17565529.2014.918868

Almeida, D. V., Kolinjivadi, V., Ferrando, T., Roy, B., Herrera, H., Gonçalves, M. V., & Van Hecken, G. (2023). The "greening" of empire: The European Green Deal as the EU first agenda. *Political Geography, 105*, 102925. DOI: 10.1016/j.polgeo.2023.102925

Amer, M., & Daim, T. U. (2010). Application of technology roadmaps for the renewable energy sector. *Technological Forecasting and Social Change, 77*(8), 1355–1370. DOI: 10.1016/j.techfore.2010.05.002

Andrews, J. C., Durvasula, S., & Akhter, S. H. (1990). A framework for conceptualizing and measuring the involvement construct in advertising research. *Journal of Advertising, 19*(4), 27–40. DOI: 10.1080/00913367.1990.10673198

Andrioli, W. (2023). EU's ESG Regulatory Framework: a Study of SFDR.

Anguelovski, I., & Carmin, J. (2011). Something borrowed, everything new – current opinion: Innovation and institutionalisation in urban climate governance. *Current Opinion in Environmental Sustainability, 3*(3), 169–175. DOI: 10.1016/j.cosust.2010.12.017

Anjanappa, J. (2024). Role of Private Sector in Driving the Green Bond Market in India. *Available at SSRN* 4959091. DOI: 10.2139/ssrn.4959091

Anurag Danda, A., Ghosh, N., Bandyopadhyay, J., & Hazra, S. (2019). Managed retreat: Adaptation to climate change in the Sundarbans ecoregion in the Bengal Delta. *Journal of the Indian Ocean Region, 15*(3), 317–335. DOI: 10.1080/19480881.2019.1652974

Anwar, G., Arianto, W., Simarmata, M., Bertham, Y. H., Yuwana, Y., Romeida, A., Ruslan, D. (2023). Floristic Composition and Species Richness of Mangrove Ecosystems in Pantai Panjang–Pulau Baai Nature Park, Bengkulu Province, Indonesia. *International Journal of Advancement in Life Sciences Research*, 6(3), 10-18. 10-18. DOI: 10.31632/ijalsr.2023.v06i03

Armitage, D. (2005). Adaptive capacity and community-based natural resource management. *Environmental Management, 35*(6), 703–715. DOI: 10.1007/s00267-004-0076-z PMID: 15940398

Armitage, D., Berkes, F., Dale, A., Kocho-Schellenberg, E., & Patton, E. (2011). Co-management and the co-production of knowledge: Learning to adapt in Canada's Arctic. *Global Environmental Change, 21*(3), 995–1004. DOI: 10.1016/j.gloenvcha.2011.04.006

Armitage, D., Plummer, R., Berkes, F., Arthur, R. I., Charles, A. T., Davidson-Hunt, I. J., Diduck, A. P., Doubleday, N. C., Johnson, D. S., Marschke, M., McConney, P., Pinkerton, E. W., & Wollenberg, E. K. (2009). Adaptive co-management for social-ecological complexity. *Frontiers in Ecology and the Environment, 7*(2), 95–102. DOI: 10.1890/070089

Arnstein, S. R. (1969). A ladder of citizen participation. *Journal of the American Institute of Planners, 35*(4), 216–224. DOI: 10.1080/01944366908977225

Ascoli, F. D. (1921). *A revenue history of the Sundarbans from 1870 to 1920.* Bengal Secretariat Book Depot, Calcutta. https://cslrepository.nvli.in//handle/123456789/7362

Aspuru-Guzik, A., Cao, Y., & Romero, J. (2022). Quantum computing for the environmental sciences. *Nature Climate Change, 12*(3), 204–211.

Atılgan, D. (2025). Çevre vergisi karbon emisyonlarını azaltmakta mı? Panel eşbütünleşme analizi. *Türk Tarim ve Doga Bilimleri Dergisi, 12*(1), 145–156. DOI: 10.30910/turkjans.1586719

Auenweide. (2024). Auenweide website: https://www.auenweide.at/

Auzepy, A., Bannier, C. E., & Martin, F. (2023). Are sustainability-linked loans designed to effectively incentivize corporate sustainability? A framework for review. *Financial Management, 52*(4), 643–675. DOI: 10.1111/fima.12437

Aydın, D. (2013). Ülkeler Arası CO2 Emisyonun Azaltılması Yönde Etkinlik Çalışması, Muğla Sıtkı Koçman Üniversitesi Sosyal Bilimler Enstitüsü Dergisi, 30, Bahar, 120-135.

Ayeb-Karlsson, S., van der Geest, K., Ahmed, I., Huq, S., & Warner, K. (2016). A people-centred perspective on climate change, environmental stress, and livelihood resilience in Bangladesh. *Sustainability Science, 11*(4), 679–694. DOI: 10.1007/s11625-016-0379-z PMID: 30174739

Ayers, J. (2011). Resolving the adaptation paradox: Exploring the potential for deliberative adaption policy-making in Bangladesh. *Global Environmental Politics, 11*(1), 62–88. DOI: 10.1162/GLEP_a_00043

Ayers, J., & Forsyth, T. (2009). Community-based adaptation to climate change, strengthening resilience through development. *Environment, 51*(4), 23–31. DOI: 10.3200/ENV.51.4.22-31

Ayers, J., Huq, S., Forer, R., Forsyth, T., Vaughan, C., & Visman, E. (2014). Ensuring climate resilient poverty alleviation. In Ensor, J., Berger, R., & Huq, S. (Eds.), *Community-based adaptation to climate change: Emerging lessons* (pp. 145–168). Practical Action Publishing., DOI: 10.3362/9781780448749.007

Ayers, J., Kaur, N., & Anderson, S. (2017). Negotiating climate resilience in Nepal. In Filho, W. L. (Ed.), *Climate change adaptation in Pacific countries* (pp. 115–132). Springer., DOI: 10.1007/978-3-319-50094-2_7

Ayvaz, E. E. (2025). Çevre vergileri, hava taşımacılığı ve CO_2 emisyonları ilişkisi: Avrupa ülkeleri üzerine bir çalışma. *Erzurum Teknik Üniversitesi Sosyal Bilimler Enstitüsü Dergisi, 21*(21), 15–33. DOI: 10.29157/etusbed.1436615

Aziz, M. A., Tollington, S., Barlow, A., Goodrich, J., Shamsuddoha, M., Islam, M. A., & Groombridge, J. J. (2017). Investigating patterns of tiger and prey poaching in the Bangladesh Sundarbans: Implications for improved management. *Global Ecology and Conservation, 9*, 70–81. DOI: 10.1016/j.gecco.2016.12.001

Bäckstrand, K., & Lövbrand, E. (2019). The road to Paris: Negotiating a stronger climate agreement. *Journal of Environmental Policy and Planning*, *21*(5), 519–532. DOI: 10.1080/1523908X.2019.1603774

Balcilar, M., Ozdemir, Z. A., & Arslanturk, Y. (2010). Economic growth and energy consumption causal nexus viewed through a bootstrap rolling window. *Energy Economics*, *32*(6), 1398–1410. DOI: 10.1016/j.eneco.2010.05.015

Bartkowiak, K. M. (2023). The European Union emission trading system and its role for green budgeting development — The case of EU member states. *Current Opinion in Environmental Sustainability*, *65*, 101390. DOI: 10.1016/j.cosust.2023.101390

Bassett, D. L. (2009). Global ruralism. Journal of Gender, Race & Justice. University of Alabama Public Law Research Paper No. 1346991.

Baxter, S., Barnes, A., Lee, C., Mead, R., & Clowes, M. (2023). Increasing public participation and influence in local decision-making to address social determinants of health: A systematic review examining initiatives and theories. *Local Government Studies*, *49*(5), 861–887. DOI: 10.1080/03003930.2022.2081551

Bayat, T., Ozsalman, E., & Derindağ, O. F. (2025). Evaluating export vulnerability through import demand elasticity in carbon border adjustment contexts: A focus on Türkiye. *Environmental Science and Pollution Research International*, *32*(5), 2238–2250. DOI: 10.1007/s11356-024-35861-8 PMID: 39762524

Bayerl, P. S. (2023). Securing the smart city: Patterns of public acceptance for integrated technological solutions. *Proceedings of the International Smart City Conference (ISC)*. DOI: 10.1109/ISC257844.2023.10293633

Bayer, P., & Aklin, M. (2020). The European Union emissions trading system reduced CO2 emissions despite low prices. *Proceedings of the National Academy of Sciences of the United States of America*, *117*(16), 8804–8812. DOI: 10.1073/pnas.1918128117 PMID: 32253304

BCBS, A. (2021). Climate-related financial risks–Measurement methodologies.

Becker, W., Norlen, H., Dijkstra, L., & Athanasoglou, S. (2020). Wrapping up the Europe 2020 strategy: A multidimensional indicator analysis. *Environmental and Sustainability Indicators*, *8*, 100075. DOI: 10.1016/j.indic.2020.100075

Beck, M., Rivers, N., Wigle, R., & Yonezawa, H. (2015). Carbon tax and revenue recycling: Impacts on households in British Columbia. *Resource and Energy Economics*, *41*, 40–69. DOI: 10.1016/j.reseneeco.2015.04.005

Beerbaum, D. (2021). Green Quadriga?-EU-Taxonomy, TCFD, Non-Financial-Reporting Directive and EBA ESG Pillar III/IFRS Foundation. *IFRS Foundation (April 11, 2021).*

Begum, F., Lobry De Bruyn, L., Amirul Islam, M., & Kristiansen, P. (2022). *The Role of Women's Participation in Co-Management of the Sundarban Mangrove Forest of Bangladesh.* University of New England. Retrieved from https://hdl.handle.net/1959.11/55652

Begum, R. A., Sarkar, M., Jaafar, A. H., & Pereira, J. J. (2014). Toward conceptual frameworks for linking disaster risk reduction and climate change adaptation. *International Journal of Disaster Risk Reduction, 10*, 362–373. DOI: 10.1016/j.ijdrr.2014.10.011

Beken, H. G., & Cebeci, A. (2024). Çatışmalı bir süreç olarak sınırda karbon düzenleme mekanizması uygulaması. *Paradigma: İktisadi ve İdari Araştırmalar Dergisi, 13*(2), 115–128.

Bellora, C., & Fontagne, L. (2023). EU in search of a carbon border adjustment mechanism. *Energy Economics, 123*, 106673. DOI: 10.1016/j.eneco.2023.106673

Benassi, D., & Morlicchio, E. (2019). New urban poverty. In Orum, A. (Ed.), *The Wiley Blackwell encyclopedia of urban and regional studies*. John Wiley & Sons Ltd., DOI: 10.1002/9781118568446.eurs0466

Bennett, N. J., & Satterfield, T. (2018). Environmental governance: A practical framework to guide design, evaluation, and analysis. *Conservation Letters, 11*(6), e12600. DOI: 10.1111/conl.12600

Berger, R. M., & Kelly, J. J. (1993). Social Work in the Ecological Crisis. *Social Work, 38*, 521–526. https://www.jstor.org/stable/23717148

Bergmann, M., Schäpke, N., Marg, O., Stelzer, F., Lang, D. J., Bossert, M., Gantert, M., Häußler, E., Marquardt, E., Piontek, F. M., Potthast, T., Rhodius, R., Rudolph, M., Ruddat, M., Seebacher, A., & Sußmann, N. (2021). Transdisciplinary sustainability research in real-world labs: Success factors and methods for change. *Sustainability Science, 16*(2), 541–564. DOI: 10.1007/s11625-020-00886-8

Berhane, G., Gilligan, D. O., Hoddinott, J., Kumar, N., & Taffesse, A. S. (2014). Can social protection work in Africa? The impact of Ethiopia's Productive Safety Net Programme. *Economic Development and Cultural Change, 63*(1), 1–26. DOI: 10.1086/677753

Berkes, F. (2009). Evolution of co-management: Role of knowledge generation, bridging organizations and social learning. *Journal of Environmental Management, 90*(5), 1692–1702. DOI: 10.1016/j.jenvman.2008.12.001 PMID: 19110363

Berkes, F. (2009). Indigenous ways of knowing and the study of environmental change. *Journal of the Royal Society of New Zealand, 39*(4), 151–156. DOI: 10.1080/03014220909510568

Berkes, F., & Ross, H. (2013). Community resilience: Toward an integrated approach. *Society & Natural Resources, 26*(1), 5–20. DOI: 10.1080/08941920.2012.736605

Besthorn, F. H., & McMillen, D. P. (2002). '*The Oppression of Women and Nature: Ecofeminism as a Framework for Social Justice Oriented Social Work*', Families and Society. *The Journal of Contemporary Human Service, 83*, 221–232. DOI: 10.1606/1044-3894.20

Bettwy, M. (2007). NASA Satellites Watch as China Constructs Giant Dam. – NASA, 14 June 2007, available at: https://www.nasa.gov/centers/goddard/news/topstory/2007/dam_construct.html

Beveridge, H. (1876). Favorite Share Flag Were the Sundarbans inhabited in ancient times. *Journal of the Asiatic Society of Bengal, 45*. https://archive.org/details/biostor-281250

Bhamra, R., Dani, S., & Burnard, K. (2011). Resilience: The concept, a literature review, and future directions. *International Journal of Production Research, 49*(18), 5375–5393. DOI: 10.1080/00207543.2011.563826

Bhattacharyya, R. (2022). Green finance for energy transition, climate action and sustainable development: Overview of concepts, applications, implementation and challenges. *Green Finance, 4*(1), 1–35. DOI: 10.3934/GF.2022001

Bhowmik, D., & Poddar, S. (2023). Impacts of Flood in India from 1953 to 2020. [AMT]. *Advancement in Management and Technology, 4*(2), 12–33.

Bhutta, U. S., Tariq, A., Farrukh, M., Raza, A., & Iqbal, M. K. (2022). Green bonds for sustainable development: Review of literature on development and impact of green bonds. *Technological Forecasting and Social Change, 175*, 121378. DOI: 10.1016/j.techfore.2021.121378

Biermann, F., Kanie, N., & Kim, R. E. (2017). Global governance by goal-setting: The novel approach of the UN Sustainable Development Goals. *Current Opinion in Environmental Sustainability, 26-27*, 26–31. DOI: 10.1016/j.cosust.2017.01.010

Biesbroek, G. R., Klostermann, J. E. M., Termeer, C. J. A. M., & Kabat, P. (2013). On the nature of barriers to climate change adaptation. *Regional Environmental Change*, *13*(5), 1119–1129. DOI: 10.1007/s10113-013-0421-y

Bimonte, S. (2002). Information access, income distribution, and the environmental kuznets curve. *Ecological Economics*, *41*(1), 145–156. DOI: 10.1016/S0921-8009(02)00022-8

Bin, D. (2021, March). Discussion on the development direction of hydropower in China. -. *Clean Energy*, *5*(1), 10–18. DOI: 10.1093/ce/zkaa025

Bioenergy, I. E. A. (2023). Biofuels Production and Development in China. - IEA Bioenergy T39 – Biofuel News, Issue 63, October 2023, available at: https://www.ieabioenergy.com/wp-content/uploads/2023/12/IEA-BIOENERGY-T39-BIOFUEL-NEWS-63-1.pdf

Birch, S. (2017). The ethics of political participation: Engagement and democracy in the 21st century. *Res Publica (Liverpool, England)*, *23*(1), 1–12. DOI: 10.1007/s11158-017-9389-7

Biswas, B. (2013). "The God of Death Takes Half of Our Children": Health of Women and Children in the Sundarbans Islands. *Journal of Health Care for the Poor and Underserved*, 24(2), 730–740. . 2013.0069DOI: 10.1353/hpu

Biswas, S., & Nautiyal, S. (2020). An assessment of socio-economic vulnerability at the household level: A study on villages of the Indian Sundarbans. *Environment, Development and Sustainability*, *23*(7), 11120–11137. DOI: 10.1007/s10668-020-01085-2

Blake, D.E., Guppy, N., & Urmetzer, P. (1997). Canadian public opinion and environmental action: evidence from British Columbia. *Canadian Journal of Political Science/Revue canadienne de science politique*, 30(3), pp.451-472.

Bo Brysch, S. L., & Czischke, D. (2021). Affordability through design: The role of building costs in collaborative housing. *Housing Studies*, *37*(10), 1800–1820. DOI: 10.1080/02673037.2021.2009778

Bock, W. J. (1980). The Definition and Recognition of Biological Adaptation. *American Zoologist*, *20*(1), 217–227. DOI: 10.1093/icb/20.1.217

Bodansky, D. (2010). *The art and craft of international environmental law*. Harvard University Press.

Bodin, Ö. (2017). Collaborative environmental governance: Achieving collective action in social-ecological systems. *Science, 357*(6352), eaan1114. DOI: 10.1126/science.aan1114 PMID: 28818915

Böhringer, C., Fischer, C., Rosendahl, K. E., & Rutherford, T. F. (2022). Potential impacts and challenges of border carbon adjustments. *Nature Climate Change, 12*(1), 22–29. DOI: 10.1038/s41558-021-01250-z

Bossuyt, D. (2021). Who Owns Collaborative Housing? A Conceptual Typology of Property Regimes. Housing. *Theory and Society*, *39*(2), 200–216. DOI: 10.1080/14036096.2021.1888788

Bovaird, T., & Loeffler, E. (2012). From engagement to co-production: The contribution of users and communities to outcomes and public value. *Voluntas*, *23*(4), 1119–1138. DOI: 10.1007/s11266-012-9309-6

Boyce, J. K. (1994). Inequality as a cause of environmental degradation. *Ecological Economics*, *11*(3), 169–178. DOI: 10.1016/0921-8009(94)90198-8

Boyce, J. K. (2018). Carbon pricing: Effectiveness and equity. *Ecological Economics*, *150*, 52–61. DOI: 10.1016/j.ecolecon.2018.03.030

BP. (2021a). China's energy market in 2020. Statistical Review of World Energy – 2021. London.

BP. (2021b). Statistical Review of World Energy 2021, 70th edition. London, available at: https://www.bp.com/content/dam/bp/business-sites/en/global/corporate/pdfs/energy-economics/statistical-review/bp-stats-review-2021-full-report.pdf

Bradford, J., & Fraser, E. D. (2008). Local authorities, climate change and small and medium enterprises: Identifying effective policy instruments to reduce energy use and carbon emissions. *Corporate Social Responsibility and Environmental Management*, *15*(3), 156–172. DOI: 10.1002/csr.151

Brandi, C. (2021). Priorities for a development-friendly EU carbon border adjustment (CBAM): Briefing Paper 20/2021. https://hdl.handle.net/10419/242609

Brandsen, T., & Honingh, M. (2016). Distinguishing different types of co-production: A conceptual analysis based on the classical definitions. *Public Administration Review*, *76*(3), 427–435. DOI: 10.1111/puar.12465

Brandsen, T., & Honingh, M. (2018). Definitions of co-production and co-creation. In *Co-production and co-creation* (pp. 9–17). Routledge. DOI: 10.4324/9781315204956-2

Breil, M., & Johnson, K. (2012). *Conceptualizing urban adaptation to climate change*. Review of Environment, Energy and Economics (Re3). Available at https://ssrn.com/abstract=2125728

Breusch, T., & Pagan, A. (1980). The lagrange multiplier test and its application to model specifications in econometrics. *The Review of Economic Studies*, *47*(1), 239–253. DOI: 10.2307/2297111

Brondizio, E. S., & Le Tourneau, F. M. (2016). Environmental governance for all. *Science*, *352*(6291), 1272–1273. DOI: 10.1126/science.aaf5122 PMID: 27284179

Broniewicz, E. (2021). Environmental protection expenditure in European Union. *Environmental Management in Practice*, CEE-M Working Paper.

Brown, L. H. (2022). Addressing the adaptation finance gap: scaling up investment in climate change adaptation using blended finance solutions.

Brown, A., Dayal, A., & Rumbaitis del Rio, C. (2012). From practice to theory: Emerging lessons from Asia for building urban climate change resilience. *Environment and Urbanization*, *24*(2), 531–556. DOI: 10.1177/0956247812456490

Brown, K. (2014). Global environmental change I: A social turn for resilience? *Progress in Human Geography*, *38*(1), 107–117. DOI: 10.1177/0309132513498837

Brulle, R. J. (2014). Institutionalizing delay: Foundation funding and the creation of U.S. climate change counter-movement organizations. *Climatic Change*, *122*(4), 681–694. DOI: 10.1007/s10584-013-1018-7

Buchholz, K. (2020). Then & now: Urban population worldwide. https://www.statista.com/chart/23349/share-of-urban-population-by-continent/

Buchner, B., Hervé Mignucci, M., Trabacchi, C., Wilkinson, J., Stadelmann, M., Boyd, R., & Micale, V. (2013). Global Landscape of Climate Finance.

Bulkeley, H., (2019). *Local Governments and Climate Change: A Discussion Note*. GSDRC.

Bulkeley, H., Schroeder, H., Janda, K., Zhao, J., Armstrong, A., Chu, S. Y., & Ghosh, S. (2011). The role of institutions, governance and urban planning for mitigation and adaptation. In Hoornweg, D., Freire, M., Lee, M. J., Bhada-Tata, P., & Yuen, B. (Eds.), *Cities and climate change: Responding to an urgent agenda* (pp. 125–159). The World Bank. DOI: 10.1596/9780821384930_CH05

Bulkeley, H., & Tuts, R. (2013). Understanding urban vulnerability, adaptation and resilience in the context of climate change. *Local Environment*, *18*(6), 646–662. DOI: 10.1080/13549839.2013.788479

Bullard, R. D. (1990). *Dumping in Dixie: Race, class, and environmental quality*. Westview Press.

Busby, J. (2018). Taking stock: The field of climate and security revisited. Social Science Research Council. https://www.ssrc.org/publications/view/taking-stock-the-field-of-climate-and-security-revisited/

C40 Cities. (2005). C40 Cities. https://www.c40.org/

C40 Cities. (2022). Copenhagen: A global leader in climate action. Retrieved from https://www.c40.org

C40 Cities. (2023). About C40 cities. Retrieved from https://www.c40.org

Cabannes, Y. (2004). Participatory budgeting: A significant contribution to participatory democracy. *Environment and Urbanization*, *16*(1), 27–46. DOI: 10.1177/095624780401600104

Calisto Friant, M. (2019). Deliberating for sustainability: Lessons from the Porto Alegre experiment with participatory budgeting. *International Journal of Urban Sustainable Development*, *11*(1), 81–99. DOI: 10.1080/19463138.2019.1570219

Callahan, K. (2007). Citizen participation: Models and methods. *International Journal of Public Administration*, *30*(11), 1179–1196. DOI: 10.1080/01900690701225366

Campiglio, E., Godin, A., Kemp-Benedict, E., & Matikainen, S. (2017). The tightening links between financial systems and the low-carbon transition. In *Economic Policies since the Global Financial Crisis* (pp. 313–356). Springer., DOI: 10.1007/978-3-319-60459-6_8

Cannon, T., & Mueller-Mahn, D. (2010). Vulnerability, resilience and development discourses in context of climate change. *Natural Hazards*, *55*(3), 621–635. DOI: 10.1007/s11069-010-9499-4

Capra, C. F. (2016). The smart city and its citizens. *International Journal of E-Planning Research*, *5*(1), 20–38. DOI: 10.4018/IJEPR.2016010102

Caragliu, A., Del Bo, C., & Nijkamp, P. (2009). Smart cities in Europe. *3rd Central European Conference in Regional Science—CERS, 2009*, 45–59.

Carbonell, P., Rodríguez Escudero, A. I., & Pujari, D. (2009). Customer Involvement in New Service Development: An Examination of Antecedents and Outcomes. *Journal of Product Innovation Management*, *26*(5), 536–550. DOI: 10.1111/j.1540-5885.2009.00679.x

Cardullo, P., & Kitchin, R. (2019). Being a 'citizen' in the smart city: Up and down the scaffold of smart citizen participation in Dublin, Ireland. *GeoJournal*, *84*(1), 1–13. DOI: 10.1007/s10708-018-9845-8

Carl, J., & Fedor, D. (2016). Tracking global carbon revenues: A survey of carbon taxes versus cap-and-trade in the real world. *Energy Policy*, *96*, 50–77. DOI: 10.1016/j.enpol.2016.05.023

Carmin, J., Anguelovski, I., & Roberts, D. (2012). Urban climate adaptation in the global South: Planning in an emerging policy domain. *Journal of Planning Education and Research*, *32*(1), 18–32. DOI: 10.1177/0739456X11430951

Carriere, Z., & Lesnikowski, A. (2024). The climate change adaptation readiness of co-operative housing in Nova Scotia, Canada. *Environmental Policy and Governance*, *34*(6), 724–739. DOI: 10.1002/eet.2116

Carter, S., & Bayer, J. L. (2006). *IR-06-043 The Projected Influence of Climate Change on the South African Wine Industry* (Interim Report IR-06-043). International Institute for Applied Systems Analysis Schlossplatz 1, Laxenburg, Austria.

Cash, D. W., (2019). Knowledge systems for sustainable development. *Proceedings of the National Academy of Sciences of the United States of America, 116*(14), 6570–6578. DOI: 10.1073/pnas.1817180116 PMID: 12777623

Castillo, R., & Purdy, C. (2022). China's Role in Supplying Critical Minerals for the Global Energy Transition: What Could the Future Hold? – Brookings, 1 August 2022, available at: https://www.brookings.edu/articles/chinas-role-in-supplying-critical-minerals-for-the-global-energy-transition-what-could-the-future-hold/

Cengiz, E., & Uğur, Ö. (Eds.) (2024). Doğal Afetler ve İklim Değişikliği, Nobel Basım Yayım, Ankara.

Cengiz, E., & Uğur, Ö. (2024). İklim değişikliği ve doğal afetler. In Cengiz, E., & Uğur, Ö. (Eds.), *Doğal afetler ve iklim değişikliği: Politik süreçler, kurumlar ve uygulamalar* (pp. 1–12). Nobel Bilimsel Eserler.

Çetin, M., & Çiftçi, Ç. (2019). Literatüre göre dünya ve ülkemizden örneklerle akıllı kent kavramının irdelenmesi. *Ulusal Çevre Bilimleri Araştırma Dergisi*, *2*(3), 134–143.

Çevre, T. C. Şehircilik ve İklim Değişikliği Bakanlığı (2025). İklim Değişikliği Azaltım Stratejisi Ve Eylem Planı (2024-2030), Ankara.

Chakrabarti, R. (2009). *Local People and the Global Tiger: An Environmental History of the Sundarbans*. Global Environment, Vol. 2, (2009),(No.3), pp. 72-95. https://www.jstor.org/stable/43201488

Chanda, A., & Akhand, A. (2023, August). Challenges towards the Sustainability and Enhancement of the Indian Sundarban Mangrove's Blue Carbon Stock. *Life (Basel, Switzerland)*, *13*(8), 1–19. DOI: 10.3390/life13081787 PMID: 37629645

Chattopadhyay, H. (1999). *The Mystery of the Sunderbans* (First ed.). A Mukherjee & Co. https://archive.org/details/the-mystery-of-the-sundarbans/page/18/mode/2up

Chaudhry, S. M., Chen, X. H., Ahmed, R., & Nasir, M. A. (2023). Risk modelling of ESG (environmental, social, and governance), healthcare, and financial sectors. *Risk Analysis*. PMID: 37480163

Chauke, H., & Pongrácz, R. (2024). The validation of climate in the wine-growing region of the Western Cape of South Africa. *GEM: International Journal on Geomathematics*, *15*(1), 4. DOI: 10.1007/s13137-023-00244-7

Cheung, M. F., & To, W. M. (2011). Customer involvement and perceptions: The moderating role of customer co-production. *Journal of Retailing and Consumer Services*, *18*(4), 271–277. DOI: 10.1016/j.jretconser.2010.12.011

Chin, N., Pinko, N., Olutoke, J., Eschalier, C., Zamarioli Santos, L., & Bendahou, S. (2024). Approaches to Meeting the Paris Agreement Goals. Options for Public Development Banks-June 2024.

China's Mid-Century Long-Term Low Greenhouse Gas Emission Development Strategy. 28 October 2021, available at: https://unfccc.int/sites/default/files/resource/China%E2%80%99s%20Mid-Century%20Long-Term%20Low%20Greenhouse%20Gas%20Emission%20Development%20Strategy.pdf

Chiu, D. (2017). The East Is Green: China's Global Leadership in Renewable Energy. - New Perspectives in Foreign Policy, Issue 13, Summer 2017, Center for Strategic and International Studies, available at: https://www.csis.org/east-green-chinas-global-leadership-renewable-energy

Chowdhury, A. N. (2019). Environment and wellbeing: Eco-psychiatry in Sundarban Delta, India. *International Journal of Management and Social Sciences*, *37*, 2. DOI: 10.21013/jmss.v14.n2sp.p5

Chu, L., Do, T. N., Le, T. H. L., Ho, Q. A., & Dang, K. (2024). Carbon border adjustment mechanism, carbon pricing, and within-sector shifts: A partial equilibrium approach to Vietnam's steel sector. *Energy Policy*, *193*, 114293. DOI: 10.1016/j.enpol.2024.114293

Chun, Z. (2015). China's New Blueprint for an "Ecological Civilization". – The Diplomat, 30 September 2015, available at: https://thediplomat.com/2015/09/chinas-new-blueprint-for-an-ecological-civilization/

Ciscar, J. C., Iglesias, A., Feyen, L., Szabó, L., Van Regemorter, D., Amelung, B., Nicholls, R., Watkiss, P., Christensen, O. B., Dankers, R., Garrote, L., Goodess, C. M., Hunt, A., Moreno, A., Richards, J., & Soria, A. (2011). Physical and economic consequences of climate change in Europe. *Proceedings of the National Academy of Sciences of the United States of America, 108*(7), 2678–2683. DOI: 10.1073/pnas.1011612108 PMID: 21282624

City of Amsterdam. (n.d.). *Circular economy*. Erişim tarihi: 24 Kasım 2024, https://www.amsterdam.nl/en/policy/sustainability/circular-economy/

Clark, R., Reed, J., & Sunderland, T. (2018). Bridging funding gaps for climate and sustainable development: Pitfalls, progress and potential of private finance. *Land Use Policy*, *71*, 335–346. DOI: 10.1016/j.landusepol.2017.12.013

Clements, K. (2021). *Kriz Dönüşümüne ve Adil Bir Barışa Doğru, Çev*. Friedrich Naumann Vakfı, Berghof Handbook For Conflict Transformation.

Cochrane, L., & Ziervogel, G. (2020). The Cape Town Drought: Lessons from the City that Nearly Ran Out of Water – Gina Ziervogel. *NokokoPod*, *3*, 1–16.

CO-HOPE. https://jpi-urbaneurope.eu/project/co-hope-collaborative-housing-in-a-pandemic-era/, 11/12/2023

Coimbra, J., & Almeida, M. (2013). Challenges and benefits of building sustainable cooperative housing. *Building and Environment*, *62*, 9–17. DOI: 10.1016/j.buildenv.2013.01.003

Çokgezen, J. (2007). Avrupa Birliği çevre politikası ve Türkiye. *Marmara Üniversitesi İ.İ.B.F. Dergisi*, *23*(2), 91–115.

Cooperative City. https://cooperativecity.org/2022/02/04/bikes-and-rails-co-housing-with-solidarity-and-an-environmental-mission/, 18/12/2023

COP 13 Local Governments Climate Roadmap. (2007). COP 13 Local Governments Climate Roadmap.

Corfee-Morlot, J., Cochran, I., Hallegatte, S., & Teasdale, P. (2011). Multilevel risk governance and urban adaptation policy. *Climatic Change*, *104*(1), 169–197. DOI: 10.1007/s10584-010-9980-9

Cornago, E., & Berg, A. (2024). Learning from CBAM's transitional phase: Early impacts on trade and climate efforts. https://www.cer.eu/publications/archive/policy-brief/2024/learning-cbams-transitional-impacts-trade

Cornwall, A., & Gaventa, J. (2017). Participation in governance. In *International development governance* (pp. 405–413). Routledge. DOI: 10.4324/9781315092577-22

Corrigan, C., & Granziera, A. (2018). Governance of protected areas from impact evaluations to celebrating success. In Worboys, G. L., Lockwood, M., Kothari, A., Feary, S., & Pulsford, I. (Eds.), *Protected area governance and management* (pp. 771–788). ANU Press., DOI: 10.22459/PAGM.04.2018.04

Cox, D., & Pawar, M. (2006). *International Social Work: Issues, Strategies and Programs*. SAGE., DOI: 10.4135/9781544308685

CPLC (Carbon Pricing Leadership Coalition). (2016). What Are the Options for Using Carbon Pricing Revenues? https://thedocs.worldbank.org/en/doc/668851474296920877-0020022016/original/CPLCWhataretheOptionsforUsingCarbonPricingRevenues092016.pdf

Cronin, J. A., Fullerton, D., & Sexton, S. (2017). Vertical and horizontal redistribution from a carbon tax and rebate. NBER Working Paper No. 23250. National Bureau of Economic Research. DOI: 10.3386/w23250

Cui, A. S., & Wu, F. (2017). The impact of customer involvement on new product development: Contingent and substitutive effects. *journal of product innovation management, 34*(1), 60-80. DOI: 10.1111/jpim.12333

Cushing, L., Blaustein-Rejto, D., Wander, M., Pastor, M., Sadd, J., Hyun, A., & Morello-Frosch, R. (2018). Carbon trading, co-pollutants, and environmental equity: Evidence from California's cap-and-trade program (2011–2015). *PLoS Medicine, 15*(7), e1002604. DOI: 10.1371/journal.pmed.1002604 PMID: 29990353

Cutter, S. L., Boruff, B. J., & Shirley, W. L. (2003). Social vulnerability to environmental hazards. *Social Science Quarterly, 84*(2), 242–261. DOI: 10.1111/1540-6237.8402002

Czischke, D. (2019). Collaborative Housing: The Resurgence of Collectively Self-organised and Self-managed Housing in Europe, 5

Czischke, D. (2018). Collaborative housing and housing providers: Towards an analytical framework of multi-stakeholder collaboration in housing co-production. *International Journal of Housing Policy, 18*(1), 55–81. DOI: 10.1080/19491247.2017.1331593

Czischke, D., Carriou, C., & Lang, R. (2020). Collaborative Housing in Europe: Conceptualizing the Field. Housing. *Theory and Society*, *37*(1), 1–9. DOI: 10.1080/14036096.2020.1703611

D'Orazio, P., & Dirks, M. W. (2022). Exploring the effects of climate-related financial policies on carbon emissions in G20 countries: A panel quantile regression approach. *Environmental Science and Pollution Research International*, *29*(5), 7678–7702. DOI: 10.1007/s11356-021-15655-y PMID: 34476706

Dağdemir, Ö. (2015). *Çevre sorunlarına ekonomik yaklaşımlar ve optimal politika arayışları (3. Baskı)*. Gazi Kitabevi.

Dagnino, E. (2003). Citizenship in Latin America: An introduction. *Latin American Perspectives*, *30*(2), 211–225. DOI: 10.1177/0094582X02250624

Dagnino, E., & Teixeira, A. C. C. (2014). The participation of civil society in Lula's government. *Journal of Politics in Latin America*, *6*(3), 39–66. DOI: 10.1177/1866802X1400600303

Dai, L., & Xiong, Y. (2023). Does the green finance development and renewable energy affect the economic recovery in Asian economies. *Renewable Energy*, *216*, 118922. DOI: 10.1016/j.renene.2023.118922

Daly, M. (2017). Quantifying the environmental impact of ecovillages and co-housing communities: A systematic literature review. *Local Environment*, *22*(11), 1358–1377. DOI: 10.1080/13549839.2017.1348342

Danda, A. A. (2019, November). *Environmental Security in the Sundarban in the Current Climate Change Era Strengthening India-Bangladesh Cooperation*. Observer Research Foundation, ORF Occasional Paper No. 220.

Das Nair, R., Chisoro, S., & Ponte, S. (2023). *Sustainability in the South African wine industry: Status, opportunities, and challenges*. CBS, Centre for Business and Development Studies.

Dasgupta, S., Wheeler, D., & Ghosh, S. (2021). Fishing in Salty Waters: Poverty, Occupational Saline Exposure, and Women's Health in the Indian Sundarban. *Journal of Management and Sustainability*, *12*(1), 1. DOI: 10.5539/jms.v12n1p1

Das, S., & Hazra, S. (2020). Trapped or resettled: Coastal communities in the Sundarbans Delta. *Climate Crisis and Local Communities*, *64*, 15–17.

Davidson, S. (1998). Spinning the wheel of empowerment. *Planning (Chicago, Ill.)*, *1262*(3), 14–15.

De Galhau, V. (2019, June). Climate change: central banks are taking action. In *FSR* (p. 7).

De Jong, M. D., Neulen, S., & Jansma, S. R. (2019). Citizens' intentions to participate in governmental co-creation initiatives: Comparing three co-creation configurations. *Government Information Quarterly, 36*(3), 490–500. DOI: 10.1016/j.giq.2019.04.003

De Koning, J. I. J. C., Crul, M. R., & Wever, R. (2016, May). Models of co-creation. In Service design geographies. Proceedings of the servdes. 2016 conference (Vol. 125, pp. 266-278). Linköping, Sweden: Linköping University Electronic Press. DOI: 10.1111/0272-4332.00064

de Moor, J., Uba, K., Wahlström, M., Wennerhag, M., & De Vydt, M. (2020). Protest for a future: Composition, mobilization and motives of the participants in Fridays For Future climate protests on 15 March, 2019 in 13 European cities. *Sustainability, 12*(6), 2191. DOI: 10.3390/su12062191

Değirmenci, T. (2020). *Türkiye'de çevre koruma harcamalarının sürdürülebilir kalkınma boyutu. H. Yavuz (Edt.), Çevre ekonomisi ve mali ilişkiler üzerine seçme yazılar*. Sakarya Üniversitesi Yayınları.

Değirmenci, T., & Aydın, M. (2020). Çevre koruma harcamaları ile gelir dağılımı ve ekonomik büyüme arasındaki dinamik ilişkiler: Seçili OECD ülkeleri için panel nedensellik yaklaşımı. *Sosyoekonomi, 28*(46), 391–406. DOI: 10.17233/sosyoekonomi.2020.04.18

Demirci, M. (2015). Urban climate change governance. *Erciyes University Faculty of Economics and Administrative Sciences Journal, 46*, 75–100.

Denchak, M. (2021). *Paris climate agreement: Everything you need to know*. National Resource Defend Council.

Dent, C. (2015). China's renewable energy development: policy, industry and business perspectives. - Asia Pacific Business Review, Volume 21, Issue 1: Managing China's Energy Sector: Between the Market and the State, pp. 26-43.

Dewulf, A. (2013). Contrasting frames in policy debates on climate change adaptation. *Wiley Interdisciplinary Reviews: Climate Change, 4*(4), 321–330. DOI: 10.1002/wcc.227

Dey, M. K., & Hazra, A. K. (2021). Abundance of microarthropods population in different sites of Midnapore east coast of West Bengal, India. *International Journal of Advancement in Life Sciences Research*, 13-25. DOI: DOI: 10.31632/ijalsr.2021.v04i03.003

Díaz-Caravantes, R. E., Zuniga-Teran, A., Martín, F., Bernabeu, M., Stoker, P., & Scott, C. (2020). Urban water security: A comparative study of cities in the arid Americas. *Environment and Urbanization, 32*(1), 275–294. DOI: 10.1177/0956247819900468

Diffenbaugh, N. S., & Burke, M. (2019). Global warming has increased global economic inequality. *Proceedings of the National Academy of Sciences of the United States of America, 116*(20), 9808–9813. DOI: 10.1073/pnas.1816020116 PMID: 31010922

Dikau, S., & Volz, U. (2021). Central bank mandates, sustainability objectives and the promotion of green finance. *Ecological Economics, 184*, 107022. DOI: 10.1016/j.ecolecon.2021.107022

Dilling, L., Daly, M. E., Travis, W. R., Wilhelmi, O. V., & Klein, R. A. (2019). The dynamics of vulnerability: Why adapting to climate variability will not always prepare us for climate change. *Wiley Interdisciplinary Reviews: Climate Change, 10*(1), e571. DOI: 10.1002/wcc.571

Dinar Wahidayat Putra, Z., & Van Der Knaap, W. G. M. (2018). Urban innovation system and the role of an open web-based platform: The case of Amsterdam smart city. *Journal of Regional and City Planning, 29*(3), 234–249. DOI: 10.5614/jrcp.2018.29.3.4

Ding, H., Veit, P. G., Blackman, A., Gray, E., Reytar, K., Altamirano, J. C., & Hodgdon, B. (2016). Climate benefits, tenure costs: The economic case for securing indigenous land rights in the Amazon. World Resources Institute. https://www.wri.org/research/climate-benefits-tenure-costs

Dışişleri Bakanlığı, T. C. (2025). İklim Değişikliğiyle Mücadelenin Önemi, https://www.mfa.gov.tr/iklim-degisikligiyle-mucadelenin-onemi.tr.mfa

Dodman, D., & Mitlin, D. (2013). Challenges for community-based adaptation: Discovering the potential for transformation. *Journal of International Development, 25*(5), 640–659. DOI: 10.1002/jid.1772

Dodman, D., & Satterthwaite, D. (2008). Institutional capacity, climate change adaptation and the urban poor. *IDS Bulletin, 39*(4), 67–74. DOI: 10.1111/j.1759-5436.2008.tb00478.x

Doğan, B., Chu, L. K., Ghosh, S., Truong, H. H. D., & Balsalobre-Lorente, D. (2022). How environmental taxes and carbon emissions are related in the G7 economies? *Renewable Energy, 187*, 645–656. DOI: 10.1016/j.renene.2022.01.077

Doğan, K. C. (2021). In Kavanoz, S. E., Uzun, A., Şengün, H., & Yayınları, A. (Eds.), *İklim Değişkliği ve Kent, Karmaşık Sorunlar ve Kent* (pp. 189–204).

Doğan, K. C., & Uğur, Ö. (Eds.). (2020). *From Efficiency Discussions to Democracy in Public Administration: A Theoretical Analysis.* Peter Lang.

Doğan, S., & Tüzer, M. (2011). Küresel iklim değişikliği ile mücadele: Genel yaklaşımlar ve uluslararası çabalar. *Istanbul Journal of Sociological Studies*, (44), 157–194.

Dominelli, L. (2010). '*Globalization, Contemporary Challenges and Social Work Practice*'. *International Social Work, 53*(5), 599–612. DOI: 10.1177/0020872810371201

Dominelli, L. (2011). '*Climate Change: Social Workers' Roles and Contributions to Policy Debates and Interventions*'. *International Journal of Social Welfare, 20*(4), 430–438. DOI: 10.1111/j.1468-2397.2011.00795.x

Doshi, R., Dadhich, M., Poddar, S., & Hiran, K. K. (2024). *Integrating generative AI in education to achieve sustainable development goals.*

Drummond, K. (2019). *The relationship between droughts and the tourism industry: A case study on Cape Town, South Africa* [Honours Thesis. 3113, Western Michigan University]. https://scholarworks.wmich.edu/honors_theses/3113

Dryzek, J. S., & Pickering, J. (2019). *The politics of the Anthropocene.* Oxford University Press.

Dubey, S., Trivedi, R., Chand, B., Mandal, B., & Rout, S. (2017). Farmers' perception of climate change, impacts on freshwater aquaculture and adaptation strategies in climatic change hotspots: A case of the Indian Sundarban delta. *Environmental Development, 21*, 38–51. DOI: 10.1016/j.envdev.2016.12.002

Duchelle, A. E., Crompton, T., Gebara, M. F., Guerra, R., Melo, T., Larson, A., & Sills, E. (2014). Linking forest tenure reform, environmental compliance, and incentives: Lessons from REDD+ initiatives in the Brazilian Amazon. *World Development, 55*, 53–67. DOI: 10.1016/j.worlddev.2013.01.014

Dunlap, R. E., & McCright, A. M. (2015). Challenging climate change: The denial countermovement. In Dunlap, R. E., & Brulle, R. J. (Eds.), *Climate change and society: Sociological perspectives* (pp. 300–332). Oxford University Press. DOI: 10.1093/acprof:oso/9780199356102.003.0010

Dunn, S., & Flavin, C. (2002). İklim Değişikliğini Gündemin Ön Sıralarına Taşımak, Tema Vakfı Yayınları, http://dogadostlari.free.fr/kuresel/ıklim%20degisikligini%20g%c3%bcndemin%20%c3%96n%20siralarina%20tasimak.pdf.

Dwivedi, Y. K., Hughes, L., Kar, A. K., Baabdullah, A. M., Grover, P., Abbas, R., Andreini, D., Abumoghli, I., Barlette, Y., Bunker, D., Chandra Kruse, L., Constantiou, I., Davison, R. M., De', R., Dubey, R., Fenby-Taylor, H., Gupta, B., He, W., Kodama, M., & Wade, M. (2022). Climate change and COP26: Are digital technologies and information management part of the problem or the solution? An editorial reflection and call to action. *International Journal of Information Management, 63*, 102456. DOI: 10.1016/j.ijinfomgt.2021.102456

Eakin, H., & Lemos, M. C. (2006). Adaptation and the state: Latin America and the challenge of capacity-building under globalization. *Global Environmental Change, 16*(1), 7–18. DOI: 10.1016/j.gloenvcha.2005.10.004

Eakin, H., & Luers, A. L. (2006). Assessing the vulnerability of social-environmental systems. *Annual Review of Environment and Resources, 31*(1), 365–394. DOI: 10.1146/annurev.energy.30.050504.144352

Eccles, R. G., & Krzus, M. P. (2019). Implementing the task force on climate-related financial disclosures recommendations: An assessment of corporate readiness. *Schmalenbach Business Review, 71*(2), 287–293. DOI: 10.1007/s41464-018-0060-4

Efşan, N. Ç., Uğur, Ö. and Aksoy, M. (2024). Assesment in The Context of Climate Migration: Hurricane Katrina, Karadeniz Teknik Üniversitesi Sosyal Bilimler Enstitüsü Sosyal Bilimler Dergisi, 27 / Haziran, 119-143.

Ekins, P. (2000). *Economic growth and environmental sustainability: The prospects for green growth.* Routledge.

Elliot, D. (2012). *Social Development and Social Work.* In Healy, L. M., & Link, R. J. (Eds.), *Handbook of International Social Work* (pp. 102–108). Oxford University Press., DOI: 10.1093/acprof:oso/9780195333619.003.0014

Elliott, D. (1993). Social Work and Social Development: Towards an Integrative Model for Social Work Practice. *International Social Work, 36*(1), 21–36. DOI: 10.1177/002087289303600103

Elliott, E., Seldon, B. J., & Regens, J. L. (1997). Political and economic determinants of individuals support for environmental spending. *Journal of Environmental Management, 51*(1), 15–27. DOI: 10.1006/jema.1996.0129

Emose, G. (2021). Sustainability bond for the Pacific feasibility study.

Energy Institute. (2024a), Statistical Review of World Energy, 73rd edition, available at: https://www.energyinst.org/__data/assets/pdf_file/0006/1542714/684_EI_Stat_Review_V16_DIGITAL.pdf

Energy Institute. (2024b), Statistical Review of World Energy – all data, 1965–2023, available at: https://www.energyinst.org/__data/assets/excel_doc/0020/1540550/EI-Stats-Review-All-Data.xlsx

Enqvist, J. P., & Ziervogel, G. (2019). Water governance and justice in Cape Town: An overview. *WIREs. Water*, *6*(4), e1354. DOI: 10.1002/wat2.1354

Erb, T., Perciasepe, B., Radulovic, V., & Niland, M. (2022). Corporate climate commitments: The trend towards net zero. In *Handbook of climate change mitigation and adaptation* (pp. 2985–3018). Springer International Publishing. DOI: 10.1007/978-3-030-72579-2_146

Eriksen, S. H., Brown, K., & Kelly, P. M. (2011). The dynamics of vulnerability: Locating coping strategies in Kenya and Tanzania. *The Geographical Journal*, *177*(4), 287–305. DOI: 10.1111/j.1475-4959.2005.00174.x

Eriksen, S. H., Nightingale, A. J., & Eakin, H. (2015). Reframing adaptation: The political nature of climate change adaptation. *Global Environmental Change*, *21*(2), 670–680. DOI: 10.1016/j.gloenvcha.2015.09.014

Eriksen, S., Aldunce, P., Bahinipati, C. S., Martins, R. D., Molefe, J. I., Nhemachena, C., O'Brien, K., Olorunfemi, F., Park, J., Sygna, L., & Ulsrud, K. (2011). When not every response to climate change is a good one: Identifying principles for sustainable adaptation. *Climate and Development*, *3*(1), 7–20. DOI: 10.3763/cdev.2010.0060

Ertürk, H. (1998). *Çevre bilimlerine giriş (3. Baskı)*. Ceylan Matbaacılık.

ESA (European Space Agency). (2023). Copernicus: Europe's eyes on Earth. Retrieved from https://www.esa.int

ESCAP. (2012). Finding a green engine for economic growth: China's renewable energy policies", Low Carbon Green Growth Roadmap for Asia and the Pacific, Economic and Social Commission for Asia and the Pacific, Bangkok, available at: https://www.unescap.org/sites/default/files/14.%20CS-China-Renewable-Energy-Policies.pdf

Esen, M., & Batuhan, T. (2024). İklim Değişikliğiyle Mücadelede Kent Yönetimleri. *EBYÜ İİBF Dergisi*, *6*(2), 140–151.

Esposito, L., & Mastromatteo, G. (2019). Sustainable finance, the good, the bad and the ugly: A critical assessment of the EU institutional framework for the green transition. In *Sustainable finance, the good, the bad and the ugly: a critical assessment of the EU institutional framework for the green transition* (Vol. 2019, No. 4, pp. 1-42). Vita e Pensiero.

European Commission. (2012). Smart cities and communities- European innovation partnership. Accessed date: 19 October 2024, https://eur-lex.europa.eu/legal-content/EN/TXT/?uri=CELEX:52012DC0470

European Commission. (2019). *The European Green Deal.* https://commission.europa.eu/publications/communication-european-green-deal_en

European Commission. (2020). A new Circular Economy Action Plan for a cleaner and more competitive Europe. https://eur-lex.europa.eu/legal-content/EN/TXT/?qid=1583933814386&uri=COM:2020:98:FIN

European Commission. (2021). Cybersecurity in the EU ETS: Lessons from the 2021 incident. Retrieved from https://ec.europa.eu

European Commission. (2021). Delivering the European Green Deal. https://commission.europa.eu/strategy-and-policy/priorities-2019-2024/european-green-deal/delivering-european-green-deal_en

European Commission. (2023). About the EU ETS, https://climate.ec.europa.eu/eu-action/eu-emissions-trading-system-eu-ets/about-eu-ets_en

European Commission. (2023). Copenhagen's smart city initiatives. Retrieved from https://ec.europa.eu

European Commission. (2025). Application User Manual. CBAM Declarant Portal ver 1.4.1.1, https://taxation-customs.ec.europa.eu/document/download/47322ae4-27c9-49f1-bdb7-2a1fef73d647_en?filename=CBAM-UMN-UI%20manual%20for%20Declarants-Release%201.4.1.1-v1.81.pdf (Accessed 20 April 2025).

European Commission. (n.d.). Circular economy action plan Accessed date: 29 November 2024, https://environment.ec.europa.eu/strategy/circular-economy-action-plan_en

European Environment Acency. (2024). Use of auctioning revenues generated under the EU Emissions Trading System, https://www.eea.europa.eu/en/analysis/indicators/use-of-auctioning-revenues-generated (Accessed 12 February 2025).

European Parliament. (2014). Mapping smart cities in the EU. Accessed date 15 November 2024, https://www.europarl.europa.eu/RegData/etudes/etudes/join/2014/507480/IPOL-ITRE_ET(2014)507480_EN.pdf

European Parliamentary Research Service. (2023). Carbon border adjustment mechanism, https://www.europarl.europa.eu/RegData/etudes/ATAG/2023/754626/EPRS_ATA%282023%29754626_EN.pdf (Accessed 25 February 2025).

Extinction Rebellion. (2019). The truth behind the UK Parliament's climate emergency declaration. https://rebellion.earth/2019/05/02/the-truth-behind-the-uk-parliaments-climate-emergency-declaration

Falkner, R. (2016). The Paris Agreement and the new logic of international climate politics. *International Affairs*, *92*(5), 1107–1125. DOI: 10.1111/1468-2346.12708

FAO. (2019). The State of Food and Agriculture 2019. Moving forward on food loss and waste reduction. https://www.fao.org/3/ca6030en/ca6030en.pdf

Fenkart, S. (2021). China's Influence in the Western Balkans: Partnership or Confrontation? – Institute of New Europe, 29 November 2021, available at: https://ine.org.pl/en/chinas-influence-in-the-western-balkans-partnership-or-confrontation/

Ferreri, M., & Vidal, L. (2021). Public-cooperative policy mechanisms for housing commons. *International Journal of Housing Policy*, *22*(2), 149–173. DOI: 10.1080/19491247.2021.1877888

Ferri, G., & Acosta, B. A. (2019). Sustainable finance for sustainable development. *Center for Relationship Banking and Economics Working Paper Series, 30.*

Few, R., Brown, K., & Tompkins, E. L. (2007). Public participation and climate change adaptation: Avoiding the illusion of inclusion. *Climate Policy*, *7*(1), 45–59. DOI: 10.1080/14693062.2007.9685637

Finance, S. U. P. (2017). National Green Banks in Developing Countries.

Fischer, M., Lindberg, T., & Lenton, T. M. (2022). Participatory governance for climate resilience: Empowering local communities with data and technology. *Climate Policy*, *22*(4), 445–460.

Fluter, (2024). Fluter website: https://fluter.de/wohnprojekte-oesterreich-zusammenleben

Folke, C., Hahn, T., Olsson, P., & Norberg, J. (2005). Adaptive governance of social-ecological systems. *Annual Review of Environment and Resources*, *30*(1), 441–473. DOI: 10.1146/annurev.energy.30.050504.144511

Ford, J. D., Maillet, M., Pouliot, V., Meredith, T., & Cavanaugh, A. (2020). Adaptation and Indigenous peoples in the United Nations framework convention on climate change. *Climatic Change*, *164*(3), 1–14. DOI: 10.1007/s10584-020-02849-6

Forino, G., Von Meding, J., & Brewer, G. J. (2015). A Conceptual Governance Framework for Climate Change Adaptation and Disaster Risk Reduction Integration. *International Journal of Disaster Risk Science*, *6*(4), 372–384. DOI: 10.1007/s13753-015-0076-z

Forster, P. M., Smith, C. J., Walsh, T., Lamb, W. F., Lamboll, R., Hauser, M., Ribes, A., Rosen, D., Gillett, N., Palmer, M. D., Rogelj, J., von Schuckmann, K., Seneviratne, S. I., Trewin, B., Zhang, X., Allen, M., Andrew, R., Birt, A., Borger, A., & Zhai, P. (2023). Indicators of Global Climate Change 2022: Annual update of large-scale indicators of the state of the climate system and human influence. *Earth System Science Data, 15*(6), 2295–2327. DOI: 10.5194/essd-15-2295-2023

Forsyth, T. (2013). Community-based adaptation: A review of past and future challenges. *Wiley Interdisciplinary Reviews: Climate Change, 4*(5), 439–446. DOI: 10.1002/wcc.231

Francesco, B., & Carlo, C. (2018). *DEFINING ADAPTATION: An Analysis of Adaptation as a Response to Climate Change*. Copenhagen Consensus Center. http://www.jstor.com/stable/resrep16322.7

Francesco, E. D., & Rahman, M. (2020, December). *Natural, social and economic regeneration in the Sundarbans of Bangladesh.* Circular Conversations. Retrieved from https://www.circularconversations.com/earth-heroes/bangladesh-environment-and-development-society

Free, C., Jones, S., & Tremblay, M. S. (2024). Greenwashing and sustainability assurance: A review and call for future research. *Journal of Accounting Literature.* DOI: 10.1108/JAL-11-2023-0201

Fu, C., Lu, L., & Pirabi, M. (2023). Advancing green finance: A review of sustainable development. *Digital Economy and Sustainable Development*, *1*(1), 20. DOI: 10.1007/s44265-023-00020-3

Ganobal, M. (2019). Save Aru: Grassroots resistance and the cancellation of a commercial project in Indonesia. [Case study].

García, C. J., Herrero, B., Miralles-Quirós, J. L., & del Mar Mirallles-Quirós, M. (2023). Exploring the determinants of corporate green bond issuance and its environmental implication: The role of corporate board. *Technological Forecasting and Social Change*, *189*, 122379. DOI: 10.1016/j.techfore.2023.122379

Garnett, S. T., Burgess, N. D., Fa, J. E., Fernández-Llamazares, Á., Molnár, Z., Robinson, C. J., Watson, J. E. M., Zander, K. K., Austin, B., Brondizio, E. S., Collier, N. F., Duncan, T., Ellis, E., Geyle, H., Jackson, M. V., Jonas, H., Malmer, P., McGowan, B., Sivongxay, A., & Leiper, I. (2018). A spatial overview of the global importance of Indigenous lands for conservation. *Nature Sustainability*, *1*(7), 369–374. DOI: 10.1038/s41893-018-0100-6

Gasper, R., Blohm, A., & Ruth, M. (2011). Social and economic impacts of climate change on the urban environment. *Current Opinion in Environmental Sustainability*, *3*(3), 150–157. DOI: 10.1016/j.cosust.2010.12.009

GEF. (2018). Least Developed Countries Fund and Special Climate Change Fund. Global Environment Facility. https://www.thegef.org/topics/least-developed-countries-fund-ldcf

Gelderman, C. J., Semeijn, J., & Vluggen, R. (2017). Development of sustainability in public sector procurement. *Public Money & Management*, *37*(6), 435–442. DOI: 10.1080/09540962.2017.1344027

Gemenne, F., Barnett, J., Adger, W. N., & Dabelko, G. D. (2014). Climate and security: Evidence, emerging risks, and a new agenda. *Climatic Change*, *123*(1), 1–9. DOI: 10.1007/s10584-014-1074-7

Georgiadis, P., Paraschivoiu, M., & Wulff, A. (2023). AI and climate resilience: Leveraging technology for a sustainable future. *Climate Policy Journal*, *18*(3), 123–140.

Germain, C. B. (1978). General-systems Theory and Ego Psychology: An Ecological Perspective. *The Social Service Review*, *52*(4), 535–550. https://www.jstor.org/stable/30015668. DOI: 10.1086/643676

Ghazouani, A., Jebli, M. B., & Shahzad, U. (2021). Impacts of environmental taxes and technologies on greenhouse gas emissions: Contextual evidence from leading emitter European countries. *Environmental Science and Pollution Research International*, *28*(18), 22758–22767. DOI: 10.1007/s11356-020-11911-9 PMID: 33423194

Ghosh, A. (2012). *Living with changing climate–impact, vulnerability and adaptation challenges in Indian Sundarbans*. Centre for Science and Environment.

Ghosh, A. (2019). *The Tides of Life*. Indira Gandhi National Centre for the Arts.

Ghosh, O. (2023). Environmental hazards and disasters - A response towards mitigating disaster management. *AIP Conference Proceedings*, *2763*(1), 050008. DOI: 10.1063/5.0158892

Giddens, A. (2015). *The politics of climate change*. Polity Press.

Gills, B. K., & Morgan, J. (2020). Global climate emergency: After COP24, climate science, urgency, and the threat to humanity. *Globalizations*, *17*(6), 885–900. DOI: 10.1080/14747731.2019.1669915

Global Covenant of Mayors for Climate and Energy. (2023). About us. Retrieved from https://www.globalcovenantofmayors.org

Global Footprint Network. *Ecological footprint data explorer*. https://data.footprintnetwork.org/#/analyzeTrends?type=EFCpc&cn=210

Göksu, E., & Somuncu, M. (2023). İklim Değişikliğine Uyum İçin İş Birliğine Dayalı Yönetişimde Kent Konseylerinin Rolü ve Önemi. *Ankara Üniversitesi Çevrebilimleri Dergisi*, *10*(2), 1–14.

Gómez-Baggethun, E., Corbera, E., & Reyes-García, V. (2012). Traditional ecological knowledge and global environmental change: Research findings and policy implications. *Ecology and Society*, *18*(4), 72. DOI: 10.5751/ES-06288-180472 PMID: 26097492

Gong, Z. (2024). China Innovates Dual-tower Solar Thermal Plant for Better Efficiency. – CGTN, 16 July 2024, available at: https://news.cgtn.com/news/2024-07-16/China-innovates-dual-tower-solar-thermal-plant-for-better-efficiency-1vhVgy6Jp2E/p.html#:~:text=The%20two%20towers%20are%20each%20200%20meters%20tall,the%20concentrated%20sunlight%20heats%20water%20to%20generate%20electricity

Gormez, K. (2018). *Environmental issues*. Nobel Publications.

Government of West Bengal. (2020). (*rep.*). State Action Plan on Climate Change.

Gray, M. (2006). The Progress of Social Development in South Africa. *International Journal of Social Welfare*, *15*(s1, Suppl. 1), S53–S64. DOI: 10.1111/j.1468-2397.2006.00445.x

Gray, M., & Coates, J. R. (2012). Environmental Ethics for Social Work: Social Work's Responsibility to the Non-human World. *International Journal of Social Welfare*, *21*(3), 239–247. DOI: 10.1111/j.1468-2397.2011.00852.x

Gray, M., Mazibuko, F., & O'Brien, F. (1996). Social Work Education for Social Development. *Journal of Social Development in Africa*, *11*(1), 33–42. https://api.semanticscholar.org/CorpusID:221926574

Greenfield, A. (2006). *Everyware: The dawning age of ubiquitous computing*. New Riders.

Green, J. F. (2021). Does carbon pricing reduce emissions? A review of ex-post analyses. *Environmental Research Letters*, *16*(4), 1–17. DOI: 10.1088/1748-9326/abdae9 PMID: 33897808

Gregory, R., Ohlson, D., & Arvai, J. (2006). DECONSTRUCTING ADAPTIVE MANAGEMENT: CRITERIA FOR APPLICATIONS TO ENVIRONMENTAL MANAGEMENT. *Ecological Applications*, *16*(6), 2411–2425. DOI: 10.1890/1051-0761(2006)016[2411:DAMCFA]2.0.CO;2 PMID: 17205914

Griese, K. M., Werner, K., & Hogg, J. (2017). Avoiding greenwashing in event marketing: An exploration of concepts, literature and methods. *Journal of Management and Sustainability*, *7*(4), 1. DOI: 10.5539/jms.v7n4p1

Griffith, E., Jepma, M., & Savini, F. (2022). Beyond collective property: A typology of collaborative housing in Europe. *International Journal of Housing Policy*, *24*(1), 121–141. DOI: 10.1080/19491247.2022.2123272

Grothmann, T., & Patt, A. (2005). Adaptive capacity and human cognition: The process of individual adaptation to climate change. *Global Environmental Change*, *15*(3), 199–213. DOI: 10.1016/j.gloenvcha.2005.01.002

Grubb, M., Vrolijk, C., & Brack, D. (1999). *The Kyoto Protocol: A guide and assessment.* Royal Institute of International Affairs/Earthscan.

GSDRC. (2015). Local Governments and Climate Change. Retrieved from https://gsdrc.org/document-library/local-governments-and-climate-change/

Guidance on Promoting Green Belt and Road. (2017). – Belt and Road Portal, 8 May 2017, available at: https://eng.yidaiyilu.gov.cn/p/12479.html

Guillén Bolaños, T., Scheffran, J., & Máñez Costa, M. (2022). Climate Adaptation and Successful Adaptation Definitions: Latin American Perspectives Using the Delphi Method. *Sustainability (Basel)*, *14*(9), 5350. DOI: 10.3390/su14095350

Gül, A., & Atak Çobanoğlu, Ş. (2017). Avrupa'da akıllı kent uygulamalarının değerlendirilmesi ve Çanakkale'nin akıllı kente dönüşümünün analizi. *Süleyman Demirel Üniversitesi İktisadi ve İdari Bilimler Fakültesi Dergisi, 22*(Kayfor 15 Özel Sayısı), 1543–1565.

Gunderson, L. (1999). Resilience, Flexibility and Adaptive Management—- Antidotes for Spurious Certitude? *Conservation Ecology*, *3*(1), art7. DOI: 10.5751/ES-00089-030107

Gupta, A., Pandey, R., & Sapatneka, S. (2024). Potential implications of the EU's carbon border adjustment mechanism. NIPFP Working Paper Series.

Gupta, H., Sharma, M., & Mishra, R. (2020). Disaster management and predictive analytics: Lessons from Cyclone Fani. *International Journal of Disaster Risk Reduction*, *44*, 101–105.

Gutierrez, K. S., & LePrevost, C. E. (2016). Climate justice in rural southeastern United States: A review of climate change impacts and effects on human health. *International Journal of Environmental Research and Public Health, 13*(2), 189. DOI: 10.3390/ijerph13020189 PMID: 26848673

Haase, I., Velten, E. K., Branner, H., & Reyneri, A. (2022). *The use of auctioning revenues from the EU ETS for climate action – An analysis based on eight case studies. Technical report.* Ecologic Institute.

HABITAT III. (2016). New urban agenda. https://habitat3.org/documents-and-archive/new-urban-agenda/

Hafiz Iqbal, M. (2020). Valuing ecosystem services of Sundarbans mangrove forest for improved conservation: Approach of randomized conjoint experiment. *Forestry Economics Review, 2*(1), 2631–3030. DOI: 10.1108/FER-04-2020-0008

Haf, S., Robins, N., & Huq, S. (2019). Defining a policy agenda to mobilize climate investment through a just transition. In Robins, N., Muguet, S., & Marquant, A. (Eds.), *The Routledge handbook of climate change and finance* (pp. 193–208). Routledge., DOI: 10.4324/9781351127318-13

Haites, E. (2018). Carbon taxes and greenhouse gas emissions trading systems: What have we learned? *Climate Policy, 18*(8), 955–966. DOI: 10.1080/14693062.2018.1492897

Hallegatte, S., Bangalore, M., Bonzanigo, L., Fay, M., Kane, T., Narloch, U., Vogt-Schilb, A. (2016). Shock waves: Managing the impacts of climate change on poverty. The World Bank. https://openknowledge.worldbank.org/handle/10986/22787

Hallegatte, S., Rentschler, J., & Rozenberg, J. (2019). *Lifelines: The resilient infrastructure opportunity.* World Bank., DOI: 10.1596/978-1-4648-1430-3

Hannam, P. (2023). Confusion Surrounds China's Energy Policies as GDP and Climate Goals Clash. – The Guardian, 28 February 2023, available at: https://www.theguardian.com/world/2023/feb/28/china-energy-policies-gdp-climate-goals-clash?_hsenc=p2ANqtz-815FZ3Nq9xcpSsG5O2AfYsdZTVvCD-43lGMIR_eeIuQ-ZKHLyBJyEsnz4oO7V4rH9nj6dy

Hanna, P., & Vanclay, F. (2013). Human rights, Indigenous peoples and the concept of Free, Prior and Informed Consent. *Impact Assessment and Project Appraisal, 31*(2), 146–157. DOI: 10.1080/14615517.2013.780373

Hao, Y., Chen, H., & Zhang, Q. (2016). Will income inequality affect environmental quality? Analysis basedon China's provincial panel data. *Ecological Indicators, 67*, 533–542. DOI: 10.1016/j.ecolind.2016.03.025

Haque, A. N., Dodman, D., & Hossain, M. M. (2014). Individual, communal and institutional responses to climate change by low-income households in Khulna, Bangladesh. *Environment and Urbanization*, *26*(1), 112–129. DOI: 10.1177/0956247813518681

Harvey, D. (1985). *Urbanisation of capital*. Blackwell.

Hashim, M., Awaluddin, Z., R, A., Sarkum, S., Sholiha, A., & Aziz, A. (2021). Architectural Design in Rethinking the Future of Co-Housing in Malaysia. Journal of Design and Built Environment. DOI: 10.22452/jdbe.vol21no3.6

Hassan, S. M. (2024). Greenwashing in ESG: Identifying and Addressing False Claims of Sustainability. *Journal of Business and Strategic Management*, *9*(8), 90–105. DOI: 10.47941/jbsm.2390

Hasselman, L. (2017). Adaptive management; adaptive co-management; adaptive governance: What's the difference? *Australasian Journal of Environmental Management*, *24*(1), 31–46. DOI: 10.1080/14486563.2016.1251857

Hawkins, J., & Turner, P. (2000). Managing foreign debt and liquidity risks in emerging economies: An overview. *BIS Policy Papers*, *8*, 3–60.

Healy, L. M. (2008). *International Social Work: Professional Action in an Interdependent World* (2nd ed.). Oxford University Press., Available at https://global.oup.com/ushe/product/international-social-work-9780190922252?cc=in&lang=

Hilbrich, S., Berensmann, K., Artmann, G., Ashman, S., Herbold, T., Lötters-Viehof, S., & Steenkamp, L. A. (2024). Implementing sustainability taxonomies to redirect capital flows: The case of South Africa. *Climate Policy*, 1–15.

Hilton, I. (2024). How China Became the World's Leader on Renewable Energy. - Yale School of the Environment, 13 March 2024, available at: https://e360.yale.edu/features/china-renewable-energy

Hiran, K. K., Khazanchi, D., Vyas, A. K., & Padmanaban, S. (Eds.). (2021). *Machine learning for sustainable development* (Vol. 9). Walter de Gruyter GmbH & Co KG., DOI: 10.1515/9783110702514

Hofer, A. (2017) Cohousing projects: The role of relationships and its dynamics, Masters thesis, J.Keppler University, Linz, Austria

Honadle, G. (1999). *How Context Matters: Linking Environmental Policy to People and Place*. Kumarian Press., https://api.semanticscholar.org/CorpusID:152578133

Hove, A. (2023). Synergies between China's Whole County PV Program and Rural Heating Electrification. – The Oxford Institute for Energy Studies, May 2023, available at: https://www.oxfordenergy.org/wpcms/wp-content/uploads/2023/05/CE6-Synergies-between-Chinas-Whole-County-PV-program-and-rural-heating-electrification.pdf

Huang, T., Liu, Z., & Zhao, T. (2022). Evolutionary Game Analysis of Responding to the EU's Carbon Border Adjustment Mechanism. *Energies*, *15*(2), 427. DOI: 10.3390/en15020427

Hu, J., Harmsen, R., Crijns-Graus, W., & Worrell, E. (2018). Barriers to investment in utility-scale variable renewable electricity (VRE) generation projects. *Renewable Energy*, *121*, 730–744. DOI: 10.1016/j.renene.2018.01.092

Hurlimann, A. C., & March, A. P. (2012). The role of spatial planning in adapting to climate change. *Wiley Interdisciplinary Reviews: Climate Change*, *3*(5), 477–488. DOI: 10.1002/wcc.183

Huttunen, S., Ojanen, M., Ott, A., & Saarikoski, H. (2022). What about citizens? A literature review of citizen engagement in sustainability transitions research. *Energy Research & Social Science*, *91*, 102714. DOI: 10.1016/j.erss.2022.102714

IEA. (2020). Sustainable Recovery: World Energy Outlook Special Report. International Energy Agency. https://www.iea.org/reports/sustainable-recovery

IHA. (2024). China's Fengning Station: World's Largest Pumped Hydro Power Plant Sets New Global Benchmark. - International Hydropower Association, 16 August 2024, available at: https://www.hydropower.org/news/chinas-fengning-station-worlds-largest-pumped-hydro-power-plant-sets-new-global-benchmark

IISD. (2020). Global Climate Change Governance: The Search for Effectiveness. Retrieved from https://www.iisd.org/articles/deep-dive/global-climate-change-governance-search-effectiveness-and-universality

ILO. (2018). World Employment and Social Outlook 2018: Greening with jobs. International Labour Organization. https://www.ilo.org/weso-greening/documents/WESO_Greening_EN_web2.pdf

IMF. (2019). *Fiscal policies for Paris Climate Strategies—From principle to practice*. International Monetary Fund. https://www.imf.org/en/Publications/Policy-Papers/Issues/2019/05/01/Fiscal-Policies-for-Paris-Climate-Strategies-From-Principle-to-Practice-46826

Inman, P. (2021). How bad is China's energy crisis? – The Guardian, 29 September 2021, available at: https://www.theguardian.com/world/2021/sep/29/how-bad-is-chinas-energy-crisis

Innes, J. E., & Booher, D. E. (2004). Reframing public participation: Strategies for the 21st century. *Planning Theory & Practice*, *5*(4), 419–436. DOI: 10.1080/1464935042000293170

Interesse, G. (2024). China's New Renewable Energy Plan: Key Insights for Businesses. - China Briefing, 26 November 2024, available at: https://www.china-briefing.com/news/chinas-new-renewable-energy-plan-key-insights-for-businesses/

Intergovernmental Panel on Climate Change (IPCC). (2021). [The Physical Science Basis.]. *Climatic Change*, 2021.

Intergovernmental Panel on Climate Change (IPCC). (2021). *Climate change 2021: The physical science basis*. Cambridge University Press.

Intergovernmental Panel on Climate Change. (2022). Climate change 2022: Mitigating climate change. https://www.ipcc.ch/report/ar6/wg3/

International Energy Agency. (2020). Sustainable Recovery. Paris, available at: https://www.iea.org/reports/sustainable-recovery

IPBES. (2019). Global assessment report on biodiversity and ecosystem services of the Intergovernmental Science-Policy Platform on Biodiversity and Ecosystem Services. IPBES Secretariat. https://ipbes.net/global-assessment

IPCC. (2018). *Global Warming of 1.5°C: An IPCC Special Report*. Intergovernmental Panel on Climate Change.

IPCC. (2022). Climate Change 2022: Impacts, Adaptation and Vulnerability. Cambridge University Press. https://www.ipcc.ch/report/ar6/wg2/

ISD. (2020). Global Climate Change Governance: The Search for Effectiveness. Retrieved from https://www.iisd.org/articles/deep-dive/global-climate-change-governance-search-effectiveness-and-universality

Islam, S. T., Dey, A., Sanyal, N., & Bandyopadhyay, S. (2022). *Troubles in The Sundarbans: A study of social and ecological issues in Hingalganj Block, North 24 Parganas West Bengal*. ActionAid International.

Islam, S., Chu, C., Smart, J. C. R., & Liew, L. (2020). Integrating disaster risk reduction and climate change adaptation: A systematic literature review. *Climate and Development*, *12*(3), 255–267. DOI: 10.1080/17565529.2019.1613217

Ismagilova, E., Slade, E. L., & Kitchin, R. (2020). Security, privacy, and risks within smart cities: Literature review and development of a smart city interaction framework. *Information Systems Frontiers, 22*, 243–264. PMID: 32837262

Jabeen, H., Johnson, C., & Allen, A. (2010). Built-in resilience: Learning from grassroots coping strategies for climate variability. *Environment and Urbanization, 22*(2), 415–431. DOI: 10.1177/0956247810379937

Jäger-Waldau, A., Kougias, I., Taylor, N., & Thiel, C. (2020). How photovoltaics can contribute to GHG emission reductions of 55% in the EU by 2030. *Renewable & Sustainable Energy Reviews, 126*, 109836. DOI: 10.1016/j.rser.2020.109836

Jahan, R. (1995). *The elusive agenda: Mainstreaming women in development*. London: ZED Books. https://www.jstor.org/stable/41260001

Jalais, A. (2007). The Sundarbans: Whose World Heritage Site? *Conservation & Society, 5*(3), 335–342. https://www.jstor.org/stable/26392892

Jamal, S., Ghosh, A., Hazarika, R., & Sen, A. (2022, July). Livelihood, conflict and tourism: An assessment of livelihood impact in Sundarbans, West Bengal. *International Journal of Geoheritage and Parks, 10*(1), 383–399. DOI: 10.1016/j.ijgeop.2022.07.004

Jandhyala, B., & Tilak, G. (2002). Education and poverty. *Journal of Human Development, 3*(2).

Jaworski, B. J., & Kohli, A. K. (2014). Co-creating the voice of the customer. In *The service-dominant logic of marketing* (pp. 109–117). Routledge.

Jeppesen, L. B. (2005). User toolkits for innovation: Consumers support each other. *Journal of Product Innovation Management, 22*(4), 347–362. DOI: 10.1111/j.0737-6782.2005.00131.x

Ji, Y. (2007). Dealing with the Malacca Dilemma: China's effort to protect its energy supply. – Strategic Analysis, Vol. 31/3/2007, Institute for Defense Studies and Analyses, available at: www.idsa.in

Johnson, K. (2009). Protectionist Breezes: Wind-Power Companies Cry Foul on China. - Wall Street Journal, 28 May 2009, available at: http://europe.wsj.com

Johnson, C. L., Bozeman, B., & He, Z. (2020). Canadian smart cities: Are we wiring new citizen-local government interactions? *The Canadian Geographer. Geographe Canadien, 64*(4), 548–561. DOI: 10.1111/cag.12623

Johnston, I. (2016). Developing world invests more in renewable energy than rich countries for first time, new study says. - The Independent, 1 June 2016, available at: https://www.independent.co.uk/climate-change/news/renewable-energy-investment-developed-world-developing-world-ren21-report-a7058436.html

Joshi, A., & Moore, M. (2004). Institutionalised co-production: Unorthodox public service delivery in challenging environments. *The Journal of Development Studies*, *40*(4), 31–49. DOI: 10.1080/00220380410001673184

Joss, S., Cook, M., & Dayot, Y. (2017). Smart cities: Towards a new citizenship regime? A discourse analysis of the British smart city standard. *Journal of Urban Technology*, *24*(4), 29–49. DOI: 10.1080/10630732.2017.1336027

Jung, H., & Song, C. K. (2023). Effects of emission trading scheme (ETS) on change rate of carbon emission. *Scientific Reports*, *13*(1), 912. DOI: 10.1038/s41598-023-28154-6 PMID: 36650382

Kahn, M. E., & Matsusaka, J. G. (1997). Demand for environmental goods: Evidence from voting patterns on California initiatives. *The Journal of Law & Economics*, *40*(1), 137–174. DOI: 10.1086/467369

Kahraman, A. C., & Acar. (2024). Notes from COP 28. Marmara Municipalities Union.

Kaijser, A., & Kronsell, A. (2014). Climate change through the lens of intersectionality. *Environmental Politics*, *23*(3), 417–433. DOI: 10.1080/09644016.2013.835203

Kalkınma Bakanlığı, T.C. (2015). *Bilgi toplumu stratejisi ve eylem planı (2015-2018).*

Kalkışım, H. M. (2019). Yönetişim Kavram ve Teorisi Üzerine Bir Değerlendirme, E-Yönetişim, Eds. B. Parlak ve K.C. Doğan, Beta Yayıncılık, İstanbul, 19-32.

Kalkışım, H. M., Doğan, K. C., & Soylu, A. C. (2023). Türk Sendikalarının 1978 Ekonomik Krizi Karşısındaki Tutumları, Sosyal. *Beşerî ve İdari Bilimler Dergisi*, *6*(5), 641–662.

Kamb, Z., Shi, L., Silva, S., & Spicer, J. (2022). Resident-Owned Resilience: Can Cooperative Land Ownership Enable Transformative Climate Adaptation for Manufactured Housing Communities?. *Housing Policy Debate*, 33, 1055 -. DOI: 10.1080/10511482.2021.2013284

Kanjilal, B., Bose, S., Patra, N., Barman, D., Ghosh, U., Mandal, A., Vardrevu, L., & Sengupta, P. (2013). *How Healthy are the Children of Indian Sundarbans?* http://r4d.dfid.gov.uk/pdf/outputs/FutureHealth_RPC/SHWforweb.pdf

Kanjilal, B., Mazumdar, P. G., Mukherjee, M., Mondal, S., Barman, D., & Singh, S. M. (2010, January). *Health care in the Sundarbans (India): Challenges. Future Health Systems Research Programme*, 21-30. Retrieved from https://assets.publishing .service.gov.uk/media/57a08b24ed915d3cfd000b68/sundarbans.pdf

Kanjilal, B., Mazumdar, P., Mukherjee, M., Mondal, S., Barman, D., Singh, S., & Mandal, A. (2010). Health care in the Sundarbans (India): Challenges and plan for a better future. *Future Health Systems Research Programme* (Issues 1–91). DOI: 10.6133/apjcn.012016.01

Karmaker, S. C., Hosan, S., Chapman, A. J., & Saha, B. B. (2021). The role of environmental taxes on technological innovation. *Energy*, *232*, 121052. DOI: 10.1016/j. energy.2021.121052

Kashif, M., Pinglu, C., Ullah, S., & Zaman, M. (2024). Evaluating the influence of financial technology (FinTech) on sustainable finance: A comprehensive global analysis. *Financial Markets and Portfolio Management*, *38*(1), 123–155. DOI: 10.1007/s11408-023-00439-w

Katona, Á. (2022). Customer involvement: A systematic literature review of the definition, measurement, and innovation outcomes. *Vezetéstudomány/Budapest. Management Review*, *53*(5), 59–69.

Kaulio, M. A. (1998). Customer, consumer and user involvement in product development: A framework and a review of selected methods. *Total Quality Management*, *9*(1), 141–149. DOI: 10.1080/0954412989333

Kaya, M. H. (2025). İklim Değişikliği ve Güvenlik: Bir Tehdit Analizi. *Nous Academy Journal*, *4*, 36–53.

Kaygısız, Ü., & Aydın, S. Z. (2017). Yönetişimde yeni bir ufuk olarak akıllı kentler. *Mehmet Akif Ersoy Üniversitesi Sosyal Bilimler Enstitüsü Dergisi*, *9*(18), 56–81.

Kaymaz, C. and Tut, G. (2020). İklim Değişikliği Politikalarına Yönelik Bir Çerçeve Oluşturma Denemesi, Kent ve Çevre Araştırmaları Dergisi, 2(1), Haziran, 78-98.

Kazancı, G., & Tezer, A. (2021). İklim Değişikliğine Uyumda Mekânsal Planlama Ve Akıllı Yönetişim Çerçevesinde Türkiye. *Planlama*, *31*(2), 302–320.

Keane, J., Mendez-Parra, M., Pettinotti, L., & Colenbrander, S. (2021). Carbon markets and standards. Policy Brief. London, www.odi.org/publications/carbon -markets-andstandards

Kemp, S. P., Whittaker, J. K., & Tracy, E. M. (1997) *Person–Environment Practice: The Social Ecology of Interpersonal Helping*. New York: Aldine De Gruyter. https://api.semanticscholar.org/CorpusID:141687227

Kemp, R., & Never, B. (2017). Green transition, industrial policy, and economic development. *Oxford Review of Economic Policy*, *33*(1), 66–84. DOI: 10.1093/oxrep/grw037

Kernaghan, S., & da Silva, J. (2013). Initiating and sustaining action: Experiences building resilience to climate change in Asian cities. *Urban Climate*, *7*, 47–63. DOI: 10.1016/j.uclim.2013.10.008

Kerzner, H. (2025). *Project management: a systems approach to planning, scheduling, and controlling*. John Wiley & Sons.

Keweloh, R. (2015). Integrating Disaster Risk Reduction and Climate Change Adaptation in Theory and Practice. *IFHV Working Paper, 5*(3).

Khandker, S. R., Samad, H. A., Ali, R., & Barnes, D. F. (2014). Who benefits most from rural electrification? Evidence in India. *The Energy Journal (Cambridge, Mass.)*, *35*(2), 75–96. DOI: 10.5547/01956574.35.2.4

Khan, S. A. R., Umar, M., Yu, Z., & Nawaz, M. T. (2023). A Recent Digitalization in Recycling Industry Attaining Ecological Sustainability: A Comprehensive Outlook and Future Trend. *Environmental Science and Pollution Research International*, *30*(47), 103760–103775. DOI: 10.1007/s11356-023-29537-y PMID: 37695483

Kim, J., Chung, J., & Suh, K. (2020). Green new deal policy and wind energy utilization in the Republic of Korea. *Sustainability*, *12*(18), 7599. DOI: 10.3390/su12187599

Kim, J., Park, J., & Glovinsky, P. L. (2018). Customer involvement, fashion consciousness, and loyalty for fast-fashion retailers. *Journal of Fashion Marketing and Management*, *22*(3), 301–316. DOI: 10.1108/JFMM-03-2017-0027

Kindon, S., Pain, R., & Kesby, M. (2007). *Participatory action research approaches and methods. Connecting people, participation and place*. Routledge. DOI: 10.4324/9780203933671

Kircher, M. (2019). Bioeconomy: Markets, implications, and investment opportunities. *Economies*, *7*(3), 73. DOI: 10.3390/economies7030073

Kiss, B., Sekulova, F., Hörschelmann, K., Salk, C. F., Takahashi, W., & Wamsler, C. (2022). Citizen participation in the governance of nature-based solutions. *Environmental Policy and Governance*, *32*(3), 247–272. DOI: 10.1002/eet.1987

Kitchin, R. (2014). The real-time city? Big data and smart urbanism. *GeoJournal, 79*(1), 1–14. DOI: 10.1007/s10708-013-9516-8

Kivilä, J., Martinsuo, M., & Vuorinen, L. (2017). Sustainable project management through project control in infrastructure projects. *International Journal of Project Management, 35*(6), 1167–1183. DOI: 10.1016/j.ijproman.2017.02.009

Klinsky, S., Roberts, T., Huq, S., Okereke, C., Newell, P., Dauvergne, P., O'Brien, K., Schroeder, H., Tschakert, P., Clapp, J., Keck, M., Biermann, F., Liverman, D., Gupta, J., Rahman, A., Messner, D., Pellow, D., & Bauer, S. (2017). Why equity is fundamental in climate change policy research. *Global Environmental Change, 44*, 170–173. DOI: 10.1016/j.gloenvcha.2016.08.002

Kónya, L. (2006). Exports and growth: Granger causality analysis on OECD countries with a panel data approach. *Economic Modelling, 23*(6), 978–992. DOI: 10.1016/j.econmod.2006.04.008

Kosenen, K., & Nicodime, G. (2009). The role of fiscal instruments in environmental policy. CESifo Working Paper, No. 2719.

Kyoto Protocol. (1997). *United Nations Framework Convention on Climate Change*.

Lannoo, K., & Thomadakis, A. (2020). Derivatives in sustainable finance. *CEPS-ECMI Study. Brussels: Centre for European Policy Studies, 3*.

Lanteigne, M. (2008). China's maritime security and the "Malacca dilemma". – Asian Security, Vol. 4, Issue 2/2008, pp. 143–161.

Lawrence, A. (2006). 'No personal motive?' Volunteers, biodiversity, and the false dichotomies of participation. *Ethics Place and Environment, 9*(3), 279–298. DOI: 10.1080/13668790600893319

Leal-Arcas, R. Faktaufon, M., & Kyprianou, A. (2022). A legal exploration of the European Union's carbon border adjustment mechanism. *European Energy and Environmental Law Review*, 223 – 240.

Leck, H., & Simon, D. (2012). Fostering multiscalar collaboration and co-operation for effective governance of climate change adaptation. *Urban Studies (Edinburgh, Scotland), 50*(6), 1221–1238. DOI: 10.1177/0042098012461675

Lee, D., Feiertag, P., & Unger, L. (2024). Co-production, co-creation or co-design of public space? A systematic review. *Cities (London, England), 154*, 105372. DOI: 10.1016/j.cities.2024.105372

Lee, H., Son, J., Joo, D., Ha, J., Yun, S., Lim, C.-H., & Lee, W.-K. (2020). Sustainable Water Security Based on the SDG Framework: A Case Study of the 2019 Metro Manila Water Crisis. *Sustainability (Basel)*, *12*(17), 6860. DOI: 10.3390/su12176860

Lehner, O. M. (2016). *Routledge handbook of social and sustainable finance*. Routledge. DOI: 10.4324/9781315772578

Leichenko, R. (2011). Climate change and urban resilience. *Current Opinion in Environmental Sustainability*, *3*(3), 164–168. DOI: 10.1016/j.cosust.2010.12.014

Leichenko, R., & Silva, J. A. (2014). Climate change and poverty: Vulnerability, impacts, and alleviation strategies. *Wiley Interdisciplinary Reviews: Climate Change*, *5*(4), 539–556. DOI: 10.1002/wcc.287

Lemos, M. C., & Agrawal, A. (2006). Environmental governance. *Annual Review of Environment and Resources*, *31*(1), 297–325. DOI: 10.1146/annurev.energy.31.042605.135621

Lim, T. (2024). Environmental, social, and governance (ESG) and artificial intelligence in finance: State-of-the-art and research takeaways. *Artificial Intelligence Review*, *57*(4), 76. DOI: 10.1007/s10462-024-10708-3

Lim, Y., & Lee, H. (2018). Citizen participation in building citizen-centric smart cities. *Geografia: Malaysian Journal of Society and Space*, *14*(4), 112–127. DOI: 10.17576/geo-2018-1404-04

Lints, T. (2010). The Essentials of Defining Adaptation. *4th Annual IEEE Systems Conference*, 113–116.

Lin, Y., Lu, Z., & Wang, Y. (2023). The impact of environmental, social, and governance (ESG) practices on investment efficiency in China: Does digital transformation matter? *Research in International Business and Finance*, *66*, 102050. DOI: 10.1016/j.ribaf.2023.102050

Local Governments and Municipal Authorities. (2025). About the LGMA. https://www.cities-and-regions.org/about-the-lgma/

Local Governments Climate Summit. (2023). Local Governments Climate Summit. https://www.cop28.com/en/lcas

Lockwood, M. (2010). Good governance for terrestrial protected areas: A framework, principles and performance outcomes. *Journal of Environmental Management*, *91*(3), 754–766. DOI: 10.1016/j.jenvman.2009.10.005 PMID: 19896262

Lockwood, M. (2018). Right-wing populism and the climate change agenda: Exploring the linkages. *Environmental Politics, 27*(4), 712–732. DOI: 10.1080/09644016.2018.1458411

Loukoianova, E., Natalucci, F., Wang, D., & Kanada, S. (2024). *Fintech Applications for Boosting Climate Finance*. International Monetary Fund. DOI: 10.5089/9798400293740.066

Lu, J., Gulzar, F., & Lai, Y. (2024). Exploring the intersection between sustainable finance and achieving carbon neutrality in the transportation sector. *Heliyon, 10*(15), e34467. DOI: 10.1016/j.heliyon.2024.e34467 PMID: 39145029

Luker, E., & Harris, L. M. (2019). Developing new urban water supplies: Investigating motivations and barriers to groundwater use in Cape Town. *International Journal of Water Resources Development, 35*(6), 917–937. DOI: 10.1080/07900627.2018.1509787

Luque-Vega, L. F., Michel-Villarreal, R., & Lopez-Neri, E. (2020). Smart cities oriented project planning and evaluation methodology driven by citizen perception—IoT smart mobility case. *Sustainability (Basel), 12*(15), 6214. DOI: 10.3390/su12177088

Macrotrends. (n.d.). World urban population. https://www.macrotrends.net/global-metrics/countries/wld/world/urban-population

Magida, N., & Makoza, F. (2023). The Effect of Water Crisis on Fine Dining Restaurant Entrepreneurs in the Cape Town Metro. *Tourism and Leisure, 12*.

Magnani, E. (2000). The environmental kuznets curve, environmental protection policy and income distribution. *Ecological Economics, 32*(3), 431–443. DOI: 10.1016/S0921-8009(99)00115-9

Mahadevia, K., & Vikas, M. (2012). Climate change—Impact on the Sunderbans: A case study. *International Scientific Journal-Environmental Science, 2*(1), 7–15.

Mahmood, F., Marwat, I. U. K., & Shah, S. A. (2024). Customer Involvement and Sustainable Product Innovations: The Role of Design Excellence and Technological Capability. *The Lighthouse Journal of Social Sciences, 3*(1), 146–160.

Manville, C., Cochrane, G., Cave, J., Millard, J., Pederson, J. K., Thaarup, R. K., Liebe, A., Wissner, M., Massink, R., & Kotterink, B. (2014). *Mapping smart cities in the EU*. [Rapor]. European Parliament, Policy Department A: Economic and Scientific Policy. https://www.europarl.europa.eu/RegData/etudes/etudes/join/2014/507480/IPOL-ITRE_ET(2014)507480_EN.pdf

Maor, M., Gilad, S., & Ben-Nun Bloom, P. (2013). Organizational reputation, regulatory talk, and strategic silence. *Journal of Public Administration: Research and Theory, 23*(3), 627–649. DOI: 10.1093/jopart/mus047

Marcatajo, G. (2023). Green claims, green washing and consumer protection in the European Union. *Journal of Financial Crime, 30*(1), 143–153. DOI: 10.1108/JFC-11-2021-0240

Marsilliani, L., & Renström, T. I. (2000). *Inequality environmental Protection and growth.* Center for Economic Research, Durham University. DOI: 10.2139/ssrn.235083

Martens, M. L., & Carvalho, M. M. (2017). Key factors of sustainability in project management context: A survey exploring the project managers' perspective. *International Journal of Project Management, 35*(6), 1084–1102. DOI: 10.1016/j.ijproman.2016.04.004

Martinsuo, M. (2013). Project portfolio management in practice and in context. *International Journal of Project Management, 31*(6), 794–803. DOI: 10.1016/j.ijproman.2012.10.013

Matikinca, P., Ziervogel, G., & Enqvist, J. P. (2020). Drought response impacts on household water use practices in Cape Town, South Africa. *Water Policy, 22*(3), 483–500. DOI: 10.2166/wp.2020.169

Mattelmäki, T., & Visser, F. S. (2011). Lost in co-x. *Proceedings of the IASDR2011.*

Mazlum, S. C. (2019). Küresel İklim Politikaları, İklim Değişikliği Eğitim Modülleri Serisi 2, WEglobal, Ankara.

Mazzarano, M. (2024). Emission trading systems, their fiscal progressivity and its limits. https://fsr.eui.eu/emission-trading-systems-their-fiscal-progressivity-and-its-limits/

McManus, S., Seville, E., Vargo, J., & Brunsdon, D. (2008). Facilitated Process for Improving Organizational Resilience. *Natural Hazards Review, 9*(2), 81–90. DOI: 10.1061/(ASCE)1527-6988(2008)9:2(81)

Meadowcroft, J., Banister, D., Holden, E., Langhelle, O., Linnerud, K., & Gilpin, G. (2019). *What next for sustainable development? Our common future at thirty.* Edward Elgar Publishing. DOI: 10.4337/9781788975209

Meijer, A., & Bolívar, M. P. R. (2016). Governing the smart city: A review of the literature on smart urban governance. *International Review of Administrative Sciences, 82*(2), 392–408. DOI: 10.1177/0020852314564308

Mellner, C., Niemi, M., Pöllänen, E., & Osika, W. (2021). *Enhancing social and individual sustainability in urban co-living*. International Journal of Housing Markets and Analysis., DOI: 10.1108/IJHMA-09-2020-0117

Mengi, A. (1999). *Global environmental issues and policies*. Mülkiyeliler Association Foundation Publications.

Mertz, O., Mbow, C., Reenberg, A., & Diouf, A. (2009). Farmers' perceptions of climate change and agricultural adaptation strategies in rural Sahel. *Environmental Management, 43*(5), 804–816. DOI: 10.1007/s00267-008-9197-0 PMID: 18810526

Meyer, P. B., & Schwarze, R. (2019). Financing climate-resilient infrastructure: Determining risk, reward, and return on investment. *Frontiers of Engineering Management, 6*(1), 117–127. DOI: 10.1007/s42524-019-0009-4

Microsoft. (2023). AI for Earth. Retrieved from https://www.microsoft.com

Midgley, J. (1994). Defining Social Development: Historical Trends and Conceptual Formulations. *Social Development Issues, 16*(3), 3–19. DOI: 10.4135/9781446221839.n2

Milbrath, L. W. (1981). Political participation. In *The Handbook of Political Behavior* (Vol. 4, pp. 197–240). Springer US. DOI: 10.1007/978-1-4684-3878-9_4

Millard, J. (2018). Open governance systems: Doing more with more. *Government Information Quarterly, 35*(4), S77–S87. DOI: 10.1016/j.giq.2015.08.003

Miller, C. A., & Wyborn, C. (2020). Co-production in global sustainability: Histories and theories. *Environmental Science & Policy, 113*, 88–95. DOI: 10.1016/j.envsci.2018.01.016

Miller, S. E., Hayward, R. A., & Shaw, T. V. (2012). Environmental Shifts for Social Work: A Principles Approach. *International Journal of Social Welfare, 21*(3), 270–277. DOI: 10.1111/j.1468-2397.2011.00848.x

Millington, N., & Scheba, S. (2021). Day Zero and The Infrastructures of Climate Change: Water Governance, Inequality, and Infrastructural Politics in Cape Town's Water Crisis. *International Journal of Urban and Regional Research, 45*(1), 116–132. DOI: 10.1111/1468-2427.12899

Mimi, R., & Haynes, L. (2018). *The role of livelihood activities and income portfolios in shaping women's perceptions of wellbeing and adaptation choices, in the Indian Bengal Delta Statement of Originality*.

Mir, F. A., & Pinnington, A. H. (2014). Exploring the value of project management: Linking project management performance and project success. *International Journal of Project Management*, *32*(2), 202–217. DOI: 10.1016/j.ijproman.2013.05.012

Mishra, M., Acharyya, T., Santos, C. A. G., da Silva, R. M., Kar, D., Mustafa Kamal, A. H., & Raulo, S. (2021). Geo-ecological impact assessment of severe cyclonic storm Amphan on Sundarban mangrove forest using geospatial technology. *Estuarine, Coastal and Shelf Science*, *260*, 107486. DOI: 10.1016/j.ecss.2021.107486

Mistri, A. (2019). Is the migration from Indian Sundarban an environmental migration? Investigating through Sustainable Livelihood Approach (SLA). *Asian Profile*, *47*(3), 195–218.

Mitlin, D., & Satterthwaite, D. (2012). Addressing poverty and inequality new forms of urban governance in Asia. *Environment and Urbanization*, *24*(2), 395–401. DOI: 10.1177/0956247812458062

Mitra, R. (2024). *Nature-based Solutions can Help Sundarbans Survive. DownToEarth*. Retrieved from https://www.downtoearth.org.in/news/climate-change/nature-based-solutions-can-help-sundarbans -survive-95230

Mizanur Rahman, M., Rakib Hossain, M., & Nazrul Islam, M. (2021). Climate Change Impact on Sundarbans: Challenges for Mitigation Strategies. *Climate Change in Bangladesh: A Cross-Disciplinary Framework*, 47-64.

Mohd, S., & Kaushal, V. K. (2018). Green finance: A step towards sustainable development. *MUDRA: Journal of Finance and Accounting*, *5*(1), 59–74. DOI: 10.17492/mudra.v5i01.13036

Molinari, N. (2017). Intensifying Insecurities: The impact of climate change on vulnerability to human trafficking in the Indian Sundarbans. *Anti Trafficking Review*, *363*(8), 1–11. DOI: 10.14197/atr.20121784

Monasterolo, I., Mandel, A., Battiston, S., Mazzocchetti, A., Oppermann, K., Coony, J., Stretton, S., Stewart, F., & Dunz, N. (2024). The role of green financial sector initiatives in the low-carbon transition: A theory of change. *Global Environmental Change*, *89*, 102915. DOI: 10.1016/j.gloenvcha.2024.102915

Mondal, M. K., Bhuiyan, S. I., & Franco, D. T. (2001). Soil salinity reduction and prediction of salt dynamics in the coastal rice lands of Bangladesh. *Agricultural Water Management*, *47*(1), 9–23. DOI: 10.1016/S0378-3774(00)00098-6

Mondal, T., Bowers, W. W., & Hossen, M. S. (2021, March 25). Sustainable Management of Sundarbans: Stakeholder Attitudes Towards Participatory Management and Conservation of Mangrove Forests. *Journal of Sustainable Development, 14*(3), 23–40. DOI: 10.5539/jsd.v14n3p23

Montreal Protocol. (1987). *United Nations Environment Programme.*

Moore, J. (2023, March 6). *PayBitoPro saves Sundarbans' child education.* Retrieved from PayBitoPro: https://www.paybito.com/lost-in -the -mangroves -paybito -steps -up -to -save -child -education -in -sundarbans/

Mora, L., & Bolici, R. (2017). How to become a smart city: Learning from Amsterdam. A. Bisello, D. Vettorato, R. Stephens, & P. Elisei (Ed.), *Smart and sustainable planning for cities and regions: Results of SSPCR 2015* (ss. 251–266). Springer.

Morse, S. (2018). Relating environmental performance of nation states to income and income inequality. *Sustainable Development (Bradford), 26*(1), 99–115. DOI: 10.1002/sd.1693

Moser, S. C. (2016). Reflections on climate change communication research and practice in the second decade of the 21st century: What more is there to say? *Wiley Interdisciplinary Reviews: Climate Change, 7*(3), 345–369. DOI: 10.1002/wcc.403

Moser, S. C., & Ekstrom, J. A. (2010). A framework to diagnose barriers to climate change adaptation. *Proceedings of the National Academy of Sciences of the United States of America, 107*(51), 22026–22031. DOI: 10.1073/pnas.1007887107 PMID: 21135232

Mozumdar, M., Shamsuzzaman, M., Un-Nabi, R., & Harun-Al-Rashid, A. (2018). Socio-economic characteristics and fishing operation activities of the artisanal fishers in the Sundarbans Mangrove Forest, Bangladesh. *Turkish Journal of Fisheries and Aquatic Sciences, 18*(6), 788–799. DOI: 10.4194/1303-2712-v18_6_05

Mukherjee, V., & Bali, A. S. (2019). Policy coherence for sustainable development in India: An analysis of sustainable development goals. NITI Aayog. https://niti.gov.in/sites/default/files/2019-07/SDG-Policy-Coherence-in-India.pdf

Mukherjee, S., Sundberg, T., Sikdar, P. K., & Schütt, B. (2022). An Integrated Quantitative Assessment of Urban Water Security of a Megacity in the Global South. *Frontiers in Water, 4*, 834239. DOI: 10.3389/frwa.2022.834239

Mukhopadhyay, A. (2009). *Cyclone Aila and the Sundarbans: An Enquiry into the Disaster and Politics of Aid and Relief.* In Research Report on CRG Annual Winter Course on Forced Migration.

Mupa, M. N., Pamful, E. E., Nnaji, J. C., & Adu-Boahen, J. O. A. N. N. A. (2024). Integrating ESG factors in investment decision-making for renewable energy projects. *Iconic Research and Engineering Journals*, *8*(2), 273–293.

Muth, D. (2023). Pathways to stringent carbon pricing: Configurations of political economy conditions and revenue recycling strategies. A comparison of thirty national level policies. *Ecological Economics*, *214*, 107995. DOI: 10.1016/j.ecolecon.2023.107995

Myllyvirta, L., & Qin, Q. (2024). Clean Energy Was Top Driver of China's Economic Growth in 2023. – Carbon Brief, 25 January 2024, available at: https://www.carbonbrief.org/analysis-clean-energy-was-top-driver-of-chinas-economic-growth-in-2023/

Myllyvirta, L., & Zhang, X. (2022). What Do China's Gigantic Wind and Solar Bases Mean for Its Climate Goals? – Carbon Brief, 3 May 2022, available at: https://www.carbonbrief.org/analysis-what-do-chinas-gigantic-wind-and-solar-bases-mean-for-its-climate-goals/

Myllyvirta, L. (2021a, October). The Real Reasons Behind China's Energy Crisis. –. *Foreign Policy*, *7*, https://foreignpolicy.com/2021/10/07/china-energy-crisis-electricity-coal-pricing-renewables

Myllyvirta, L. (2021b, September). Why Xi's Coal Pledge Is a Big Deal. –. *Foreign Policy*, *28*, https://foreignpolicy.com/2021/09/28/china-xi-jinping-coal-pledge-unga-clean-energy developing-countries/

Nabatchi, T., Sancino, A., & Sicilia, M. (2017). Varieties of Participation in Public Services: The Who, When, and What of Participation in Service Design and Delivery. *Public Administration Review*, *77*(5), 766–776. DOI: 10.1111/puar.12765

Naeem, M. A., Karim, S., Rabbani, M. R., Bashar, A., & Kumar, S. (2023). Current state and future directions of green and sustainable finance: A bibliometric analysis. *Qualitative Research in Financial Markets*, *15*(4), 608–629. DOI: 10.1108/QRFM-10-2021-0174

Naess, L. O. (2013). The role of local knowledge in adaptation to climate change. *Wiley Interdisciplinary Reviews: Climate Change*, *4*(2), 99–106. DOI: 10.1002/wcc.204

Nambisan, S. (2002). Designing virtual customer environments for new product development: Toward a theory. *Academy of Management Review*, *27*(3), 392–413. DOI: 10.2307/4134386

NASA, & NOAA. (2024). Annual Global Climate Report 2023. U.S. Government Printing Office.

NASA. (2020). *The causes of climate change.* https://climate.nasa.gov/causes/

NASA. (2023). Globe observer app. Retrieved from https://www.nasa.gov

NASA. (2023). NASA's Earth Observing System (EOS). Retrieved from https://www.nasa.gov

NASA. (2023). Orbiting Carbon Observatory-2. Retrieved from https://www.nasa.gov

Nature Climate Change. (2016). Nature Climate Change. https://www.nature.com/articles/nclimate3169

Naude, R. T., & Naude, M. J. (2019). Impact of climate change and extreme weather conditions on wine growing within the Stellenbosch region. *Journal of Contemporary Management, 16*(2), 111–134. DOI: 10.35683/jcm19041.0029

Navigant ve The Generation Foundation. (2018). Raising the acceptability and effectiveness of carbon pricing: The crucial role of carbon revenue recycling, Carbon Pricing Unlocked Partnership. https://guidehouse.com/-/media/www/site/downloads/energy/2018/cpu2018carbonrevenuerecycling.pdf

Nelson, D. R. (2011). Adaptation and resilience: Responding to a changing climate. *Wiley Interdisciplinary Reviews: Climate Change, 2*(1), 113–120. DOI: 10.1002/wcc.91

Nelson, K., Gillespie-Marthaler, L., Baroud, H., Abkowitz, M., & Kosson, D. (2020). An integrated and dynamic framework for assessing sustainable resilience in complex adaptive systems. *Sustainable and Resilient Infrastructure, 5*(5), 311–329. DOI: 10.1080/23789689.2019.1578165

Newell, S. J., & Green, C. L. (1997). Racial differences in consumer environmental concern. *The Journal of Consumer Affairs, 31*(1), 53–69. DOI: 10.1111/j.1745-6606.1997.tb00826.x

Newman, N. S. (2022). The European Union opean Union's CBAM: Is ıt effective economic climate policy? *Pepperdine Policy Review*, 14 (3), https://digitalcommons.pepperdine.edu/ppr/vol14/iss1/3

Nhamo, G., & Agyepong, A. O. (2019). Climate change adaptation and local government: Institutional complexities surrounding Cape Town's Day Zero. *Jàambá, 11*(3). DOI: 10.4102/jamba.v11i3.717 PMID: 31308900

Nicholls, R. J., Adger, W. N., Hutton, C. W., & Hanson, S. E. (2020). *Deltas in the Anthropocene*. Palgrave Macmillan Springer Nature., DOI: 10.1007/978-3-030-23517-8

Nicol, M., Shishlov, I., & Cochran, I. (2018). Green Bonds: Improving their contribution to the low-carbon and climate resilient transition. *I4CE–Institute for Climate Economics, Paris, 217*.

Nilsson, M., Chisholm, E., Griggs, D., Howden-Chapman, P., McCollum, D., Messerli, P., Neumann, B., Stevance, A.-S., Visbeck, M., & Stafford-Smith, M. (2018). Mapping interactions between the sustainable development goals: Lessons learned and ways forward. *Sustainability Science, 13*(6), 1489–1503. DOI: 10.1007/s11625-018-0604-z PMID: 30546483

Nilsson, M., Griggs, D., & Visbeck, M. (2016). Policy: Map the interactions between sustainable development goals. *Nature, 534*(7607), 320–322. DOI: 10.1038/534320a PMID: 27306173

Noll, D. (2020). *Day Zero: Communicating the water crisis in Cape Town, 2015-2018—A qualitative study about the communication during the water crisis and how the public responded to the messages* [Bachelor's Thesis]. Karlstads Universiteit.

Norgaard, K. M. (2017, September). The Sociological Imagination in a Time of Climate Change. *Global and Planetary Change, Elsevier*. DOI: DOI: 10.1016/j.gloplacha.2017.09.018

Norris, F., Stevens, S., Pfefferbaum, B., Wyche, K., & Pfefferbaum, R. (2008). Community resilience as metaphor, theory, set of capacities, and strategy for disaster readiness. *American Journal of Community Psychology, 41*(1–2), 127–150. DOI: 10.1007/s10464-007-9156-6 PMID: 18157631

Novaković, I. (2020). China in the Western Balkans Energy Sector: Between Development and Ecological Concerns. - European Western Balkans, 2 November 2020, available at: https://europeanwesternbalkans.com/2020/11/02/china-in-the-western-balkans-energy-sector-between-development-and-ecological-concerns/

NREL. (2004). Renewable Energy in China. - National Renewable Energy Laboratory, U.S. Department of Energy - Office of Energy Efficiency and Renewable Energy, April 2004, Colorado, available at: www.nrel.gov

Nwariaku, H., Fadojutimi, B., Lawson, L. G. L., Agbelusi, J., Adigun, O. A., Udom, J. A., & Olajide, T. D. (2024). Blockchain technology as an enabler of transparency and efficiency in sustainable supply chains. *International Journal of Science and Research Archive, 12*(2), 1779–1789. DOI: 10.30574/ijsra.2024.12.2.1454

Nyong, A., Adesina, F., & Elasha, B. O. (2007). The value of indigenous knowledge in climate change mitigation and adaptation strategies in the African Sahel. *Mitigation and Adaptation Strategies for Global Change*, *12*(5), 787–797. DOI: 10.1007/s11027-007-9099-0

O'Brien, K. (2012). Global environmental change II: From adaptation to deliberate transformation. *Progress in Human Geography*, *36*(5), 667–676. DOI: 10.1177/0309132511425767

O'Brien, K., Eriksen, S., Nygaard, L. P., & Schjolden, A. (2007). Why different interpretations of vulnerability matter in climate change discourses. *Climate Policy*, *7*(1), 73–88. DOI: 10.1080/14693062.2007.9685639

Obaid, T., & Abu-Naser, S. S. (2023). Big data analytics in project management: a key to success. International Journal of Academic Engine e ring Re search (IJAER), ISSN: 2643-9085, Vol. 7 Is s ue 7, July - 2023, Page s: 1-8

Ocaklı, Ş. N. (2021). İklim Değişikliği Yönetişimi ve Türkiye Üzerine Bir Değerlendirme, III. Kent Konseyleri Sempozyumu Geleceğin Kentlerinde İşbirliği ve Dayanışma (Bildiri Kitabı), 18-20 Kasım, Balıkesir, 561-576.

OECD Data Explorer. *OECD environmental and carbon data*. https://data-explorer.oecd.org/

OECD. (2011). *Environmental taxation: A guide for policy makers*. OECD Publishing.

OECD. (2021). *Carbon pricing and climate policy: How to align policies and achieve net zero*. OECD Publishing.

OECD. (2021). Carbon pricing in action. Retrieved from https://www.oecd.org

OECD. (2021). Engaging citizens in climate policy. Retrieved from https://www.oecd.org

OECD. (2021). Enhancing transparency in climate action: Balancing privacy and accountability. Retrieved from https://www.oecd.org

OECD. (2024). Climate Adaptation: Why Local Governments Cannot Do It Alone | OECD Policy Paper.

OECD. (n.d.). Climate Adaptation: Why Local Governments Cannot Do It Alone. Retrieved from https://www.oecd.org/en/publications/climate-adaptation-why-local-governments-cannot-do-it-alone_be90ac30-en.html

Okoye, K., Nganji, J. T., Hiran, K. K., & Hosseini, S. (2024, May). Impact and implications of AI methods and tools for the future of education. [). Frontiers Media SA.]. *Frontiers in Education*, *9*, 1434052. DOI: 10.3389/feduc.2024.1434052

Olhoff, A., & Schaer, C. (2010). *Screening tools and guidelines to support the mainstreaming of climate change adaptation into development assistance a stocktaking report.* New York: UNDP. http://www.undp.org/climatechange/library.shtml

Oliveira, D. J. S., & Ckagnazaroff, I. B. (2023). Citizen participation as one of the principles of open government. *Cadernos Gestão Pública e Cidadania*, *28*, e84867. DOI: 10.12660/cgpc.v28.84867

Olovsson, C. (2018). Is climate change relevant for central banks? *Sveriges Riksbank Economic Commentaries*, *13*(13), 1–8.

Olsen, L. (2009). The employment effects of climate change and climate change responses: A role for international labour standards? Global Union Research Network Discussion Paper No. 12. https://www.gurn.info/en/discussion-papers/no12-may09-the-employment-effects-of-climate-change-and-climate-change-responses-a-role-for-international-labour-standards

Olsson, P., Folke, C., & Berkes, F. (2004). Adaptive co-management for building resilience in social–ecological systems. *Environmental Management*, *34*(1), 75–90. DOI: 10.1007/s00267-003-0101-7 PMID: 15383875

Omer, S. (1979). *'Social Development'. International Social Work*, *22*(3), 11–26. DOI: 10.1177/002087287902200303

Öner, Ş. (2023). İklim Değişikliği Sorununun Uluslararası Gelişmeler Eşliğinde Türkiye'nin Politika ve Kurumlarına Yansıması, Ombudsman Akademik, 18, Ocak-Haziran, 13-47.

OpenSecrets. (2025). Exxon Mobil: Lobbying profile. https://www.opensecrets.org/federal-lobbying/clients/summary?cycle=2024&id=D000000129

Oppenheimer, M., Glavovic, B. C., Hinkel, J., van de Wal, R., Magnan, A. K., Abd-Elgawad, A., Sebesvari, Z. (2019). Sea level rise and implications for low-lying islands, coasts and communities. In H.-O. Pörtner, D. C. Roberts, V. Masson-Delmotte, P. Zhai, M. Tignor, E. Poloczanska, ... A. Okem (Eds.), IPCC Special Report on the Ocean and Cryosphere in a Changing Climate. https://www.ipcc.

Organisation de coopération et de développement économiques (OCDE). (2023). *Exploring new frontiers of citizen participation in the policy cycle.* https://www.oecd.org/governance/exploring-new-frontiers-citizen-participation.pdf

Organisation for Economic Co-operation and Development (OECD). (2021). *Blockchain technologies as enablers of sustainable development goals*. OECD Publishing.

Örnek, İ., & Türkmen, S. (2019). Gelişmiş ve yükselen piyasa ekonomilerinde çevresel kuznets eğrisi hipotezinin analizi. *Journal of the Cukurova University Institute of Social Sciences*, *28*(3), 109–129.

Orr, S. K. (2013). *Environmental policymaking and stakeholder collaboration: Theory and practice*. Taylor & Francis. DOI: 10.1201/b16131

Osborne, S. P., Radnor, Z., & Nasi, G. (2013). A new theory for public service management? Toward a (public) service-dominant approach. *American Review of Public Administration*, *43*(2), 135–158. DOI: 10.1177/0275074012466935

Ostrom, E. (1996). Crossing the great divide: Coproduction, synergy, and development. *World Development*, *24*(6), 1073–1087. DOI: 10.1016/0305-750X(96)00023-X

Oswald, J. (2016). China Turns to Ecology in Search of "Civilisation". – Asian Studies Association of Australia, 03 August 2016, available at: https://asaa.asn.au/china-turns-to-ecology-in-search-of-civilisation/

Overton, J., Murray, W. E., & Howson, K. (2019). Doing good by drinking wine? Ethical value networks and upscaling of wine production in Australia, New Zealand, and South Africa. *European Planning Studies*, *27*(12), 2431–2449. DOI: 10.1080/09654313.2019.1628181

Ozer, M. A. (Ed.). (2023). *Local governments and the future tomorrows*. Ekin Kitabevi Publications.

Ozer, Y. E. (2017). Analysis of actors in climate change governance and the situation of Türkiye. *International Journal of Management Economics and Business*, *13*(4), 833–851.

Özer, Y. E. (2017). İklim Değişikliği Yönetişimindeki Aktörlerin Analizi ve Türkiye. *Uluslararası Yönetim İktisat ve İşletme Dergisi*, *13*(4), 833–851.

Özsoy, E. C. (2022). Sınırda karbon düzenleme mekanizması: Türkiye için yeni bir risk mi fırsat mı? VI. Anadolu International Conference on Economics, 13-15 Mayıs, Eskişehir, Türkiye.

Paidakaki, A., & Lang, R. (2021). Uncovering Social Sustainability in Housing Systems through the Lens of Institutional Capital: A Study of Two Housing Alliances in Vienna, Austria. *Sustainability (Basel)*, *13*(17), 9726. DOI: 10.3390/su13179726

Pal, P., & Poddar, S. (2024). Applications of Smart Farming Sensors: A Way Forward. *Data Science for Agricultural Innovation and Productivity*, 175.

Pana, T. N. (2021). Context & Challenges of the 2021 EU Green Bond Standard. *Available at SSRN* 3920262. DOI: 10.2139/ssrn.3920262

Panda, S., Sadhu, C., Pramanik, G., Pahari, S., & Hossain, J. (2016). Concerning public health situation of under-nutrition in children and anemia in women in Indian Sundarbans delta: A community based cross sectional investigation. *BMC Nutrition*, *2*(1), 1–10. DOI: 10.1186/s40795-016-0105-3

Pan, J., Cross, J. L., Zou, X., & Zhang, B. (2024). To tax or to trade? A global review of carbon emissions reduction strategies. *Energy Strategy Reviews*, *55*, 101508. DOI: 10.1016/j.esr.2024.101508

Park, C. Y., Yamamoto, Y., & Doong, M. A. L. (2023). European Union carbon border adjustment mechanism: Economic impact and implications for Asia. *ADB Briefs*, (276). DOI: 10.22617/BRF230561-2

Parks, R., Mclaren, M., Toumi, P. R., & Rivett, P. U. (2019). Experiences and lessons in managing water from Cape Town. *Grantham Institute Briefing Paper No.29.*

Park, Y., & Fujii, H. (2023). Civic engagement in a citizen-led living lab for smart cities: Evidence from South Korea. *Urban Planning*, *8*(2), 269–282. DOI: 10.17645/up.v8i2.6361

Parsons, D. (2010). Organisational Resilience. *Australian Journal of Emergency Management*, *25*(2), 18–20.

Pearce, W., Holmberg, K., Hellsten, I., & Nerlich, B. (2019). Climate change on Twitter: Topics, communities and conversations about the 2013 IPCC Working Group 1 report. *PLoS One*, *9*(4), e94785. DOI: 10.1371/journal.pone.0094785 PMID: 24718388

Peeters, J. (2012). The Place of Social Work in Sustainable Development: Towards Ecosocial Practice. *International Journal of Social Welfare*, *21*(3), 287–298. DOI: 10.1111/j.1468-2397.2011.00856.x

Pelling, M. (2003). *The vulnerability of cities: Natural disasters and social resilience.*

Pelling, M. (2011). *Adaptation to climate change: From resilience to transformation*. Routledge., DOI: 10.4324/9780203889046

Pelling, M., & Manuel-Navarrete, D. (2011). From resilience to transformation: The adaptive cycle in two Mexican urban centers. *Ecology and Society*, *16*(2), 11. DOI: 10.5751/ES-04038-160211

Perdana, S., & Vielle, M. (2022). Making the EU carbon border adjustment mechanism acceptable and climate friendly for least developed countries. *Energy Policy, 170*, 113245. DOI: 10.1016/j.enpol.2022.113245

Pervin, M., Sultana, S., Phirum, A., Camara, I., Nzau, M., Phonnasane, V., & Anderson, S. (2013). *A framework for mainstreaming climate resilience into development planning*. IIED Working Paper, IIED, London https://www.iied.org/10050iied

Pesaran, M. H. (2004). *General diagnostic tests for cross section dependence in panels*, Cambridge Working Papers in Economics, 435, London: BP.

Pesaran, M. H. (2004). General diagnostic tests for cross section dependence in panels. SSRN *Electronic Journal*. DOI: 10.2139/ssrn.572504

Pesaran, M. H., Ullah, A., & Yamagata, T. (2008). A bias-adjusted LM test of error cross-section independence. *The Econometrics Journal, 11*(1), 105–127. DOI: 10.1111/j.1368-423X.2007.00227.x

Pesaran, M. H., & Yamagata, T. (2008). Testing slope homogeneity in large panels. *Journal of Econometrics, 142*(1), 50–93. DOI: 10.1016/j.jeconom.2007.05.010

Pfaff, N., Utermarck, S., Altun, O., & Egorov, S. (2023). *Market integrity and greenwashing risks in sustainable finance*. International Capital Market Association.

Pham, H., & Saner, M. (2021). A Systematic Literature Review of Inclusive Climate Change Adaption. *Sustainability (Basel), 13*(19), 10617. DOI: 10.3390/su131910617

Pigou, A. C. (Ed), (1920). The economics of welfar. McMillan & Co, London.

Pinto, J. K. (2020). *Project management: achieving competitive advantage*. Pearson.

Pirlot, A. (2022). Carbon border adjustment measures: A straightforward multipurpose climate change instrument? *Journal of Environmental Law, 34*(1), 25–52. DOI: 10.1093/jel/eqab028

Pitchaikani, J. S. (2020, March). Integrated coastal zone management practices for Sundarbans, India. *Indian Journal of Geo-Marine Sciences, 49*(03), 352–356. https://core.ac.uk/download/pdf/326041666.pdf

Pla-Castells, M., Laredo, J. L. G., & Pujol, A. V. (2015). Use of ICT in smart cities: A practical case applied to traffic management in the city of Valencia. *Proceedings of the International Conference on Smart Cities and Smart Infrastructure*. DOI: 10.1109/SCSP.2015.7181559

Polin, F., & Alam, D. (2020, March 13). *Enhancing Community Participation in Forest Management: Case Study of the Sundarbans, Bangladesh.* South-West Region (Khulna Division). Retrieved from https://www.researchgate.net/publication/339916369_Enhancing_Community_Participation_in_Forest_Management_Case_Study_of_the_Sundarbans_Bangladesh

Prahalad, C. K., & Ramaswamy, V. (2000). Co-Opting Customer Competence. *Harvard Business Review*, *78*(1), 79–87.

Pramanik, M., Szabo, S., Pal, I., Udmale, P., Pongsiri, M., & Chilton, S. (2021). Population health risks in multi-hazard environments: Action needed in the Cyclone Amphan and COVID-19 - hit Sundarbans region, India. *Climate and Development*, *14*(2), 99–104. DOI: 10.1080/17565529.2021.1889948

Prasad, M. A., Loukoianova, M. E., Feng, A. X., & Oman, W. (2022). *Mobilizing private climate financing in emerging market and developing economies.* International Monetary Fund.

Preston, J., Waugh, L., & Jones, T. (2020). Citizen engagement for co-creating low carbon smart cities: Practical lessons from Nottingham City Council in the UK. *Energies*, *13*(24), 6615. DOI: 10.3390/en13246615

Pretty, J. N. (1995). Participatory learning for sustainable agriculture. *World Development*, *23*(8), 1247–1263. DOI: 10.1016/0305-750X(95)00046-F

Project Management Institute. (2021). *A guide to the project management body of knowledge (PMBOK® guide)* (7th ed.).

Purtill, J. (2024). China Is Installing the Wind and Solar Equivalent of Five Large Nuclear Power Stations per Week. – ABC News, 15 July 2024, available at: https://www.abc.net.au/news/science/2024-07-16/chinas-renewable-energy-boom-breaks-records/104086640

Puskás, N., Abunnasr, Y., & Naalbandian, S. (2021). Assessing deeper levels of participation in nature-based solutions in urban landscapes–A literature review of real-world cases. *Landscape and Urban Planning*, *210*, 104065. DOI: 10.1016/j.landurbplan.2021.104065

Qazi, A. L. (2022). *Karachi Bachao Tehreek: Housing rights, climate justice, and urban activism in Pakistan.* Peabody College, Vanderbilt University.

Qing, L., Abbas, J., Najam, H., Ma, X., & Dagestani, A. A. (2024). Investment in renewable energy and green financing and their role in achieving carbon-neutrality and economic sustainability: Insights from Asian region. *Renewable Energy*, *221*, 119830. DOI: 10.1016/j.renene.2023.119830

Rahman, M., Giedraitis, V. R., Lieberman, L. S., Akhtar, T., & Taminskienė, V. (2013). Shrimp cultivation with water salinity in Bangladesh: The implications of an ecological model. *Universal Journal of Public Health, 1*(3), 131–142. DOI: 10.13189/ujph.2013.010313

Raihan, A., Muhtasim, D. A., Farhana, S., Hasan, M. A. U., Pavel, M. I., Faruk, O., Rahman, M., & Mahmood, A. (2023). An econometric analysis of Greenhouse gas emissions from different agricultural factors in Bangladesh. *Energy Nexus, 9*, 100179. DOI: 10.1016/j.nexus.2023.100179

Rajamani, L. (2011). The Cancun Climate Agreements: Reading the text, subtext and tea leaves. *The International and Comparative Law Quarterly, 60*(2), 499–519. DOI: 10.1017/S0020589311000078

Ramaswamy, V., & Ozcan, K. (2018). What is co-creation? An interactional creation framework and its implications for value creation. *Journal of Business Research, 84*, 196–205. DOI: 10.1016/j.jbusres.2017.11.027

Ravallion, M. (2007). *Urban poverty. Finance & Development.* International Monetary Fund.

Ravallion, M., Heil, M., & Jalan, J. (2000). Carbon emissions and income inequality. *Oxford Economic Papers, 52*(4), 651–669. DOI: 10.1093/oep/52.4.651

Ray, P. S. (2020). Socio-Economic History in Sundarbans. *Journal of Critical Reviews, 7*(2), 1030–1034.

Reardon, T., & Vosti, S. A. (1995). Links between rural poverty and the environment in developing countries: Asset categories and investment poverty. *World Development, 23*(9), 1495–1506. DOI: 10.1016/0305-750X(95)00061-G

Rees, W. E. (2003). A blot on the land. *Nature, 421*(6926), 898. DOI: 10.1038/421898a PMID: 12606978

Reid, H., Alam, M., Berger, R., Cannon, T., Huq, S., & Milligan, A. (2009). *Community-based adaptation to climate change: An overview.* Participatory Learning and Action: Community Based Adaptation to Climate Change, 60, 11–33. Available at https://www.researchgate.net/publication/272791313_Community-based_adaptation_to_climate_change_An_overview

Re, L. L., Lee, C., Cassisa, C., Zhang, W., & Moarif, S. (2020). *Implementing Effective Emissions Trading Systems: Lessons from International Experiences.* IEA Publications.

REW. (2018). China Sets New Renewables Target of 35 Percent by 2030. - Renewable Energy World, 26 September 2018, available at: https://www.renewableenergyworld.com/baseload/china-sets-new-renewables-target-of-35-percent-by-2030/#gref

Reyer, C., Guericke, M., & Ibisch, P. L. (2009). Climate change mitigation via afforestation, reforestation and deforestation avoidance: And what about adaptation to environmental change? *New Forests*, *38*(1), 15–34. DOI: 10.1007/s11056-008-9129-0

Ribot, J. C., Najam, A., & Watson, G. (2009). *Climate Variation, Vulnerability and Sustainable Development in the Semi-arid Tropics*. In Schipper, E. L. F., & Burton, I. (Eds.), *The Earthscan Reader on Adaptation to Climate Change* (pp. 117–160). Earthscan., Available at https://www.researchgate.net/publication/291177311_Climate_variation_vulnerability_and_sustainable_development_in_the_semi-arid_tropics

Richardson, L. (2005). Social and political participation and inclusion. In *A more equal society?* (pp. 93–16). Policy Press., DOI: 10.2307/j.ctt9qgqd0.10

Richter, J., Basten, D., Michalik, B., Rosenkranz, C., & Smolnik, S. (2023). Opening the black box of knowledge management mechanisms: Exploring knowledge flows at a consultancy. *Kybernetes*, *52*(13), 1–28. DOI: 10.1108/K-08-2022-1118

Ridzuan, S. (2019). Inequality and the environmental Kuznets curve. *Journal of Cleaner Production*, *228*, 1472–1481. DOI: 10.1016/j.jclepro.2019.04.284

Roberts, D. (2008). Thinking globally, acting locally institutionalising climate change at the the local government level in Durban, South Africa. *Environment and Urbanization*, *20*(2), 521–537. DOI: 10.1177/0956247808096126

Rockwell, K. (2025). How China is Weaponising Its Dominance in Critical Minerals Trade. – East Asia Forum, 19 February 2025, available at: https://eastasiaforum.org/2025/02/19/how-china-is-weaponising-its-dominance-in-critical-minerals-trade/

Rodina, L. (2019). Water resilience lessons from Cape Town's water crisis. *WIREs. Water*, *6*(6), e1376. DOI: 10.1002/wat2.1376

Rogers, P. P., Jalal, K. F., & Boyd, J. A. (2007). *An Introduction to Sustainable Development*. Harvard University Press., DOI: 10.4324/9781849770477

Rolnick, D., Donti, P. L., & Kaack, L. H.. (2022). Tackling climate change with machine learning. *Nature Climate Change*, *12*(3), 203–210.

Romano, A. A., Scandurra, G., Carfora, A., & Ronghi, M. (2017). *Climate finance as an instrument to promote green growth in developing countries*. Springer.

Roostaie, S., Nawari, N., & Kibert, C. J. (2019). Sustainability and resilience: A review of definitions, relationships, and their integration into a combined building assessment framework. *Building and Environment*, *154*, 132–144. DOI: 10.1016/j.buildenv.2019.02.042

Rowe, G., & Frewer, L. J. (2000). Public participation methods: A framework for evaluation. *Science, Technology & Human Values*, *25*(1), 3–29. DOI: 10.1177/016224390002500101

Roy, D. (2020). 'On the Horns of a Dilemma'! Climate change, forest conservation and the marginal people in Indian Sundarbans. *Forum for Development Studies*, 47(2), 307–326. . 1786452DOI: 10.1080/08039410.2020

Roy, S. (2018). Livelihood resilience of the indigenous Munda community in the Bangladesh Sundarbans Forest. *Handbook Of Climate Change Resilience*, 10(1), 1–21. DOI: 10.1007/978-3-319-71025-9_10-1

Roy, M., Hulme, D., & Jahan, F. (2013). Contrasting adaptation responses by squatters and low-income tenants in Khulna, Bangladesh. *Environment and Urbanization*, *25*(1), 157–176. DOI: 10.1177/0956247813477362

Roy, S. (2019). *Women's Use of the Sundarbans Forest Resources*. Springer Briefs in Environment, Security. *Development and Peace*, *29*, 59–71. DOI: 10.1007/978-981-13-6776-2_4

Rudra, A., & Chattopadhyay, A. (2018). *Environmental vulnerabilities and risks in population: A comparison of two Sundarban Delta*. Population Association of America, May, 1–14. https://www.researchgate.net/publication/326201083

Rudra, A., & Chattopadhyay, A. (2019). Environmental change of coastal Sundarbans: Impact on livelihood and standard of living status of indigenous people. *Environmental Quality Management*, *29*(2), 77–84. DOI: 10.1002/tqem.21667

Ruiz-Martin, C., Lopez-Paredes, A., & Wainer, G. (2018). What we know and do not know about organizational resilience. *International Journal of Production Management and Engineering*, *6*(1), 11. DOI: 10.4995/ijpme.2018.7898

Sabio, N. R. (2023). How EU ETS in Portugal supports cyclists and helps birds to fly again. Carbon Market Watch, https://carbonmarketwatch.org/2023/10/09/how-eu-ets-in-portugal-supports-cyclists-and-helps-birds-to-fly-again/

Sahana, M., Rehman, S., Ahmed, R., & Sajjad, H. (2021). Analyzing climate variability and its effects in Sundarban Biosphere Reserve, India: Reaffirmation from local communities. *Environment, Development and Sustainability*, *23*(2), 2465–2492. DOI: 10.1007/s10668-020-00682-5

Saif, S., Russell, A. M., Nodie, S. I., Inskip, C., Lahann, P., Barlow, A., Greenwood Barlow, C., Islam, M. A., & MacMillan, D. C. (2016). Local usage of tiger parts and its role in tiger killing in the Bangladesh Sundarbans. *Human Dimensions of Wildlife*, *21*(2), 95–110. DOI: 10.1080/10871209.2015.1107786

Saito, O., Subramanian, S., Hashimoto, S., & Takeuchi, K. (2020). *Managing socio-ecological production landscapes and seascapes for sustainable communities in Asia mapping and navigating stakeholders, policy and action.* Springer https://www.springer.com/series/11884

Salvagnin, C., Glielmo, A., Giulis, M. E., & Mira, A. (2024). Investigating the price determinants of the European Emission Trading System: A non-parametric approach. *Quantitative Finance*, *24*(10), 1529–1544. DOI: 10.1080/14697688.2024.2407895

Sanders, E. B. N., & Stappers, P. J. (2008). Co-creation and the new landscapes of design. *CoDesign*, *4*(1), 5–18. DOI: 10.1080/15710880701875068

Sandoval-Almazan, R., & Gil-Garcia, J. R. (2012). Are government internet portals evolving towards more interaction, participation, and collaboration? Revisiting the rhetoric of e-government among municipalities. *Government Information Quarterly*, *29*, S72–S81. DOI: 10.1016/j.giq.2011.09.004

Sara, S., (2012). China's Renewable Energy Law and its impact on renewable power in China: Progress, challenges and recommendations for improving implementation. -. *Energy Policy*, *51*(December), 89–109.

Sarkar, A. N., & Dash, S. (2011). Emissions trading and carbon credit accounting for sustainable energy development with focus on India. *Asia Pacific Business Review*, *7*(1), 50–80. DOI: 10.1177/097324701100700104

Sarkar, S., & Sanyal, P. (2019). Community perception towards tourism development in the Indian Sundarbans: Case study of Jharkhali Island. *International Journal of Engineering Research & Technology (Ahmedabad)*, *8*(12), 526–540. www.ijert.org

Sarwar Hossain, M., Eigenbrod, F., Amoako Johnson, F., & Dearing, J. A. (2016). Unravelling the interrelationships between ecosystem services and human wellbeing in the Bangladesh delta. *International Journal of Sustainable Development and World Ecology*, *24*(2), 120–134. DOI: 10.1080/13504509.2016.1182087

Sathre, R., Antharam, S. M., & Catena, M. (2022). Water Security in South Asian Cities: A Review of Challenges and Opportunities. *CivilEng*, *3*(4), 873–894. DOI: 10.3390/civileng3040050

Sato, M., Rafaty, R., Calel, R., & Grubb, M. (2022). Allocation, allocation, allocation! The political economyof the development of the European Union EmissionsTrading System. *Wiley Interdisciplinary Reviews: Climate Change*, *13*(5), 1–19. DOI: 10.1002/wcc.796

Sattayapanich, T., Janmaimool, P., & Chontanawat, J. (2022). Factors affecting community participation in environmental corporate social responsibility projects: Evidence from mangrove forest management project. *Journal of Open Innovation*, *8*(4), 209. DOI: 10.3390/joitmc8040209

Satterthwaite, D., Huq, S., Reid, H., Pelling, M., & Romero Lankao, P. (2007). *Adapting to climate change in urban areas: the possibilities and constraints in low-and middle-income nations*. Human Settlements Working Paper, IIED, London. https://www.iied.org/10549iied

Satterthwaite, D., & Mitlin, D. (2014). *Reducing urban poverty in the global south*. Routledge., DOI: 10.4324/9780203104316

Saunders, C., Doherty, B., Hayes, G., & McCarthy, J. (2020). The impact of Extinction Rebellion in the UK: Disruptive protest and climate emergency declarations. *Environmental Politics*, *29*(5), 856–875. DOI: 10.1080/09644016.2020.1786331

Savelli, E., Mazzoleni, M., Di Baldassarre, G., Cloke, H., & Rusca, M. (2023). Urban water crises driven by elites' unsustainable consumption. *Nature Sustainability*, *6*(8), 929–940. DOI: 10.1038/s41893-023-01100-0

Schipper, E. L., Ayers, J., Reid, H., Huq, S., & Rahman, A. (Eds.), *Community-based adaptation to climate change: scaling it up* (pp. 214–225). Routledge., DOI: 10.4324/9780203105061

Schoeman, J., Allan, C., & Finlayson, C. M. (2019). Exploring the Multiple Meanings of Adaptive Management: A Case Study of the Lachlan Catchment in the Murray–Darling Basin. *Environmental Management*, *64*(4), 470–482. DOI: 10.1007/s00267-019-01203-5 PMID: 31511922

Scottish Government. (2021). Scotland's Climate Assembly: Research findings. https://www.gov.scot/publications/scotlands-climate-assembly-research-findings

Sen, H. S. (2019). *The Sundarbans: A disaster-prone ecoregion*. Berlin: Springer. 10, 978-3 https://link.springer.com/content/pdf/10.1007/978-3-030-00680-8.pdf

Sengers, F., Wieczorek, A. J., & Raven, R. (2019). Experimenting for sustainability transitions: A systematic literature review. *Technological Forecasting and Social Change*, *145*, 153–164. DOI: 10.1016/j.techfore.2016.08.031

Shao, S., Hu, Z., Cao, J., Yang, L., & Guan, D. (2020). Environmental regulation and enterprise innovation: A review. *Business Strategy and the Environment, 29*(3), 1465–1478. DOI: 10.1002/bse.2446

Sharma, D., & Tomar, S. (2010). Mainstreaming climate change adaptation in Indian cities. *Environment and Urbanization, 22*(2), 451–465. DOI: 10.1177/0956247810377390

Sharma, R., Webster, P., & Bhattacharyya, S. (2014). Factors affecting the performance of community health workers in India: A multi-stakeholder perspective. *Global Health Action, 7*(1), 25352. DOI: 10.3402/gha.v7.25352 PMID: 25319596

Shaw, V. (2024). SDIC Gansu Commissions 750 MW CSP-PV Plant in China. – PV Magazine, 18 December 2024, available at: https://www.pv-magazine.com/2024/12/18/sdic-gansu-commissions-750-mw-csp-pv-plant-in-china/

Shcherbakova, O. (2023). The role of green bonds in financing sustainable development. *Economy of Ukraine, 66*(12), 745.

Shirai, S. (2023). *Green central banking and regulation to foster sustainable finance* (No. 1361). ADBI Working Paper.

Sholarin, E. A., & Awange, J. L. (2016). *Environmental project management: principles, methodology, and processes*. Springer.

Shrimali, G., & Jena, L. P. (2023). Enabling Finance for Net-Zero Energy Transitions. *Available at SSRN* 4659874.

Silvius, A. G., & Schipper, R. (2015). A conceptual model for exploring the relationship between sustainability and project success. *Procedia Computer Science, 64*, 334–342. DOI: 10.1016/j.procs.2015.08.497

Simonet, G., & Duchemin, E. (2010). The concept of adaptation: Interdisciplinary scope and involvement in climate change. *Surveys and Perspectives Integrating Environment and Society, 3*(1), 1–9.

Sivaramakrishnan, K. C. (2007). *Democracy in urban India*. Urban Age. https://urbanage.lsecities.net/essays/democracy-in-urban-india

Smit, B., & Pilifosova, O. (2003). *From adaptation to adaptive capacity and vulnerability reduction.* In Climate Change, Adaptive Capacity and Development (pp. 9-28). Imperial College Press. DOI: 10.1142/9781860945816_0002

Smit, B., Burton, I., Klein, R. J. T., & Street, R. (1999). THE SCIENCE OF ADAPTATION: A FRAMEWORK FOR ASSESSMENT. *Mitigation and Adaptation Strategies for Global Change, 4*(3-4), 199–213. DOI: 10.1023/A:1009652531101

Smit, B., & Skinner, M. W. (2002). Adaptation options in agriculture to climate change: A typology. *Mitigation and Adaptation Strategies for Global Change*, *7*(1), 85–114. DOI: 10.1023/A:1015862228270

Smit, B., & Wandel, J. (2006). Adaptation, adaptive capacity and vulnerability. *Global Environmental Change*, *16*(3), 282–292. DOI: 10.1016/j.gloenvcha.2006.03.008

Solwa, I. (2019). Renewable Energy IPPs in SSA–Effective use of Blended Finance.

Somers, E. (2016). *City-zen*. Amsterdam Smart City. Erişim tarihi: 2 Aralık 2024, https://amsterdamsmartcity.com/updates/project/city-zen

Somerville, R. C., & Hassol, S. J. (2011). Communicating the science of climate change. *Physics Today*, *64*(10), 48–53. DOI: 10.1063/PT.3.1296

Sontiwanich, S., Tuan, A., & Chaisawat, M. (2022). An unsustainable smart city: Lessons from uneven citizen education and engagement in Thailand. *Sustainability (Basel)*, *14*(20), 13315. DOI: 10.3390/su142013315

Sovacool, B. K. (2011). An international comparison of four polycentric approaches to climate and energy governance. *Energy Policy*, *39*(6), 3832–3844. DOI: 10.1016/j.enpol.2011.04.014

Sovacool, B. K. (2018). Bamboo beating bandits: Conflict, inequality, and vulnerability in the political ecology of climate change adaptation in Bangladesh. *World Development*, *102*, 183–194. DOI: 10.1016/j.worlddev.2017.10.014

Soylu, A. C. (2022). Sürdürülebilir kalkınma ve gıda güvenliği ilişkisi. *Paradigma: İktisadi ve İdari Araştırmalar Dergisi*, *11*(2), 100–111.

Stanway, D. (2021). China starts Baihetan hydro project, biggest since Three Gorges. – Reuters, 28 June 2021, available at: https://www.reuters.com/business/energy/chinas-giant-baihetan-hydro-plant-begins-generating-power-cctv-2021-06-28/

Stavins, R. N. (2019). *Carbon taxes vs. cap and trade: Theory and practice*. Harvard Project on Climate Agreements.

StC. (2008). D. E. *Convention on the Conservation of Migratory Species of Wild Animals*.

Steele, J., Stoeckl, N., & McDonald, G. (2022). Blockchain applications for carbon markets and climate change mitigation. *Nature Climate Change*, *12*(3), 207–215.

Stern, N. (2007). *The economics of climate change: The Stern Review*. Cambridge University Press. DOI: 10.1017/CBO9780511817434

Steuer, S., & Tröger, T. H. (2022). The role of disclosure in green finance. *Journal of Financial Regulation*, *8*(1), 1–50. DOI: 10.1093/jfr/fjac001

Stockholm Conference. (1972). *United Nations Conference on the Human Environment*.

Stringer, L. C., Dougill, A. J., Fraser, E., Hubacek, K., Prell, C., & Reed, M. S. (2006). Unpacking “Participation” in the Adaptive Management of Social–ecological Systems: A Critical Review. *Ecology and Society*, *11*(2), art39. DOI: 10.5751/ES-01896-110239

Sundarbans National park. (1987). *UNESCO World Heritage Centre*. https://whc.unesco.org/en/list/452

Suness. (2024). China's Dual Tower CSP Innovation. – Suness Blog, 22 July 2024, available at: https://www.suness.com/Blog/china-dual-tower-csp-innovation.html

Surayya, T. (2012). Innovative Financial Instruments and mechanisms for financing forest restoration and mitigating climate change: Select cases from India. *European Journal of Sustainable Development*, *1*(2), 361–361. DOI: 10.14207/ejsd.2012.v1n2p361

Susha, I., & Grönlund, Å. (2014). Context clues for the stall of the Citizens' Initiative: Lessons for opening up e-participation development practice. *Government Information Quarterly*, *31*(3), 454–465. DOI: 10.1016/j.giq.2014.02.005

Swamy, P. A. (1970). Efficient inference in a random coefficient regression model. *Econometrica*, *38*(2), 311–323. DOI: 10.2307/1913012

Szabo, S., Hossain, M. S., Adger, W. N., Matthews, Z., Ahmed, S., Lázár, A. N., & Ahmad, S. (2016). Soil salinity, household wealth and food insecurity in tropical deltas: Evidence from south-westcoast of Bangladesh. *Sustainability Science*, *11*(3), 411–421. DOI: 10.1007/s11625-015-0337-1 PMID: 30174734

Szczubełek, G. (2022). Perspectives for energy development in Poland and in the European Union in the context of the European Green Deal: Renewable energy and energy efficiency. *Olsztyn Economic Journal*, *17*(2), 175–190. DOI: 10.31648/oej.8934

Taing, L., Chang, C. C., Pan, S., & Armitage, N. P. (2019). Towards a water secure future: Reflections on Cape Town's Day Zero crisis. *Urban Water Journal*, *16*(7), 530–536. DOI: 10.1080/1573062X.2019.1669190

Tandon, S., & Merle, K. (2024). Evaluatıng the impact of cbam on developıng countries: The EU's new approach to multilateralism and trade. Policy Study, https://feps-europe.eu/wp-content/uploads/2024/11/Impact-of-CBAM.pdf

Thakur, S., Maity, D., Mondal, I., Basumatary, G., Ghosh, P. B., Das, P., & De, T. K. (2021). Assessment of changes in land use, land cover, and land surface temperature in the mangrove forest of Sundarbans, northeast coast of India. *Environment, Development and Sustainability*, *23*(2), 1917–1943. DOI: 10.1007/s10668-020-00656-7

The Belt and Road Ecological and Environmental Cooperation Plan. (2017). – Belt and Road Portal, 14 May 2017, available at: https://eng.yidaiyilu.gov.cn/p/13392.html

Theron, H., & Hunter, J. J. (2022). Mitigation and Adaptation Practices to the Impact of Climate Change on Wine Grape Production, with Special Reference to the South African Context. *South African Journal of Enology and Viticulture*. DOI: 10.21548/43-1-4735

Thomas, J. C. (1993). Public involvement and governmental effectiveness: A decision-making model for public managers. *Administration & Society*, *24*(4), 444–469. DOI: 10.1177/009539979302400402

Tietenberg, T., & Lewis, L. (2016). *Environmental and natural resource economics* (10th ed.). Pearson.

Tippett, J., Handley, J. F., & Ravetz, J. (2007). Meeting the challenges of sustainable development—A conceptual appraisal of a new methodology for participatory ecological planning. *Progress in Planning*, *67*(1), 9–98. DOI: 10.1016/j.progress.2006.12.004

Toprak, D. (2006). Sürdürülebilir kalkınma çevresinde çevre politikalari ve mali araçlar. *Süleyman Demirel Üniversitesi Sosyal Bilimler Enstitüsü Dergisi*, (2), 146–169.

Torras, M., & Boyce, J. (1998). Income, inequality, and pollution: A reassessment of the environmental kuznets curve. *Ecological Economics*, *25*(2), 147–160. DOI: 10.1016/S0921-8009(97)00177-8

Tóth, A. (2022). Planning and designing green infrastructure across landscapes and scales. *Acta Horticulturae et Regiotecturae*, *25*(1), 1–7. DOI: 10.2478/ahr-2022-0001

Tripodi, T., & Potocky-Tripodi, M. (2007). *International Social Work Research: Issues and Prospects*. Oxford University Press., DOI: 10.1093/acprof:oso/9780195187250.001.0001

Twigg, J. (2007). *Characteristics of a disaster-resilient community, a guidance note*. Available at https://www.researchgate.net/publication/315799865_Characteristics_of_a_Disaster-resilient_Community_A_Guidance_Note_Characteristics_of_a_Disaster-resilient_Community_A_Guidance_Note

Tyagi, S. K. S., Mukherjee, A., Pokhrel, S. R., & Hiran, K. K. (2020). An intelligent and optimal resource allocation approach in sensor networks for smart agri-IoT. *IEEE Sensors Journal, 21*(16), 17439–17446. DOI: 10.1109/JSEN.2020.3020889

Tyler, S., & Moench, M. (2012). A framework for urban climate resilience. *Climate and Development, 4*(4), 311–326. DOI: 10.1080/17565529.2012.745389

U.S. Department of Energy. (2023). Open energy data initiative. Retrieved from https://www.energy.gov

U.S. Department of Energy. (2023). Smart water management. Retrieved from https://www.energy.gov

U.S. Environmental Protection Agency. (2014). Inventory of U.S. greenhouse gas emissions and sinks. Retrieved from https://www.epa.gov/ghgemissions

U.S. House Committee on Oversight and Reform. (2022). Fueling the climate crisis: Exposing big oil's disinformation campaign to prevent climate action. https://oversight.house.gov/legislation/hearings/fueling-the-climate-crisis-exposing-big-oil-s-disinformation-campaign-to

Uçar, A., Şemşit, S., & Negiz, N. (2017). Avrupa Birliği akıllı kent uygulamaları ve Türkiye'deki yansımaları. *Süleyman Demirel Üniversitesi İktisadi ve İdari Bilimler Fakültesi Dergisi, 22*(Kayfor 15 Özel Sayısı), 1785–1798.

Uddin, S. M. (2019). Religion, Nature, and Life in the Sundarbans. *Asian Ethnology, 78*(2), 289–310. https://www.jstor.org/stable/26845329

Uğur, Ö. (2019). Climate change in global politics in the postmodern eera: A new challenge and responses. Ömer Uğur ve Kadir Caner Doğan (Ed). *Contemporary Debates on Politics and Public Administration in the Postmodern Era*, 83-99.

Uğur, Ö., & Aksoy, M. (2023). İklim Değişikliğinin Göç Üzerindeki Etkisi, 2. *Bilsel International Truva Scientific Researches and Innovation Congress*, 16-17 December, 2023, Çanakkale /Türkiye.

Uğur, Ö., & Cengiz, E. (2022). İklim değişikliği siyaseti ve ekolojik diplomasi: Küresel sorunlara karşı küresel mücadele, *1 St International Conference on Scientific and Academic Research*, 318 – 322, Konya.

Uğur, Ö., & Cengiz, E. (2023). İklim değişikliği siyaseti. In Uğur, Ö., & Cengiz, E. (Eds.), *İklim değişikliği siyaseti: Riskler, fırsatlar ve yeni vizyonlar* (pp. 1–12). Nobel Bilimsel Eserler.

Uğur, Ö., & Cengiz, E. (Eds.). (2023). *İklim Değişikliği Siyaseti*. Nobel Basım Yayım.

Uğur, Ö., & Doğan, K. C. (Eds.). (2019). *Contemporary Debates on Politics and Public Administration in the Postmodern Era*. Peter Lang. DOI: 10.3726/b16122

Uğur, Ö., & Efşan, N. Ç. (2024). Avrupa Birliği İklim Değişikliği Politikası: Politik Fırsatlar İçin İtici ve Destekleyici Faktörler. *Uluslararası Ekonomi İşletme ve Politika Dergisi, 8*(2), 305–327. DOI: 10.29216/ueip.1474284

Uğur, S. (2014). Sera gazı emisyonlarının azaltımında karbon-enerji vergilerinin rolü. *FSM İlmî Araştırmalar İnsan ve Toplum Bilimleri Dergisi, 0*(3), 341–358. DOI: 10.16947/fsmiad.58262

Ulucak, R., Danish, , & Kassouri, Y. (2020). An assessment of the environmental sustainability corridor: Investigating the non-linear effects of environmental taxation on CO_2 emissions. *Sustainable Development (Bradford), 28*(4), 1010–1018. DOI: 10.1002/sd.2057

UN DESA. (2024). Multilevel Governance for Climate Change Mitigation: Effective Policy Frameworks at All Levels.

UNDP. (2022). What is Climate Governance? Retrieved from https://www.undp.org/sites/g/files/zskgke326/files/2022-09/Volume%20III%20What%20is%20climate%20governance.pdf

UNEP (United Nations Environment Programme). (2022). The emissions gap report 2022. Retrieved from https://www.unep.org

Unequal Scenes. (n.d.). https://www.millefoto.com/unequalscenes

UNFCCC. (1998). *Kyoto Protocol to the United Nations Framework Convention on Climate Change*. https://unfccc.int/resource/docs/convkp/kpeng.pdf

UNFCCC. (2015). *The Paris Agreement*. https://unfccc.int/sites/default /files/english_paris_ agreement.pdf

UNFCCC. (2023). Paris agreement. Retrieved from https://unfccc.int

UNFCCC. (n.d.). Climate Change Adaptation Strategies for Local Impact. Retrieved from https://unfccc.int/resource/docs/2009/smsn/igo/054.pdf

UN-Habitat. (2010). The state of Asian cities 2010/11.

UN-Habitat. (2022). Smart cities and inclusivity: Challenges and opportunities. Retrieved from https://www.unhabitat.org

UNICEF. (2020). What is Climate Governance? Retrieved from https://www.unicef.org/lac/en/reports/what-climate-governance

Union of Concerned Scientists. (2015). The climate deception dossiers: Internal fossil fuel industry memos reveal decades of corporate disinformation. https://www.ucsusa.org/resources/climate-deception-dossiers

United Nations Conference on Trade and Development [UNCTAD]. (2017). *Handbook of statistics*. https://unctad.org/system/files/official-document/tdstat42_en.pdf

United Nations Department of Economic and Social Affairs. (2024). *The 17 goals*. Sustainable Development. Erişim tarihi: 20 Kasım. https://sdgs.un.org/goals

United Nations Framework Convention on Climate Change (UNFCCC). (2015). *The Paris Agreement*. Retrieved from https://unfccc.int

United Nations Framework Convention on Climate Change. (1992). *UNFCCC*. https://unfccc.int/resource/docs/convkp/conveng.pdf

United Nations Framework Convention on Climate Change. (2025). Climate change conferences of the parties (COP). https://unfccc.int/process/bodies/supreme-bodies/conference-of-the-parties-cop

United Nations Office for Disaster Risk Reduction (UNDRR). (2022). *Global Assessment Report on Disaster Risk Reduction 2022*. UNDRR.

United Nations. (1997). Kyoto Protocol.

United Nations. (2016). Paris Agreement.

United Nations. (2016). Sustainable development goal 13: Climate action. https://sdgs.un.org/goals/goal13

United Nations. (n.d.). Climate action and support trends. Retrieved from https://www.un.org/en/climatechange

United Nations. (n.d.). Scenario modeling for climate action. Retrieved from https://www.un.org/en/climatechange

United Nations. (n.d.). Statement of commitment for action to eradicate poverty adopted by Administrative Committee on Coordination [Press release ECOSOC/5759].

Unogwu, O. J., Doshi, R., Hiran, K. K., Mijwil, M. M., Catherine, A. T., & Abotaleb, M. (2024). *Exploring the Implications of Emerging Artificial Intelligence Technologies at Edge Computing in Higher Education.* In Integrating Generative AI in Education to Achieve Sustainable Development Goals (pp. 169-182). IGI Global. DOI: 10.4018/979-8-3693-2440-0.ch009

Usta, A. E. (2024). Sınırda karbon düzenleme mekanizması: Türkiye'yi neler bekliyor?. İktisadi Kalkınma Vakfı, No: 277.

Uyar, B., & Bilgici, Y. (2021). Karbon fiyatlandırmasında emisyon ticaret sistemi ve önemi. *Kırklareli Üniversitesi İktisadi ve İdari Bilimler Fakültesi Dergisi, 10*(1), 47–72.

Uyduranoğlu, A., Gevrek, Z. E., & Acar, S. (2023). Avrupa Yeşil Mutabakatı'na uyum çerçevesinde Türkiye'deki işletmelerin emisyon ticaretine verdikleri desteği etkileyen faktörler. *Ankara Avrupa Çalışmaları Dergisi, 22*(1), 257–289. DOI: 10.32450/aacd.1327099

Van der Veen, E. (2016, 15 Ocak). *IoT living lab.* Amsterdam Smart City. Erişim tarihi: 25 Kasım 2024, https://amsterdamsmartcity.com/updates/project/iot-living-lab

Van Ginkel, K. C. H., Hoekstra, A. Y., Buurman, J., & Hogeboom, R. J. (2018). Urban Water Security Dashboard: Systems Approach to Characterizing the Water Security of Cities. *Journal of Water Resources Planning and Management, 144*(12), 04018075. DOI: 10.1061/(ASCE)WR.1943-5452.0000997

Van Winden, W., Oskam, I., van den Buuse, D., Schrama, W., & Van Dijck, E. J. (2016). *Organising smart city projects: Lessons from Amsterdam.* Hogeschool van Amsterdam.

Vanolo, A. (2014). Smartmentality: The smart city as disciplinary strategy. *Urban Studies (Edinburgh, Scotland), 51*(5), 883–898. DOI: 10.1177/0042098013494427

Vargas, C., Whelan, J., Brimblecombe, J., & Allendera, S. (2022). Co-creation, co-design and co-production for public health: A perspective on definitions and distinctions. *Public Health Research & Practice, 32*(2). DOI: 10.17061/phrp3222211 PMID: 35702744

Vink, N., Deloire, A., Bonnardot, V., & Ewert, J. (2012). Climate change and the future of South Africa's wine industry. *International Journal of Climate Change Strategies and Management, 4*(4), 420–441. DOI: 10.1108/17568691211277746

Vishnoi, J. N. (2022). *The Role of the financial institutions in transition to low carbon economy* (Doctoral dissertation, Wien).

Vision and Actions on Energy Cooperation in Jointly Building Silk Road Economic Belt and 21st-Century Maritime Silk Road. (2017). – Belt and Road Portal, 25 December 2017, available at: http://obor.nea.gov.cn/detail/669.html

Visser, G., & Cloete, J. (2021). Guest Houses, Drought and Water Management Strategies in the Cape Winelands, South Africa. *African Journal of Hospitality, Tourism and Leisure*, *10*(10(5)), 1549–1561. DOI: 10.46222/ajhtl.19770720.178

Visser, W. P. (2018). A perfect storm: The ramifications of Cape Town's drought crisis. *The Journal for Transdisciplinary Research in Southern Africa*, *14*(1). DOI: 10.4102/td.v14i1.567

Voci, D., Bruns, C. J., Lemke, S., & Weder, F. (2020). Framing the End: Analyzing Media and Meaning Making During Cape Town's Day Zero. *Frontiers in Communication*, *5*, 576199. DOI: 10.3389/fcomm.2020.576199

Vona, F., & Patriarca, F. (2011). Income inequality and the development of environmental technologies. *Ecological Economics*, *70*(11), 2201–2213. DOI: 10.1016/j.ecolecon.2011.06.027

Voorberg, W. H., Bekkers, V. J., & Tummers, L. G. (2015). A systematic review of co-creation and co-production: Embarking on the social innovation journey. *Public Management Review*, *17*(9), 1333–1357. DOI: 10.1080/14719037.2014.930505

Vorster, S., Winkler, H., & Jooste, M. (2011). Mitigating climate change through carbon pricing: An emerging policy debate in South Africa. *Climate and Development*, *3*(3), 242–258. DOI: 10.1080/17565529.2011.598367

Wahlström, M., Kocyba, P., De Vydt, M., & de Moor, J. (2022). *Youth activism and the Fridays for Future movement: Mobilizing for climate justice*. Routledge.

Wang, Z., (2014). The Ecological Civilization Debate in China: The Role of Ecological Marxism and Constructive Postmodernism—Beyond the Predicament of Legislation. - Monthly Review, 01 November 2014, available at: https://monthlyreview.org/2014/11/01/the-ecological-civilization-debate-in-china/#en73

Wang, G., & Zhang, Q. (2023). Transparency, Governance, and Innovation: Unveiling the Impact of Environmental Information Disclosure on Corporate Value and Green Innovation Capabilities. *International Journal of Operations and Quantitative Management*, *29*(4), 54–70.

Warner, J. F., & Meissner, R. (2021). Cape Town's "Day Zero" water crisis: A manufactured media event? *International Journal of Disaster Risk Reduction*, *64*, 102481. DOI: 10.1016/j.ijdrr.2021.102481

Washburn, D., Sindhu, U., Balaouras, S., Dines, R. A., Hayes, N., & Nelson, L. E. (2010). *Helping CIOs understand "smart city" initiatives*. Forrester Research. http://public.dhe.ibm.com/partnerworld/pub/smb/smarterplanet/forr_help_cios_und_smart_city_initiatives.pdf

We Are Social. (2024). Digital 2024: 5 billion social media users. https://wearesocial.com/uk/blog/2024/01/digital-2024-5-billion-social-media-users/

Webler, T., Kastenholz, H., & Renn, O. (1995). Public participation in impact assessment: A social learning perspective. *Environmental Impact Assessment Review*, *15*(5), 443–463. DOI: 10.1016/0195-9255(95)00043-E

Webler, T., & Renn, O. (1995). A brief primer on participation: philosophy and practice. In *Fairness and competence in citizen participation: Evaluating models for environmental discourse* (pp. 17–33). Springer Netherlands. DOI: 10.1007/978-94-011-0131-8_2

Wei, L. (2023). China Set to Open World's Largest Hybrid Energy Plant in 2023. – CGTN, 1 October 2022, available at: https://news.cgtn.com/news/2022-10-01/China-set-to-open-world-s-largest-hybrid-energy-plant-in-2023-1dLLxnhKDbW/index.html

Werrell, C. E., & Femia, F. (Eds.). (2015). Climate change as threat multiplier: Understanding the broader nature of the risk. The Center for Climate and Security. https://climateandsecurity.org/climate-change-as-threat-multiplier/

Wijnhoven, F., Ehrenhard, M., & Kuhn, J. (2015). Open government objectives and participation motivations. *Government Information Quarterly*, *32*(1), 30–42. DOI: 10.1016/j.giq.2014.10.002

Wilbanks, T. J., Romero Lankao, P., Bao, M., Berkhout, F., Cairncross, S., Ceron, J.-P., Kapshe, M., Muir-Wood, R., & Zapata-Marti, R. (2007). Industry, settlement and society. In Parry, M. L., Canziani, O. F., Palutikof, J. P., van der Linden, P. J., & Hanson, C. E. (Eds.), *Climate change 2007: Impacts, adaptation and vulnerability. Contribution of Working Group II to the Fourth Assessment Report of the Intergovernmental Panel on Climate Change*. Cambridge University Press.

Wilker, J., Rusche, K., & Rymsa-Fitschen, C. (2016). Improving participation in green infrastructure planning. *Planning Practice and Research*, *31*(3), 229–249. DOI: 10.1080/02697459.2016.1158065

Williams, B. K., & Brown, E. D. (2012). *Adaptive management: the US Department of the Interior applications guide*. US Department of the Interior, Adaptive Management Working Group.

Williams, H. T. P., McMurray, J. R., Kurz, T., & Lambert, F. H. (2015). Network analysis reveals open forums and echo chambers in social media discussions of climate change. *Global Environmental Change*, *32*, 126–138. DOI: 10.1016/j.gloenvcha.2015.03.006

Wiser, R., Lantz, E., Mai, T., Zayas, J., DeMeo, E., Eugeni, E., Lin-Powers, J., & Tusing, R. (2015). Wind vision: A new era for wind power in the United States. *The Electricity Journal*, *28*(9), 120–132. DOI: 10.1016/j.tej.2015.09.016

World Bank. (2020). Reference guide to climate change framework legislation. https://openknowledge.worldbank.org/handle/10986/34593

World Bank. (2022). *Harnessing technology for climate-smart development*. World Bank Group.

World Bank. (2022). State and Trends of Carbon Pricing. https://documents1.worldbank.org/curated/en/099045006072224607/pdf/P1780300092e910590acb201757ecd54322.pdf

World Bank. (2023). Climate change knowledge portal. Retrieved from https://climateknowledgeportal.worldbank.org

World Bank. (2023). Disaster resilience initiatives. Retrieved from https://www.worldbank.org

World Bank. (2024). State and Trends of Carbon Pricing 2024, https://documents1.worldbank.org/curated/en/099081624122529330/pdf/P50228315fd8d1050186341ea02e1c107bc.pdf

World Economic Forum. (2023). Blockchain for sustainable and transparent carbon markets. *Global Future Council Reports*.

World Meteorological Organization (WMO). (2023). *State of the Global Climate 2022*. WMO.

World Resources Institute. (2023). Climate data explorer. Retrieved from https://www.wri.org

World Resources Institute. (2023). Data challenges in climate governance. Retrieved from https://www.wri.org

Wu, Y. (2024). Embracing Sustainability: How Businesses Can Contribute to the 'Beautiful China' Initiative. – China Briefing, 30 January 2024, available at: https://www.china-briefing.com/news/embracing-sustainability-how-businesses-can-contribute-to-the-beautiful-china-initiative/

Xiaoyun, W., & Chuanmin, Z. (2024). Does Chinese environmental policy affect income inequality? Evidence from the central environmental protection inspection. *China & World Economy. Institute of World Economics and Politics*, *32*(3), 130–161.

Xinyu, G., Bo, J., Bin, L., Kai, Y., Hongguang, Z., & Boyuan, F. (2011). Study on Renewable Energy Development and Policy in China. -. *Energy Procedia, 5/2011*, 1284–1290. DOI: 10.1016/j.egypro.2011.03.224

Xun, L. (2022). China's First Solar-tidal Photovoltaic Power Plant Fully Operational. – CGTN, 02 June 2022, available at: https://news.cgtn.com/news/2022-05-30/China-s-first-solar-tidal-photovoltaic-power-plant-fully-operational-1asqPv0xC2k/index.html

Xu, W., & Zhu, Q. (2021). Evaluating the impact mechanism of citizen participation on citizen satisfaction in a smart city. *Environment and Planning. B, Urban Analytics and City Science*, *48*(8), 2419–2436. DOI: 10.1177/2399808320980746

Yadav, A., Sharma, H., & Venkataramani, V. (2023). Mobilising Finance for Climate Action and Disaster-Resilient Infrastructure in Developing Countries.

Yadav, A., Gyamfi, B. A., Asongu, S. A., & Behera, D. K. (2024). The role of green finance and governance effectiveness in the impact of renewable energy investment on CO2 emissions in BRICS economies. *Journal of Environmental Management*, *358*, 120906. DOI: 10.1016/j.jenvman.2024.120906 PMID: 38636419

Yang, F., Liu, W., Zhang, Y., Yang, G., & wala, T. (2024). Money grows on green energy: Financing a sustainable power future. *Heliyon*, *10*(7), e28353. DOI: 10.1016/j.heliyon.2024.e28353 PMID: 38590910

Yilan, G., Cordella, M., & Morone, P. (2022). Evaluating and managing the sustainability of investments in green and sustainable chemistry: An overview of sustainable finance approaches and tools. *Current Opinion in Green and Sustainable Chemistry*, *36*, 100635. DOI: 10.1016/j.cogsc.2022.100635

Yılmaz Uğur, C. (2021). In Dogan, K. C., Ugur, Ö., & de Lyon, L. (Eds.), *Struggling With Covid-19 Outbreak in The Framework of Global Public Goods, Public Administration and Public Finance Research* (pp. 292–311).

Yılmaz Uğur, C. (2022). The Polluter Pays Principle within the Framework of the European Union Emissions Trading System. *Abant Sosyal Bilimler Dergisi*, *22*(2), 862–872. DOI: 10.11616/asbi.1102031

Yimsek, F. S., & Yakar, M. (2023). A general glimpse into the smart cities. *Türkiye Arazi Yönetimi Dergisi*, *5*(1), 49–56. DOI: 10.51765/tayod.1270504

Yohe, G., & Tol, R. S. J. (2002). Indicators for social and economic coping capacity moving toward a working definition of adaptive capacity. *Global Environmental Change, 12*(1), 25–40. DOI: 10.1016/S0959-3780(01)00026-7

You, X. (2022). Why China's Provinces Are So Important for Action on Climate Change. – Carbon Brief, 10 October 2022, available at: https://www.carbonbrief.org/explainer-why-chinas-provinces-are-so-important-for-action-on-climate-change/

Zhang, Z., & Baranzini, A. (2004). What do we know about carbon taxes? An inquiry into their impacts on competitiveness and distribution of income. *Energy Policy, 32*(4), 507–518. DOI: 10.1016/S0301-4215(03)00152-6

Ziervogel, G. (2019). *UNPACKING THE CAPE TOWN DROUGHT:LESSONS LEARNED* [Report for Cities Support Programme undertaken by African Centre for Cities].

Ziervogel, G., & Morgan, G. (2022). The Adaptive Governance Capacities of the City of Cape Town Built in Response to Extreme Events. *Journal of Extreme Events, 09*(01), 2250005. DOI: 10.1142/S2345737622500051

Zimmermann, M., Van Begin, G., & Cristobal, I. V. (2010). The international local government climate roadmap. In Van Staden, M., & Musco, F. (Eds.), *Local governments and climate change* (pp. 79–89). Springer.

Ziolo, M., Bak, I., & Cheba, K. (2021). The role of sustainable finance in achieving sustainable development goals: Does it work? *Technological and Economic Development of Economy, 27*(1), 45–70. DOI: 10.3846/tede.2020.13863

Zirek, D., & Unsal, O. (2023). Green bonds: Do investors benefit from third-party certification? *Global Finance Journal, 58*, 100872. DOI: 10.1016/j.gfj.2023.100872

About the Contributors

Ömer Uğur is an associate professor of at the Faculty of Economics and Administrative Scienfces, Gumushane University, Turkey. He works on The European Union, global environmental governace, and environmental law.

Kadir Caner Doğan received his bachelor's degree in econometrics from Bursa Uludağ University in Turkey in 2006. In 2013, he received his bachelor's degree in public administration from Anadolu University. In 2009, he received his master's degree in public administration from Bolu Abant İzzet Baysal University. In 2014, he received his doctorate degree in public administration from Bursa Uludağ University. His research interests include ombudsman, natural law, ethics, governance and postmodern public administration. He is currently working as a professor at Gümüşhane University in Turkey.

***

Naboshree Bhattacharya is an assistant professor with a specialization in healthcare economics and a broader research focus on development economics. With a rich academic background, she holds alumni status from prestigious institutions including St. Xavier's College, Ranchi University, and Jadavpur University. Over the course of her career spanning 12 years, Naboshree has garnered extensive experience in both academic and industrial settings. Currently serving as an Assistant Professor at Amity University Jharkhand, she brings a wealth of knowledge and expertise to her role. Naboshree's contributions to the field extend beyond the classroom. She is a prolific author, having penned six books on various topics within economics, collaborating with both national and international authors. Additionally, her research output includes numerous papers and book chapters published in esteemed journals and publications, both domestically and internationally.

Özge Yalçıner Ercoşkun is a graduate of the Department of City and Regional Planning at Istanbul Technical University. She completed her master's degree in Geographic Information Systems at Middle East Technical University and earned her PhD in City and Regional Planning at Gazi University. She is currently a faculty member at the Department of City and Regional Planning at Gazi University. Her research interests include sustainable urban planning, smart cities, eco-tech urban design, resilient cities, and geographic information systems. She has participated in numerous panels, exhibitions, and conferences, both nationally and internationally, particularly in the fields of ecological urban planning and GIS. She has also contributed to various research projects. She has authored and edited many books and over one hundred publications on sustainable urban design, ecological and smart city planning, and information technologies. She has received awards for her work on sustainability, urban development, and sustainable tourism.

Oindrilla Ghosh is a Research Scholar hailing from the School of Law at Christ (Deemed to be University) Bangalore, India, who focuses her research on the health of tribal communities. Her work includes Neuroscience on Organizational Behavior publications at IGI Global and Environmental Disasters, of which 19 publications are indexed in SCOPUS and Web of Sciences conference proceedings. She completed her early education at Loreto Day School, Bowbazar, and achieved distinction in her undergraduate (B.A/LL.B) and postgraduate (LL.M) studies at Calcutta University. Having served as an Assistant Professor of Law at Kingston Law College, Barasat, Kolkata, Ghosh has actively participated in numerous National Conferences, International Conferences, UN Global Summits, webinars, and seminars organized by prestigious academic institutions. Her expertise lies in health perspectives, environmental studies, human behavior, and related areas. Additionally, she has conducted various workshops and lectures in educational institutions, and is engaged with NGO's to help the needy

Binod Kumar is a graduate student known for his academic excellence, with his 14 scholarly works indexed in Scopus database. Hailing from Ranchi, Jharkhand in India, Binod is currently pursuing a Master of Laws (LL.M) at the prestigious Radha Govind University. His passion for technology and innovation along with climate changes is evident through his high academic performance and numerous projects. Binod has a keen interest in artificial intelligence, machine learning, neuropsychology, climate changes and disaster management, fields in which he has undertaken several research projects. He is particularly fascinated by the potential of AI to solve real-world problems and is dedicated to learning and applying new techniques in these areas. He volunteers at local NGOs, teaching underprivileged children basic computer skills. With a strong foundation in law and a commitment to making a positive impact, Binod Kumar is poised to become a future leader in the legal industry.

Lucky Diliza Madikiza works in the University of South Africa's Research, Postgraduate Studies, Innovation and Commercialisation section. He has research interests in sustainability, media and communication studies and public finance. He holds postgraduate degrees in Communication Studies and International Business from the University of South Africa and the University of Liverpool.

Manel Masmoudi is an assistant professor at Institut of High Commercial Studies Institute at the University of Sfax. Her areas of expertise include emerging trends in consumer behavior, with a particular focus on online behavior, tourist behavior, culinary heritage preservation and consumer behavior during crises. She is dedicated to understanding how these behaviors evolve and adapt in various contexts, using different methodologies that include both qualitative and quantitative approaches.

Ananya Pandey is a Research Scholar in School of Law at Christ University, Bangalore. She has her research interests in the field of Constitutional Law, Environment Law, Human Rights, Food Security, Intellectual Property Law and Jurisprudence.

Latifa Trabelsi is a permanent Master Technologist at the Higher Institute of Technological Studies of Bizerte (Tunisia) and she is a member of the Marketing Research Laboratory (Sfax). She holds a PhD in marketing from the Faculty of Economics and Management of Sfax and a master's degree in marketing from the Higher Institute of Management of Tunis. CRM, CRM performance, Customer Knowledge Management, Knowledge Discovery from Data, Web Analytics, and visual rhetoric in advertising are just a few of Dr. Trabelsi's research topics. The international Journal of Knowledge Management Studies and La Revue Tunisienne du Marketing, have published her works. She also published a chapter in the book « Social Customer Relationship Management (Social-CRM) in the Era of Web 4.0 » (pp. 169-195), IGI Global. She has papers published in the proceedings of international and national academic conferences (ACFAS, Marketing Trends, ATM, AGECSO, and so on).

Elizabeth Yoneva holds a Ph.D. degree in Political Science. She is Associate Professor in the International Relations Department at the Faculty of International Economics and Politics, University of National and World Economy, Sofia (Bulgaria). Her teaching activity and research interests include energy security, European integration and regional cooperation in the Black Sea zone and Central

Asia. She is a member of the editorial board of the "Bulgarian Journal of International Economics and Politics".

Index

A

B

C

D

E

F

G

H

I

J

L

M

P

R

S

T

U

V

www.ingramcontent.com/pod-product-compliance
Lightning Source LLC
LaVergne TN
LVHW080349011026
849161LV00009B/2241